WATER CONTAMINATION AND HEALTH

Environmental Science and Pollution Control Series

Series Advisors

Dr. Suzanne Lesage
Environment Canada
Burlington, Ontario, Canada

Dr. Hans van Bergen
International Flavors and Fragrances
Hilversum, The Netherlands

Dr. Willy J. Masschelein
Consultant
Brussels, Belgium

Dr. Constantine Yapijakis
The Cooper Union
New York, New York, U.S.A.

1. Toxic Metal Chemistry in Marine Environments, *Muhammad Sadiq*
2. Handbook of Polymer Degradation, *edited by S. Halim Hamid, Mohamed B. Amin, and Ali G. Maadhah*
3. Unit Processes in Drinking Water Treatment, *Willy J. Masschelein*
4. Groundwater Contamination and Analysis at Hazardous Waste Sites, *edited by Suzanne Lesage and Richard E. Jackson*
5. Plastics Waste Management: Disposal, Recycling, and Reuse, *edited by Nabil Mustafa*
6. Hazardous Waste Site Soil Remediation: Theory and Application of Innovative Technologies, *edited by David J. Wilson and Ann N. Clarke*
7. Process Engineering for Pollution Control and Waste Minimization, *edited by Donald L. Wise and Debra J. Trantolo*
8. Remediation of Hazardous Waste Contaminated Soils, *edited by Donald L. Wise and Debra J. Trantolo*
9. Water Contamination and Health: Integration of Exposure Assessment, Toxicology, and Risk Assessment, *edited by Rhoda G. M. Wang*
10. Pollution Control in Fertilizer Production, *edited by Charles A. Hodge and Neculai N. Popovici*

Additional Volumes in Preparation

Groundwater Contamination and Control, *edited by Uri Zoller*

Handbook of Toxic Properties of Pesticides, *Nicholas P. Cheremisinoff and John A. King*

Combustion and Incineration Processes: Applications in Environmental Engineering. Second Edition, Revised and Expanded, *Walter R. Niessen*

WATER CONTAMINATION AND HEALTH

INTEGRATION OF EXPOSURE ASSESSMENT, TOXICOLOGY, AND RISK ASSESSMENT

EDITED BY
RHODA G. M. WANG
Human Risk Assessment Branch
U.S. Environmental Protection Agency
Washington, D.C.

Marcel Dekker, Inc. New York • Basel • Hong Kong

Library of Congress Cataloging-in-Publication Data

Water contamination and health: integration of exposure assessment, toxicology, and risk assessment / edited by Rhoda G.M. Wang.
p. cm.—(Environmental science and pollution control series: 9)
Includes index.
ISBN 0-8247-8922-9
1. Water—Pollution—Health aspects. 2. Water—Toxicology.
3. Health risk assessment. 4. Water—Pollution—Health aspects—United States. I. Wang, Rhoda G.M. II. Series:
Environmental science and pollution control: 9.
RA591.W32 1994
363.73'94—dc20 93-48586
CIP

The publisher offers discounts on this book when ordered in bulk quantities. For more information, write to Special Sales/Professional Marketing at the address below.

This book is printed on acid-free paper.

Marcel Dekker, Inc.
270 Madison Avenue, New York, New York 10016

Current printing (last digit):
10 9 8 7 6 5 4 3 2 1

PRINTED IN THE UNITED STATES OF AMERICA

Preface

We all realize the water we drink should be clean and safe, free of disease-causing organisms and harmful chemicals, and that minute quantities of contaminants present in water should meet the drinking water standards set by the U.S. Environmental Protection Agency (EPA) for protecting public health. Another consideration is concern about routes of exposure to water contaminants other than ingestion. Should we be concerned about additional doses of water contaminants (e.g., volatile organic chemicals) that may be attributed to inhalation and dermal routes of exposure, such as while showering or bathing? What kind of health hazard do we face when we swim in a pool treated with chlorinated chemicals for disinfection purposes? It is quite obvious that an added dose of a water contaminant to the ingested dose through inhalation and dermal exposure may exceed the "safe dose" allowable, which is defined by the maximum contamination levels (MCL).

This book addresses the various concerns and the pros and cons of including all routes of exposure to water contaminants for health risk assessment. The unique feature of this book is the integration of exposure assessment, toxicology, and risk assessment in a single volume, with prominent contributors applying their expertise in an array of diversified yet closely related fields to deal with the intricate problems of water contaminants and health. The diversification of techniques applied in this book was consolidated into five major sections, as discussed here, which clearly indicates this is not just another book about "water." Rather, it addresses the profound questions dealing with water contaminants and the various scientific approaches entailed.

The first section, "Water Contaminants: Monitoring, Treatment, and Health Impact," covers both U.S. EPA and State of California efforts in the monitoring and treatment of water contaminants, including the various regulatory mandates. The discussion on the historical development of public health awareness toward water contaminants and the most current application of health risk assessment is presented in two chapters to provide readers with a rather amusing view of how far modern technology has come.

The second section, "Exposure Assessment of Water Contaminants: Specific Dosimetry on Radon and Chloroform," emphasizes exposure assessment and dose, an extremely important component of modern health risk assessment. The importance of incorporating various routes of exposure to water contaminants is introduced in this section for the evaluation of potential health risks. Also introduced are more recent viewpoints at the federal and state levels on this subject. Chapters include indoor exposure to radon from water sources and exposure to chloroform in swimming pools. These chapters contain actual monitoring data and incorporate all potential routes of exposure in computing a theoretical total dose. Inhalation exposure is an indisputable major route of exposure to radon in domestic water, and both inhalation and dermal routes of exposure to chloroform are relevant to swimming pool exposure monitoring.

The next section, "Methodology Development in Exposure Assessment and Dose Estimates," is considered the focal point of this book. It discusses the most up-to-date techniques for dose estimation. The internal dose concept using a physiologically based pharmacokinetic model is emphasized here, with various examples and different approaches. Some chapters may appear highly mathematical and difficult to appreciate by audiences with less mathematical background. However, they represent some of the most advanced scientific approaches of which we all should be aware and that invariably will have impact on the computation of exposure in deriving "dose" under a variety of exposure scenarios. In dealing with many water contaminants, a screening approach may be needed initially to sort out a large number of chemicals of exposure concern. Attempts were made to develop a firsthand screening strategy that may be of some use to risk assessors.

The last sections, "Human Health Risk Assessment and Water Contaminants" and "Epidemiological Evaluation of Water Contaminants and Toxicological End Points," deal strictly with modern risk assessment techniques applicable to water contaminants. Each chapter was contributed by one or more experts in the field and can be regarded truly as masterpiece documents. The epidemiologic section emphasizes an emerging need to pinpoint the actual correlation of chemical exposure and human health effects. Since the ultimate goal of risk assessment in humans is to discern the type of adverse health effects that may be evoked in a population exposed to certain chemicals, no other results are more representative and accurate than those derived from human epidemiology studies.

For any given population, however, numerous chemical exposures may coexist. Even when a single chemical source can be identified, the dose derived from such an exposure is often unknown. Other confounding factors render such studies less successful. Because of the scope of this book, no further discussion is made here. It is clearly recognized that a concerted effort is needed to generate a high-quality and defensible epidemiological background to fill the data gap of human exposure to toxic chemicals and the adverse health effects that may result. We are obliged to provide the public with solid data whether a positive or negative association of an exposure exists and whether an adverse effect has been identified. Chapter 22, "Are There Carcinogens in Water for Human Consumption?" is highly recommended to readers. The peer review panels of Marcel Dekker, Inc. singled out this chapter as the most outstanding paper.

The audience for this book may include scientists in federal and state government agencies and in the private sector dealing with water and waste products, as well as consultants for both government and private industry. Because of the highly technical nature of this book, it is recommended for university-level professionals, which includes students who may use it as a reference book, both in this country and abroad. When this topic was first presented on a much smaller scale at the 1992 annual meeting of the SOT, the response was overwhelming. Surely this book does not provide all the answers, or even attempt to solve such a controversial problem. It is certain, however, that the book provides a challenge and the means to deal with problems associated with water contaminants and health.

Rhoda G. M. Wang

Contents

EXPOSURE ASSESSMENT OF WATER CONTAMINANTS: SPECIFIC DOSIMETRY ON RADON AND CHLOROFORM

METHODOLOGY DEVELOPMENT IN EXPOSURE ASSESSMENT AND DOSE ESTIMATES

Contributors

Charles O. Abernathy Human Risk Assessment Branch, U.S. Environmental Protection Agency, Washington, D.C.

A. John Bailer Department of Mathematics and Statistics, Miami University, Oxford, Ohio, and Division of Standards Development and Technology Transfer, National Institute for Occupational Safety and Health, Cincinnati, Ohio

Andrew Battin Office of Information Resources Management, U.S. Environmental Protection Agency, Washington, D.C.

David L. Bayliss Office of Research and Development, U.S. Environmental Protection Agency, Washington, D.C.

Jerry N. Blancato Exposure Assessment Division, U.S. Environmental Protection Agency, Las Vegas, Nevada

Robert S. Boethling Economics, Exposure, and Technology Division, Office of Pollution Prevention and Toxics, U.S. Environmental Protection Agency, Washington, D.C.

Kenneth T. Bogen Health and Ecological Assessment Division, Lawrence Livermore National Laboratory, University of California, Livermore, California

Joseph P. Brown Office of Environmental Health Hazard Assessment, California Environmental Protection Agency, Berkeley, California

Ronald Brown International Life Sciences Institute, Risk Science Institute, Washington, D.C.

Annette L. Bunge Department of Chemical Engineering and Petroleum Refining, Colorado School of Mines, Golden, Colorado

Arthur O. S. Chiu Office of Research and Development, U.S. Environmental Protection Agency, Washington, D.C.

Nancy Chiu U.S. Environmental Protection Agency, Washington, D.C.

Bill W. Colston, Jr. Environmental Sciences Division, Lawrence Livermore National Laboratory, University of California, Livermore, California

Douglas J. Crawford-Brown Department of Environmental Science and Engineering, University of North Carolina, Chapel Hill, North Carolina

James Darr Office of Pollution Prevention and Toxics, U.S. Environmental Protection Agency, Washington, D.C.

Anna M. Fan Office of Environmental Health Hazard Assessment, California Environmental Protection Agency, Berkeley, California

Jeffery W. Fisher Occupational and Environmental Health Directorate, Armstrong Laboratory, Wright-Patterson Air Force Base, Ohio

Gordon L. Flynn College of Pharmacy, The University of Michigan, Ann Arbor, Michigan

John M. Fraizer The Johns Hopkins University, Baltimore, Maryland

Richard H. Guy School of Pharmacy, University of California, San Francisco, California

L. E. Haskell University of Maine, Orono, Maine

C. T. Hess University of Maine, Orono, Maine

Kim Hoang Office of Research and Development, U.S. Environmental Protection Agency, Washington, D.C.

Russell Kinerson Office of Water, U.S. Environmental Protection Agency, Washington, D.C.

Richard H. F. Lam Office of Environmental Health Hazard Assessment, California Environmental Protection Agency, Berkeley, California

Lesia K. Machicao Environmental Sciences Division, Lawrence Livermore National Laboratory, University of California, Livermore, California

Howard I. Maibach Department of Dermatology, University of California School of Medicine, San Francisco, California

Jackie Moya Office of Research and Development, U.S. Environmental Protection Agency, Washington, D.C.

Kenneth A. Poirier Environmental Criteria and Assessment Office, Office of Research and Development, U.S. Environmental Protection Agency, Cincinnati, Ohio

Christopher J. Portier Division of Intramural Research, National Institute of Environmental Health Sciences, Research Triangle Park, North Carolina

Lorenz Rhomberg Office of Health and Environmental Assessment, U.S. Environmental Protection Agency, Washington, D.C.

John Schaum Office of Research and Development, U.S. Environmental Protection Agency, Washington, D.C.

Frank E. Scully, Jr. Department of Chemistry and Biochemistry, Old Dominion University, Norfolk, Virginia

Theresa A. Shepard College of Pharmacy, Rutgers—The State University of New Jersey, Piscataway, New Jersey

David L. Storm Office of Drinking Water, California Department of Health Services, Sacramento, California

Robert R. Vanderslice Rhode Island Department of Health, Providence, Rhode Island

Susan F. Velazquez Environmental Criteria and Assessment Office, Office of Research and Development, U.S. Environmental Protection Agency, Cincinnati, Ohio

Rhoda G. M. Wang* Office of Science and Technology, Human Risk Assessment Branch, U.S. Environmental Protection Agency, Washington, D.C.

Clifford P. Weisel Department of Environmental and Community Medicine, University of Medicine and Dentistry of New Jersey—Robert Wood Johnson Medical School, Piscataway, New Jersey

Ronald C. Wester Department of Dermatology, University of California School of Medicine, San Francisco, California

---

**Present affiliation*: Department of Pesticide Regulation, State of California Environmental Protection Agency, Sacramento, California

William N. White Department of Chemistry, University of Vermont, Burlington, Vermont

David Young Department of Pharmaceutics, University of Maryland at Baltimore, Baltimore, Maryland

1

A Retrospective on Drinking Water

Charles O. Abernathy

U.S. Environmental Protection Agency, Washington, D.C.

I. INTRODUCTION

Water is essential for all organisms; it comprises 60–95% of the weight of living cells (1), and losses through various metabolic and excretory processes must be balanced by an adequate intake. However, water may contain substances, whether natural or anthropogenic, that can affect the quality and existence of life.

It is important to distinguish between pure water and safe water. *Pure water* may be defined as water that is free of extraneous substances (2), whether harmless or not, and, from a practical standpoint, impossible to produce. On the other hand, *safe water* is water that is not likely to cause undesirable or adverse effects (2), although it may contain various contaminants. The crux is embodied in the definition of "safe." In the regulatory sense, *safe* means that, although the "purified" water may contain some contaminants, the risks imposed by those contaminants are of an acceptable nature. For example, chlorination is used extensively to disinfect water in the United States. This process introduces trihalomethanes (THM) into the finished product, however, and THM pose potential health risks (3). As expected, differences of opinion exist about the "safety" or "acceptable nature" of contaminants in drinking and surface water, and there is considerable public awareness of this problem, as evidenced by recent articles in the *New York Times* (4), *Washington Post* (5), and *U.S. News and World Report* (6).

It is also informative to consider the business aspects of water purification. The production of potable water in the United States is a very large industry. Moeller (7) estimates that it costs $5 billion/year to operate the 60,000 water treatment facilities and that these plants have required a capital investment of approximately $250 billion. These facilities produce 40–50 billion gallons/day (approximately 180 gallons per capita). Over one-half is used in the home, and nearly one-third is consumed by industry. The rest is used by public services or lost in the distribution process (Figure 1). In 1985, approximately 60% of the 36,500 million gallons of water withdrawn per day in the United States came from surface water; the remainder was from groundwater sources (8).

The objectives of this chapter are to provide a brief history of water use and purification, to give a few examples of the types of adverse effects associated with water contaminants, to mention some risk assessment methods, and briefly to consider drinking water regulations. Details of risk assessment methodology for specific end points or for various exposure routes are covered in subsequent chapters.

II. HISTORY

Since many writings have been lost, our knowledge of the early history of drinking water treatment is stochastic in nature (Table 1). The following is a brief overview; those wishing a more detailed history are referred to Baker (9).

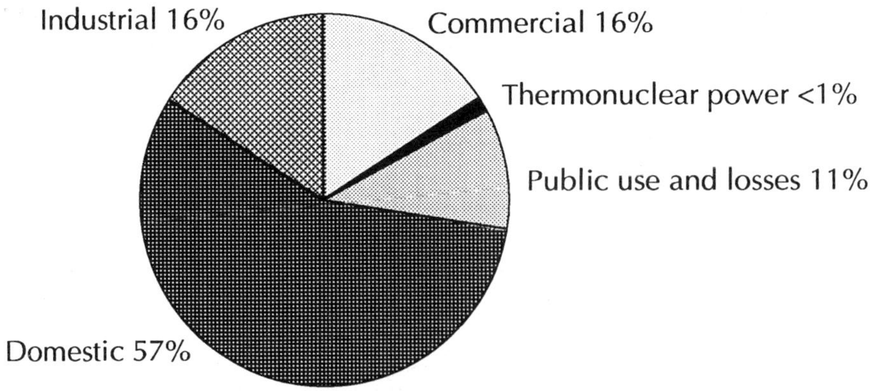

Figure 1 Percentage use by various sectors of municipally treated water in the United States in 1985. (From Reference 8.)

Table 1 Important Dates in the History of Drinking Water Treatment

Date	Comment
2000 BCE, India	Exposure to sunlight, boiling, and filtration through charcoal or sand
1500 to 1300 BCE, Egypt	Use of wick siphons
600 BCE, Persia	Cyrus the Great took boiled water for his troops
400 to 200 BCE, Greece	Boiled and filtered water through cloth and porous vessels
97 CE, Rome	Frontius was first water engineer
Eighth century	Geber used distillation to purify water
Eleventh century	Avicenna recommended boiling to purify water while traveling
Fifteenth century	Venice had filter-cisterns
1675	Walcott given the first water purification patent (distillation)
1804	Gibb used filtration to provide purified water to Paisley, Scotland
1806	Happey opened a plant in Paris for purifying water
1807	Glasgow was first city to pump purified water to inhabitants
1855	Snow demonstrated that cholera was spread by contaminated water
1894	Chlorination used to halt typhoid epidemic in Adriatic area
1896	Chlorine gas used to purify water in Louisville, Kentucky
1902	First permanent chlorination plant established in Belgium
1908	Large-scale chlorination of water in Boonton, New Jersey
1941	Over 85% of U.S. water treatment by chlorination

Source: Compiled from References 9 and 10.

The earliest records of water purification are from about 2000 BCE (before the common era) in India. Early Sanskrit writing mentions (1) keeping water in copper pots, exposing it to the sun, and filtering through charcoal, and (2) heating over a fire and filtering through gravel and sand (9). During the fifteenth to thirteenth centuries BCE, Egyptian records mention using wick siphons to clarify water. In the ninth century BCE,

Lycurgus of Sparta devised a drinking cup. It was dark and porous, which apparently served two purposes. The darkness hid the color of the water, and the porosity caused the mud in the water to stick to the sides. From a military as well as hygienic standpoint, it is of interest to note that Cyrus the Great took boiled water for his troops during the sixth century BCE (9). Hippocrates (460 to 377 BCE) recognized that water was important to health but concentrated on finding "good" water. He mentioned that rainwater should be boiled and strained through a cloth bag. This cloth bag has been designated "Hippocrates' sleeve" (10). The writings of Socrates, Plato, and Aristotle (fifth to fourth centuries BCE) mention the use of siphons and filtration through porous vessels (9).

In 97 CE, Sextus Julius Frontius became the water commissioner of Rome. He developed settling reservoirs for water, wrote a treatise on public water supply, and deserves to be called the first water engineer (9).

The earliest use of distillation to purify water was by the Arabian alchemist, Geber, in the eighth century. In the eleventh century, Avicenna, a Persian physician, recommended boiling over distillation or sedimentation as a means for travelers to purify water (9).

Although it is impossible to determine when filtration was first used, the Venetians are known to have operating filter-cisterns in the mid-1500s. Their use means that they were probably the first city to have a form of "treated" water (9).

The possibility of purifying water for towns brought business opportunities. On October 25, 1675, William Walcott of England was given the first patent for water purification by distillation, and in 1683, Robert Fitzgerald was also given a British patent for distillation. Not unexpectedly, a legal battle between the two parties ensued and the Walcott patent was set aside (9). During this period, Luc Antonio Porzio, an Italian physician concerned with soldiers' health, described the first sand filters. In 1685, he used multiple filtration through sand after straining and sedimentation. The first known depiction of water purification for industrial use is from a 1698 French manuscript on papermaking (9). In the early 1700s, the French and English proposed different types of filters and the use of sulfuric acid to purify water. By midcentury, Dr. Francis Horne was experimenting with water softening, and later, Lowitz recommended using charcoal to purify and preserve water. In France, Joseph Amy was issued the first water filter patent in 1749. He used sponges and sand as filters and described his filters in a promotional booklet. Of note during this period is Johanna Hempel, a potter, who was granted the first patent for the manufacture of household filters. She is the only woman inventor mentioned in the early history of water filtration (9).

In 1804, J. Gibb completed the first mechanical filter system to supply water to the town of Paisley, Scotland. However, the water was carted to consumers. In 1807, Glasgow became the first city to distribute purified water to its population (9). Subsequently, more communities started receiving filtered or treated water, albeit of varying quality. The safety of water was uncertain since "clear" water could cause disease, but "cloudy" water did not. Dr. John Snow of London, in 1855, was the first to prove clearly that the spread of a disease (cholera) was due to water contamination. Making this discovery even more impressive is that he did it before the advent of the germ theory of disease. Subsequently, in 1884, Robert Koch isolated *Vibrio cholerae*, the causative agent of cholera (10).

The importance of water filtration in disease reduction was also clearly shown in Germany. Although Altona and Hamburg both obtained their water from the Elbe, Altona, situated below Hamburg, used filtration because its water was more contaminated. During an 1892 cholera epidemic, Koch reported that the cholera incidence in Altona was much lower than that of Hamburg. Since the cholera bacteria were present in feces and contaminated the Altona water, it was accepted that filtration decreased the incidence by removing bacteria from the water (10).

In the United States, typhoid fever was the major concern. Hazen (11) reported that the death rate from typhoid fever in Lawrence, Massachusetts dropped 79% in the first 5 years after the introduction of sand filters. In a comparison of two New York cities, it was reported that the death rates in Albany due to cholera and diarrhea and for children under 5 decreased 54% after the introduction of sand filtering (1900–1904); in Troy, during the same years, it decreased only 16% (2).

During the middle to late nineteenth century, a variety of water treatment procedures were tried. They included voltaic action, electrolysis, and addition of various oxidizing agents, including chlorine products. It is of interest that chlorination of water was proposed in 1835 by Dr. R. Dunglinson even though there was no bacterial theory of disease (see earlier). Evidently, relatively large quantities were used and the resulting taste and odor were problems. In 1894, chlorination with a bleaching powder was employed to halt a typhoid epidemic in Pola on the Adriatic Sea. Chlorine gas was used by Jewell in Louisville, Kentucky in 1896 to purify water. He was requested to stop because it was decided that the process would not generally be useful (9). The first permanent chlorination plant was established in Belgium in 1902, and the second was placed in operation in England in 1905. In the United States, chlorination on a large scale first occurred in New Jersey in 1908, when Dr. R. Leal proposed the use of chloride of lime as a purifying agent. The use of liquid chlorine first occurred in New York in 1912 at the Niagara Falls filter plant, but the first permanent liquid chlorine treatment facility was in Philadelphia 1 year later. The use of chlorination rapidly increased, and by 1941, over 85% of water treatment facilities in the United States used chlorination (9).

Ozonation was first used for water purification in 1893 at Oudshoorn, the Netherlands. The initial use of ozone in the United States for taste and odor control occurred 13 years later at New York City's Jerome Park Reservoir (12). In the United States, ozonation was used to approximately the same extent as chlorination in the early twentieth century. The cost of installation and operation are greater than for chlorination, however, and its use diminished (9). In Europe, ozonation is still widely used, and most of the more than 2000 ozonation plants are located there (13). Because of growing concern over the potential adverse health effects of THM and new rules and regulations on disinfection by-products and organic chemicals, interest in ozonation has grown in the United States (13).

In recent years, much concern has centered around the possible toxicological effects of disinfection agents and their by-products (3). The nature of the adverse effects of chlorine or any disinfectant that may be proposed as a substitute for chlorine needs to be better characterized, however, before any final judgments can be adequately made. At the present time, it appears that there is little concern for the amounts of residual chlorine in finished water (3), but less is known about chloramine, which has been proposed as a substitute in water purification. Data on the by-products of ozonation are scant, but ozonation can cause an increase in bromine-containing by-products, and the production of bromates is cause for concern (3). In addition, ozone does not afford residual disinfection in pipelines and pipes, as does chlorine.

III. TYPES OF CONTAMINANTS

To facilitate discussion, contaminants are divided into three groups: (1) pathogenic organisms, (2) inorganic chemicals, and (3) organic chemicals. Since this is an overview, only a few specific examples are described; those requiring more information are referred to several reviews and books (3, 10,14–24).

A. Pathogenic Organisms

It is evident from the water purification attempts throughout history that humans realized that drinking water could be unsafe. As mentioned, however, Dr. John Snow was the first to demonstrate a causal link between drinking water and a disease (cholera). Several other infectious diseases may be transmitted by contaminated water. Bacterial diseases include typhoid fever, campylobacteriosis, salmonellosis, and shigellosis, in addition to cholera; hepatitis A, Norwalk virus, and rotavirus gastroenteritis are examples of water-transmitted viral diseases. Parasitic diseases, such as amebiasis and giardiasis, are also transmitted via water (25).

Before 1920, data on waterborne disease outbreaks were not recorded in the United States, and it is impossible to assess the number and seriousness of waterborne diseases during that period. From 1920 to 1980, 1405 outbreaks of infectious disease attributable to contamination of water were reported. Associated with these occurrences were over 386,000 illnesses and about 1100 deaths. Most of the deaths were attributable to typhoid fever and occurred before 1936. Since 1960, no deaths attributed to typhoid fever have been reported in the United States (26). Between 1981 and 1983, over 20,000 cases of illness associated with 112 disease outbreaks were reported in the United States. The most frequently identified causative organism was *Giardia lamblia*, however, a noticeable difference from the 1920–1980 era (26,27). In the 1980s, the use of contaminated groundwater and water treatment errors accounted for over 70% of the outbreaks (27). Compared with the eighteenth and nineteenth centuries, waterborne disease outbreaks of infectious diseases in the latter half of the twentieth century have been minor. However, we must continue to treat water because problems could recur. For example, consider the recent cholera epidemic in Peru and elsewhere in South and Central America. Not properly chlorinating the water supply permitted cholera bacteria to survive in the water distribution system. Since January 1991, over 300,000 cases of cholera and 3516 deaths have been attributed to this outbreak (28). When considering the potential or projected health effects of chlorination or other treatment procedures, one must place the risk into proper perspective and factor in the real risk of illness and death due to infectious agents.

B. Chemicals

It has been estimated that approximately 4 billion m^3 (1.5 trillion gallons) of contaminants enter the soil each year to pose a threat to groundwater (29). These compounds arise from industrial wastes, agricultural uses, and domestic refuse. The following discussion focuses on a few chemicals to illustrate potential hazards.

1. Inorganic Chemicals

Several inorganics found in drinking water are of concern (19,21,30). They include such compounds as arsenic (As), cadmium, chromium, lead (Pb), nitrite/nitrate, and sulfate (SO_4). Their effects range from cancer (As) to neuropathy (Pb) to a diarrhea (SO_4) that is potentially harmful, especially in infants (21).

Arsenic is used for illustration. Exposure to this chemical in drinking water can result in cancer, peripheral neuropathies, vascular effects, and dermal lesions (31–34). At the present time, an assessment of the carcinogenicity and toxicity of As is underway at the U.S. Environmental Protection Agency (EPA) to promulgate a new standard for drinking water. Complicating this assessment is that As may be an essential trace element (ETE) (35), although at the present time the U.S. EPA does not consider it to be essential (34).

With known ETE, however, other problems may arise. The "threshold" concept for the reference dose (RfD; see later) is based on the idea that exposures from zero to some finite level do not cause untoward effects in organisms (36). With ETE, deficient, as well as excessive, intakes can cause adverse effects (35,37). When deriving RfD for the ETE, the U.S. EPA must consider, along with the toxicity studies, the nutritional requirements, dietary intake levels, and bioavailability of each ETE (37). However, when comparing an RfD for a specific ETE with its recommended dietary allowance (RDA), it is necessary to realize that the U.S. EPA uses RfD to regulate chemicals and the RDA is a recommended intake level by the Food and Nutrition Board of the National Research Council (35).

2. Organic Chemicals

In a 1980 survey, over 400 separate organic chemicals were found in drinking water (38). Other than THM resulting from water purification processes, the most commonly occurring compounds were trichloroethylene, 1,1,1-trichloroethane, tetrachloroethylene, *cis*- and *trans*-1,2-dichloroethylene, 1,1-dichloroethane, and carbon tetrachloride (39,40).

After chlorination, chloroform, bromoform, dichlorobromomethane, and dibromochloromethane are the main THM detected in finished drinking water (3). In general, the primary short- and long-term health effects of the THM are on the liver and kidney (41,42). In various cancer studies, administration of various THM in oil has been reported to cause hepatic, renal, and intestinal tumors (43–47). However, when chloroform was given in drinking water, it did not induce hepatic tumors in mice but renal tumors in male rats were seen (48). Additional research is needed in this area.

In addition to these chemicals, various pesticides have been reported to contaminate drinking water sources. For example, 1,2-dibromo-3-chloropropane (DBCP) has been found in many wells in California (30), and ethylene dibromide (EDB) has been detected in water supplies in Florida, Georgia, and other states (21). Both are potent carcinogens in animals species, and DBCP has caused sterility and testicular injury in persons manufacturing or using it (21,30). Both DBCP and EDB have been banned from all or most uses.

IV. RISK ASSESSMENT

Risk may be considered the probability of an agent causing an adverse effect after an organism has been exposed to the agent. Each human risk assessment scenario is specific for the chemical and is composed of a hazard identification, a dose-response correlation, a human exposure assessment, and a risk characterization. The risk may be described in qualitative terms (high or low) or quantitated by use of a mathematical equation (21). Although human data are occasionally available, for example for arsenic, the hazard identification usually relies on animal data, and it is important to realize that the adverse effects noted in animals occur, or are expected to occur, in humans exposed to the same agent. The dose-response curve gives a measure of the potency of the compound in causing the effect(s). However, most dose-response curves involve animals that are exposed to relatively high doses. Accordingly, when applying such data to humans, it is necessary to consider whether high and low doses exert the same or similar effects. In the absence of conflicting data, it is assumed that they do. In addition, one must account for variations between animals and humans and intrahuman variations. In evaluating human exposure, one must estimate the number of people exposed and variations, such as magnitude and duration, of exposure (21). Since adequate human data are generally scant, this exercise usually employs exposure models. The results of the hazard identification, dose-response curve, and human exposure estimates are combined to characterize the risk for humans. For regulatory purposes, this characterization is divided into a noncancer (*threshold*) or a cancer (*nonthreshold*) scenario.

A. Noncancer

For risks involving noncarcinogens, the U.S. EPA and other organizations, such as the World Health Organization (WHO), use the threshold approach. Underlying this concept is the theory that exposure to some finite amount of an agent is not harmful (36). To derive an RfD at the EPA, a no observed adverse effect level (NOAEL) or lowest observed adverse effect level (LOAEL) obtained from an animal study is divided by uncertainty factors (UF) and/or modifying factors (MF). Generally, UF are an order of magnitude and are used to account for unknowns, such as intra- and interspecies variations, conversion of a LOAEL to a NOAEL, or the use of a subchronic study. The MF has been used for such differences as bioavailability, lack of a sensitive end point, or tolerance to a chemical. The RfD (basically equivalent to the acceptable daily intake used by the WHO and other regulatory or advisory groups) is calculated from the formula

$$\text{RfD} = \frac{\text{NOAEL (or LOAEL)}}{\text{UF} \times \text{MF}} \tag{1}$$

For a more detailed description of the RfD, the reader is directed to several reviews (36,37,49).

B. Cancer

In the risk assessment setting, carcinogens are treated as though they have no threshold; that is, a single molecule can cause the development of a cancer (21). This does not mean that each exposure results in cancer, but rather that any exposure increases the risk of developing cancer. In other words, the probability of developing cancer increases as the magnitude of exposure increases. As a general rule, only data from the exposure of relatively few animals to relatively high doses are available. Accordingly, it is necessary to utilize mathematical models to approximate the potential risk to humans. The U.S. EPA generally uses the linearized multistage (LMS) model for this purpose. The LMS is a conservative model and gives an estimate of the plausible upper limit for such risk. Within 95% confidence limits, the actual risk is unlikely to be higher than that predicted by the LMS model (21,50).

V. DRINKING WATER REGULATIONS

The Safe Drinking Water Act of 1974 (51) as amended in 1986 (52) requires the EPA to establish regulations for 83 drinking water contaminants. Selection is based on the potential for causing adverse health effects and for known, or potential, occurrence in drinking water (53). The act mandates that, for each chemical, the EPA must develop a maximum contaminant level goal (MCLG) based solely on potential for causing adverse health effects. The MCLG is a nonenforceable health goal. A maximum contaminant level (MCL) is then established as close to the MCLG as feasible, after considering the economic and regulatory impacts, treatment technology, and analytical capabilities. The MCL is an enforceable standard. For a detailed account of this process, see Reference 21.

VI. CONCLUSIONS

The occurrence of diseases transmitted by pathogens in drinking water in past (2,9,25,26) and recent times (27,28) demonstrates the necessity of water purification. However, present methods of removing these pathogens involve various chemical treatments, primarily chlorination in the United States, which introduce potentially harmful compounds. In addition, chemicals from industry and agriculture also add to the contamination. Public health organizations must not only balance the beneficial and adverse effects of chemicals involved in drinking water purification but also limit the levels of other contaminants in water supplies. To accomplish this task, the development and use of appropriate risk assessment methodologies are required.

ACKNOWLEDGMENTS

The author thanks Drs. N. Chiu, K. Khanna, and J. Murphy for their comments and the U.S. EPA headquarters library staff for their assistance in the literature search and retrieval of references. The opinions and conclusions expressed in this article are those of the author and do not necessarily reflect those of the U.S. Environmental Protection Agency.

REFERENCES

1. F.B. Armstrong, *Biochemistry*, 3d ed., Oxford University Press, New York, 1989, p. 19.
2. G.C. Whipple, *The Value of Pure Water*, John Wiley and Sons, New York, 1907, p. 58.
3. R.J. Bull and F.C Kopfler, *Health Effects of Disinfectants and Disinfection By-Products*, AWWA Research Foundation, Denver, CO, 1991.
4. K. Schneider, U.S. pushing states to curb water pollution, *New York Times*, November 7, 1991, p. A23.
5. M. Weisskopf, EPA's bind: Disinfecting water may cause cancer, *Washington Post*, October 10, 1991, p. A43.
6. B. Carpenter, S.J. Hedges, C. Crabb, M. Reilly, and M.C. Bounds, Is your water safe? *U.S. News & World Rep.*, July 29, 1991, pp. 50–55.
7. D.W. Moeller, *Environmental Health*, Harvard University Press, Cambridge, MA, 1992, pp. 60–62.
8. W.B. Solley, C.F. Merck, and R.R. Pierce, *Estimated Use of Water in the United States in 1985*, U.S. Geological Survey, Circular No. 1004, Washington, D.C., 1988.
9. M.N. Baker, *The Quest for Pure Water*, American Water Works Assn., Inc., 1949.

10. National Research Council, *Drinking Water and Health*, Vol. 1, National Academy Press, Washington, D.C., 1977, pp. 1–8.
11. A. Hazen, *Clean Water and How to Get It*, 1st ed., John Wiley, New York, 1904.
12. C.N. Haas, *Disinfection in Water Quality and Treatment. A Handbook of Community Water Supplies* (F.W. Pontius, ed.), McGraw-Hill, New York, 1990, pp. 877–932.
13. C.H. Tate, Survey of ozone installations in North America, *J. Am. Water Works Assn. 83*:40–47 (1991).
14. L.J. McCabe, J.M. Symons, R.D. Lee, and G.G. Robeck, Survey of community water supply systems, *J. Am. Water Works Assn. 62*:760–787 (1970).
15. World Health Organization, *Human Viruses in Water, Wastewater and Soil*, WHO Technical Report Series No. 639, WHO, Geneva, 1979.
16. United States Environmental Protection Agency, *Assessment of Microbiology and Turbidity Standards for Drinking Water*, U.S. EPA 570-9-83-001, Washington, D.C., 1983.
17. S.A. Esrey and J.P. Habicht, Epidemiologic evidence for health benefits from improved water and sanitation in developing countries, *Epidemiol. Rev. 8*:117–128 (1986).
18. G. Craun, *Waterborne Diseases in the United States*, CRC Press, Boca Raton, FL, 1986.
19. I.M. Sayre, International standards for drinking water, *J. Am. Water Works Assn. 80*:53–60 (1988).
20. R.A. Larson, *Biohazards of Drinking Water Treatment*, Lewis Publishers, Chelsea, MI, 1989.
21. U.S. Environmental Protection Agency, *Risk Assessment, Management and Communication of Drinking Water Contamination*, U.S. EPA 625/4-89/024, Washington, D.C., 1989.
22. N.M. Ram, R.F. Christman, and K.P. Cantor (eds.), *Significance and Treatment of Volatile Organic Compounds in Water*, Lewis Publishers, Chelsea, MI, 1990.
23. J.A. Goodrich, B.W. Lykins, Jr., and R.M. Clark, Drinking water contamination from agriculturally contaminated groundwater, *J. Environ. Qual. 20*:707–717 (1991).
24. U.S. Environmental Protection Agency, *Drinking Water Health Advisory: Pesticides*, Lewis Publishers, Chelsea, MI, 1989.
25. J.R. Harris, Clinical and epidemiological characteristics of common infectious diseases and chemical poisonings caused by ingestion of contaminated drinking water, In *Waterborne Diseases in the United States* (G. Craun, ed.), CRC Press, Boca Raton, FL, 1986, pp. 74–160.
26. G. Craun, Statistics of waterborne disease in the U.S. (1920–1980), In *Waterborne Diseases in the United States* (G. Craun, ed.), CRC Press, Boca Raton, FL, 1986, pp. 73–159.

27. G. Craun, Recent statistics of waterborne disease outbreaks (1981–1983), In *Waterborne Diseases in the United States* (G. Craun, ed.), CRC Press, Boca Raton, FL, 1986, pp. 161–168.
28. C. Anderson. Cholera epidemic traced to risk miscalculation, *Nature 354*:255 (1991).
29. M. Liston, Understanding well water problems, *Water Technol. 12*:36–37 (1989).
30. H.F. Russell, R.J. Jackson, D.P. Spath, and S.A. Book. Chemical contamination of California drinking water, *West. J. Med. 147*:615–622 (1987).
31. C.O. Abernathy and E.V. Ohanian, Noncarcinogenic effects of inorganic arsenic, *Environ. Geochem. Health 14*:35–41 (1992).
32. W.P. Tseng, Effects of dose-response relationships on skin cancer and blackfoot disease with arsenic, *Environ. Health Perspect. 19*:109–119 (1977).
33. J.T. Hindmarsh and R.F. McCurdy, Clinical and environmental aspects of arsenic toxicity, *CRC Crit. Rev. Lab. Sci. 23*:315–347 (1986).
34. U.S. Environmental Protection Agency, *Special Report on Inorganic Arsenic, Skin Cancer, Nutritional Essentiality*, U.S. EPA/625/3-87/013, Washington, D.C., 1988.
35. National Research Council, *Recommended Dietary Allowances*, 10th ed., National Academy Press, Washington, D.C., 1989.
36. D.G. Barnes and M. Dourson, Reference dose (RfD): Description and use in health risk assessments, *Regul. Toxicol. Pharmacol. 8*:471–486 (1986).
37. C.O. Abernathy, R. Cantilli, J.T. Du, and O.A. Levander, Essentiality versus toxicity: Some considerations in the risk assessment of essential trace elements, In *Hazard Assessment of Chemicals* (J. Saxena, ed.), Vol. 8, Hemisphere, New York, 1993, pp. 81–113.
38. W.E. Coleman, R.G. Mitton, F.C. Kopfler, K.A. Barone, T.A. Aurand, and M.G. Jellison, Identification of organic compounds in a mutagenic extract of surface drinking water by a computerized gas chromatography/mass spectrometry system (GC/MS/COM), *Environ. Sci. Technol. 14*:576–588 (1980).
39. J.J. Westrick, J.W. Mello, and R.F. Thomas, The groundwater supply survey, *J. Am. Water Works Assn. 76*:52–59 (1984).
40. J. Cotruvo, S. Goldhaber, and C. Vogt, Regulatory significance of organic contamination in the decade of the 1980s, In *Organic Carcinogens in Drinking Water* (N.M. Ram, E.J. Calabrese, and R.F. Christman, eds.), John Wiley and Sons, New York, 1986, pp. 511–530.
41. U.S. Environmental Protection Agency, *Drinking Water Criteria Document for Trihalomethanes (Final Draft)*, Office of Science and Technology, Office of Water, U.S. EPA, Washington, D.C., 1992.
42. T.A. Jorgenson and C.J. Rushbrook, SRI International, *Effects of Chloroform in the Drinking Water of Rats and Mice*: *Ninety-Day Subacute Toxicity Study*, U.S. Environmental Protection Agency, Menlo Park, CA, Contract No. 68-03-2616, Publication No. EPA-600/1-80-030, 1980.

43. T.A. Jorgenson, C.J. Rushbrook, and D.C.L. Jones, Dose-response study of chloroform carcinogenesis in the mouse and rat: Status report, *Environ. Health Perspect. 46*:141–149 (1982).
44. National Cancer Institute, *Report on Carcinogenesis Bioassay of Chloroform*, NTIS PB-264018, Springfield, VA, 1976.
45. National Toxicology Program, *Toxicology and Carcinogenesis Studies of Chlorodibromomethane in F344/N Rats and B6C3F1 Mice (Gavage Studies)*, U.S. Department of Health and Human Services, Research Triangle Park, NC, Technical Report Series No. 282, 1985.
46. National Toxicology Program, *Toxicity and Carcinogenesis Studies of Bromodichloromethane (CAS No. 75-27-4) in F344/N Rats and B6C3F1 Mice (gavage studies)*, U.S. Department of Health and Human Services, Research Triangle Park, NC, Technical Report Series No. 321, 1987.
47. National Toxicology Program, *Toxicology and Carcinogenesis Studies of Tribromomethane (Bromoform) in F344/N Rats and B6C3F1 Mice* (*Gavage Studies*), U.S. Department of Health and Human Services, Research Triangle Park, NC, Technical Report Series No. 350, 1989.
48. T.A. Jorgenson, E.F. Meierhenry, C.J. Rushbrook, R.J. Bull, and M. Robinson, Carcinogenicity of chloroform in drinking water to male Osborne-Mendel rats and female B6C3F1 mice, *Fund. Appl. Toxicol. 5*:760–769 (1985).
49. J. Patterson, S. Lutkenhoff, and W. Rish, Use of health risk estimates in U.S. EPA activities, In *The Analysis, Communication, and Perception of Risk* (B.J. Garrick and W. C. Gekler, eds.), Plenum Press, New York, 1991, pp. 547–554.
50. U.S. Environmental Protection Agency, EPA approach for assessing the risks associated with chronic exposures to carcinogens, Integrated Risk Information System (IRIS), Online, Intra-agency carcinogen risk assessment verification endeavor (CRAVE) Work Group, Office of Health and Environmental Assessment, Environmental Criteria and Assessment Office, Cincinnati, OH, 1992.
51. Safe Drinking Water Act, 42 U.S.C., 300f *et seq*., 1974.
52. Safe Drinking Water Act Amendments of 1986, P.L. 99-339, 1986.
53. *Federal Register 50*(219):46941 (1985).

2

Chemicals in California Drinking Water: Source of Contamination, Risk Assessment, and Drinking Water Standards

Richard H. F. Lam, Joseph P. Brown, and Anna M. Fan

California Environmental Protection Agency, Berkeley, California

I. INTRODUCTION

California has a population of 30 million, which continues to increase by an estimated 500,000 persons a year. This fast population growth rate has made the task of providing high-quality drinking water to all residents an increasingly challenging one. More than 40% of the total population is served by groundwater supplies, the rest obtaining their drinking water from surface waters from the Sierra snow melt and/or from the Delta watershed. Most rural areas depend on groundwater as the source of drinking water.

The responsibility of protecting California's ground and surface waters against toxic chemical or hazardous waste pollution is divided among various local, state, and federal agencies. The state agencies with major roles are the State Environmental Protection Agency (C/EPA), the State Health and Welfare Agency, and the State Resources Agency. Within the C/EPA are two organizations that are responsible for various water protection programs:

1. The State Water Resources Control Board (SWRCB) controls the quality of surface and groundwater under the Porter-Cologne Water Quality Control Act of 1969. The SWRCB assists and oversees the nine regional water quality control boards that formulate, adopt, and

implement water quality control policies within their own jurisdictions, as specified in the California Water Code, Section 13240.

2. The Office of Environmental Health Hazard Assessment (OEHHA) performs risk assessment of chemicals for evaluation of actual or potential contamination situations and establishes health-based permissible levels in drinking water, which are enforced by the Office of Drinking Water (ODW), California Department of Health Services (CDHS).

The ODW, CDHS in the State Health and Welfare Agency, pursuant to Title 22 of the California Administrative Code and the Safe Drinking Water Act of 1989, has the responsibility for quality control of drinking water supplies by the promulgation of drinking water standards. In the State Resources Agency, the Department of Water Resources (CDWR) manages the water resources of the state. It works closely with SWRCB and CDHS in the investigation of water contamination in both ground and surface waters.

The Department of Toxic Substances Controls (DTSC) and the Department of Pesticide Regulation (CDPR) within C/EPA are involved in activities that relate to and have impact on water quality. DTSC controls a number of chemical-handling activities that impact on surface and groundwater quality under the California Hazardous Waste Control Act of 1972, the Federal Resources Conservation and Recovery Act of 1976 (RCRA), and the state and federal superfund cleanup programs. CDPR regulates and administers controls on pesticide applications, pursuant to the California Food and Agricultural Code.

Water development in the state is controlled by its geography and climate. Highly populated areas and agriculture, the two principal users of water, are mostly centered in water-deficient areas. Many of these are in areas underlain by aquifers, which were once believed to provide unlimited water supplies. The contamination of many aquifers with industrial and agricultural chemicals has made water from these aquifers unsuitable for drinking purposes, making it necessary to import water from the Sacramento-San Joaquin Delta and other uncontaminated sources. Although state law requires water utilities to use the best available water source, in times of shortage there may be a need to use lower quality water from ground and surface water supplies. However, providing lower quality water even during shortage would not be in the best public interest.

Contamination by both natural and synthetic chemicals has been found in many of the state ground and surface water supplies. To assess the health risks associated with exposures to chemical contaminants present in treated drinking water, a review of drinking water contaminants must be determined. This report reviews the quality of drinking water from surface and groundwater sources. Risk assessments of many of the current derived contaminants in California drinking water are presented with supporting rationales on their use in the development of water standards.

II. DRINKING WATER QUALITY AND CONTAMINANT SOURCES

The quality of drinking water in the state depends on the source from which it is obtained. Most residents of the San Francisco Bay area obtain drinking water that originates as rain or snow in the Sierra Nevada. This water is brought to consumers by way of the Hetch Hetchy and Mokelumne Aqueducts, bypassing the Delta. According to traditional criteria, this water is considered pristine, the only treatment required being that against disease-causing organisms, such as *Giardia lamblia*, *Cryptosporidium*, and coliform bacteria, as required by the state surface water filtration and disinfection treatment rule. Southern California depends on surface waters from the Sacramento-San Joaquin River Delta and the Colorado River. Many communities in the Central Valley depend on groundwater as the principal source of drinking water. The blending of water from surface and groundwater sources is common among water utilities. Drinking water quality problems in some parts of the state are further complicated by some water systems withdrawing water from sources originally planned for agricultural and industrial uses. Further, during times of drought, when surface water is in short supply, the dependence on groundwater may become even more critical. Issues concerning the remediation of contaminated surface and groundwater and the control of pollutant release into these water sources is beyond the scope of this review.

A. Groundwater

1. Pesticides (AB 2021)

Groundwater resources of the state had been thought to be relatively safe from pollution by pesticides, until dibromochloropropane (DBCP) contamination was discovered in hundreds of wells in the San Joaquin Valley in the 1970s. Subsequently, DBCP has been found in over 2000 wells (both water supply and irrigation wells) statewide (1–3), and more than 50 other pesticides were found in groundwater of 23 countries (2,3). The widespread contamination of groundwater led to the enactment of AB 2021 (Article 15 of the Food and Agricultural Code), the Pesticide Contamination Prevention Act in 1985. The legislation requires the CDPR (formerly a division of the California Department of Food and Agriculture, CDFA) to maintain a statewide data base of wells sampled for pesticide active ingredients, which may contaminate groundwater by the agricultural use of pesticides. Before this, CDFA maintained a similar data base, but only from nonpoint sources (contamination that cannot be traced to a single location). Since the enactment of AB 2021, both nonpoint and point sources are monitored. The sampled water supply wells in the data base did not differentiate between those wells used as a source of drinking water and those used for irrigation. Many water supply wells, especially in rural areas, are known to have dual purposes.

To fulfill the requirements of the act, CDPR is required to provide a yearly update on the sampling for pesticide residues in California groundwater to the legislature, CDHS, and SWRCB. Yearly well sampling data are obtained from various federal, state, county, and private agencies (4–10). Between 1975 and 1991, the well inventory data base showed more than 15,000 wells sampled for pesticides, with more than 3000 showing the presence of one or more pesticides (Table 1). The most frequently detected pesticides in groundwater are shown in Table 2. DBCP is found in 2480 contaminated wells, followed by simazine (287 wells), diuron (155 wells), and atrazine (126 wells). Registrations have been canceled or suspended for several of the pesticides impacting groundwater supplies, and many others have been restricted or canceled in areas that were determined to be vulnerable to groundwater contamination by their continual use. DBCP,

Table 1 Sampling for Contaminants in Well Water Under Various California State Programs

	AB 1803 CDHS/ODW (1984–1990): large and small public water systems[a]	AB 2021 CDPR (1975–1991): well inventory[b]	SWRCB (1975–1987): well inventory[b]	CDHS/ODW (1984–1992): public water system[a]
Analytes	Organics	Pesticides	Nitrates	Nitrates
No. wells sampled	7712	15,238	38,144	NA[c]
No. positive wells	921	3,021	—	5131
No. wells over standard[d]	321	1,700 (estimated)	3,800 (estimated)	390

[a]Water supply wells.
[b]Water supply wells and irrigation wells.
[c]Not available.
[d]When no MCL is available, the AL or the U.S. EPA lifetime health advisory (HAL) is used.

Table 2 Most Frequently Detected Pesticides in Water Wells Under CDPR Well Inventory Data Base (1975–1991)

Chemical	Total no. contaminated wells	Maximum concentration detected (μg/L)
DBCP	2480	60
Simazine	287	19
Diuron	155	4
Atrazine	126	10
1,2-Dichloropropane (propylene dichloride)	91	62
Bromacil	90	15
Bentazon	64	20
Ethylene dibromide	54	380
Aldicarb	30	47
Aldicarb sulfone	32	5
Aldicarb sulfoxide	25	13
TPA (2,3,5,6-tetrachloroterephthalic acid)[a]	26	15
Prometon	22	1

[a]TPA is a breakdown product of the active ingredient chlorthal-dimethyl.

which is no longer registered for use in California, since 1979, is still the most frequently detected pesticide in groundwater supplies.

2. Organic Pollutants (AB 1803)

Public drinking water wells throughout California have been sampled by the CDHS since 1984, as specified under AB 1803 (Chapter 881, Statutes of 1983) to determine how many wells have been contaminated by organic pollutants. Monitoring was first conducted on large supply systems (more than 200 connections) and later on small systems (5–199 connections) (11,12). Between 1984 and 1990 7712 wells were sampled, and 921 of them had been found to be polluted with one or more chemicals (Table 1). Of these, 321 drinking water wells were found to exceed one or more state drinking water standards. Table 3 shows the organic chemicals most frequently detected in the monitoring of drinking water wells under AB 1803 (combined large and small public water systems). DBCP is the most frequent organic chemical found, followed closely by tetrachloroethylene (PCE) and trichloroethylene (TCE). Chloroform, which is found in many wells, is probably formed from the chlorination of well water. Wells with concentrations of organics above the maximum contaminant level (MCL) or action level (AL) were either closed or had remedial action taken to reduce the level below the drinking water standard. In all cases, water delivered

Table 3 Most Frequently Detected Organic Contaminants in Water Supply Wells Sampled Under AB 1803

Chemical	Total no. contaminated wells	Maximum concentration (μg/L)
Dibromochloropropane	275	7.4
Tetrachloroethylene	264	166
Trichloroethylene	217	538
Chloroform	169	120
1,1,1-Trichloroethane	99	202
1,1-Dichloroethylene	72	78
Carbon tetrachloride	45	29
Bromodichloromethane	43	40
Atrazine	42	2.4
Dibromochloromethane	39	30
1,2-Dichloroethane	36	24
1,2-Dichloroethylene[a]	36	100
Simazine	32	28
Bromoform	27	78
1,1-Dichloroethane	28	30
1,2-Dichloropropane	26	53
Trichlorofluoromethane	20	24

[a]Combined *cis*- and *trans*-1,2-DCE.

to consumers meets drinking water standards, although this is obtained in some instances by the blending of water from slightly contaminated wells with water from clean sources (13).

With the completion of the small public water system monitoring effort in 1990, the requirements of AB 1803 program were met. This survey sampling program is now being replaced with regulatory monitoring requirements pursuant to federal and state laws. New state regulations require that all public water systems found contaminated pursuant to AB 1803 be retested.

3. Nitrate

Although much attention has been focused on the contamination of groundwater supplies in the state by organic chemicals, the greatest threat to this source of drinking water is from contamination by nitrate. The principal concern for excessive nitrate levels in drinking water is the development of methemoglobinemia in infants. After ingestion, nitrate is converted to nitrite in the gut of the infant. The absorbed nitrite reacts with hemoglobin in the blood, forming methemoglobin, which significantly reduces the oxygen-carrying capacity of the blood. This gives rise to the "blue baby" syndrome. More detailed environmental and toxicological assessment of nitrate can be obtained from other publications (14,15).

Data compiled by the SWRCB (16) in a report to the legislature showed that of 38,144 wells (water supply and irrigation wells) sampled for nitrate, about 10% were found to have levels exceeding the current drinking water standard of 45 mg/L (Table 1). Monitoring by the CDHS (17) indicated that the same level of contamination was found in drinking water wells in the state, where 390 of 5131 nitrate-contaminated wells are found to be above the current MCL (Table 1). Contamination in wells were found in the coastal counties of Orange, Los Angeles, San Diego, Ventura, Monterey, Santa Clara, and Contra Costa and in an agricultural zone stretching from Kern County to Sacramento County.

More public water supply wells in California have been closed because of violations of the nitrate standard than any other contaminant or class of contaminants (18). Nitrate contamination and the loss of drinking water wells can be especially devastating to small water systems with less than 200 service connections. Many small water systems, after losing a principal source of drinking water supply to nitrate contamination, could not afford the cost of developing a new source or installing costly water treatment. The common remedial action for many of these water systems is to construct a new, deeper well or deepening an existing contaminated well. Larger water systems with multiple sources usually close the highly contaminated wells and have been able to meet the nitrate drinking water standard by the blending of water from various wells. Some water utilities have also installed ion-exchange nitrate treatment facilities.

Agricultural practice using nitrogen fertilizers resulted in the leaching of nitrate to groundwater. When the amount of nitrogen applied to the soil exceeds the crop utilization capacity and with water application rates resulting in deep percolation, increased nitrate loading to the groundwater can occur. The situation is aggravated by the decrease in the capacity of the groundwater volume to dilute this load. Other sources of nitrate are from dairy waste, septic tank leachate, and wastewater disposal.

4. Radon

Radon can enter the indoor environment via the drinking water supply, and inhabitants can be exposed to radon by the direct ingestion of drinking water or, more significantly, through the inhalation of radon and its daughter isotopes after radon is volatilized from portable water used for domestic purposes within the confines of the house (e.g., bathing). However, most of the radon present in household living spaces comes from soil and bedrock located directly beneath the foundation, and the major route of radon exposure is by inhalation.

In 1991, the U.S. EPA proposed a new maximum contaminant level goal (MCLG) of zero for radon and a MCL of 300 pCi/L_w. In this proposal, radon is classified as a group A human carcinogen causing lung cancer in humans and other animals, with estimated unit risks via the inhalation and ingestion routes as 4.9×10^{-7}/pCi/L_w and 1.5×10^{-7}/pCi/L_w, respectively (19). These estimates for lung cancer risk from radon exposure were based on human epidemiological data obtained from groups of underground, metal ore miners who smoked (19). In reviewing the relative risk from radon, the U.S. EPA estimated that for smokers, the estimated risks range up to 450% (average of 180%) of the overall average population risk. Because smokers were used in estimating the lung cancer risk from radon exposure, the risks of cancer due to radon exposure in the general population are probably overestimated.

Radon in drinking water was estimated to contribute between 1 and 5% of indoor airborne radon. The U.S. average radon level in public groundwater supplies is estimated as 650 pCi/L_w, which corresponds to approximately 200 lung cancer deaths per year. In comparison, radon in homes may account for as many as 40,000 lung cancer deaths annually (19,20). At present, radon in indoor air is not being regulated. The U.S. EPA uses a nonenforceable action level of 4 pCi/L_a in indoor air as a level above which corrective steps should be taken to reduce the risk of exposure. The individual lifetime cancer risk at the 4 pCi/L_a air action level is of the order of 10^{-2}. Using a 1:10,000 transfer efficiency ratio to determine the average radon concentration in indoor air contributed by the radon in the water supply, 300 pCi/L_w (proposed MCL) would generate 0.03 pCi/L_a radon in the air (19,21). Under the 1986 amendments to the Safe Drinking Water Act, the U.S. EPA is required to regulate radon and other radionuclides in drinking water but no legal authority was given to regulate the much higher risk of radon in indoor air. Water utilities are concerned that the proposed regulation, which will be expensive to implement, is only reducing a minor portion of the overall risks from radon exposure.

In California, radon contamination in groundwater is found in wells adjacent to granitic rocks or alluvial soils derived from granite, as in the San Joaquin Valley (22). Cretaceous-Tertiary sediments that contain sandstone or feldspar are also related to high radon contamination. Overall, the coastal areas of California contain few locations that have elevated radon contamination. Based on aquifer geology, the greatest potential of radon in groundwater (1000–10,000 pCi/L_w) are in the counties of Alpine, Amador, Calaveras, El Dorado, Fresno, Kern, Madera, Mariposa, Riverside, San Diego, Tulare, and Tuolumne (22). Monitoring recently conducted by the CDHS (22) on wells chosen on the basis of aquifer geology, found high radon concentrations (>2000 pCi/L_w) in wells located in high-radon rock formations. When large system wells were sampled on a random basis, however, more than 64% of the wells exceeded 500 pCi/L_w and 95% of the wells were above 200 pCi/L_w of radon. This study shows that at the proposed radon standard of 300 pCi/L_w, geological criteria cannot be used to determine a well's vulnerability for selective radon monitoring. It is estimated that in California, more than 7000 large and small water system wells will be out of compliance with the proposed radon rule.

5. Arsenic

Arsenic, a natural element, is widely found in the environment in both organic and inorganic forms. The inorganic forms of arsenic that occur as either arsenate [As(V)] or arsenite [As(III)] are much more toxic than the organic forms. Inorganic forms of arsenic rock deposits in underground strata are the main source of arsenic in drinking water.

In 1976, the U.S. EPA established an MCL of 50 ppb for inorganic arsenic based on skin cancer incidences in humans. This standard was subsequently adopted by the State of California. Since 1985, new human studies on the consumption of arsenic-contaminated drinking water have provided persuasive evidence that arsenic can also cause liver and lung cancer and increases the risk of bladder, kidney, and nasal cavity cancers in both sexes and prostate cancer in males (23–25). In a recent analysis of the human data, Smith et al. (25,26) estimated that the cancer potency of arsenic, based on skin cancer, to be 5.3×10^{-3}/μg/kg per day, with a 10^{-6} lifetime risk, which is equivalent to drinking water consumption of 2 L/day at the 6.2 ppt (0.0062 ppb) level. Comparable risks on other tumor sites, ranges from 0.2 in 1000 (liver cancer in males) to 17.2 in 1000 (lung cancer in females). The current drinking water standard of 50 ppb carries a risk of approximately 10^{-2}, or 7 in 1000. Detailed description of the quantitative risk assessment is available from other publications (25–27).

Arsenic is not a widespread contaminant of drinking water in California. Most public water systems have less than 25 ppb arsenic in their drinking water. A total of 15 water systems in six counties have arsenic concentrations above the 50 ppb MCL. In a monitoring study of water from large systems in the state, a significant number of arsenic detections (i.e., above the 10 ppb reporting limit) in drinking water wells occurred in the counties of Fresno, San Joaquin, Riverside, Kern, Stanislaus, Monterey, Tulare, Sonoma, Kings, Los Angeles, Merced, and Sacramento. In one area around North Edwards in the Mojave Desert, several drinking water wells have been found to have arsenic levels between 65 and 200 ppb, one with 406 ppb (17). Problematic water systems have been able to meet the 50 ppb arsenic water standard by the closing of highly contaminated wells and blending water from various wells. The reevaluation of arsenic using newly available data on internal cancers will see a lowering in the MCL, probably closer to the 1–2 ppb limit of detection (27). A new lower MCL will probably cause many water systems to be out of compliance.

B. Surface Water

1. California State Water Project

The California State Water Project (SWP) provides drinking water to over 20 million people in north and Southern California. The Sacramento and San Joaquin rivers are the two major rivers providing water to the Sacramento-San Joaquin Delta, but numerous smaller rivers also feed into the system. The SWP has 27 lakes and reservoirs, which impound 6.8 million acre-feet of water, and approximately 700 miles of canals and pipelines. The purposes of the project include providing municipal and industrial (M&I) and agricultural water supply, flood control, hydroelectrical power generation, recreation, fish and wildlife preservation, and water quality control. Drinking water destined for Southern California is pumped from the South Delta into the California Aqueduct. Four terminal reservoirs store water for delivery to southern water supply agencies.

2. Central Valley Project

The Central Valley Project (CVP), operated by the U.S. Bureau of Reclamation, also pumps water from the Delta, but its primary purpose is to provide water for agriculture in the Central Valley, not for drinking water purposes. However, this may soon change because of legislation (the Reclamation Projects Authorization and Adjustment Act, Title 34 of HR 429) signed into law by President Bush in 1992. In addition to changes in CVP operation, the act authorizes the transfer of CVP water outside its service area. This could benefit California's growing metropolitan areas, giving them an alternative source of drinking water.

3. Contaminant Sources

The quality of water in the Delta is affected by discharges in the watersheds of the Sacramento and San Joaquin Rivers and seawater intrusion from the San Francisco Bay. There are 149 M&I discharges into the Delta, the majority complying with current National Pollutant Discharge Elimination System (NPDES) permit requirements. Contaminants discharged from treatment plants are pathogens, nutrients, organics, and metals. Bacteria die off rapidly in receiving waters relative to viruses and *Giardia* cysts, which survive longer. Nutrients can stimulate algae and plant growth downstream of the discharge, leading to high concentrations of organic carbon at downstream water intakes (13). Organic carbon combined with disinfectants used at water treatment plants can produce trihalomethanes (THM) and other disinfection by-products (DBP). Organics and metals are reduced in raw water by adsorption to particulate matter and sedimentation (28,29).

With increasing urbanization of the Central Valley, especially in areas near the Delta, urban runoff discharges into the Delta have increased in recent years. Key contaminants in urban runoff are sediment, heavy metals, and petroleum hydrocarbons. Agricultural drainage contributes pesticides, herbicides, metals (including selenium), and nutrients to the SWP system. Most agricultural discharges are seasonal and/or episodic and are related to specific crop practices. There is concern that periodic discharges by some of the 450 or more agricultural drains in the Delta may go undetected. This can be a potential public health problem for small communities that take their drinking water in close proximity to the drain pipes. Other sources of contamination to the SWP include seasonal mine discharges (heavy metals, asbestos, mercury, and cyanide), seawater intrusion (sodium, chloride, and bromide), and others, such as highway and canal roadside drainage, pipeline overcrossings, illegal dumping, and accidental spills (28,29).

To assess the water quality in the Delta, various monitoring programs were carried out by the CDWR to sample for various contaminants in the water that may affect human health. Contaminants determined included trihalomethane formation potential, pesticides, asbestos, metals, and synthetic organic pollutants (28,29). Because of the importance of the Delta as a source of drinking water, the quality of Delta water has been under intense scrutiny in recent years (30,31). In 1991, in a report to the legislature (as required by California Senate Concurrent Resolution No. 55), the SCWRB reported that drinking water obtained from the Delta does not pose a serious health risk to California's population but poses a serious problem to the utilities charged with treating the water to meet current state and federal standards (32). It has been anticipated by water utilities that stricter standards for THM and new standards for other disinfectant by-products may make it difficult for utilities to achieve compliance.

a. Trihalomethanes

One major issue of concern with regard to the use of Delta water for drinking is the formation of trihalomethanes. THM are formed when certain organic substances dissolved in the water combine with chlorine used to disinfect drinking water. The most common THM compounds formed during chlorination include chloroform, bromodichloromethane, chlorodibromomethane, and bromoform. With the exception of chlorodibromomethane (placed in group C), the other THM compounds are classified by the U.S. EPA as probable human carcinogens (group B2). All four

THM compounds are currently treated by the state as probable human carcinogens. Other chlorinated by-products identified include various haloacids (dichloroacetic acid, trichloroacetic acid), the haloaldehydes (chloroacetaldehyde, dichloroacetaldehyde, and trichloroacetaldehyde), the haloketones (1,1,1-trichloroacetone, 1,1,3,3-tetrachloroacetone, and hexachloroacetone), the haloacetonitriles (dichloroacetonitrile, trichloracetonitrile, trichloromethylnitrile, bromochloroacetonitrile, and dibromoacetonitrile), the chloropicrins (nitrotrichloromethane and trichloronitromethane), and the chlorophenols (monochlorophenols, dichlorophenols, and trichlophenols). Only a few of these chemicals have been sufficiently characterized toxicologically as to their potential human health effects. Others have received limited testing or available data are of little use in risk quantification. Currently available toxicological data for these chlorinated by-products are summarized by Bull and Kopfler (33) and the National Academy of Sciences (NAS) (34).

Water utilities at present must reduce the total concentration of THM to 100 μg/L to meet current state and federal drinking water standards. The current standard was placed in effect in 1979 and applies only to water treatment systems serving more than 10,000 people. The new state surface water treatment rule, which came into effect on September 1, 1991, requires all approved surface water supplied by utilities to the public for drinking purposes to be treated with a continuous disinfection process. The process must inactivate 99.9% of *Giardia* cysts and 99.99% of viruses and must maintain residual disinfectant concentrations of not less than 0.2 mg/L for at least 4 h/day. Most utilities are able to comply with the new rule and still meet the total THM standard. The U.S. EPA is in the process of developing comprehensive regulation to control disinfectants and disinfection by-products. These new standards will apply to all public water systems and will include standards for individual THM and possibly also a standard for total THM and other DBP. The new standards arise primarily from an increasing concern about the reported risk of bladder and rectal cancer in humans associated with THM and DBP in drinking water treated with chlorine (35). In establishing the new rule, U.S. EPA has chosen to use the negotiated rule-making process to achieve a new workable standard. By this process, the U.S. EPA is trying to balance the risks involved with disinfection and its by-products with the additional costs and risks of waterborne disease.

There is concern about the technical feasibility and cost of meeting a more stringent standard when using Delta water because of its high THM formation potential. In the Delta, the total THM formation potential for water is generally lower at freshwater stations (e.g., American River and

Greene's Landing), than in the central Delta stations (e.g., Rock Slough, Clifton Court intake, Banks Headworks, Delta-Mendota Canal intake, Middle River, and Mallard Island) (30,31). The peat soils of the Delta islands is a major contribution to organic precursors for the formation of THM. Organic THM precursors also come from a variety of sources, including agricultural drainage, surface runoff, wastewater treatment plant discharge, seawater intrusion, and algae (30–32,36). Studies by the CDWR (30,31,36) indicated that agricultural drainage and urban storm water runoff can contribute significantly to the source of THM precursors in the Delta. In the Delta, bromides from seawater or estuarine water are important contributors to THM formation. During chlorination of bromine-rich Delta water, bromide ions compete with chlorine to form THM compounds. Because of its higher atomic weight, more brominated THM compounds than chloroform are formed, substantially increasing the concentrations of THM in the water. When surface water low in bromine is chlorinated, chloroform is the major THM compound formed. In general, untreated Delta water supplies have total THM formation potential values three to nine times higher than the THM standard for treated water.

The major challenge for water utilities is to sustain a high level of disinfection while minimizing the formation of by-products of disinfection that may be potentially harmful. Treatment processes have been modified, and many utilities have began to use alternative disinfectants to chlorine, such as chloramines, chlorine dioxide, and ozone. Studies on the use of other treatment options to reduce THM formation have shown that other DBP of health concern are being formed (34,37,38). THM are not formed when chlorine dioxide is employed as a disinfectant. However, chlorine dioxide is reduced to chlorite as it oxidizes organic matter in water, and part of it is then transformed to chlorate (38). Because of concerns about the toxicity of chlorite and chlorate (hematological effects in both humans and laboratory animals), the U.S. EPA has limited the use of chlorine dioxide to a concentration of 1 ppm, which renders its use as a primary disinfectant impractical. Chloramines, a significantly weaker disinfectant than chlorine, produces low levels of chlorinated and other organic by-products. It is unsuitable as a primary disinfectant when virus and/or parasites are a concern (34). Many water utilities are using chloramines to maintain a disinfectant residual in the distribution system following treatment. Ozone is the most reactive of the oxidants used in water treatment. Relatively little information exists about the resulting by-products formed. When bromide ion is present, as in Delta water, ozonation causes the formation of bromate. Bromate is at present being evaluated by the U.S. EPA as a probably human carcinogen. All alternative disinfecting chemicals are being studied, but current treatment technologies may be inadequate to reduce the level of DBP in treated Delta water to meet their anticipated drinking water standards (29).

Instead of searching for disinfectants that do not produce undesirable by-products, water utilities are also studying ways to reduce the formation of DBP. Options include the removal of precursors by pretreatment of source waters, modifying water conditions (e.g., pH), preventing overtreatment with disinfectant, and the removing of the DBP themselves.

b. Pesticides and Synthetic Organic Chemicals

Pesticides and synthetic organic chemicals have not been a problem in Delta water. In a monitoring program by the CDWR (28,29), 24 pesticides were detected at concentrations marginally above detection levels and none were above drinking water standards (Table 4). Paraquat was detected in one location (Vernalis) at 74 μg/L. There is currently no drinking water standard for paraquat, but this is higher than the U.S. EPA lifetime health advisory of 30 μg/L. Paraquat was not detected in other locations downstream from Vernalis. In this monitoring program, pesticide sampling periods were selected to coincide with summer pesticide application, winter surface water runoff, and spring preemergent herbicide application, so that there is a higher likelihood of the pesticides being found at the monitoring sites.

c. Selenium

Selenium is a known contaminant discharged from agricultural drainage. It is monitored in Delta water because of concern arising from the 1984 Kesterson National Wildlife Refuge episode in which deformities in young aquatic birds were attributed to high selenium levels in the refuge (39,40). The source of selenium was traced to the San Luis Drain, which emptied agricultural drain water containing selenium leached from soils in the western part of the San Joaquin Valley. Selenium is an essential nutrient to animals and humans at low levels of intake but can be toxic at high levels of exposure. Symptoms characteristics of overexposure in humans are nail changes and hair loss. Selenium concentrations in the Delta are meeting the 10 μg/L drinking water standard. Selenium-laden waters enter the San Joaquin River between February and March each year from Mud Slough and Salt Slough. During this period, elevated levels can be traced down the San Joaquin River and through the southern Delta. In monitoring studies, the highest concentrations are found in the lower San Joaquin River: 3 μg/L was measured in the Delta-Mendota intake, Banks Delta pumping plant, and Vernalis (28). Because of dilution and natural removal processes, selenium does not currently appear to constitute a health threat

Table 4 Pesticide Detected in Sacramento-San Joaquin Delta Water

Pesticide	Highest concentration detected (μg/L)	Drinking water standard (μg/L)[a]
2,4-D	1.0	70
4,4-DDD	0.004	NA
4,4-DDE	0.007	NA
Atrazine	0.18	3.0
Bentazon	2.8	18
BHC-α	0.003	0.7
BHC-β	0.006	0.3
BHC-γ (lindane)	0.006	NA
Carbofuran	1.33	18
Dacthal (DCPA)	0.15	400[b]
Diazinon	0.1	14[c]
Dieldrin	0.005	0.05[d]
Dimethoate	0.046	140[c]
Endosulfan	0.01	NA
Glyphosate	10.0	700
Gluthion	0.02	NA
Methyl parathion	2.5	30[c]
Molinate	1.5	20
Paraquat	74.0	30[c]
Parathion	0.035	30[c]
Simazine	0.36	10 (4[e])
Thiobencarb	1.7	70

[a]MCL unless specified. NA = none available.
[b]U.S. EPA Health Advisory, lifetime for 70 kg adult.
[c]State action level.
[d]Limit of quantification.
[e]New federal MCL, 1992.
Source: Adapted from Table 5, Reference 28.

to consumers of Delta water, although the possibility of future impacts should not be dismissed.

d. Asbestos

Asbestos (chrysotile) is a naturally occurring mineral in water that arises from the erosion of serpentine rocks, which are common in California (41). Although asbestos has been shown to be a human carcinogen through inhalation exposure, based on evidence from the National Toxicology Program (NTP) dietary ingestion studies (42,43) of asbestos, asbestos in water is classified by the U.S. EPA as a category II contaminant (limited evidence of carcinogenicity via ingestion). The U.S. EPA proposed a drinking water standard for asbestos of 7.1 million fibers per L water for fibers 10 μm or greater in length. This standard will be adopted by the state. The range of total (small and long fibers combined) asbestos concentrations in Delta water varied from 12 million to 7500 million fibers per L water (28,44). About 1% of asbestos fibers in Delta water are greater than 10 μm in length. Although the concentration of asbestos in Delta (raw) water is high, it bears little resemblance to that in treated Delta water. Treated water rarely contains asbestos concentrations exceeding the drinking water standard. Normal water treatment processes for raw water, such as coagulation, sedimentation, and filtration, generally remove 99% or more of the asbestos (45).

III. TOXICOLOGICAL RISK ASSESSMENT

The assessment of health risk and safe exposure to substances in drinking water has been described in various publications (46–51). The purpose of this section is to outline the scientific process by which risk to humans can be estimated and to provide information on how this is used for the setting of drinking water standards. A detailed description of the risk assessment process is available from various publications (52–56). The Office of Environmental Health Hazard Assessment, California Environmental Protection Agency, is given responsibility for performing risk assessment for drinking water contaminants. The OEHHA also determines recommended public health levels (RPHL; see Section IV) and action levels for occasional chemical contaminants in drinking water. The guidelines currently used by the OEHHA to identify and assess the risks of carcinogens are documented in the CDHS *Guidelines for Chemical Carcinogen Risk Assessment and Their Scientific Rationale* (1985) (57). The CDHS guidelines are similar in many ways to those adopted by the U.S. EPA in 1986 (54). The risk assessment of noncarcinogens in general follows guidelines that were used by the U.S. EPA and the NAS (47,58).

In California, the toxicological risk assessment of chemical contaminants in drinking water is conducted as part of an evaluation program to establish "safe" concentrations for human exposure through drinking water standards. The review and evaluation of toxicology, epidemiology, metabolism, and exposure data are conducted in-house by OEHHA staff and consultants. All contracted health risk assessments, usually within the University of California system or affiliated national laboratories, are subjected to two external peer reviews. In all cases the development of rationales supporting the health-based drinking water standards are conducted by OEHHA staff.

A. Noncarcinogens

For noncarcinogens the evaluations resulted in identification of suitable no observable adverse effect level (NOAEL) or lowest observable adverse effect level (LOAEL) values. For volatile organics the evaluation includes an assessment of exposure via noningestion routes, such as inhalation, and dermal exposures resulting from typical household activities (showering, bathing, flushing toilets, and others) (59,60). The initial proposed MCL (PMCL) or RPHL values were usually derived by the relation

$$\text{PMCL or RPHL (mg/L)} = \frac{\text{NOAEL/LOAEL (mg/kg-day)} \times 70\ \text{kg} \times \text{RSC}}{\text{UF} \times W\ \text{(L/day)}}$$

where NOAEL or LOAEL values are in mg/kg-day; 70 kg is the standard human body weight; RSC is the relative source contribution, usually assumed to be 20% or 0.2 if adequate data on other sources of human exposure are absent; and UF is the overall uncertainty factor or the product of 10-fold uncertainty factors reflecting the quality of data on which the NOAEL or LOAEL is based. Guidelines for the selection of uncertainty factors are similar to those suggested by Dourson and Stara (61) and used by the U.S. EPA (46,58). *W* is the estimated daily water consumption of 2 L/day for drinking water (oral) exposure only and larger L equivalents per day for volatile organics subject to inhalation and dermal exposure estimates.

To determine the maximum contaminant level goal for noncarcinogens, the U.S. EPA has determined that the contribution from other sources of exposure, including air and food, should be taken into account. The U.S. EPA is now utilizing a 20% default assumption in calculating drinking water contributions when relative source contribution data are lacking or when data indicate that drinking water contributes 20% or less of the total exposure. When data indicate drinking water exposure is between 80 and 100% of total exposure to a contaminant, a relative source contribution for drinking water of up to 80% is assigned. Although 80% may be a more realistic ceiling since it is difficult to be certain that 100% of exposure to any chemical comes solely from drinking water, 20% may be too high for a minimum contribution and its use could lead to numerically excessive MCLG. The OEHHA has remained flexible on setting the RSC and will evaluate each contaminant on its own merits.

In 1991, after analyzing the recommendations of its risk assessment forum, the U.S. EPA concluded that exposure to volatile compounds in contaminated water supplies can occur not only through ingestion but also through inhalation after transfer to the air or through dermal absorption during various household uses (62). It issued informal guidance recommending that "when time, resources, or other constraints prevent the use of monitoring or modelling data, risk assessors should assume that exposure to volatile chemicals during showering is equivalent to exposures from ingesting two liters of same water per day." OEHHA has advocated the use of monitoring data and modeling techniques to estimate likely human exposures via alternative routes of exposure to volatile organic compounds (VOC) in drinking water and has used these estimates in establishing human health-based drinking water standards for California for many years. In our experience, estimated human exposures from these additional routes for VOC could range from 3 to 10 times that by the ingestion route, depending on the individual chemical and other environmental and human behavioral factors. Thus, although it provides a useful default assumption, the agency's conclusion that contribution from other routes of exposure is equivalent to exposure from ingesting 2 L/day of the contaminated water may underestimate human exposure and the associated risk of adverse human health effects in some cases.

B. Carcinogens

For water contaminants that have the potential to cause cancer in humans, the relationship between risk and exposure can be determined. In studies in which excess cancer risk has been associated with exposure to a chemical, the estimated human potency can be derived directly from the epidemiological data. This type of data is rarely available or adequate for cancer risk assessment, however. When animal studies are used, the human potency estimate can be derived using the linearized multistage model fitted to the animal tumor incidence data (63).

This default approach uses the 95% upper bound estimate of the low-dose slope extrapolated from dose-response data and assumes that the upper bound on slope or the potency is linear at very low doses. To extrapolate potency q_1^* derived from animal data $q_1^*(A)$ to humans $q_1^*(H)$, a conversion based on the surface area or (body weight)$^{2/3}$ is used:

$$q_1^*(H) = q_1^*(A)\left(\frac{W_H}{W_A}\right)^{1/3}$$

where:

W_H = standard human body weight, kg
W_A = animal weight, kg
$q_1^*(A)$ = animal potency, (mg/kg per day)$^{-1}$

$q_1^*(H)$ = human potency, (mg/kg per day)$^{-1}$

The cube root of the body weight ratio adjusts for species differences with the assumption that metabolic rate is proportional to the two-thirds power of body weight (63,64). In fitting the linearized multistage model to the animal cancer data, actual dosing data are converted to daily average lifetime doses. When data are available, other dose metrics, which include pharmacokinetic behavior, are also taken into account. For reduced study duration the potency values are converted for intercurrent mortality by the factor $(L/Le)^3$, where Le is the duration of the experiment and L is the life span of the animal, usually 104 weeks.

The PMCL or RPHL derived for a carcinogen is determined by the equation

$$\text{PMCL or RPHL (mg/L)} = \frac{\text{risk} \times 70 \text{ kg} \times \text{RSC}}{q_1^*(H)\ (\text{mg/kg per day})^{-1}\ W(\text{L/day})}$$

where:

risk = the minimum individual extra lifetime cancer risk of 10^{-6}

RSC = relative source contribution, often but not always assumed to be 1.0

W = water consumption, L/day or L equivalent per day, if addition exposure routes are considered

The potency estimates calculated by the OEHHA occasionally differed from those adopted by the U.S. EPA. Differences are mainly attributed to study or data set selection, averaging of potencies by arithmetical or geometrical means, and the use of different exposure estimates, particularly with regard to volatile organic compounds.

In Table 5 are presented new and proposed drinking water standards for 67 chemicals. For each chemical, from left to right, are listed by PMCL, based only on health considerations, the final MCL, resulting from evaluation of technical and economic feasibility, and the proposed or preliminary RPHL. Also listed is a summary of the toxicological end point used to set the standard, species, exposure route, and overall uncertainty factor. For carcinogens, the cancer potency values q_1^* are also listed.

IV. CALIFORNIA'S DRINKING WATER PROGRAM

The 1974 Federal Safe Drinking Water Act (SDWA) and its amendments require the U.S. EPA to regulate contaminants in drinking water at levels at which no known or anticipated adverse effects will occur. In promulgating the National Primary Drinking Water Regulations (NPDWR), the

Table 5 California Maximum Contaminant Levels and Proposed Recommended Public Health Levels

Chemical	RPHL[a] (ppb)	MCL[b] (ppb)	PMCL[c] (ppb)	Health effects end point(s)	Species route(s)	Overall uncertainty factor	q_1^* (mg/kg per day)$^{-1}$ [d]
Aluminum	1000	1000	1,000	Increased Al concentration in the blood, changes in phosphate absorption in gastrointestinal tract	Human diet	10	None
Antimony	6[e]	—	—	Shorten life span, increased blood cholesterol, decreased blood glucose	Rat drinking water	1,000	None
Arsenic	0.002[e]	50	—	Skin cancer; lung, kidney, liver, bladder cancer	Human drinking water	None	5.3
Atrazine	3	3	3	Increased heart, liver weight	Beagle dog diet	100	None
Bentazon	18	18	18	Prostatitis, weight loss	Beagle dog diet	1,000	None
Benzene	0.35	1	0.05–1.63	Leukemia	Human inhalation	None	0.1–0.01 (ppm^{-1})
Beryllium	4[e]	—	—	Decreased body weight gains	Male and female (M/F) rat drinking water	1,000	None
Bromodichloromethane	—	—	0.06–38.9	Liver nodule, carcinoma	F mouse, rat gavage	None	0.0009–0.013
Cadmium	5.0[e]	10.0	—	Renal dysfunction	Human oral	10	None
Carbofuran	18	—	18	Plasma cholinesterase inhibition	Dog diet	100	None
Carbon tetrachloride	0.5	0.5	0.04–2.3	Hepatocellular carcinoma	Mouse, rat gavage	None	0.014–0.18
Chlordane	0.03	0.1	0.03	Liver carcinoma	M/F mouse diet	None	1.2 (geometric mean)
Chloroform	1.0[e]	—	—	Kidney tumors	Rat, mouse gavage/drinking water/toothpaste	None	0.031 (weighted average)
Chloropicrin	—	—	12	Decreased body weight	F rat gavage	10,000	None

Chromium (total)	2.0[e]	50	—	Forestomach carcinoma [Cr(VI)]	Mouse diet (K_2CrO_4)	None	0.42
Cyanide	200[e]	—	—	Body weight loss, thyroid effects, myelin degeneration	Rat diet	500	None
Dalapon	200[e]	—	—	Increased kidney weight	Rat diet	300	None
DBCP	0.002	0.2	0.015	Forestomach carcinoma	M mouse gavage	None	7.0
1,1-Dichloroethane (1,1-DCA)	5.0	5.0	2.5–20	Decreased body weight, increased death, intoxication	M rat gavage	100,000	None
1,2-DCA	0.3	0.5	0.004–0.3	Mammary adenocarcinoma or fibroadenoma	F rat gavage	None	0.02–1.9
1,1-Dichloroethylene (1,1-DCE)	6.0	6.0	6.3	Midzonal hepatocellular fatty degeneration	Rat water	10,000	None
Cis-1,2-DCE	6.0	6.0	1–18	Changes in hepatic xenobiotic metabolism	F rat inhalation	10,000	None
Trans-1,2-DCE	10.0	10.0	1–10	Increased serum glucose levels	M mouse water	10,000	None
1,2-Dichloropropane	5.0	5.0	0.08–12.5	Liver adenoma/carcinoma	M/F mouse gavage	None	0.063
1,3-Dichloropropene	0.2	0.5	0.05–0.2	Forestomach, squamous cell papillomas, carcinomas (rat); bladder transitional cell carcinoma (mouse)	M rat F mouse gavage	None	0.18 (potency derived from either study)
2,4-D	70[e]	100	—	Blood, liver, kidney effects	Rat oral	100	None
Di(2-ethylhexyl)adipate (DEHA)	400[e]	—	—	Reduced maternal body weight, reduced fetal ossification	Rat diet	300	None
Di(2-ethylhexyl)-phthalate (DEHP)	4.0	4.0	0.84–10.5	Hepatocellular carcinoma	Rat, mouse diet	None	0.0084
1,4-Dichlorobenzene	1.0	5.0	0.2–8.1	Hepatocellular adenoma or carcinoma, adrenal gland pheochromocytoma	M mouse gavage	None	0.04
Dinoseb	7[e]	—	—	Decreased body weight gains, decreased thyroid weights	M rat	1,000	None
Diquat	15[e]	—	—	Lens opacity, cataracts	Rat diet	100	None
Endothall	100[e]	—	—	Increased stomach, small intestine weights	Beagle dog diet	100	None
Endrin	2[e]	2	—	Increased liver weights, vacuolization of liver cells	Dog diet	100	None
Ethylbenzene	680	680	680	Increased liver, kidney weight	Rat gavage	1,000	None
Ethylene dibromide	0.01	0.02	0.02	Forestomach squamous cell carcinoma	M/F rat, mouse gavage	None	3.6 (geometric mean)

Table 5 Continued

Chemical	RPHL[a] (ppb)	MCL[b] (ppb)	PMCL[c] (ppb)	Health effects end point(s)	Species route(s)	Overall uncertainty factor	q_1^* (mg/kg per day)$^{-1}$ [d]
Glyphosate	700	700	700	Bilateral renal tubular dilation	Rat diet	100	None
Heptachlor	0.01	0.01	0.0005–0.018	Liver carcinoma	M/F mouse diet	None	5.7 (geometric mean)
Heptachlor epoxide	0.007	0.01	0.0002–0.007	Liver carcinoma	M/F mouse diet	None	13 (geometric mean)
Hexachlorobenzene	0.02[e]	—	—	Hepatocellular carcinoma	F rat drinking water	None	1.6
Hexachlorocyclopenta-diene (HEX)	50[e]	—	—	Stomach lesions	Rat gavage	1,000	None
Lindane	0.04[e]	4.0	—	Liver carcinoma	Mouse diet	None	1.1
Methoxychlor	4.0[e]	100	—	Reproductive toxicity	Rabbit	1,000	None
Molinate	20	20	20	Decreased number of female pups, body weight of pups	Rat diet	100	None
Monochlorobenzene	70	30	30	Hepatic neoplastic nodule formation	M rat diet	1,000	None
Nickel	100[e]	—	—	Decreased body, organ weights	Rat diet	300	None
Oxamyl	200[e]	—	—	Decreased body weight gain	Rat diet	100	None
PCE	0.7	5.0	2.0	Hepatocellular carcinoma leukemia	Mouse, rat gavage, inhalation	None	0.001–0.085
PCP	0.25[e]	—	—	Liver adenoma/carcinoma	Mouse diet	None	0.15
Picloram[e]	400[e]	—	—	Increased liver weight	Dog diet	100	None
Selenium	10[e]	10	—	Hair loss, nail changes, skin lesions, nervous system effects	Human diet	1.5	None
Silvex (2,4,5-TP)	10[e]	10	—	Changes in liver histopathology	Dog diet	100	None
Simazine	4[e]	10[f]	35	Decreased body weight gain, hematological changes, reproductive and developmental effects	Rat diet	3,000	None

1,1,1-Trichloroethane (1,1,1-TCA)	200	200	200	Fetal mortality, increased concentration of glial fibrillary acid protein, hepatic changes (gerbil)	Rat water, gerbil inhalation, mouse inhalation	100	None
1,1,2-TCA	1[e]	32[f]	32–100	Liver effects, decreased prothrombin time, increased fibrinogen levels	Mouse gavage	10,000	None[g]
1,2,4-Trichlorobenzene	70[e]	—	—	Increased adrenal gland weight in offspring (P_0, F_1 generations)	Rat drinking water	1,000	None
TCE	2.5	5.0	3.2	Liver tumor	Mouse gavage	None	0.011
Tetrachloroethane	1.0	1.0	1.0	Transient body weight depression, changes in liver fat content	Rat inhalation	10,000	None
Thallium	0.5[e]	—	—	Increased blood SGOT, LDH, sodium; decreased blood glucose	Rat diet	3,000	None
Thiobencarb	70	70	70	Increased packed red cell volume, hemoglobin concentration, red cell count, dermal lesions, serum urea concentration (rat); increased kidney weight, decreased erythrocytes, hemoglobin, hematocrit values, plasma cholinesterase inhibition (dog)	Rat diet Dog gelatin capsule	100	None
Toluene	150[e]	—	—	Increased liver and kidney weights (rat) neurotoxicity	Rat gavage, human inhalation	200	None
Toxaphene	0.03[e]	5.0	—	Hepatocellular carcinoma	Mouse water	200	1.2
Trichlorofluoromethane	150	150	150–2,000	Increased serum urea nitrogen	Dog inhalation	10,000	None
Trichlorotrifluoroethane	1200	1200	1200–18,000	Increased liver weight	Rat inhalation	1,000	None
Uranium	30[e]	30	30	Decreased content nucleic acids renal hepatic tissues	Rabbit diet	100	None
Vinyl chloride	0.15	0.5	0.03–0.5	Hepatocellular carcinoma	Rat diet, gavage	None	0.07–0.4
Xylene(s)	1750	1750	1,750	Reduced body weight	Rat gavage	1,000	None

[a]Recommended public health levels are new health-based drinking water standards mandated by the Safe Drinking Water Act of 1989.
[b]Maximum contaminant levels for California drinking water, California Department of Health Services.
[c]Hazard evaluation section proposed maximum contaminant levels: health-based numbers used as basis for MCL setting (pre-1989).
[d]Cancer potency value(s) used in deriving RPHL and PMCL. Several of the values are historical; for current OEHHA values contact Dr. D. Siegel, current listing at 916-324-2829.
[e]Preliminary values. Remaining values have been proposed but are not yet in regulation.
[f]State MCL higher than federal MCL (final rules June 17,1992). New state MCL under review.
[g]Based on CDHS guidelines, 1,1,2-TCA is not a carcinogen.

U.S. EPA must establish maximum contaminant level goals, which are nonenforceable health goals, and maximum contaminant levels, the enforceable standards. For MCL the technical feasibility of removing the chemical, the analytical detection limit, and the economic impact of regulating the contaminant are taken into account. The MCLG are theoretical goals for contaminant levels in drinking water, set at levels at which no adverse human health effects are anticipated. It also includes a margin of safety. The U.S. EPA bases the calculation of the MCLG on the weight of evidence classification given to the chemical. For group A (known human carcinogens) and B chemicals (probably human carcinogens), the MCLG is set at 0. The MCL must be set as close to the MCLG as is technically and economically feasible. Secondary MCL are also set for esthetic water quality, such as taste, odor, and appearance (65).

With the exception of Indiana and Wyoming, all other states are given the authority of enforcing the NPDWR. To maintain primacy, the states must within 18 months of promulgation or within a specified time either adopt the federal drinking water standards or set more stringent standards of their own. In California, the CDHS has primary enforcement responsibilities. For water systems with fewer than 200 service connections, this authority has been subdelegated to the counties. The federal government still retains jurisdiction for the water program for Native American lands.

Historically, California has taken the enforcement of drinking water standards very seriously, promulgating its own regulations under the California Safe Drinking Water Act (CSDWA), which are typically more stringent than the federal regulations. Section 4026(a) of the CSDWA requires the CDHS to promulgate drinking water standards, MCL, and action levels. The state MCL are legally enforceable and are based on health effects with adjustments made for technical feasibility and treatment cost. AL are health-based interim standards for contaminants found in California water supplies for which an MCL has yet to be established. Although AL are not legally enforceable, they are often adhered to by water providers. MCL are calculated for single chemicals and do not take into consideration the toxicological interactions of chemical mixtures. Further, MCL are derived for drinking water only and are not to be used as target levels for the cleanup of contaminated environmental waters. The 1989 revision to the CSDWA (AB 21) requires the development of recommended public health levels in addition to MCL for drinking water contaminants. RPHL are based solely on health effects without regard to cost impacts or other factors. It must also consider possible effects resulting from the interaction of two or more contaminants, sensitive individuals, pharmacokinetics, and the contribution of the resulting body burden from media other than drinking water. The act requires MCL to be reviewed every 5 years after adoption and the RPHL to be "periodically revised as necessary based upon the availability of new scientific data" (66). RPHL are aimed primarily at public water systems that have more than 10,000 service connections. These system must reduce the level of contaminants in drinking water to as close to the RPHL as feasible. All "reasonable measures" available to the water system to reduce the level of the contaminant, including costs to consumers and the water system, must be submitted in a water quality improvement plan to the CDHS. At the moment, analytical method detection limits (MLD) for many of the contaminants appear to be one of the important limiting factors for the implementation of RPHL. In general, MLD need to be about 5–10 times lower than the enforceable levels for monitoring to be conducted effectively (67). The basic underlying philosophy behind the AB 21 legislation is for all water providers to meet the enforceable MCL, but since this is a risk management level, they should in addition try to remove as much of the contaminants from drinking water as they can, taking into account economical and technical feasibility. The new regulation will establish RPHL monitoring requirements and compliance evaluations, water quality improvement plans (WQIP), and the best available technology for mitigation purposes. Water utilities will incur costs from the development of the WQIP and from mitigation efforts. RPHL are similar in many ways to the federal MCLG, except that for carcinogens they are not set at zero but at the 10^{-6} risk level. Initially, 34 RPHL (32 organics and 2 inorganics) will be proposed (Table 5). Of the RPHL 9 will be lower than existing MCL: benzene, chlordane, DBCP, 1,3-dichloropropene (1,3-DCP), ethylene dibromide (EDB), heptachlor epoxide, tetrachloroethylene, trichloroethylene, and vinyl chloride. For 3 chemicals, chloroform, pentachlorophenol (PCP), and toluene, RPHL have been proposed but MCL have not yet been promulgated.

V. SUMMARY

California depends on both surface and groundwater for its source of drinking water. Issues relating to the chemical contaminants in drinking water and the potential human health effects in exposure to these chemicals have received increasing public attention, especially in recent years. With its rapid population growth, many utilities may be forced to use lower quality water for drinking water purposes. Chemical contaminants found in drinking water are due to the presence of salts, nitrate, pesticides, herbicides, metals, and organic chemicals that enter drinking water supplies from both natural and human-engineered sources. In addition, water obtained from the Sacramento-San Joaquin Delta contains high concentrations of organic carbon, which can combine with disinfectants used at the water treatment plants to produce THM and other disinfection by-products. These pose a serious problem to utilities using Delta water because they many not be able to comply with the stricter standards for THM and DBP that the U.S. EPA is in the process of developing.

The responsibilities of protecting the quality of drinking water in the state are divided among various state, federal, and local agencies. To ensure the safety of drinking water, the state enacted legislation that implemented water quality to detect, monitor, control, and clean up contaminants affecting drinking water supplies. Major ongoing activities include the sampling of water supply wells for the active ingredients of pesticides which contaminate groundwater by the agricultural use of pesticides, and the monitoring of public drinking water wells throughout California for chemical contamination. These programs are not centrally controlled or coordinated, but they represent a major effort to protect the state's drinking water sources, and this approach appears to be meeting its goal.

REFERENCES

1. State Water Resources Control Board (SWRCB), *Water Quality and Pesticides: A California Risk Assessment Program*, SWRCB Pub. 84-62P, 1984.
2. D.B. Cohen, Ground water contamination by toxic substances: A California assessment, *Evaluation of Pesticides in Ground Water* (W.Y. Garner, R.C. Honeycut, and H.N. Niggs, eds.), ACS Symposium Series 315, American Chemical Society, Washington, D.C., 1986, pp. 499–529.
3. H.H. Russell, R.J. Jackson, D.P. Spath, and S.A. Book, Chemical contamination of California drinking water, *West. J. Med. 147*:615–622 (1987).
4. California Department of Food and Agriculture, *Agricultural Pesticide Residues in California Well Water: Development and Summary of a Well Inventory Data Base for Non-point Sources*, 1985.
5. California Department of Food and Agriculture, *Sampling for Pesticide Residues in California Well Water: 1986 Well Inventory Data Base*, 1986.
6. California Department of Food and Agriculture, *Sampling for Pesticide Residues in California Well Water: 1987 Update, Well Inventory Data Base*, 1987.
7. California Department of Food and Agriculture, *Sampling for Pesticide Residues in California Well Water: 1988 Update, Well Inventory Data Base*, 1988.
8. California Department of Food and Agriculture, *Sampling for Pesticide Residues in California Well Water: 1989 Update, Well Inventory Data Base*, 1989.
9. California Department of Food and Agriculture, *Sampling for Pesticide Residues in California Well Water: 1990 Update, Well Inventory Data Base*, 1990.
10. California Department of Pesticide Regulation, *Sampling for Pesticide Residues in California Well Water: 1991 Update, Well Inventory Data Base*, 1991.
11. California Department of Health Services, *Organic Chemical Contamination of Large Public Water Systems in California*, 1986.

12. California Department of Health Services, *Organic Contamination of Small Public Water Systems in California*, 1990.
13. L.A. Smith, K.P. Green, and D.M. Mackay, Quality of ground water in California: Overview and implications, *Coping with Water Scarcity: The Role of Ground Water*, Seventeenth Biennial Conference on Ground Water, San Diego, CA, Report No. 72, ISSN 0575-4968, 1990, pp. 93–107.
14. National Academy of Sciences, *Nitrates: An Environmental Assessment*, Washington, D.C., 1978.
15. A.M. Fan, C.C. Willhite, and S.A. Book, Evaluation of nitrate drinking water standard with specific references to methemoglobinemia and reproductive toxicity, *Regul. Toxicol. Pharmacol. 7*:135–148 (1987).
16. State Water Resources Control Board, *Nitrate in Drinking Water, Report to the Legislature*, Report No. 88-11WQ, 1988.
17. A. Milea, *Drinking Water Quality Monitoring for Selected Chemicals, January 1984 to May 1992*, California Department of Health Services, Office of Drinking Water, 1992 (personal communication).
18. D.P. Spath, The nitrate threat, in *Coping with Water Scarcity: The Role of Ground Water*, Seventeenth Biennial Conference on Ground Water, San Diego, CA, Report No. 72, ISSN 0575-4968, 1990, pp. 149–155.
19. Federal Register, U.S. Environmental Protection Agency, National Primary Drinking Water Regulation: Radionuclides, Proposed Rules, *Federal Register*, *56* (138):33050, 1991.
20. National Academy of Sciences, *Health Effects of Exposure to Low Levels of Ionizing Radiation*, BEIR V. National Academy Press, Washington, D.C., 1990.
21. U.S. Environmental Protection Agency, *Radon in Drinking Water: Assessment of Exposure Pathways*, Life Systems, Inc., June 1991.
22. R.H. Sakaji, *Radon in California's Groundwater: A Water Quality Survey and Exposure Assessment*, Office of Drinking Water, California Department of Health Services, Berkeley, CA, October 1991.
23. C.J. Chen, Y.C. Chuang, S.L. You, T.M. Lin, and H.Y. Wu, A retrospective study of malignant neoplasms of bladder, lung and liver in blackfoot disease endemic area in Taiwan, *Br. J. Cancer 53*:399–405 (1986).
24. C.J. Chen and C.J. Wang, Ecological correlation between arsenic level in well water and age-adjusted mortality from malignant neoplasms, *Cancer Res. 50*:5470–5474 (1990).
25. A.H. Smith, C. Hopenhayn-Rich, M.N. Bates, H.M. Goeden, I. Hertz-Picciotto, H.M. Duggan, R. Wood, M.J. Kosnett, and M.T. Smith, Cancer risks from arsenic in drinking water, *Environ. Health Perspect. 97*:259–267 (1992).
26. A.H. Smith, H. Goeden, V. Shearn, M. Bates, and M.T. Smith, *Health Risk Assessment for Arsenic Ingestion*, Health Risks Associates, Berkeley, CA, 1990.
27. J.P. Brown and A.M. Fan, *Arsenic: Recommended Public Health Level for Drinking Water*, Office of Environmental Health Hazard Assessment, Berkeley, March 1992.

28. California Department of Water Resources, *The Delta as a Source of Drinking Water, Monitoring Results 1983 to 1987*, Interagency Delta Health Aspects Monitoring Program, 1989.
29. California Department of Water Resources, *Delta Island Drainage Investigation Report of the Interagency Delta Health Aspects Monitoring Program*, 1990.
30. California Water Resources Center, *Protecting Drinking Water Quality at the Source*, Proceedings of a Conference (J.J. DeVries, ed.), University of California Report No. 76, ISSN 0575-4968, 1991.
31. State Water Resources Control Board, *Water Quality Assessment 1991*, 1992.
32. State of California, *Delta Water Quality: A Report to the Legislature on Trihalomethanes and the Quality of Drinking Water Available from the Sacramento-San Joaquin Delta*, SWRCB, CDHS, and CDWR, 1991.
33. R.J. Bull and F.C. Kopfler, *Health Effects of Disinfectants and Disinfection By-Products*, American Water Works Association Research Foundation, Denver, CO, 1991.
34. National Academy of Sciences, Disinfectants and disinfectant by-products, in *Drinking Water and Health*, Vol. 7, National Academy Press, Washington, D.C., 1987.
35. R.D. Morris, A.M. Audet, I.F. Angelillio, T.C. Chalmers, and F. Mosteller, Chlorination, chlorination by-products, and cancer: A meta-analysis, *Am. J. Public Health 82*:955–963 (1992).
36. California Department of Water Resources, *Trihalomethane Formation Potential in the Sacramento-San Joaquin Delta: Mathematical Model Development*, 1991.
37. International Agency for Research on Cancer, *IARC Monographs on the Evaluation of Carcinogenic Risks to Humans: Chlorinated Drinking Water, Chlorination By-products, and Some Other Halogenated Compounds*, Vol. 52, IARC Scientific Publications, Lyons, France, 1991, pp. 45–362.
38. R.G. Rice and M. Gomez-Taylor, Occurrence of by-products of strong oxidants reacting with drinking water contaminants—scope of the problem, *Environ. Health Perspect. 69*:31–44 (1986).
39. T.S. Pesser and I. Barnes, *Dissolved Constituents Including Selenium in Waters in the Vicinity of Kesterson National Wildlife Refuge and the West Grassland, Fresno and Merced Counties, California*, U.S. Geological Survey Water Resources Investigations Report 85-4200, 1985.
40. State Water Resources Control Board, *Selenium in California*, Vol. 1, *History, Chemistry, Biology, Uses, Management*, Report No. 88-10-I-WR, 1988.
41. S.B. Hayward, Field monitoring of chrysotile asbestos in California waters, *J. Am. Water Works Assn. 76*:66–72 (1984).
42. National Toxicology Program, *Toxicology and Carcinogenesis Studies of Chrysotile Asbestos in F344/N Rats (Feed Studies)*, NTP Technical Report Series No. 295, 1985.
43. National Toxicology Program, *Lifetime Carcinogenesis Studies of Chrysotile Asbestos in Syrian Golden Hamsters (Feed Studies)*, NTP Technical Report Series No. 246, 1983.

44. R.C. Bales, D.D. Newkirk, and S.B. Hayward, Chrysotile asbestos in California surface water: From upstream rivers through water treatment, *J. Am. Water Works Assn.* *76*:76–82 (1984).
45. M.J. McGuire, A.E. Bowers, and D.A. Bowers, Optimizing large-scale water treatment plants for asbestos fiber removal. *J. Am. Water Works Assn.* *75*:364–369 (1983).
46. U.S. Environmental Protection Agency, *Workshops on Assessment and Management of Drinking Water Contamination*, U.S. EPA/Office of Water, Washington, D.C., EPA/600/M-86/026, 1987.
47. National Academy of Sciences, Chemical contaminants: Safety and risk assessment, in *Drinking Water and Health*, Vol. 1, National Academy Press, Washington, D.C., 1977, pp. 19–62.
48. National Academy of Sciences, Problems of risk estimation, in *Drinking Water and Health*, Vol. 3, National Academy Press, Washington, D.C., 1980, pp. 25–66.
49. National Academy of Sciences, Risk assessments, in *Drinking Water and Health*, Vol. 6, National Academy Press, Washington, D.C., 1986, pp. 250–293.
50. National Academy of Sciences, Pharmacokinetics in risk assessment, in *Drinking Water and Health*, Vol. 8, National Academy Press, Washington, D.C., 1987.
51. National Academy of Sciences, Selected issues in risk assessment, in *Drinking Water and Health*, Vol. 9, National Academy Press, Washington, D.C., 1989.
52. Office of Technology Assessment, *Identifying and Regulating Carcinogens*, Lewis Publishers, Chelsea, MI, 1988.
53. National Academy of Sciences, *Risk Assessment in the Federal Government: Managing the Process*, National Academy Press, Washington, D.C., 1983.
54. U.S. Environmental Protection Agency, *The Risk Assessment Guidelines of 1986*, U.S. EPA, Office of Health and Environmental Assessment, Washington, D.C., EPA/600/8-87/045, 1987.
55. U.S. Environmental Protection Agency, Guidelines for exposure assessment, *Federal Register 57*(104):22888, 1992.
56. D.J. Paustenbach, *The Risk Assessment of Environmental and Human Health Hazards: A Textbook of Case Studies*, Wiley-Interscience, New York, 1989.
57. California Department of Health Services, *Guidelines for Chemical Carcinogen Risk Assessment and Their Scientific Rationale*, Sacramento, 1985.
58. J.A. Cotruvo, Drinking water standards and risk assessment, *Regul. Toxicol. Pharmacol.* *8*:288–299 (1988).
59. T.B. McKone, Human exposure to volatile organic compounds in household tap water: The indoor inhalation pathway. *Environ. Sci. Technol.* *21*:1194–1201 (1987).
60. H.S. Brown, D.R. Bishop, and C.A. Rowan, The role of skin absorption as a route of exposure for volatile organic compounds (VOCs) in drinking water, *Am. J. Public Health* *74*:479–484 (1984).
61. M.L. Dourson and J.F. Stara, Regulatory history and experimental support of uncertainty (safety) factors, *Regul. Toxicol. Pharamcol.* *3*:224–238 (1983).

62. U.S. Environmental Protection Agency, *Guidance on Estimating Exposure to VOCs During Showering*, Memo from Risk Assessment Forum/Office of Research and Development, July 10, 1991.
63. K.S. Crump, An improved procedure for low dose carcinogenic risk assessment from animal data, *J. Environ. Pathol. Toxicol. Oncol. 5*:339–348 (1984).
64. J.F. Stara, J. Patterson, and M.L. Dourson, Risk assessment approaches for ground water contamination by pesticides and other organic substances, in *Evaluation of Pesticides in Ground Water* (W.Y. Garner, R.C. Honeycut, and H.N. Niggs, eds.), ACS Symposium Series 315, American Chemical Society, Washington, D.C., 1986, pp. 445–461.
65. U.S. Environmental Protection Agency, Guidelines and methodology used in preparation of health effects assessment chapters of the consent decree water criteria documents, *Federal Register 45*:79347, 1980.
66. California Department of Health Services, *California Safe Drinking Water Act, Laws and Standards Relating to Domestic Water Supply*, Section 4010-4095, 1990.
67. Y.Y. Wang, R.H.F. Lam, J.P. Brown, A.M. Fan, and A.M. Milea, Development of recommended public health levels for drinking water in California, *Toxicologist 11*:198 (1991).

3

Reactions of Drinking Water Contaminants with Aqueous Chlorine and Monochloramine

Frank E. Scully, Jr.
Old Dominion University, Norfolk, Virginia

William N. White
University of Vermont, Burlington, Vermont

Robert S. Boethling
U.S. Environmental Protection Agency, Washington, D.C.

I. INTRODUCTION

Unwarranted exposure of humans and animals to pesticides and toxic substances can elicit adverse health effects. Of major concern is the presence of these substances in ground and surface waters used as drinking water supplies. Based solely on available dose-response data on adverse health effects of these compounds in humans or animals, exposure levels can be identified at which significant long- or short-range harmful effects are not expected. However, most drinking waters in the United States are prepared from natural waters that have been subjected to a series of physical and chemical processes designed to disinfect the water and make it pleasing to drink. Because chemical disinfectants are by their very nature highly reactive, organic contaminants in raw water may undergo transformation to more or less toxic substances.

Aqueous chlorine and monochloramine are the two disinfectants most widely used in drinking water treatment in the United States. The objective of this chapter is to review the available information on the reactions of aqueous chlorine and monochloramine with various organic functional groups, to predict which classes of chemical substances are most likely to undergo transformation under drinking water disinfection conditions. For those reactions for which kinetic data are available, half-lives of substances containing these functional groups are estimated. When possible, reaction mechanisms are also identified and structure-reactivity relationships discussed.

II. WATER TREATMENT PROCESSES

The major sources of drinking water in the United States are groundwater, rivers, and freshwater lakes. These waters generally contain suspended particles, chemical substances, and microbiological agents that make them undesirable to drink without treatment. Therefore, municipal treatment systems subject raw water to combinations of physical and chemical processes to improve their quality. A description of one combination of these processes illustrates the variety of treatments that can affect the fate of a chemical substance in the raw water. Often a coagulating agent (either alum or ferric chloride) and powdered activated carbon are added to the raw water as it enters the plant. The powdered activated carbon adsorbs and removes a variety of organic chemicals. Alum and ferric chloride hydrolyze in water to colloidal aluminum and iron hydroxides. These hydroxides are charged and bind oppositely charged particles, such as clay particles and bacterial cells, as well as organic and inorganic ions in the water. At the same time the hydroxides aggregate, become heavier, and settle over several hours in a settling basin. The clarified water is next filtered through sand to remove particles that have not settled. At this stage, the pH of the water is adjusted upward with lime, chlorine is added as a disinfectant, and sometimes fluoride is added. Utilities that employ monochloramine as the disinfectant add chlorine and ammonia either simultaneously or sequentially. Following disinfection, the treated water is generally held in a clear well for several hours to allow time for the disinfectant to act before the water is allowed to enter the distribution system.

Not all utilities employ all of these processes; others utilize additional processes, such as lime softening. Although a U.S. Environmental Protection Agency (EPA) report indicated that in 1984 only about half of the surface water utilities in the United States employed disinfection, the recent surface water treatment rule (1) now requires all public water systems that use either surface waters or groundwaters under the direct influence of surface water to disinfect their water. A variety of disinfectants is used in drinking water treatment, but aqueous chlorine and monochloramine are by far the most common. In this chapter, only the reactions of these two disinfectants are considered.

Sufficient disinfectant is added to the water at the plant to ensure that there is at least a minimum concentration of disinfectant at every tap in the distribution system. The actual concentration of disinfectant in the water at the tap varies with the amount added to the water at the treatment plant, the temperature of the water, the reactions that might dissipate the disinfectant while it is in the pipes, and the length of time the water has remained in the pipes before it is withdrawn. Many states have established a required minimum residual chlorine concentration of 0.2 mg/L of free chlorine or 1.0 mg/L of combined chlorine (chloramines) at every point in the distribution system. The EPA found that most of the drinking waters in a survey of 80 localities contained residual aqueous chlorine concentrations between 0.4 and 2.8 mg/L (as Cl_2), with an average of approximately 2×10^{-5} M (2).

The residence time of water in the pipes, which is the time between when it leaves the plant and when it reaches the consumer, varies with the size of the distribution system. Consumers living near a treatment plant receive water on the same day it left the plant, but the detention time may be greater than 5 days in large systems. For the purposes of this chapter, the typical distribution system is assumed to have a 5 day residence time. Clearly, this time can affect the extent of reaction of a chemical substance with a disinfectant if the reaction has a half-life of days or less.

III. NATURE OF REACTIONS OF CHLORINE-BASED DISINFECTANTS WITH ORGANIC SUBSTRATES

The most common reactions result in substitution (replacement of a hydrogen by chlorine), oxidation (usually implying increased bonding of an atom to oxygen), and addition (saturation of a multiple bond). Initial reactions are often followed by others that result in carbon skeleton cleavage, decarboxylation, elimination, and so on. These subsequent reactions may or may not involve additional disinfectant. It is important in considering the nature of reactions of organic compounds with aqueous chlorine and chloramine to have a clear notion of the actual molecular or ionic species involved. The reactive molecules in aqueous chlorine and chloramine may seem obvious, but complex interactions of these substances with water or with themselves actually produce a variety of species in solution.

A. Reactive Species in Aqueous Chlorine

When chlorine is added to water, it hydrolyzes:

$$Cl_2 + H_2O \rightleftharpoons HOCl + HCl \qquad (1)$$

The product HOCl is a weak acid, however, and the absolute concentration of HOCl is affected dramatically by changes in pH. HOCl has a pK_a of

7.5 (3).

$$HOCl \rightleftharpoons OCl^- + H^+ \qquad (2)$$

At pH values above 7.5, the aqueous chlorine is primarily in the form of OCl^-. At pH values below 7.5, HOCl predominates. Consequently, there are a variety of chlorinating species in chlorinated drinking water: Cl_2, HOCl, and OCl^-. The initial amounts of each depend on the initial concentration of chlorine and the pH.

Typical chlorinated drinking water may contain about 1×10^{-5} M "free available chlorine" and 3×10^{-4} M Cl^- (4) and has a pH of about 7.5. Free available chlorine (FAC) is the sum of Cl_2, OCl^-, and HOCl, expressed as Cl_2. Under these conditions but at pH 1, the actual Cl_2 concentration is about 7% of the FAC. As the pH increases to 7, this concentration falls by a factor of 10 per pH unit until it is 0.000006% of the total. Above pH 7, the molarity of Cl_2 decreases by a factor of 100 per pH unit. From pH 1 to 7, hypochlorous acid (HOCl) accounts for almost all the active chlorine, but at higher pH (>8) its molarity drops by a factor of 10 per pH unit. The concentration of hypochlorite ion (OCl^-) is very low at pH 1 (0.00003% of total chlorine) but increases by a factor of 10 per pH unit until pH 7, and above pH 8 almost all the original Cl_2 is present in this form. The approximate concentrations of the various active chlorine species at pH 7.5 are given in Table 1.

Although the concentration of Cl_2 in chlorinated drinking water is likely to be very low (only 0.0000013% of FAC at pH 7.5), it is not necessarily insignificant. Molecular chlorine is generally more reactive than HOCl. Furthermore, since it can be produced from the other species by the preceding equilibria if it is consumed in a reaction, it is regenerated. If this happens fast enough, reactions with Cl_2 can account for the reactivity of aqueous chlorine solutions toward some organic substrates.

Table 1 Concentrations of Active Chlorine Species in Typical Chlorinated Drinking Water

Species	Concentration (M)
Cl_2	1.27×10^{-13}
HOCl	5.00×10^{-6}
OCl^-	5.00×10^{-6}

B. Reactive Species in Aqueous Chloramine

The situation with aqueous chloramine is even more complex. Potential equilibria in these solutions are as follows (5–7):

$$NH_2Cl + H_2O \rightleftharpoons HOCl + NH_3 \quad (3)$$

$$NH_2Cl + H^+ \rightleftharpoons NH_3Cl^+ \quad (4)$$

$$2NH_2Cl + H^+ \rightleftharpoons NHCl_2 + NH_4^+ \quad (5)$$

$$NHCl_2 + HOCl \rightleftharpoons NCl_3 + H_2O \quad (6)$$

$$HOCl \rightleftharpoons H^+ + OCl^- \quad (7)$$

All these chlorine-containing molecules (NH_2Cl, NH_3Cl^+, $NHCl_2$, NCl_3, HOCl, and OCl^-) have been shown to be reactive species. The relative amounts of each of these in a particular solution depend on the concentrations and the pH.

If a drinking water contains low initial concentrations of both FAC and NH_3 (about 10^{-5} M), NCl_3 predominates at very low pH, at which most of the NH_3 is tied up as NH_4^+, but the amount of NCl_3 drops rapidly as the pH increases. The concentration of $NHCl_2$ is low at low pH, rises to a maximum at pH 5, where it accounts for most of the active chlorine, and diminishes again at higher pH. NH_2Cl shows similar behavior except that it peaks at pH 10. The concentration of inorganic chlorammonium ion (NH_3Cl^+) is very low at neutral pH, but it is also a more potent chlorinating agent than NH_2Cl. As expected, HOCl is important only at low pH and OCl^- at high pH. Table 2 gives typical concentrations for the various species in chloramine-treated drinking water at pH 7.5.

The concentrations of NH_3Cl^+, NCl_3, HOCl, and OCl^- are rather low, and it might be questioned whether they can serve as effective reactants in drinking water. The various species in aqueous chloramine generally interconvert much more slowly than those in aqueous chlorine. However, a reaction that is dependent upon a highly reactive species that is present in very low concentration may still proceed at a significant rate if the reactive species can be regenerated rapidly enough through the shift in an equilibrium. In general, it appears that any of the active chlorine species (NCl_3, $NHCl_2$, NH_2Cl, NH_3Cl^+, HOCl, or OCl^-) in chloramine-treated drinking waters may be kinetically significant chlorinating agents, either directly or through hydrolysis to HOCl.

C. Reactions with Organic Substrates

Reaction of organic compounds with chlorinated or chloraminated water can result in substitution, addition, oxidation, and so on. Despite this

Table 2 Concentrations of Ammonia and Active Chlorine Species in NH_3- and Cl_2-Treated Drinking Water

Species	Concentration (M)
NH_2Cl	0.80×10^{-6}
$NHCl_2$	4.53×10^{-6}
NCl_3	6.14×10^{-8}
HOCl	9.21×10^{-11}
OCl^-	9.21×10^{-11}
NH_3	5.96×10^{-8}
NH_4^+	4.53×10^{-6}
NH_3Cl^+	2.5×10^{-13}

diversity of reaction results, the initiating step in most of the reactions has a common mechanistic characteristic. It is a nucleophilic substitution by an electron-rich group (the nucleophile) on a halogen:

$$\text{Nu:}^- + \text{Cl—}X \rightarrow \text{Nu—Cl} + X\text{:}^- \qquad (8)$$

$$\text{H—Nu:} + \text{Cl—}X \rightarrow \text{:Nu—Cl} + \text{H}^+ + X\text{:}^- \qquad (9)$$

These reactions are characterized by the transfer of an electrophilic (electron-deficient) chlorine from Cl—X to the nucleophile and are often acid or base catalyzed. In essence, a Cl^+ is exchanged. Reaction (9) can also be described as an electrophilic substitution or chlorination. Both reactions are usually oxidations in the theoretical sense, since they most often result in an increase in the oxidation state of the atom or group Nu. Cl—X may be Cl—Cl, Cl—OH, Cl—O^-, Cl—NH_2, Cl—NHCl, or Cl—NCl_2.

The reactivity of the different electrophilic chlorine donors that have been mentioned varies greatly. Nevertheless, a general order of reactivity may be constructed as $Cl_2 > HOCl, OCl^-, NCl_3 > NH_2Cl > NHCl_2$. When such Cl^+ donors react with a nucleophile, a substitution or association product is formed (8–10). For example,

$$\text{Me}_2\text{N—H} + \text{HO—Cl} \rightarrow \text{Me}_2\text{N—Cl} + \text{H—OH} \qquad (10)$$

$$\text{H—C}_6\text{H}_4\text{—OH} + \text{HO—Cl} \rightarrow o\text{-and } p\text{-Cl—C}_6\text{H}_4\text{—OH} + \text{H—OH} \qquad (11)$$

$$\text{Ph}_2\text{S:} + \text{Cl—Cl} \rightarrow \text{Ph}_2\text{S}^+\text{—Cl} + \text{Cl}^- \qquad (12)$$

In some cases, the initial product is not stable and undergoes further reaction. For example, the electrophilic chlorine may simply be transferred to another nucleophile (11). Alternatively, the products of chlorination may undergo rearrangement or may suffer elimination, and carbon-carbon bond cleavage may occur. A simple example of the latter is the well-studied haloform reaction (12), in which chloroform is generated by the chlorination of methyl ketones in drinking water:

$$C_{10}H_7\text{—}\underset{\underset{O}{\|}}{C}\text{—}CH_3 + 3HO\text{—}Cl \rightarrow 3HO\text{—}H + C_{10}H_7\text{—}\underset{\underset{O}{\|}}{C}\text{—}CCl_3$$

$$\rightarrow C_{10}H_7\text{—}COOH + CHCl_3 \qquad (13)$$

Frequently, this type of transformation involves highly chlorinated intermediates. Finally, Cl^+ donors may react with carbon-carbon multiple bonds, since the π electrons serve as an electron-rich center to combine with the putative Cl^+.

In general, "ionic" reactions are favored for the reactive species found in aqueous chlorine and chloramine. As we have seen, these almost always involve the transfer of an electrophilic Cl^+ from the chlorine donor to a nucleophile [Equation (8)]. However, nucleophilic substitution on the X of X—Cl is also possible under certain conditions. In this reaction, Cl^- is expelled and X is transferred to the substrate:

$$Nu:^- + X\text{—}Cl \rightarrow Nu\text{—}X + Cl^- \qquad (14)$$

For example, reactions in which an amino group is transferred from chloramine (NH_2Cl) to a nucleophilic organic substrate are well known in organic synthesis, but in most cases they are slow compared with Cl^+ transfer in drinking water.

IV. SURVEY OF REACTIONS OF FUNCTIONAL GROUPS

In this section we summarize the known reactions of aqueous chlorine and monochloramine with various organic functional groups. The purpose of this survey is to provide a compendium of reactions and their rates that might be used to assess the relative susceptibility of water contaminants to chemical transformation in a drinking water treatment system. Theory suggests that different functional groups that may be present in an organic chemical can be treated independently. For example, the aromatic ring and the carbamate moiety (half ester and half amide) present in the pesticide carbaryl can undergo reactions independent of one another. Thus, the primary reactions of carbaryl with aqueous chlorine or monochloramine can be modeled by the reactions of these disinfectants with aromatic compounds, esters, and amides.

A. Alkanes

Alkanes are comparatively inert. Halogens react with alkanes, but the reaction requires that either light or heat initiate the formation of halogen free radicals and that the concentrations of reactants are sufficient to sustain a chain reaction. These conditions are unlikely to exist in a water distribution system.

B. Haloalkanes

This class of compounds contains a number of pesticides, such as dieldrin, aldrin, chlordane, endrin, heptachlor, lindane, toxaphene, mirex, and Kepone (chlordecone). Several of these (e.g., dieldrin and heptachlor) are also alkenes; Kepone contains a ketone functional group. In general, haloalkanes are electrophilic, as are HOCl and aqueous Cl_2, and are not chlorinated further. For example, chloroform is produced in drinking water treatment systems when methyl ketones and polyhydroxylated aromatic residues of natural humic substances are chlorinated, but it is not further chlorinated to carbon tetrachloride.

C. Alcohols

Alcohols react with aqueous chlorine to form alkyl hypochlorites (13). Primary and secondary alkyl hypochlorites decompose to aldehydes and ketones (14,15). However, these reactions are too slow to be of significance in drinking waters treated with aqueous chlorine or monochloramine.

D. Ketones and Aldehydes

Ketones and aldehydes are halogenated at the α-carbon atoms on reaction with aqueous chlorine. The halogenation of methyl ketones results in the elimination of a trihalomethane molecule in a reaction referred to as the haloform reaction. This reaction has been thought to be partially responsible for the formation of chloroform on chlorination of natural humic substances during water treatment processes. The α-halogenation of ketones and aldehydes is acid or base catalyzed (16–21) and involves initial slow enolization of the ketone. For example, the mechanism for the base-catalyzed reaction of acetone is as follows:

$$HO^- + CH_3{-}\overset{\overset{\displaystyle O}{\|}}{C}{-}CH_3 \underset{k_{-3}}{\overset{k_3}{\rightleftharpoons}} CH_3{-}\overset{\overset{\displaystyle O^-}{|}}{C}{=}CH_2 \qquad (15)$$

$$Cl_2 + CH_3\text{—}\overset{\overset{\displaystyle O^-}{|}}{C}\text{=}CH_2 \xrightarrow{k_4} CH_3\text{—}\overset{\overset{\displaystyle O}{\|}}{C}\text{—}CH_2Cl + Cl^- \qquad (16)$$

The rate of the chlorination reaction is apparently dependent only on the rate of enolization; that is, the first step is rate limiting.

Kinetic considerations suggest that the α-halogenation of simple aliphatic ketones and aldehydes is not likely to be a significant reaction for these compounds under water treatment conditions. In contrast, 1,3-diketones, which enolize to a much greater extent than acetone, are much more reactive with aqueous chlorine and produce one molecule of chloroform per molecule of ketone (22). Using values of the rate constants k_3 for the reaction of hydroxide ion with ketones of varying structure and other compounds that readily form enolates, half-lives at pH 7.5 can be calculated (23,24). A selection of these is listed in Table 3. It is obvious from inspection of Table 3 that chemical structure can dramatically affect the half-lives of ketones and other compounds that form enolates.

Since monochloramine is a much less effective chlorinating agent than aqueous chlorine, it is not expected to chlorinate the α-carbon of ketones or aldehydes to any significant extent under typical treatment conditions.

E. Amines

Chlorination of aliphatic amino nitrogens by either HOCl or monochloramine is one of the most rapid and significant reactions that can take place in chlorinated drinking water. Both primary and secondary aliphatic amines react rapidly over a wide range of pH values. The mechanism involves the reaction of a free amino group with hypochlorous acid, as illustrated here

Table 3 Half-lives for Reaction of Compounds That Form Enolates in Water Treated with Aqueous Chlorine

Compound	k_3	$k_3\,[HO^-]$	$t_{1/2}$
Acetone	0.25	7.9×10^{-8}	102 days
Hexan-2,5-dione	1.67	5.2×10^{-7}	15.4 days
Chloroacetone	9.3	2.9×10^{-6}	2.7 days
1,1-Dichloroacetone	450	1.4×10^{-4}	1.4 h
Diethyl malonate	10^4	3.2×10^{-3}	3.6 minutes
Acetylacetone	2×10^6	6.3×10^{-1}	1.1 s

for the amino acid glycine:

$$H_2N{-}CH_2{-}COO^- + HOCl \rightarrow ClNH{-}CH_2{-}COO^- + H_2O \qquad (17)$$

The concentration of free amino acid is dependent on pH, and the fraction of unprotonated amino acid present at pH 7.5 is considerably lower than the total amino acid concentration. Because unprotonated amino acid is consumed in the reaction, however, the equilibrium shifts and additional unprotonated amino acid is generated. Free amines react even more rapidly with Cl_2 (25).

N-Chlorinated aliphatic amines without a good leaving group attached to the α-carbon appear to be stable and do not decompose rapidly at pH 7.5 (26). If a good leaving group is attached to the α-carbon, the compound may undergo an elimination reaction to form an unstable imine that subsequently hydrolyzes to an aldehyde (27–29). For example, α-amino acids decompose by the scheme

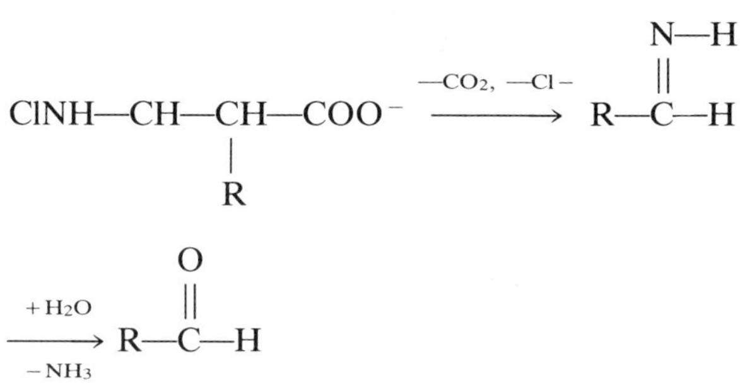

(18)

Because there is an excess of FAC under drinking water treatment conditions, primary amines can be chlorinated further to form *N*,*N*-dichloramines:

$$Cl{-}HN{-}CH_2{-}CH_2{-}COO^- + HOCl \rightarrow Cl_2N{-}CH_2{-}CH_2{-}COO^- + H_2O \qquad (19)$$

The rate of this reaction is considerably slower than the rate of the first chlorination, yet sufficiently fast that *N*,*N*-dichloroamines are the major products of chlorination of primary amines in drinking water. *N*,*N*-dichlorinated α-amino acids are unstable and decay rapidly to *N*-chloroaldimines and nitriles (27,30,31):

$$\underset{\displaystyle \mathrm{R}}{\mathrm{Cl_2N{-}\underset{|}{C}H{-}COO^-}} \xrightarrow{-CO_2,\ -Cl^-} \mathrm{R{-}\overset{\displaystyle N{-}Cl}{\overset{\|}{C}}{-}H} + \mathrm{R{-}C{\equiv}N} \qquad (20)$$

At neutral pH, *N*-chloroaldimines, such as *N*-chloroisobutyraldimine, *N*-chlorophenylacetaldimine, and *N*-chloro-2-methylbutyraldimine, have half-lives of about 40 h and decompose to the corresponding aldehydes, presumably by hydrolysis (F.E. Scully, Jr. and B. Conyers, unpublished data).

Monochloramine has been shown to transfer its chlorine directly to the amino group of organic amines and amino acids (32,33):

$$NH_3Cl^+ + H_2N\text{-}CH_2\text{-}COO^- \rightarrow NH_4^+ + ClNH\text{-}CH_2\text{-}COO^- \qquad (21)$$

Unlike reaction with hypochlorous acid, however, reaction with monochloramine is not likely to result in dichlorinated amines. The stability of an organic N-chloramine depends on the structure of the compound.

Aromatic amines react with aqueous chlorine to form ring chlorinated products. The mechanism involves electrophilic aromatic substitution and is discussed in Section IV.G.

F. Amides

Reactions of amides (and probably carbamates as well) are too slow to be of significance in a drinking water disinfected with aqueous chlorine or monochloramine (34).

G. Aromatic Compounds

The reactivity of homocyclic aromatic compounds varies widely and depends on the type of substituents attached to the ring. Aromatic compounds substituted with simple alkyl groups, such as toluene and xylene, do not react to any significant extent in chlorinated drinking water. In contrast, aromatics substituted with more strongly electron-donating groups, such as the methoxy group (e.g., anisole), are quite reactive, as are phenolic compounds. For compounds with activating (electron-donating) groups, the reaction rate increases with the number of groups attached to the ring. On the other hand, most aromatic compounds are not expected to undergo significant transformation in chloraminated water.

The presence of chlorinated phenols is frequently associated with undesirable taste and odor in chlorine-treated water. The reactivity of phenols is markedly different from that of other aromatics because they can ionize at higher pH values. Lee (35) and Soper and Smith (36) studied the rate of reaction of phenol with aqueous chlorine at pH values from 5 to 12 and found that the reaction mechanism involves chlorination of the phenoxide ion by HOCl. The products of the chlorination of phenol, primarily *o*- and *p*-chlorophenol, can react further in drinking water to form dichlorinated products. The observed rates of chlorination vary widely (35); for example, *p*-chlorophenol is four times less reactive than phenol, but the maximum observed rate constant for the reaction of *o*-chlorophenol is approximately equal to that for phenol. Because of the lower pK_a of the chlorinated phenols, however, the pH at which the maximum rate is observed is lower than that for phenol. Thus at pH 7 *o*-chlorophenol reacts with aqueous chlorine about 50% faster than does phenol.

Ring substitution in some phenols may involve replacement of other functional groups on the ring. Chlorination of phenolic acids, such as 4-hydroxybenzoic acid, results in replacement of the carboxyl group by a chlorine atom.

The reactions of aqueous chlorine with polyhydroxylated aromatics can yield numerous products. 1,3-Dihydroxylated aromatics are believed to be one of the main sources of chloroform produced on chlorination of humic substances in drinking water (37).

Other aromatic compounds that have been shown to react with aqueous chlorine include 4-nitrophenol and 3-hydroxybenzoic acid (38) and a series of polynuclear aromatics (39). Both chlorinated and oxidation products were observed for the latter. In contrast, Burleson et al. (40) found no ring substitution products from phenylalanine upon superchlorination of wastewater, only the expected products from chlorination of the terminal amino group. Chlorinated aromatics, such as polychlorinated biphenyls and the herbicide 2,4-dichlorophenoxyacetic acid (2,4-D), are also unlikely to be reactive because of deactivation of the ring by the halogen substituents.

H. Heterocyclic Compounds

No mechanistic studies of the reactions of heterocyclic compounds have been conducted that allow prediction of half-lives for these compounds. Nevertheless, chlorination products from heterocyclic aromatics have been identified as contaminants of chlorinated wastewaters, and studies employing conditions designed to mimic chlorinated and chloraminated drinking waters suggest that some heterocyclics may undergo significant transformation.

Gould et al. (41–43) studied the chlorination of purines and pyrimidines. Seven products from the chlorination of caffeine were identified, and 5-chlorouracil was identified as a product from the chlorination of uracil. At high chlorine-substrate ratios, dichlorouracil and oxidation products were formed, but it is unclear whether these reactions would occur under actual treatment conditions. The aminopyrimidines cytosine and 5-methylcytosine formed remarkably stable N-chloramino products. Jolley (discussed in Reference 44) identified several chlorinated heterocyclic compounds that were present at significant concentrations in a chlorinated wastewater: 5-chlorouracil (4.3 μg/L), 5-chlorouridine (1.7 μg/L), 8-chlorocaffeine (1.7 μg/L), 6-chloroguanine (0.9 μg/L), and 8-chloroxanthine (1.5 μg/L).

Lin and Carlson (45) examined the reactivity of a group of environmentally important heterocyclic compounds under disinfection conditions. More than 95% of the indole, 3-methylindole, 2-phenylindole, *N*-phenylpyrrole, dibenzothiophene, and carbazole initially present at concentrations of 10^{-5}–10^{-6} M reacted with aqueous chlorine (10^{-4} M) within 15 minutes at pH 7. Other heterocyclic compounds were also reactive but required somewhat longer for the reaction to be 95% complete: acridine (45 minutes); phenanthridine (1 h); benzo[*b*]thiophene (1.5 h); and isoquinoline (4 h). Reaction times were somewhat longer with chloraminated water, but still short enough for the heterocyclics to be considered reactive: indole, 21.5 h; 3-methylindole, 21.5 h; 2-phenylindole, 3.5 h; and *N*-phenylpyrrole, 50 h.

I. Alkenes and Alkynes

In general, alkenes and alkynes are considered electron-rich species and react with electrophilic reagents. For example, alkenes react with HOCl to form halohydrins, and these may react further to form expoxides. This reaction is illustrated here for ethylene:

$$H_2C{=}CH_2 + HOCl \rightarrow H_2\overset{\overset{HO}{|}}{C}{-\!\!-}\overset{\overset{Cl}{|}}{C}H_2 \rightarrow H_2C \overset{\overset{O}{/\quad\backslash}}{-\!\!-\!\!-} CH_2 + HCl \qquad (22)$$

The reactivity of an alkene depends on the substituents attached to the double bond and is greatly diminished by electron-withdrawing groups. Unfortunately, to our knowledge there have been no detailed rate studies of the reactions of alkenes with aqueous chlorine or chloramine. Predictions of the rate of chlorohydrin formation based on theoretical arguments (46) range from minutes to weeks, but more information is needed to determine the true reactivity of these compounds.

J. Carboxylic Acids

Reactions of carboxylic acids in drinking water treated with chlorine or ammonia and chlorine do not appear to occur very rapidly and therefore are probably not very significant. However, acids with "activated" α-carbons may be reactive enough to be partially chlorinated under these conditions. This category includes compounds that have keto, carboxyl, cyano, nitro, and similar groups attached to the α(2)-carbon.

K. Carboxylic Esters

The acyl group of esters is subject to nucleophilic addition. The resulting intermediate can decompose by loss of the alkoxyl group and formation of a new acid derivative:

$$\mathrm{R{-}C({=}O){-}OR} + \mathrm{Nu{:}^-} \rightarrow \mathrm{R{-}C(O^-)(OR){-}Nu} \rightarrow \mathrm{R{-}C({=}O){-}Nu} + \mathrm{RO^-} \tag{23}$$

The principal nucleophiles in aqueous chlorine are water and HO^- and OCl^- ions. Jencks and Carriuolo (47) studied the reaction of these species with the ester *p*-nitrophenyl acetate. At the concentrations typical of chlorinated drinking water (Table 1), the OCl^--catalyzed hydrolysis of the ester is about 30 times faster than HO^- catalysis and about 250 times faster than the water reaction. The calculated half-life for the OCl^--catalyzed hydrolysis is 1.4 h, which is well within the residence time of drinking water distribution systems. However, *p*-nitrophenyl acetate is very reactive hydrolytically because *p*-nitrophenoxide is an excellent leaving group. Simple esters, such as ethyl acetate, generally have lower reactivities. Thus, ethyl acetate is about 120 times less reactive than *p*-nitrophenyl acetate, and ethyl benzoate is more than 10-fold less reactive than ethyl acetate (48,49). The calculated half-reaction time for the reaction of OCl^- with ethyl acetate is about 35 h, however, still fast enough for the reaction to be of significance.

The esters derived from β-ketoacids, 2-cyanoacids, 2-nitroacids, and malonic acid may possess hydrogens on the α-carbons that can be easily replaced by halogens, just as was the case with the corresponding carboxylic acids. For example, diethyl malonate (Table 3) reacts with halogens to undergo displacement of the methylene hydrogens by halogens. The reaction is catalyzed by base, the first step being conversion to the enolate and/or enol, followed by reaction of these intermediates with the halogen. This pathway is the same as that described for simple aldehydes and ketones (see Section IV.D). The second halogen is more easily introduced than the first in base-catalyzed halogenation because the enolate anion of the monohalo compound is stabilized by the electronegative halogen. In the chlorination of diethyl malonate, the rate constant for the substitution of the second chlorine is about 13 times larger than that for introduction of the first chlorine.

Other carboxylic esters that have been studied include ethyl acetoacetate (50), ethyl cyanoacetate, and ethyl nitroacetate (51). The former has a calculated half-life in aqueous chlorine of about 22 minutes, and the others are expected to undergo chlorination even more rapidly. In general, the chlorinated products formed in these reactions are themselves quite reactive, since halogens situated on carbons next to carbonyl groups are easily replaced by nucleophiles. Thus, these compounds are good alkylating agents.

Unfortunately, not enough information is available on the behavior of aqueous chloramine to make any quantitative conclusions about the fate of carboxylic esters in this medium.

L. Sulfur Compounds

Reduced organic sulfur compounds, such as alkyl sulfides and thiols, are generally quite reactive toward oxidizing agents, and several different products may be formed. For example, thiols can be oxidized to sulfinic or sulfonic acids in a multistep oxidation, as follows:

$$\mathrm{R{-}S{-}H} \xrightarrow{[\mathrm{O}]} \mathrm{R{-}}\overset{\overset{\displaystyle \mathrm{O}}{\|}}{\mathrm{S}}\mathrm{{-}H} \xrightarrow{[\mathrm{O}]} \mathrm{R{-}}\underset{\underset{\displaystyle \mathrm{O}}{\|}}{\overset{\overset{\displaystyle \mathrm{O}}{\|}}{\mathrm{S}}}\mathrm{{-}H} \xrightarrow{[\mathrm{O}]} \mathrm{R{-}}\underset{\underset{\displaystyle \mathrm{O}}{\|}}{\overset{\overset{\displaystyle \mathrm{O}}{\|}}{\mathrm{S}}}\mathrm{{-}O^{-}} \tag{24}$$

To our knowledge there have been no mechanistic studies of the reactions of alkyl sulfides with aqueous chlorine or monochloramine. Based on theoretical considerations, however, it seems likely that the reactions are extremely rapid, with half-lives of less than a second for the initial step. More highly oxidized sulfur compounds are slower to react with aqueous chlorine (52), but their reactions are still very rapid, with predicted half-lives of the order of a few minutes. Studies with chloramine-T suggest that even monochloramine reacts extremely rapidly with alkyl sulfides (53,54).

A recent paper discussed the reaction of the sulfide-containing pesticide aldicarb in chlorinated water (55). The pesticide is oxidized rapidly to aldicarb sulfoxide and aldicarb sulfone and eventually to more highly oxidized fragments. The true environmental significance of the rates of the reported secondary reactions are difficult to determine from the studies reported.

M. Organophosphorus Compounds

Hypochlorite solutions have long been used to detoxify equipment contaminated by the phosphoramide insecticides schradan and dimefox, providing prima facie evidence that organophosphorus compounds may be unstable in aqueous chlorine (56). Epstein et al. (57) studied the reaction of the nerve gas sarin (isopropyl methylphosphonofluoridate) with hypochlorite. Sarin hydrolyzes much more rapidly in slightly acid or slightly basic chlorine solutions than in water. The tentative reaction mechanism is

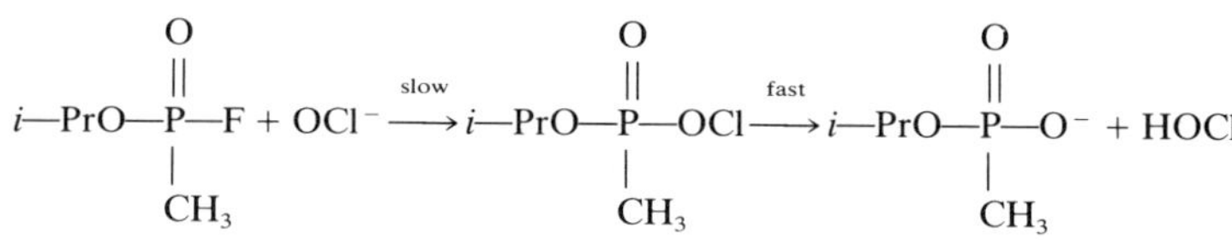

(25)

An unstable phosphoryl hypochlorite intermediate is proposed, which rapidly hydrolyzes. Hypochlorite acts as a true catalyst and is regenerated in the reaction. The estimated half-life in chlorine-treated drinking water is approximately 4 h, making this reaction highly significant under these conditions.

Aqueous chlorine also reacts with S-alkylphosphorothioates. Lordi and Epstein studied the hydrolysis of triethylphosphorothioate (TETP) in aqueous chlorine (58) and observed the following reaction:

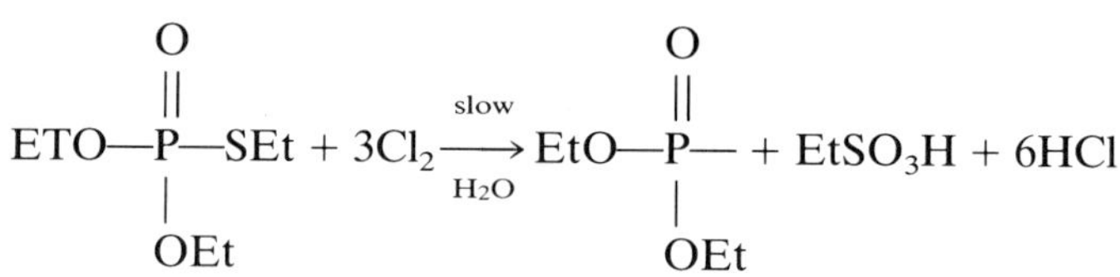

(26)

The calculated half-life for this reaction is 16.9 days, which suggests that the hydrolysis of S-alkylphosphorothioates would proceed, but only to a very limited extent at typical residence times. TETP may be considered a good model for the reaction of S-alkylated thiophosphates, such as the pesticide malathion, but it is not clear whether this is true for O-alkylated thiophosphates, such as parathion.

Organic chloramines have been reported to undergo reaction with a wide variety of organophosphorus compounds (59), but most of these reactions have been carried out in nonaqueous media and few rate studies are available. Epstein et al. (57) also examined the rate of detoxification of sarin with chloramine-T. They found no enhancement of the rate of aqueous hydrolysis in the presence of 3×10^{-3} M chloramine-T. Therefore, monochloramine is not expected to react with phosphonofluoridates in a drinking water at significant rates.

V. SUMMARY

A treated drinking water spends several hours in a treatment plant and as much as 5 days in a typical municipal distribution system. Based on this,

Table 4 Overall Reactivities of Chemical Classes and Functional Groups

Reactivity	Aqueous chlorine	Monochloramine
Very reactive	Amines Alkyl sulfides β-Diketones β-Ketoesters Malonic ester (diethyl malonate) analogs *Some* nitrogen heterocyclic aromatics	Alkyl sulfides
Reactive	Phenols Alkoxybenzenes β-Ketoacids Phosphonofluoridates *p*-Nitrophenyl acetate	Amines *Some* nitrogen heterocyclic aromatics
Somewhat reactive	α,α-Diketones Aliphatic esters *S*-alkylphosphonofluoridates	
Slightly reactive	Aromatic esters Amides Carbamates Alcohols Simple ketones and aldehydes	Alcohols
Unreactive	Alkanes Alkyl halides Toluene and xylene Simple carboxylic acids	Alkanes Alkyl halides Aromatic compounds Phenols Alkenes Amides Carboxylic acids
Unknown	Alkenes	Esters Aldehydes and ketones

the various structural classes and functional groups discussed here can be classified according to reactivity using the following scheme:

Very reactive: >50% reacted in <5 minutes
Reactive: >50% reacted in <1 day
Somewhat reactive: >50% reacted in <5 days
Slightly reactive: >50% reacted in <1 year
Unreactive

Unknown: compounds for which reactions with disinfectants are known but for which information about their reactivity in dilute aqueous solution is not available

Table 4 summarizes this information for both aqueous chlorine- and monochloramine-treated drinking water. Table 4 is intended to serve only as a general guide to reactivity, since (1) within any chemical class there may be exceptions to the rule, and (2) for some classes the reactivity assignment is based on relatively few data.

Chemicals classified as "very reactive" can be expected to undergo stoichiometric conversion to products in drinking water treatment. Foremost among these for aqueous chlorine treatment (of the chemicals that have been studied) are the amines, alkyl sulfides, ketones, and esters with active methylene groups (such as diethyl malonate and acetylacetone), and *some* *N*-heterocyclic aromatics. Chemicals classified as "reactive" can be expected to undergo extensive but not necessarily stoichiometric conversion to products at typical residence times.

REFERENCES

1. F.W. Pontius, Complying with the new drinking water quality regulations, *J. Am. Water Works Assn. 82*:32–52 (1990).
2. J.A. Symons, T.A. Bellar, J.K. Carswell, J. DeMarco, K.L. Kropp, G.G. Robeck, D.R. Seeger, C.J. Slocum, B.L. Smith, and A.A. Stevens, National organics reconnaissance survey for halogenated organics, *J. Am. Water Works Assn. 67*:634–647 (1975).
3. W.P. Jencks and J. Regenstein, Ionization constants of acids and bases, in *Handbook of Biochemistry* (H.A. Sober, ed.), Chemical Rubber Co., Cleveland, OH, 1970, p. J-190.
4. V.L. Snoeyink and D. Jenkins, *Water Chemistry*, Wiley, New York, 1980.
5. E.T. Gray, D.W. Margerum, and R.P. Huffman, Chloramine equilibria and the kinetics of disproportionation in aqueous solution, in *Organometals and Organometalloids, Occurrence and Fate in the Environment* (F.E. Brinckman and J.M. Belloma, eds.), American Chemical Society, ACS Symposium Series 82, Washington, D.C., 1978, pp. 264–277.
6. V.C. Hand and D.W. Margerum, Kinetics and mechanism of the decomposition of dichloramine in aqueous solution, *Inorg. Chem. 22*:1449–1456 (1983).
7. K. Kumar, R.W. Shinness, and D.W. Margerum, Kinetics and mechanism of the base decomposition of nitrogen trichloride in aqueous solution, *Inorg. Chem. 26*:3430–3434 (1987).
8. G.H. Coleman, The reaction of alkylchloramines with Grignard reagents, *J. Am. Chem. Soc. 55*:3001–3005 (1933).
9. C.K. Ingold, *Structure and Mechanism in Organic Chemistry*, 2nd ed., Cornell University Press, Ithaca, NY, 1969, pp. 341–344.

10. K. Fries and W. Vogt, Chlorides and bromides of diphenyl sulfide, *Liebigs Ann. Chem. 381*:337–346 (1911).
11. L.O. Brown and F.G. Soper, Studies of N-halogeno-compounds. Part V. The interaction of dialkylchloramines with phenols, *J. Chem. Soc.* 3576–3580 (1953).
12. M.S. Newman and H.L. Holmes, Beta-naphthoic acid, in *Organic Syntheses* (A.H. Blatt, ed.), Wiley, New York, 1943, Coll. Vol. II, pp. 428–430.
13. M. Anbar and I. Dostrovsky, A kinetic study of the formation and hydrolysis of tert-butyl hypochlorite, *J. Chem. Soc.* 1094–1104 (1954).
14. D.C. Chattaway and O.G. Backeberg, Alkyl hypochlorites, *J. Chem. Soc.* 2999–3003 (1923).
15. R.V. Stevens, K.T. Chapman, and H.N. Weller, Convenient and inexpensive procedure for oxidation of secondary alcohols to ketones, *J. Org. Chem. 45*:2030–2032 (1980).
16. A. Lapworth, The action of halogens on compounds containing the carbonyl group, *J. Chem. Soc.* 30–36 (1904).
17. F.O. Rice and C.F. Fryling, The kinetics of the reaction between halogens and saturated aliphatic ketones in dilute aqueous solution, *J. Am. Chem. Soc. 47*:382–386 (1925).
18. P.D. Bartlett and J.R. Vincent, Enolization as directed by acid and basic catalysts. I. The acid-catalyzed enolization of menthone, *J. Am. Chem. Soc. 55*:4992–4997 (1933).
19. P.D. Bartlett, Enolization as directed by acid and basic catalysts. II. Enol mechanism of the haloform reaction, *J. Am. Chem. Soc. 56*:967–969 (1934).
20. P.D. Bartlett and J.R. Vincent, The rate of the alkaline chlorination of ketones, *J. Am. Chem. Soc. 57*:1596–1600 (1935).
21. R.P. Bell and K. Yates, Kinetics of the chlorination of acetone in aqueous solution, *J. Chem. Soc.* 1927–1933 (1962).
22. S.D. Boyce and J.F. Hornig, Formation of chloroform from the chlorination of diketones and polyhydroxybenzenes in dilute aqueous solution, in *Water Chlorination: Environmental Impact, and Health Effects*, Vol. 3 (R.L. Jolley et al., eds.), Ann Arbor Science, Ann Arbor, MI, 1980, pp. 131–140.
23. R.P. Bell and O.M. Lidwell, The base catalyzed prototropy of substituted acetones, *Proc. R. Soc. [A] 176*:88–113 (1940).
24. J.C. Morris, Reaction dynamics in water chlorination, in *Water Chlorination: Chemistry, Environmental Impact, and Health Effects*, Vol. 5 (R.L. Jolley et al., eds.), Lewis Publishers, Chelsea, MI, 1985, pp. 701–711.
25. D.W. Margerum, E.T. Gray, and R.P. Huffman, Chlorination and the formation of N-chloro compounds in water treatment, in *Organometals and Organometalloids, Occurrence and Fate in the Environment* (F.E. Brinckman and J.M. Belloma, eds.), American Chemical Society, ACS Symposium Series 82, Washington, D.C., 1978, pp. 278–291.
26. F.E. Scully, Jr., and M.A. Bempong, Stability of aqueous solutions of *N*-chloropiperidine and *N*-chlorodiethylamine with varying pH, in *Water Chlorination: Environmental Impact, and Health Effects*, Vol. 3 (R.L. Jolly et al., eds.), Ann Arbor Science, Ann Arbor, MI, 1980, pp. 203–208.

27. A. Nweke and F.E. Scully, Stable *N*-chloraldimines and other products of the chlorination of isoleucine in model solutions and in a wastewater, *Environ. Sci. Technol. 23*:989–994 (1989).
28. W.D. Stanbro and W.D. Smith, Kinetics and mechanism of the decomposition of *N*-chloroalanine in aqueous solution, *Environ. Sci. Technol. 13*:446–451 (1979).
29. V.C. Hand, M.P. Snyder, and D.W. Margerum, Concerted fragmentation of *N*-chloro-α-amino acid anions, *J. Am. Chem. Soc. 105*:4022–4025 (1983).
30. T. Stelmaszynska and J.M. Zgliczynski, *N*-(2-oxoacyl)amino acids and nitriles as final products of dipeptide chlorination mediated by the myeloperoxidase/H_2O_2/Cl^- system, *Eur. J. Biochem. 92*:301–308 (1978).
31. W.D. Stanbro and M.J. Lenkovitch, Kinetics and mechanisms of the decomposition of N,N-dihalopeptides, *Int. J. Chem. Kinetics, 17*:401–411 (1985).
32. R.A. Isaac and J.C. Morris, Transfer of active chlorine to nitrogenous organic compounds. 2. Mechanism, *Environ. Sci. Technol. 19*:810–814 (1985).
33. M.P. Snyder and D.W. Margerum, Kinetics of chlorine transfer from chloramine to amines, amino acids, and peptides, *Inorg. Chem. 21*:2545–2550 (1982).
34. R.P. Mauger and F.G. Soper, Acid catalysts in the formation of chloramides from hypochlorous acid, *J. Chem. Soc.* 71–75 (1946).
35. C.F. Lee, Kinetics of reactions between chlorine and phenolic compounds, in *Principles and Applications of Water Chemistry* (S.D. Faust and J.V. Hunter, eds.), Wiley, New York, 1967, pp. 55–74.
36. F.G. Soper and G.F. Smith, Halogenation of phenols, *J. Chem. Soc.* 1582–1591 (1926).
37. J.J. Rook, Possible pathways for the formation of chlorinated degradation products during chlorination of humic acids and resorcinol, in *Water Chlorination: Environmental Impact and Health Effects*, Vol. 3 (R.L. Jolley et al., eds.), Ann Arbor Science, Ann Arbor, MI, 1980, pp. 85–98.
38. L.H. Nowell and D.G. Crosby, Photodegradation of water pollutants in chlorinated water, in *Water Chlorination: Chemistry, Environmental Impact and Health Effects*, Vol. 5 (R.L. Jolley et al., eds.), Lewis, Chelsea, MI, 1985, pp. 1055–1062.
39. R.M. Carlson, R.E. Carlson, H.L. Kopperman, and R. Caple, Facile incorporation of chlorine into aromatic systems during aqueous chlorination processes, *Environ. Sci. Technol. 9*:674–675 (1975).
40. J.L. Burleson, G.R. Peyton, and W.H. Glaze, Gas-chromatographic/mass spectrometric analysis of derivatized amino acids in municipal wastewater products, *Environ. Sci. Technol. 14*:1354–1359 (1980).
41. J.P. Gould and T.R. Hay, The nature of the reactions between chlorine and purine and pyrimidine bases: Products and kinetics, *Water Sci. Technol. 14*:629–640 (1982).
42. J.P. Gould, J.T. Richards, and M.G. Miles, The kinetics and primary products of uracil chlorination, *Water Res. 18*:205–212 (1984).
43. J.P. Gould, J.T. Richards, and M.G. Miles, The formation of stable organic chloramines during the aqueous chlorination of cytosine and 5-methylcytosine, *Water Res. 18*:991–999 (1984).

44. D.H. Rosenblatt, Chlorine and oxychlorine species reactivity with organic substances, in *Disinfection, Water and Wastewater* (J.D. Johnson, ed.), Ann Arbor Science, Ann Arbor, MI, 1975, pp. 249–276.
45. S. Lin and R.M. Carlson, Susceptibility of environmentally important heterocycles to chemical disinfection: Reactions with aqueous chlorine, chlorine dioxide, and chloramine, *Environ. Sci. Technol. 18*:743–748 (1984).
46. G.C. Israel, J.K. Martin, and F.G. Soper, The kinetics of chlorohydrin formation. Part I. The reaction between hypochlorous acid and allyl alcohol in aqueous solution, *J. Chem. Soc.* 1282–1285 (1950).
47. W.P. Jencks and J. Carriuolo, Reactivity of nucleophilic reagents toward esters, *J. Am. Chem. Soc. 82*:1778–1786 (1960).
48. J.E. Potts and E.S. Amis, The alkaline hydrolysis of ethyl acetate from the standpoint of ion-dipole theory, *J. Am. Chem. Soc. 71*:2112–2116 (1949).
49. E. Tommila and L. Ketonen, Kinetic studies on ester hydrolysis. X. The alkaline hydrolysis of ethyl benzoate and certain substituted ethyl benzoates in methanol-water mixtures, *Suomen Kemistilehti 18B*:24–28 (1945).
50. K.J. Pedersen, The velocity of bromination of acetoacetic ethyl ester. II. The general base catalysis, *J. Phys. Chem. 38*:601–621 (1934).
51. R.G. Pearson and R.L. Dillon, Rates of ionization of pseudo acids. IV. Relation between rates and equilibria, *J. Am. Chem. Soc. 75*:2439 (1953).
52. J.L. Kice and A.R. Puls, The reaction of hypochlorite with various oxidized derivatives of disulfides and with sulfinate ions, *J. Am. Chem. Soc. 99*:3455–3460 (1977).
53. J.G. Jacangelo and V.P. Olivieri, Aspects of the mode of action of monochloramine, in *Water Chlorination: Chemistry, Environmental Impact and Health Effects*, Vol. 5 (R.L. Jolley et al., eds.), Lewis, Chelsea, MI, 1985, pp. 575–586.
54. H. Yuki, T. Kishikawa, Y. Tohira, S. Suzuki, F. Sano, and S. Hayakawa, Antiviral agents. V. Synthesis and antiviral activity of *N*-chloro compounds, *Chem. Pharm. Bull. 15*:1123–1128 (1967).
55. C.J. Miles, Degradation of aldicarb, aldicarb sulfoxide, and aldicarb sulfone in chlorinated water, *Environ. Sci. Technol. 25*:1774–1779 (1991).
56. H. Tsuyuki, M.A. Stahmann, and J.E. Casida, Preparation, purification, isomerization, and biological properties of octamethylpyrophosphoramide *N*-oxide, *J. Agr. Food Chem. 3*:922–930 (1955).
57. J. Epstein, V.E. Bauer, M. Saxe, and M.M. Demek, The chlorine-catalyzed hydrolysis of isopropyl methylphosphonofluoridate (Sarin) in aqueous solution, *J. Am. Chem. Soc. 78*:4068–4071 (1956).
58. N. Lordi and J. Epstein, Kinetics and mechanism of chlorination of triethylphosphorothiolate in dilute aqueous media at 25°, *J. Am. Chem. Soc. 80*:509–515 (1958).
59. P. Kovacic, M.K. Lowery, and K.W. Field, Chemistry of *N*-bromamines and *N*-chloramines, *Chem. Rev. 70*:646–647, 649–651, 658–661 (1970).

4

Chemical Monitoring of California's Public Drinking Water Sources: Public Exposures and Health Impacts

David L. Storm

California Department of Health Services, Sacramento, California

I. INTRODUCTION

The drinking water resources in California were for many years considered pristine and plentiful. Groundwater from wells was assumed to be generally pure and wholesome as a result of natural filtration in the ground. Surface water from lakes and rivers was readily purified by filtration and chemical disinfection with chlorine. In recent decades, however, it has become apparent that California's increasing population, industrialization, and agricultural development have impacted on the quality of a portion of its drinking water supplies. Pesticides and nitrate fertilizers have been detected in groundwater and in rivers of areas subject to heavy agricultural activities. The groundwater of many urbanized, industrial areas of the state have been found to contain industrial organic solvents. Concerns are now growing about the effects of certain natural contaminants, such as heavy metals and radionuclides, on human health.

The estimated urban demand for water in California, including drinking water, is 7 million acre-feet annually (1) (one acre-foot of water is approximately equal to 326,000 gallons water). This accounts for about 17% of the total water demand in the state. Approximately 11,000 public water systems provide drinking water to persons in California. A public water system is defined in Section 64411 of the California Code of Regulations (CCR) as a system providing piped water to the public for domestic use and that has at least 5 service connections or regularly serves an average of at least five individuals daily at least 60 days of the year. About one-tenth of these systems are large water systems having 200 or more service connections. They supply drinking water to about 97% of the state's population. Most of the remaining 3% of the population obtains its water from the 10,000 small water systems having 200 or fewer service connections. In all, these public water systems draw their water supplies from approximately 20,000 groundwater and surface sources throughout the state. Approximately 61% of the large water systems and 80% of the small water systems utilize groundwater (2).

The quality of water served by public drinking water systems has been monitored for various physical, chemical, and biological parameters for many years. The federal and state safe drinking water acts that were passed in recent years now require that public water systems monitor on a regular basis their water sources and water being served for a wide variety of chemicals and parameters. California's water quality monitoring requirements are specified in Title 22, Chapter 15 of the California Code of Regulations. These regulations include drinking water standards for heavy metals, minerals, radionuclides, and other inorganics; organic chemicals, such as industrial solvents and pesticides; and coliform bacteria. The regulations specify schedules to which water system operators must conform for sampling and resampling for each of the parameters.

Drinking water quality has also been investigated through special studies conducted by public agencies. In the late 1970s and early 1980s frequent discoveries of organic chemical contamination were made in large water system and small water system wells throughout the state. Few of these organic chemicals were regulated through drinking water standards or regulatory monitoring requirements. Accordingly, the state legislature passed Assembly Bill 1803 (AB 1803, Chapter 881, Statutes of 1983). That legislation mandated the California Department of Health Services to evaluate all public water system groundwater wells for potential contamination by organic chemical contaminants. The large water system phase of the AB 1803 program commenced in 1984. During 1984–1985, 2947 of the 8800 large water system wells were sampled and tested for organic chemical contamination. It was found that over 18% of the wells tested were contaminated with organic chemicals, and 5.6% exceeded one of the department's action levels (3). The small water system phase of AB 1803 was carried out between 1986 and 1989. Approximately 4800 of the 10,000 small water systems were sampled by local health departments. Approximately 8% of them were found to have contaminated sources, and 3% exceeded an action level or a maximum contaminant level (4). Federal and state regulations were adopted in the late 1980s that require periodic monitoring of drinking water sources for "unregulated" organic chemicals that have no enforceable drinking water standards.

II. SCOPE AND LIMITATIONS

This chapter describes the major sources of drinking water in California and the important contaminants that can adversely affect the quality of that water and have impact on public health. Data and information are provided that address our current state of knowledge about drinking water source quality in California, and some of the more serious chemically related water quality problems existing at present in the state are discussed. It is beyond the scope of this chapter to report on the monitoring of the many chemicals tested in public drinking water sources in a state the size of California. With the exception of radon, the discussion is limited to those chemicals for which there currently are primary drinking water standards or maximum contaminant levels (MCL) in California. The emphasis is placed on drinking water sources that have exceeded the MCL. The department's comprehensive drinking water quality report addresses all chemicals tested, along with more information about positive detections not exceeding the MCL (2). The primary goal here is to identify the important chemical contamination problems in California and examine some approaches for prioritizing them for more in-depth, targeted exposure and risk assessments in the future.

All drinking water analyses reported here were carried out by California state-certified laboratories using standardized and approved methods and procedures. These laboratories are certified and regulated through the department's Environmental Laboratory Accreditation Program. Currently the Office of Drinking Water quality-monitoring database (WQM) contains 247 laboratories certified to conduct chemical analyses of various kinds. Most source sampling in large drinking water systems are carried out by personnel of the systems or by contracted firms. Small water system samples were collected usually by state or local county personnel. This chapter does not describe the multitude of the regulatory monitoring requirements and analytical methods for the chemicals discussed here. They have been covered in detail by Wong (5).

An introductory discussion is provided that describes drinking water sources and causes of their contamination. A detailed evaluation of the specific causes and approaches for its prevention and mitigation, however, is not addressed here. The California Environmental Protection Agency (CEPA) is responsible for this aspect of water contamination.

All the data presented in this chapter, with the exception of some radon data, were extracted from the Department's Office of Drinking Water

WQM data base. The data base, containing more than 2 million analyses, covers almost entirely monitoring results for drinking water sources, such as wells and surface intakes from rivers, lakes, reservoirs, and aqueducts. Little information is available in the WQM about monitoring within public drinking water utility distribution systems. None is available for monitoring of consumer taps. Accordingly, the data presented here must be interpreted with some discretion with regard to potential human exposures. The caveats and cautions necessary in using them are detailed in the Section VI. Most of the data in WQM cover sampling from January 1984, small amounts covering sampling before 1984. This chapter reports data for water samples collected through December 1992.

For organic chemicals, positive detections and cases when they exceed MCL are reported here only if they were confirmed by at least one more independent sampling and testing of the source. The purpose of this validation of organic chemical results is to prevent the use of false positive findings, which have frequently been observed as a result of outside contamination of water samples during transit or laboratory handling.

III. SOURCES OF CALIFORNIA'S DRINKING WATER

Table 1 summarizes the number of ground and surface public drinking water sources currently recorded in the WQM according to their operational status and the population served by the parent water systems. The sources listed for small water systems having 5–199 water connections (generally less than 200 persons) include those regulated by the department, such as those operated by state and federal institutions, and about one-half of the approximate 10,000 regulated by local health jurisdictions. As of December 31, 1992, WQM had 15,444 drinking sources on record. This is approximately three-quarters of the total public sources in the state. More than 94% of the sources on record are wells; only 870 are surface intakes.

Most of the sources are on active status (over 83%). The remainder are on standby or nonoperational or are other kinds of water sampling points not classified as drinking water sources. The number of drinking water sources in the state and the status of many is continually changing as new sources are opened and others are closed.

A. Groundwater Sources

More than 90% of the water systems in California utilize groundwater to serve a population of about 9 million (about 30%). In addition, many of the systems serving surface water depend on groundwater as a source of

Table 1 Public Drinking Water Sources in Office of Drinking Water Data Base

Source type	Population served by system	Number of systems: Active[a]	Standby[b]	Non-operational[c]	Other[d]	Totals
Ground	<200	5,465	10	51	7	5,533
	200–2999	1,200	38	273	13	1,524
	3000–9999	1,113	45	318	33	1,509
	>9999	4,125	225	1542	116	6,008
	Totals	11,903	318	2184	169	14,574
Surface	<200	127	1	6	0	134
	200–2999	225	1	5	2	233
	3000–9999	115	0	1	2	118
	>9999	330	1	11	43	385
	Totals	797	3	23	47	870
	Total sources	12,700	321	2207	216	15,444

[a]Sources currently in operation and serving water.
[b]Sources in operation less than 15 days each year, usually in times of emergency or high demand.
[c]Sources that are destroyed, inactive for a period of 1 year or greater, or abandoned with no intent of future use.
[d]Nondrinking water sources including monitoring wells, agricultural wells, and sampling points within water utility distribution systems.

reserve supply during emergencies and as augmentation of the surface water during peak demand periods. In rural areas, where small water systems are prevalent, over 90% of the population relies on groundwater as the only source of drinking water.

Approximately 40% of California overlays groundwater basins (6). The California Department of Water Resources has defined 449 groundwater basins in nine hydrological study areas. These hydrological study areas store about 850 million acre-feet of water; however, less than one-half of that water lies close enough to the earth's surface to be economically pumped (1). Groundwater basins are discrete geological-hydrological units where political boundary lines are also considered whenever practical (7).

Currently 14,574 of the sources recorded in WQM are wells. Their physical locations in the state are illustrated in Figure 1.

B. Surface Water

During an average year approximately 71 million acre-feet of water derived from precipitation drains from the state's land surface. An additional 1.4

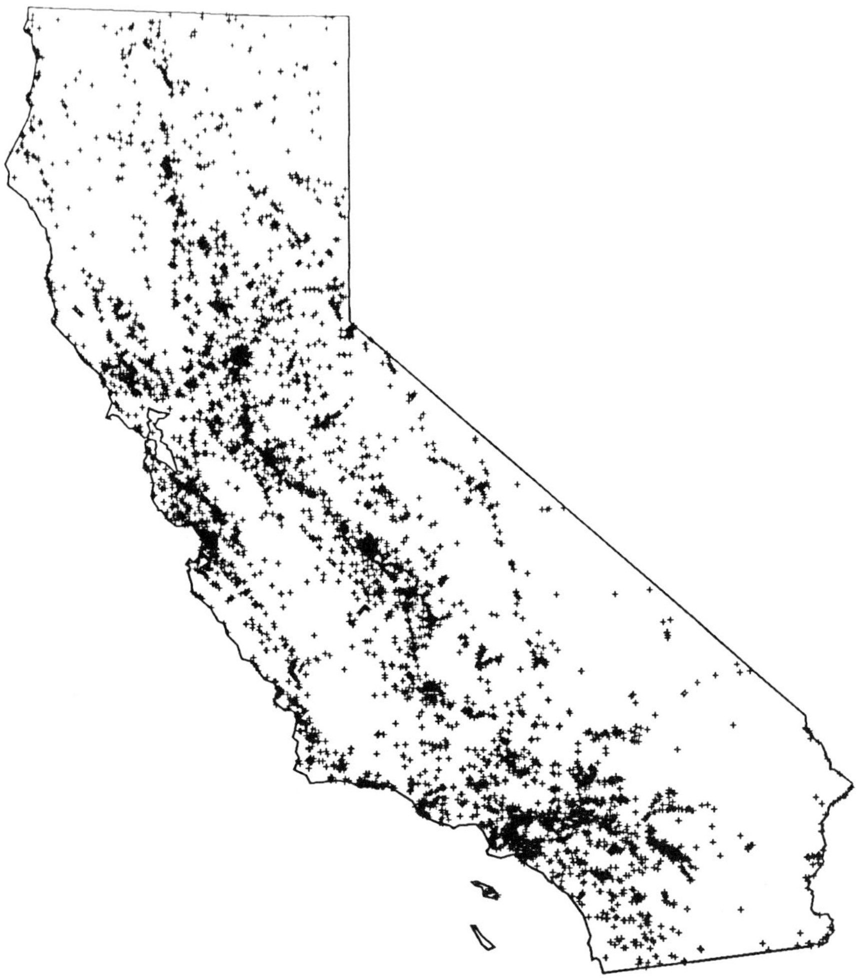

Figure 1 Public drinking water wells in California.

million acre-feet flows into the state from streams that have all or part of their watershed in Oregon, and 4.8 million acre-feet enter the state by way of the Colorado River. California drinking water systems using surface water obtain their supplies from the resulting lakes, streams, and reservoirs. Only about 900 water systems (less than 10%) utilize surface water as a source for their supplies, but they serve approximately 21 million people, about 70% of the population. Figure 2 shows the locations of public drinking water surface intakes.

California's surface water is managed through an extensive system of local, state, and federal dams, reservoirs, and aqueducts to provide the maximum benefit for agricultural and urban usage. There are a total of 1313 reservoirs in and adjacent to California with a combined capacity of

Figure 2 Public drinking water surface sources in California.

43 million acre-feet. The State Water Project (SWP), including the California Aqueduct, dams, and reservoirs, is the most extensive water development and conveyance system in the state. It has the capacity to transport annually up to 3.6 million acre-feet of surplus water not subject to prior rights from northern California's Sacramento Valley over 600 miles to urban and agricultural water consumers in the Central Valley, San Francisco Bay area, and Southern California. Approximately one-half of this water is runoff collected by Lake Oroville from the Feather River watershed and the remainder from surplus flows in the Sacramento-San Joaquin Delta. During the 10 year period of 1978 to 1988 the SWP delivered almost 10 million acre-feet of water to municipal water suppliers for delivery to domestic customers. This represents an estimated average of 100 gallons per day for each person served from the SWP (1). The Central Valley Project (CVP), operated by the U.S. Bureau of Reclamation, has three elements (8). The northern element includes Shasta Dam and Lake, which collects water from the Sacramento River watershed for delivery to farmers in the Sacramento Valley. The second pumps water from the Sacramento-San Joaquin Delta to users in the San Joaquin Valley. The third element is Friant Dam and Millerton on the San Joaquin River. It provides water to users on the east side of the San Joaquin Valley as far south as Kern County.

IV. TYPES AND SOURCES OF DRINKING WATER CONTAMINATION

The waters of California are subject to a multitude of natural and human-caused threats to their quality (Table 2). Natural threats, occurring in all parts of the state, include geological structures, such as mineral springs, steam geysers, natural hot-water areas, storm runoff, soluble mineral deposits, and peat- or organic-based formations. It is not possible to avoid the effects of these types of contaminants on water quality. Sources so contaminated must be avoided, treated, or blended with uncontaminated water to lower the contaminant's concentration before use.

Contamination of human derivation on the other hand, can often be controlled. The CEPA is the primary California state agency responsible for controlling chemical waste and effluent discharges to minimize water contamination. In addition, the U.S. Environmental Protection Agency (EPA) oversees the cleanup of hazardous contaminants. Many are point sources, and responsible individuals can be identified. These include underground tanks, leaking pipelines, solid and hazardous waste disposal sites, waste discharges, and hazardous spills. Point sources are those that

Table 2 Chemical Threats to Drinking Water Quality

Source of contamination	Contaminant	Typical sites
Natural (occurs statewide)	Dissolved minerals	Mineral deposits, mineralized waters, hot springs, seawater intrusion
	Asbestos	Mine tailings and treatment wastes, serpentine formations
	Hydrogen sulfide	Subsurface organic deposits, such as Delta Islands and San Joaquin Valley trough; present in most geological formations
	Radon	
Commercial businesses	Gasoline	Service station underground storage tanks
	Solvents	Dry cleaners, machine shops
	Toxic metals	Photographic processors, laboratories, metal plating works
Municipal	Toxic metals, nutrients, and miscellaneous liquid wastes	Contributions from industrial dischargers, households, and septic tanks
Industrial	VOC, industrial solvents, toxic metals, acids	Electronics manufacturing, metal fabricating and plating, transporters, storage facilities, on-site hazardous waste storage and disposal
	Pesticides and herbicides	Chemical manufacture, storage, and formulating plants; refineries
	Wood preservatives	Pressure waste treating power poles, wood pilings, railroad ties
Waste disposal	Solvents, pesticides, toxic metals, organics, petroleum wastes, any chemicals used in industrial and commercial activities	Solid and liquid waste disposal, storage, and treatment sites located statewide that receive waste from a variety of industries, municipal solid wastes, waste petroleum products, household waste
Agricultural	Pesticides, herbicides, fumigants, fungicides, fertilizers, concentrated mineral salts, nitrates, microbial agents	Irrigated farm runoff, agricultural chemical applications, airborne migration, fertilizer usage, chemical storage at farms and applicators' airstrips, agricultural produce packing sheds and processing plants, meat-processing plants, dairies and feedlots
Disasters	Solvents, petroleum products, pesticides, any other chemical shipped or stored	Earthquake-caused pipeline and storage tank failures and damage to sewage treatment and containment facilities; major spills of hazardous materials in storage or transit; flood water contamination of storage reservoirs and groundwater sources

cause localized contamination affecting a relatively small portion of a groundwater basin or stream area.

Surface and groundwater contamination can differ significantly in their impact, longevity, and ease of cleanup. When a river, lake, reservoir, or aqueduct is contaminated by a chemical discharge or spill, the immediate, local effects may be acute but are generally short-lived. The remedy often entails merely stopping the discharge or spill release and discontinuing water use while the contaminant flows downstream. If the contaminated water enters a larger body, such as a lake or river, adverse effects are often diminished through dilution, precipitation, or degradation.

Groundwater contamination, on the other hand, can be difficult or impossible to correct for many years. Groundwater migrates at a very slow rate, usually measured in feet per year, if it is undisturbed by pumping. When a large utility or agricultural well is in operation, the direction of flow and the rate of flow can be significantly changed. When several such wells are active within the basin, the combined effect can greatly increase the rate and extent of spreading of the contamination. Wells drawing water from a contaminated basin can do so for many years, resulting in potential chronic exposure of consumers to the contaminants.

The following is a more detailed discussion of the various types of contamination in groundwater and surface water and the extent of their influence.

A. Groundwater Contamination

The subsurface environment is a complex system subject to contamination from many sources. Groundwater quality is often affected by waste disposal and inappropriate handling of chemicals. Water-soluble wastes and other materials that are dumped, spilled, or stored on the surface of the land or in burial pits and lagoons can be dissolved by precipitation, irrigation waters, or liquid wastes and eventually move through the soil in the unsaturated zone to pollute groundwater. Additionally, waste fluids disposed in injection wells can eventually migrate directly into drinking water aquifers. Groundwater can also be contaminated by agricultural activities, such as pesticide and fertilizer applications to the land. Excess irrigation water or annual precipitation can dissolve the applied chemicals or their degradation products and leach them into underlying groundwater. The potential for agricultural chemicals to contaminate groundwater depends on the individual properties of the chemicals, the quantities applied, and manner of application. Livestock waste produced from large-scale operations and deposited on the land can result in the infiltration of leachates high in nitrogen content into groundwater supplies. These leachates can contaminate the water supplies with nitrates and nitrites. Groundwater can also be contaminated by natural sources. As groundwater slowly migrates through subsurface strata, it can dissolve heavy metals, selenium, fluorides, and radionuclides from subsurface minerals.

The potential for various natural and artificial chemicals to contaminate groundwater is dependent on many factors, including the following:

Chemical and physical characteristics of the soil
Rainfall and percolation rates
Distance from source of contamination
Water solubility of the chemical
Volatility of the chemical
Mobility of the chemical in soil
Resistance of the chemical to degradation
Nature of any degradation products
Nature of any physical encapsulation of the chemical
Amount of chemical

Because groundwater generally moves very slowly, there is little mixing and dilution of contaminants. A contaminant in a plume of groundwater may remain highly concentrated for decades as it moves from the recharge to the discharge area or to wells in its path.

1. Types of Groundwater Contamination

Groundwater contamination is typically categorized as either point or non-point source contamination.

a. *Point Source Contamination*

Point source contamination is generally localized and is caused by recognized, defined sources. Virtually every area of the state has experienced some form of contamination from a point source discharge. Many commercial establishments, such as dry cleaners, have in the past practiced on-site storage and disposal of cleaning agents and solvents, causing numerous cases of localized groundwater contamination. Industries have been responsible for contamination of community supplies through the inadequate handling and storage of chemicals and disposal of wastes. Businesses involved in the storage and disposal of hazardous wastes, such as chemical drum recyclers, often create potential contamination through unsafe practices. Leakage from undergound gasoline storage tanks, tanks for pressure treating of utility poles, exploding and leaking transformers, brine waste ponds at food-processing plants, and numerous other activities have contributed to the contamination of the groundwater resource. Contamination has not always been the result of irresponsible actions on the part of an individual or of an industry. In many cases contamination occurred years before controls or regulations existed.

b. Nonpoint Source Contamination

Nonpoint discharges have no easily identified points of discharge. They are the most common causes of contamination seen in public drinking water wells. Examples are storm runoff from heavily populated areas, surface and subsurface agricultural drainage from irrigated areas, and percolation of pesticides applied in agricultural areas. Nonpoint sources can cause contamination and affect entire groundwater basins. The contamination of the San Gabriel and San Fernando Valley water basins with volatile organic chemicals (VOC) in Southern California and the San Joaquin Valley groundwater basin with dibronochloropropane (DBCP) are examples of nonpoint contamination. These are described in Section V.

2. Types of Groundwater Contaminants

a. Organic Chemicals

Volatile organic chemicals are the most common types of industrial chemicals found contaminating groundwater systems. VOC are typically liquid chemical substances utilized in a wide variety of commercial and manufacturing processes to clean equipment and other materials and in chemical production activities as solvents. Volatile halocarbons are by far the most frequently occurring VOC contaminants detected in groundwater. These are synthetic VOC that contain halogens, such as chlorine, fluorine, and bromine, and include such solvents as trichloroethylene, tetrachloroethylene, methylene chloride, and chloroform. Another category of VOC found often in groundwater are volatile aromatic compounds, such as benzene, toluene, xylenes, and ethylbenzene. In addition to being important chemical intermediates and solvents, these chemicals are natural components of petroleum products, such as gasoline. The U.S. EPA has declared many volatile halocarbons and several volatile aromatics as priority pollutants (9). VOC having regulatory MCL in California are listed in Table 3. All the industrial chemicals and solvents in Table 3 are VOC with the exception of diethylhexylphthalate (DEHP), which belongs to the class of BNA (base, neutral, and acid extractable chemicals; see later).

VOC have been detected in many drinking water wells in California. The AB 1803 studies found that the majority of wells in which organic chemical contamination were found contained VOC (3,4). Wells in the urban, industrialized areas of the state were the most frequently contaminated. In most cases the sources of the contaminants are not known but are generally believed to be the result of industrial and commercial operations. VOC have been spilled and dumped illegally onto the land surface, disposed in sanitary landfills, and discharged into sewers. In the latter

Table 3 Organic Chemicals Having Maximum Contaminant Levels in California

Chemical	MCL (μg/L)
Industrial chemicals and solvents	
Benzene	1.0
Carbon tetrachloride	0.5
1,4-Dichlorobenzene	5.0
1,1-Dichloroethane	5.0
1,2-Dichloroethane	0.5
1,1-Dichloroethylene	6.0
Cis-1,2-dichloroethylene	6.0
Trans-1,2-dichloroethylene	10.0
Di-(2-ethylhexyl)phthalate (DEHP)	4.0
Ethylbenzene	680
Monochlorobenzene	30
1,1,2,2-Tetrachloroethane	1.0
Tetrachloroethylene	5.0
1,1,1-Trichloroethane	200
1,1,2-Trichloroethane	32
Trichloroethylene	5.0
Trichlorofluoromethane	150
1,1,2-Trichloro-1,2,2-trifluoroethane	1200
Vinyl chloride	0.5
Xylenes[a]	1750
Agricultural chemicals	
Atrazine	3.0
Bentazon	18
Carbofuran	18
Chlordane	0.1
2,4-D	100
Dibromochloropropane (DBCP)	0.2
1,2-Dichloropropane	5.0
1,3-Dichloropropene	0.5
Endrin	0.2
Ethylene dibromide (EDB)	0.0
Glyphosate	700
Heptachlor	0.0
Heptachlor epoxide	0.0
Lindane	4.0
Methoxychlor	100
Monlinate	20
Simazine	10
Thiobencarb	70
2,4,5-TP silvex	10
Toxaphene	5.0
Trihalomethanes (THM)[a]	
Bromodichloromethane	100
Bromoform	100
Dibromochloromethane	100
Chloroform	100

[a]MCL is for either single chemical or the sum of all of them.

cases, the sewer lines have sometimes leaked and contaminated the groundwater with illegally discharged chemicals. In some cases, leaking storage tanks and lines at industrial operations have resulted in contamination of groundwater supplies by VOC. The chemicals, because they are moderately soluble in water, can then be leached by precipitation into the groundwater. The high volatilities of VOC and the fact that in vapor form they are more dense than air also facilitate their contamination of subsurface aquifers. VOC vapors have a tendency to migrate downward through the unsaturated zone to reach groundwater. Some organic chemicals can also undergo chemical transformation in subsurface strate and aquifers to produce additional contaminants. For example, tetrachloroethylene can be biotransformed to trichloroethylene, dichloroethylene, vinly chloride, and carbon dioxide (10).

Once VOC have reached a groundwater supply, the contamination is likely to remain in the water for very long time periods. Since the subsurface water is essentially a closed, migrating system not open to the atmosphere, the VOC cannot evaporate from the water. Because of the great potential for VOC to contaminate groundwater systems, public water wells are now routinely monitored for them. Most VOC can be analyzed with a single scanning procedure.

The less volatile organic chemicals, many categorized as base, neutral, and acid extractable chemicals, are seen very seldom in groundwater. The BNA generally have a lower water solubility than the VOC. They consist of a diverse collection of chemicals, which include industrial synthetic intermediates, such as 2-chlorophenol, plasticizers, such as the phthalate esters, chlorinated pesticides, and polychlorinated biphenyls. As a class they are not as ubiquitous in industry and commerce as VOC and are not as likely to contaminate groundwater because of their generally lower water solubilities and lower mobilities in soil. The only BNA detected in public drinking water sources have been diethylhexylphthalate, endrin, and methoxychlor. DEHP was found through the AB 1803 study in several small water system wells in San Benito and Monterey counties. Since DEHP exhibits very low water solubility and does not migrate readily through soil, it has been speculated that the contamination is derived from construction materials used within the wells. Even though BNA do not readily contaminate groundwater, the U.S. EPA has identified selected BNA as priority pollutants because of the potential hazards they pose to the environment through other routes (9). Those having California MCL are included in Table 3.

b. Trihalomethanes (THM)

THM are a subset of volatile organic chemicals that are produced when drinking water is disinfected with chlorine. THM of concern are listed in

Table 3. They are produced when naturally occurring organic humic substances, such as humic, fulvic, and hymatomelanic acids, and naturally occurring bromide salts in the water react with the added chlorine. As such THM are not normally direct groundwater contaminants but are rather undesirable disinfection by-products. Although THM formation is a more serious problem for surface waters, which contain much more humic substances than well water, the formation of small amounts of THM are common when well water is chlorinated. Chloroform is the only THM that is a common industrial manufacturing intermediate and solvent and may therefore occasionally appear as a groundwater contaminant.

c. *Pesticides*

Agriculture is a major industry in California. As a result, a wide variety of agricultural pesticide chemicals have been used in large quantities throughout the state. Pesticides are used to control insects, rodents, mites, snails, slugs, fungi, weeds, and soil organisms that might otherwise damage crops or compete with them.

There are thousands of different pesticide chemicals, and their physical and chemical properties differ considerably. For example, the insecticide DDT is of low volatility and low water solubility and has never been confirmed as a contaminant in a public drinking water well. Dibromochloropropane, on the other hand, is a volatile, moderately soluble soil fumigant that has been found in hundreds of public drinking water wells in California. The manner in which pesticides are applied can affect their potential for groundwater contamination. Depending on their functions and target organisms, pesticides can be applied directly to plant foliage by spraying or dusting, to the land surface by spraying, to the land dissolved in irrigation water, or to the subsurface by injection. Direct application onto or under the land surface increases the potential for groundwater contamination. DBCP, the most frequently found pesticide well contaminant, was usually injected directly into the land subsurface to control nematodes until it was deregistered in the late 1970s. The number of different pesticide chemicals utilized in California is too large to list all of them in this report. Table 3 lists those that have California MCL. A number of the agricultural chemicals have been detected in private domestic wells by the Department of Pesticide Regulation, CEPA (formerly Department of Food and Agriculture), but have not been found in public drinking water wells (11).

There is no single scanning analytical method that covers all the pesticides in Table 3. As a consequence, several methods would have to be carried out by a laboratory at considerable cost to analyze a water sample for all the pesticides. Public drinking water sources have been selectively analyzed for pesticides according to the amounts used in the vicinity of the well. Use patterns are usually obtained from local county agricultural commissioners. Additionally, the pesticides in Table 3 with regulatory MCL are subject to periodic testing in public drinking water wells pursuant to regulations. Several pesticide chemicals shown in Table 3 are analyzed by the group scanning methods utilized for VOC and BNA.

d. Inorganic Chemicals

The inorganic chemicals of concern in drinking water include the water-soluble compounds of metals and anion salts. Those having MCL are listed in Table 4. Contamination of groundwater by inorganic chemicals can result from natural processes or from the human activities listed in Table 2. Two important natural sources are chemical weathering of mineral deposits and soil leaching. The abundance of minerals containing the chemicals often vary across the state; therefore, the contamination of groundwater from natural sources is regional.

The release of inorganic chemicals from natural or human-derived materials is a complex process influenced by solubility, pH (degree of acidity), adsorption characteristics, hydration, coprecipitation, colloidal dispersion, and formation of chemical complexes. The chemistry of each substance

Table 4 Inorganic Chemicals Having Maximum Contaminant Levels in California

Chemical	MCL (mg/L)
Arsenic	0.05
Barium	1
Cadmium	0.010
Chromium	0.05
Lead	0.05
Mercury	0.002
Selenium	0.01
Silver	0.05
Fluoride	
(annual average maximum daily air temperatures, °F)	
≥53.7F	2.4
53.8–58.3	2.2
58.4–63.8	2.0
63.9–70.6	1.8
70.7–79.2	1.6
79.3–90.0	1.4
Nitrate (as NO_3)	45

listed in Table 4 varies considerably, along with its potential for contaminating groundwater. Although all the listed metals are insoluble in their elemental form, they can form soluble complex anions that can contaminate groundwater. The compounds of arsenic, barium, fluoride, and selenium are frequent groundwater contaminates. Fluoride, one of the most common, occurs naturally as calcium fluoride in the mineral fluorspar (fluorite) and as sodium aluminum fluoride in cryolite.

Nitrate (NO_3^-) is a highly soluble anion of nitrogen that can readily contaminate groundwater by percolation. An important constituent of the nitrogen cycle in nature, it is a very commonly occurring substance and the most frequently observed contaminant in drinking water wells. There are several important sources of nitrate. Biological nitrogen fixation and lightning are two natural processes that convert elemental nitrogen in the atmosphere to nitrate. Human activities causing nitrate contamination include application of fertilizers, waste disposal, wood burning and fossil fuel combustion, industrial processing of nitrogen, release by domestic livestock and humans, and growing such crops as legumes, which fix nitrogen (12).

e. *Radiological Chemicals*

Minute traces of radionuclides, isotopic forms of elements that exhibit radioactivity, exist in all drinking water. A large number of natural and artificial radionuclides have been found in water supplies. For groundwater, the nature of the contaminants and their concentration vary regionally depending on the composition of soils and mineral deposits through which the water has moved. Most radioactivity found in drinking water, however, has been the result of a small number of radionuclides. The low linear energy transfer (LET) radionuclides seen in drinking water have included potassium 40 (^{40}K), tritium, (^{3}H), carbon 14 (^{14}C), and rubidium 87 (^{87}R) (13). Tritium, a β-emitting isotope of hydrogen, is the only low-LET radionuclide of sufficient concern to result in the establishment of an MCL. Natural tritium is produced in the atmosphere as a result of the interactions of cosmic rays with oxygen and nitrogen. When it is oxidized, it forms tritiated water, which after precipitation can contaminate ground and surface water. As a consequence, background tritium concentrations in drinking water range from 10 to 25 pCi/L (13). Tritium, produced and utilized by humans, is of the greatest concern. Tritium is used for research, in self-luminous phosphors, and in nuclear power plants. Because of its short half-life of 12.5 days, the greatest potential threat in the event of an accidental release or improper disposal is to surface water rather than groundwater.

The high-LET α-particle-emitting radionuclides that have been the most important drinking water contaminants are radium 226 (^{226}Ra), the daughter isotopes (decay products) of radium 228 (^{228}Ra), polonium (^{210}Po), uranium (U), thorium (Th), radon 220 (^{220}Rn), and radon (^{222}Rn). The major nuclides of concern are uranium and radon. Contamination of groundwater by both these elements, like other inorganic substances, is often regional, depending on the occurrence of subsurface minerals containing uranium and other radon precursors (14). The current MCL addressing radionuclides and radio emissions are summarized in Table 5.

B. Surface Water Contamination

California's surface water supplies reside in a vast network of rivers, streams, lakes, aqueducts, and reservoirs. With the exception of covered reservoirs and enclosed pipelines, surface waters are open to the environment. As such, they are potentially exposed directly to contaminants that reside in the atmosphere or on the earth's surface. Waste material contaminated with chemicals, pesticides, or pathogenic organisms can pollute surface water through direct discharge, surface runoff, spills, or atmopsheric drift into water bodies.

The quality of California's surface water supplies is generally good. Surface water movement in rivers, streams and aqueducts is usually measured in terms of feet per second, contrasted to less than 1000 feet per year for groundwater. As a result, the dilution and dispersion of contaminants occurs much more rapidly in surface waters, and elevated concentrations of chemical contaminants are rare. VOC, so commonly seen in groundwater, are very rare in surface waters because volatile chemicals rapidly evaporate from the water into the atmosphere.

1. Types of Surface Water Contamination

a. Point Source Contamination

Most of the sources of point source contaminations described for groundwater sources can potentially affect surface waters. Under certain circum-

Table 5 Radiological Chemicals Having Maximum Contaminant Levels in California

Constituent	MCL (pCi/L)
Combined radium 226 and radium 228	5
Gross α particle (including radium 226 but excluding and uranium)	15
Tritium	20,000
Strontium 90	8
Gross β particle activity	50
Uranium	20

stances groundwater can migrate to and contaminate surface water supplies. Additionally, surface supplies can be directly contaminated by point source incidents, such as spills, landfill and mine tailing leachates, and direct discharges.

b. *Nonpoint Source Contamination*

Again, nonpoint source contamination sources that affect groundwater can also potentially contaminate surface waters. An example of nonpoint source contamination of surface water was the contamination of the Sacramento River by rice herbicides, described later, as a result of agricultural return flows.

2. Type of Surface Water Contaminants

a. *VOC*

Little information is available about the types and quantities of VOC discharge and spills into surface supplies. Once in a surface water body, however, such chemicals quickly evaporate into the atmosphere. Accordingly, VOC are seldom detected in surface water supplies.

b. *Trihalomethanes*

The potential for microbiological contamination of surface waters is much higher than for groundwater. As a consequence, disinfection is an essential part of providing a safe, wholesome, and potable drinking water supply. Surface waters contain much more suspended and dissolved organic material than groundwater. This material, largely naturally occurring humic substances generated from decaying plants, reacts with disinfection agents. Chlorine, the most commonly used disinfection agent, reacts with the humic substances and naturally occurring bromide in source water to produce THM. Although much of the precursor humic materials are removed by filtration of surface water before disinfection, enough remains to produce concentrations of THM often approaching and exceeding the THM MCL of 100 μg/L.

All water systems in California that serve a population of 10,000 or more and that disinfect all or part of the water served are required to conduct quarterly monitoring for THM within the distribution network in conformance with U.S. EPA monitoring criteria. Little of the THM distribution system testing has been recorded in WQM.

c. *Pesticides*

Pesticides are frequently applied near surface waters. They can enter surface water through runoff from precipitation, spray drift, agricultural return flow, or infiltration. The detection of pesticides in drinking water derived from surface waters is rare; however, surface water supplies have not been monitored as thoroughly for pesticides as groundwater supplies (15). The most notable case involving a drinking water surface source contamination by pesticides is in the Sacramento River watershed. As a result of agricultural return flows from rice-growing activities into the Sacramento River during April–June each year, drinking water supplies in Sacramento have been periodically contaminated by the rice herbicides molinate and bentazon. Consequently, MCL have been established for these chemicals.

d. Inorganic Chemicals

Although traces of heavy metals and other inorganic substances listed in Table 4 are common in surface waters, serious contamination of drinking water derived from surface water is rare. Nitrate has been reported in several rivers in Southern California in excess of its MCL (16). This contamination is attributed to one or more of the following: (1) effluent from sewage treatment plants; (2) agricultural effluent runoff; (3) the stream receiving exfiltration from a high nitrate content groundwater body; and (4) industrial and municipal runoff.

e. Radiological Chemicals

A low background concentration of tritium exists in all surface water supplies as a result of its natural formation in the atmosphere. Contamination of surface water supplies by radionuclides occurring naturally in geological deposits is less of a potential problem than for groundwater supplies. Surface water seldom contains significant amounts of radium or uranium, and the treatment processes normally required, such as flocculation and filtration, removes most of these contaminants.

Contamination of surface supplies by artificial radionuclides is somewhat greater than for groundwater. As discussed, the potential for surface water contamination by tritium from manufacturing and nuclear power generation has been recognized. During the period of atmospheric testing of nuclear weapons, contamination of surface water by airborne fission products was observed; however, that contamination has sharply decreased since implementation of the Nuclear Test Ban Treaty of 1963 (13).

V. RESULTS

This section of the report presents available information about the quality of public drinking water supplies in California. According to WQM, public drinking water sources have been tested for over 375 different chemicals since 1984. In the following discussion, data for those 60 chemicals and parameters having MCL are presented.

A. Contaminants

1. VOC

Over 11,000 drinking water sources have been sampled for VOC. Much of the early sampling between 1984 and 1989 was carried out pursuant to the AB 1803 program. Subsequent sampling was compliance monitoring conducted in response to recent regulations requiring the monitoring of organic chemicals. Chlorinated solvents have been the most frequently detected group of VOC and those most often found to exceed their MCL (Table 6). Of the approximately 11,700 sources sampled for chlorinated VOC, 2266 (19%) and 487 (4.1%) were found to be positive and to exceed an MCL, respectively. They have been found to exceed their MCL most frequently in the Los Angeles area and San Joaquin Valley (Figures 3 and 4). The statewide concentration means for 10 of the VOC, including benzene, tetrachloroethylene, trichloroethylene, and vinyl chloride, have exceeded their MCL.

2. Pesticides

As with VOC, most of the pesticide monitoring reported in the data base are the result of the AB 1803 program and subsequent compliance monitoring. The most numerous seriously contaminated sources have been in the San Joaquin Valley. The pesticide of primary concern has been dibromochloropropane, as illustrated in Table 7 and Figure 5. Thus, 641 (10.9%) of the 5855 sources sampled for DBCP have been found to contain detectable levels of the chemical, and 282 (4.8%) had levels in excess of the MCL. Ethylene dibromide (EDB), although tested in about the same number of sources, was detected in only 0.8% and exceeded its MCL in only 0.7%. The statewide means for DBCP and EDB exceed their MCL.

3. Inorganic Chemicals

The inorganic chemicals listed in Table 4 have had MCL for several years, and nearly all the results on record for them were collected pursuant to compliance monitoring requirements. Fluoride and nitrate are the most frequently detected inorganic chemicals (Table 8). Of the approximately 7000 sources sampled for the chemicals, 5679 (81%) and 6200 (87%) were found to be contaminated by fluoride and nitrate, respectively. Nitrate has a record number of positive observations of 18,445 as a result of repeat testing of many sources. Sources exceeding the nitrate MCL have been concentrated in Southern California and the San Joaquin Valley (Figure 6). Fluoride, on the other hand, has been more dispersed across the state. The MCL of fluoride and nitrate have been exceeded in 195 (2.8%) and 451 (6.4%) of the sources sampled, respectively. The fluoride MCL of 1.4

Table 6 Organic Chemical Monitoring of Public Drinking Water Sources

Contaminant	Number of sources			Concentration data on positives (μg/L)				Number observations
	Total sampled	Positives	>MCL	Mean	Median	Maximum	Minimum	
Trichloroethylene	11,761	486	194	9.57	3.3	643.40	0.40	7665
Tetrachloroethylene	11,775	593	144	5.39	2.00	301.00	0.04	7424
Carbon tetrachloride	11,701	77	71	3.27	2.00	29.00	0.20	739
1,2-Dichloroethane	11,686	45	43	3.03	1.40	24.00	0.20	236
1,1-Dichloroethylene	11,696	120	14	2.45	1.00	99.05	0.10	1437
Benzene	11,642	10	8	83.96	4.10	1,300.00	0.39	113
Cis-1,2-dichloroethylene	8,270	87	7	2.23	1.60	69.00	0.10	1083
1,1-Dichloroethane	11,680	39	6	2.19	1.30	30.00	0.47	159
Diethylhexylphthalate	1,955	9	6	19.57	8.90	93.00	3.3	20
Total trihalomethanes	9,151	395	4	17.11	3.50	670.00	0.05	1976
1,1,2,2-Tetrachloroethane	11,665	3	3	3.23	2.55	5.80	1.40	6
Vinyl chloride	11,649	3	3	13.66	12.00	44.00	1.60	11
Trans-1,2-dichloroethylene	11,669	19	2	13.77	2.00	190.00	0.58	71
Chloroform	11,848	536	2	6.20	1.50	620.00	0.05	3588
1,1,1-Trichloroethane	11,704	157	0	2.67	1.00	202.20	0.20	1399
Bromoform	11,765	145	0	7.63	3.13	90.50	0.03	705
Bromodichloromethane	11,773	167	0	6.99	3.20	100.00	0.20	994
Dibromochloromethane	11,782	171	0	8.95	3.27	89.0	0.12	828
1,4-Dichlorobenzene	11,659	11	0	1.15	0.96	4.20	0.30	34
Dichlorodifluoromethane	11,646	58	0	17.86	3.30	404.90	0.50	259
Ethylbenzene	11,636	4	0	21.24	6.30	120.00	0.54	29
Monochlorobenzene	11,645	5	0	1.89	1.55	5.40	0.50	24
1,1,2-Trichloroethane	11,665	8	0	2.75	1.75	6.40	0.60	26
Trichlorofluromethane	11,664	50	0	10.81	2.05	1,444.00	0.100	246
Trichlorotrifluoroethane	8,161	39	0	1.55	0.85	27.90	0.26	579
Xylenes	11,589	23	0	22.91	3.80	573.00	0.20	76

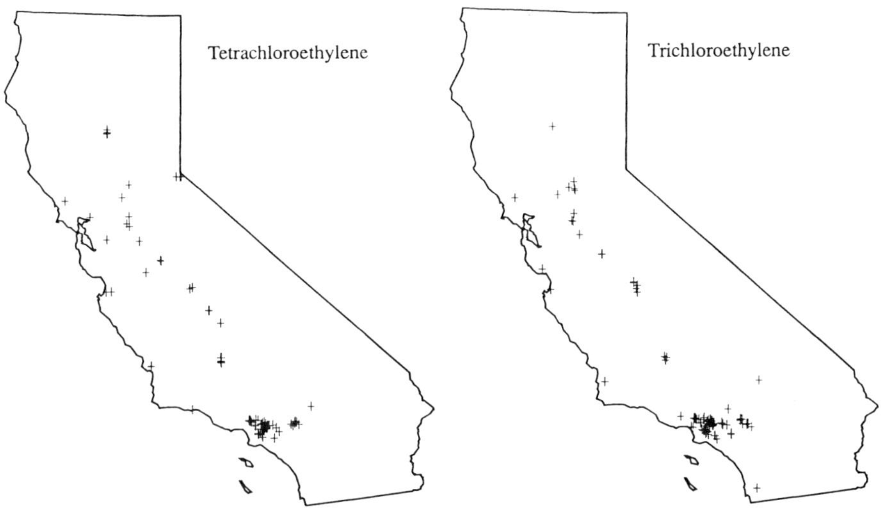

Figure 3 Public drinking water sources exceeding tetrachloroethylene and trichloroethylene MCL.

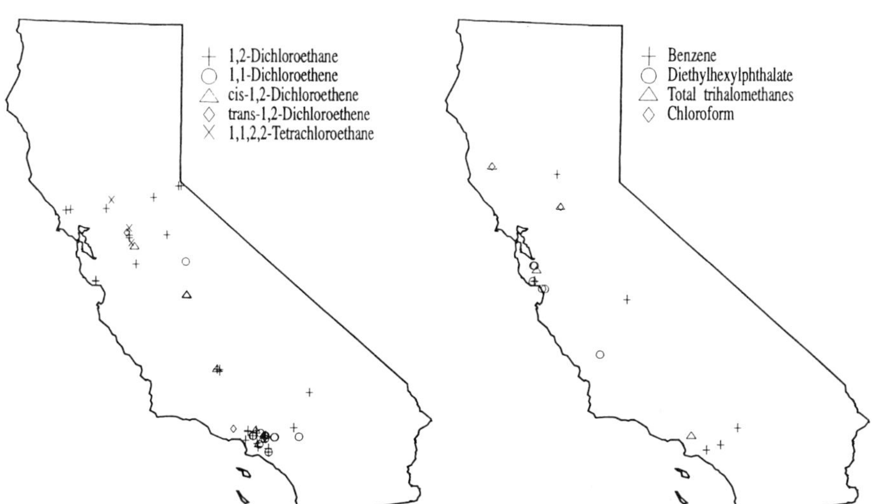

Figure 4 Public drinking water sources exceeding organic chemical MCL.

Table 7 Pesticide Monitoring of Public Drinking Water Sources

	Number of sources			Concentration data on positives (μg/L)				
Contaminant	Total sampled	Positives	>MCL	Mean	Median	Maximum	Minimum	Number observations
Dibromochloroethane (DBCP)	5,855	641	282	0.40	0.160	74.00	0.001	6579
Ethylene dibromide (EDB)	5,571	45	37	0.08	0.04	1.10	0.004	328
1,2-Dichloropropane	11,660	49	5	2.79	1.25	53.00	0.10	202
Alachlor	2,375	1	0	1.00	1.00	1.10	0.90	2
Atrazine	5,798	51	0	0.66	0.50	2.40	0.06	219
Bentazon	1,767	2	0	1.69	1.60	3.00	0.60	11
2,4-D	4,775	3	0	5.31	6.00	10.90	0.10	6
Endrin	5,109	2	0	0.06	0.06	0.10	0.01	4
γ-BHC (lindane)	5,128	2	0	0.08	0.10	0.10	0.01	4
Methoxychlor	4,600	1	0	0.10	0.10	0.10	0.10	2
Molinate	3,672	1	0	3.37	3.10	6.50	0.60	23
Simazine	6,158	66	0	0.75	0.50	28.00	0.06	228
Thiobencarb	2,988	2	0	1.75	1.30	3.40	1.00	4
Toxaphene	5,279	2	0	0.30	0.30	0.50	0.10	4
2,4,5-TP (silvex)	4,378	1	0	0.10	0.10	0.10	0.10	2
Carbofuran	2,893	0	0	—	—	—	—	—
1,3-Dichloropropene	8,395	0	0	—	—	—	—	—
Glyphosate	1,772	0	0	—	—	—	—	—
Heptachlor	3,652	0	0	—	—	—	—	—
Heptachlor epoxide	3,400	0	0	—	—	—	—	—

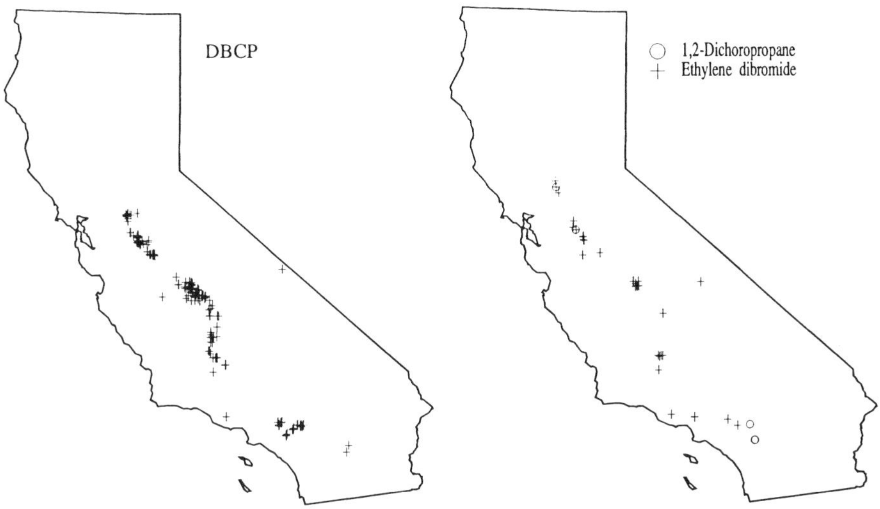

Figure 5 Public drinking water sources exceeding pesticide MCL.

mg/L, for the highest temperature value (Table 4), was used here as the limit. Although arsenic and selenium have been found to exceed their MCL much less frequently statewide, they are of concern because of their high frequencies in certain water systems and localities (Figure 7), as discussed later. Nearly all inorganic chemical testing reported here has been conducted on sources operated by state-regulated large water systems. Very few inorganic chemical monitoring data are available for small water system sources regulated by local county health agencies.

4. Radiological Chemicals

Total, or gross, α-particle emission is the most frequently detected radiological contaminant or parameter in public drinking water sources (Figure 8). Approximately 68% of the 6250 sources tested for total α were found contaminated, and 3.4% exceeded its MCL (Table 9). Natural uranium, although sampled less, has been observed at about the same frequencies as α. Again, very little radiological monitoring has been conducted on small water system sources.

During 1989, Sakaji of the Office of Drinking Water conducted a special drinking water sampling study to provide an exposure assessment of risk associated with radon in groundwater (14). Two rounds of sampling were carried out on large water system wells, selected because they were likely to have elevated levels of radon because of the geology of their surround-

Table 8 Inorganic Chemical Monitoring of Public Drinking Water Sources

	Number of sources			Concentration data on positives (μg/L)				
Contaminant	Total sampled	Positives	>MCL	Mean	Median	Maximum	Minimum	Number observations
Nitrate[a]	7075	6208	451	17.081	0.30	630.00	0.002	18,445
Fluoride[a]	7009	5679	195	0.58	0.30	375.00	0.11	13,717
Selenium	6860	279	84	112.23	8.00	200.00	2.10	563
Arsenic	6857	678	51	22.66	15.00	460.00	5.10	1,033
Aluminum	5453	546	44	397.72	210.00	12,000	0.01	666
Cadmium	6866	333	40	10.65	3.00	1000.00	1.10	389
Mercury	6856	225	27	6.50	0.62	300.00	0.21	255
Chromium (total)	6862	484	19	23.02	19.00	330.00	10.10	718
Lead	6867	525	18	21.55	14.00	620.00	5.10	572
Silver	6860	83	6	35.85	20.00	470.00	11.00	99
Barium	6856	1539	5	301.68	160.00	280,280	101.00	2,513

[a]Concentration data in mg/L.

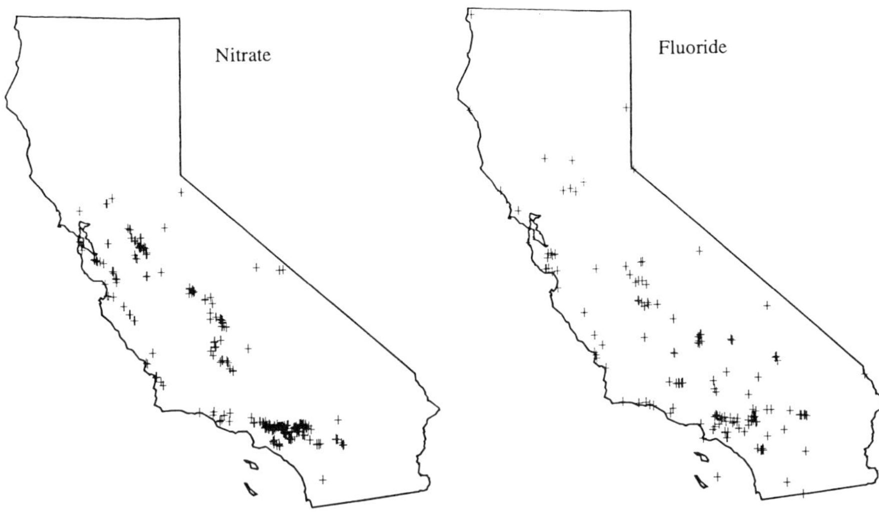

Figure 6 Public drinking water sources exceeding nitrate and fluoride MCL.

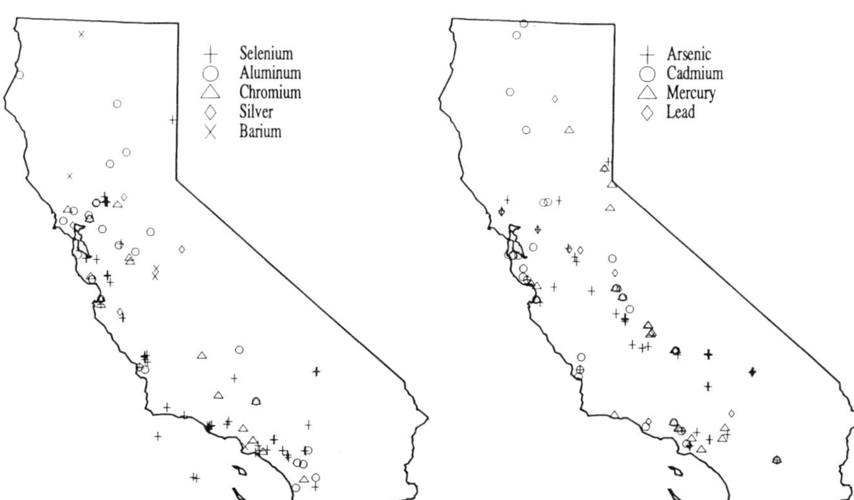

Figure 7 Public drinking water sources exceeding inorganic chemical MCL.

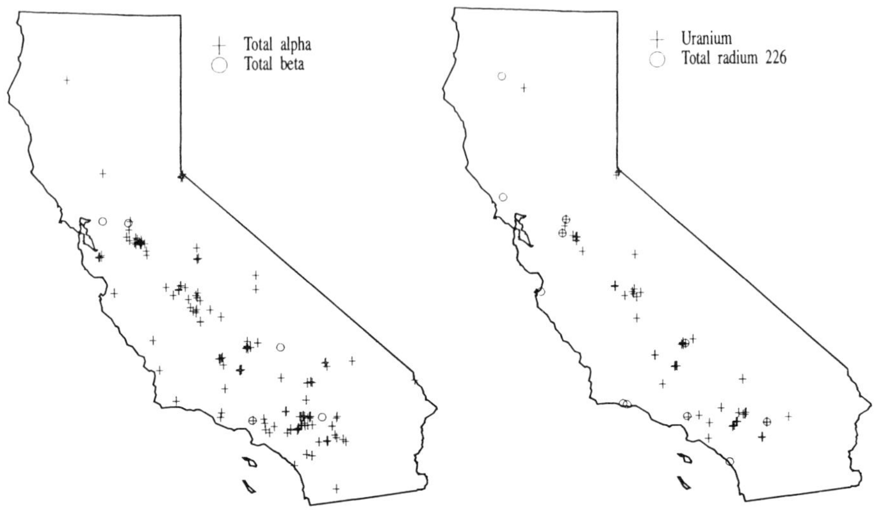

Figure 8 Public drinking water sources exceeding radiological MCL.

ings. In a third round, large system wells were selected randomly. Table 10 summarizes the results of the study.

Sakaji conducted a GIS (geographical information system) analysis of the sampling results to ascertain whether the location of a well and a consideration of the geology in the area could predict the potential for the well to have elevated levels of radon. GIS was shown to be a viable tool to predict potential contamination of drinking water wells by radon.

In 1986 the U.S. EPA announced its intent to establish an MCL for radon (17). The GIS projection indicated that if an MCL of 200 pCi/L is established for radon, the majority of public drinking water wells in California would be vulnerable to radon. This is supported by data in WQM. Although the data base contains sampling data for only 20 wells, 14 (70%) of which had detectable radon, the minimum concentration observed was 190 pCi/L and the mean was 960 pCi/L.

B. Source Status

The quality of water provided by sources is summarized according to their operational status in Table 11. Approximately 25% of the drinking water sources found to exceed an MCL have been placed on standby or nonoperational status. No accurate historical information is available about the reasons for placing sources in a standby or nonoperational status; however, it was frequently done as a result of contamination. Table 11 and

Table 9 Radiological Chemical Monitoring of Public Drinking Water Sources

	Number of sources			Concentration data on positives (pCi/L)				
Contaminant	Total sampled	Positives	>MCL	Mean	Median	Maximum	Minimum	Number observations
Total α	6250	4256	212	5.99	2.90	9,999.00	1.02	10,872
Natural uranium	1253	787	96	12.59	8.00	129.00	2.10	1,847
Radium 226 + radium 228	67	32	32	0.26	0.08	3.30	0.02	26
Total tritium	181	29	29	5138.58	1014.00	20,000.00	26.00	31
Total radium 226	399	77	12	6.53	1.10	444.00	0.60	93
Total strontium 90	169	16	9	104.75	20.00	330.00	8.00	16
Total β	1234	594	5	10.03	7.0	154.90	4.10	937
Total radon 222	78	48	0	570.64	595.50	1,199.00	106.00	52

Table 10 Radon Contamination of Public Drinking Water Wells

	Number of sources				
Round	<200 pCi/L	200–499 pCi/L	500–1000 Ci/L	>1000 pCi/L	Totals
1 and 2	26	36	25	96	213
3	3	19	24	14	60
Totals	29	55	49	110	373

Source: From Reference 14.

Figure 9 support this assertion: they indicate that the percentage of analyses that have exceeded MCL for major problem contaminants is higher for standby and nonoperational sources. Overall, 1.0% of the analyses of samples from active sources have been found to exceed an MCL, and 2.0 and 2.8% of those from standby and nonoperational sources, respectively, have exceeded an MCL. Despite this, most of the sources exceeding MCL remain on active status.

C. Locations of Contamination

There are some areas in the state where contamination of drinking water sources for certain contaminants is much more prevalent than in others. Figure 10 shows the regions of the state discussed in this section. Figure 11 illustrates the frequencies of occurrences where MCL have been exceeded for the regions. This graph is expressed in terms of the percentage of analyses for the chemical categories that have found MCL to be exceeded. The following describes the contamination in terms of major regions within the state.

1. San Joaquin Valley

The San Joaquin Valley includes portions of San Joaquin, Stanislaus, Merced, Madera, Fresno, Tulare, Kings, and Kern counties. It is an area of widespread groundwater contamination, showing the greatest frequency of analyses exceeding MCL for pesticides and radiological chemicals (Figure 11). The San Joaquin Valley is an important agricultural area of the state. DBCP, before its ban, was heavily used in the valley. Consequently, it is the primary groundwater contaminant in the area and is the major concern. DBCP was originally detected in 1979 in the vicinity of the Occidental Chemical Company at Lathrope, California. Before 1984, DBCP had already been detected in about 125 large water system wells and 1800 small water system and private wells in the region (3). Currently, 11.1% of the

Table 11 Drinking Water Quality Results According to Source Status

Chemical	Active[a]			Standby[b]			Nonoperational[c]			Other[d]		
	Sources sampled	Sources positive	Sources >MCL	Sources sampled	Sources positive	Sources >MCL	Sources sampled	Sources positive	Sources >MCL	Sources sampled	Sources positive	Sources >MCL
Aluminum	4,943	479	37	185	28	2	328	27	3	87	12	2
Arsenic	5,934	607	40	240	22	1	550	44	9	133	5	1
Barium	5,940	1329	4	238	77	1	546	114	0	132	19	0
Benzene	10,587	8	6	258	0	0	650	2	2	147	0	0
Cadmium	5,944	284	33	240	10	2	550	31	3	132	8	2
Carbon tetrachloride	10,630	54	50	260	7	6	661	11	11	150	5	4
Chloroform	10,725	445	1	264	14	0	670	43	1	189	34	0
Chromium (total)	5,940	427	15	240	14	3	550	37	1	132	6	0
Dibromochloropropane	5,254	516	195	145	25	19	347	94	66	109	6	2
1,1-Dichloroethane	10,617	32	5	258	1	0	659	6	1	146	0	0
1,2-Dichloroethane	10,620	36	35	259	2	2	660	7	6	147	0	0
1,1-Dichloroethylene	10,622	90	11	260	6	0	664	20	3	150	4	0
Cis-1,2-dichloroethylene	7,513	71	6	229	3	0	397	13	1	131	0	0
1,2-Dichloropropane	10,608	43	4	257	1	0	648	5	1	147	0	0
Diethylhexylphtalate	1,709	8	5	74	0	0	158	1	1	14	0	0
Ethylene dibromide	5,026	33	27	142	1	1	296	9	7	107	2	2
Fluoride	6,007	4863	173	249	213	2	609	493	17	144	110	3
Lead	5,942	466	15	241	18	0	552	34	3	132	7	0
Mercury	5,935	203	25	240	9	0	549	9	2	132	4	0
Natural uranium	1,109	682	75	43	31	6	77	56	13	24	18	2
Nitrate (as NO_3)	6,048	5315	304	256	223	25	624	552	113	147	118	9

Table 11 (continued)

Chemical	Active[a]			Standby[b]			Nonoperational[c]			Other[d]		
	Sources sampled	Sources positive	Sources >MCL	Sources sampled	Sources positive	Sources >MCL	Sources sampled	Sources positive	Sources >MCL	Sources sampled	Sources positive	Sources >MCL
Ra 226 + Ra 228	51	24	24	2	1	1	12	7	7	2	0	0
Selenium	5,937	240	70	241	11	4	550	20	7	132	8	3
Silver	5,938	70	4	241	4	0	550	6	2	131	3	0
Tetrachloroethylene	10,663	450	85	269	31	16	688	96	38	155	16	5
Total α	5,535	3787	172	227	141	10	387	265	26	101	63	4
Total β	1,099	531	5	32	13	0	77	38	0	26	12	0
Total radium 226	354	70	8	12	2	1	29	4	2	4	1	1
Total radon 222	75	47	47	1	0	0	2	1	1	0	0	0
Total strontium 90	137	16	9	1	0	0	22	0	0	9	0	0
Total trihalomethanes	8,336	333	3	235	10	0	400	19	1	180	33	0
Total tritium	150	28	28	1	1	1	21	0	0	9	0	0
1,1,2,2-Tetrachloroethane	10,605	3	3	258	0	0	655	0	0	147	0	0
Trans-1,2-dichloroethylene	10,606	14	1	258	1	0	658	4	1	147	0	0
Trichloroethylene	10,654	361	128	269	27	15	681	85	46	157	13	5
Vanadium	9	5	5	0	0	0	0	0	0	0	0	0
Vinyl chloride	10,600	2	2	258	0	0	644	1	1	147	0	0

[a]Currently in operation.
[b]On standby and used only during times of emergency or unusually high demand.
[c]Destroyed, abandoned, or permanently inactive.
[d]Not drinking water sources. Agricultural or monitoring wells or sample points within a utility distribution system.

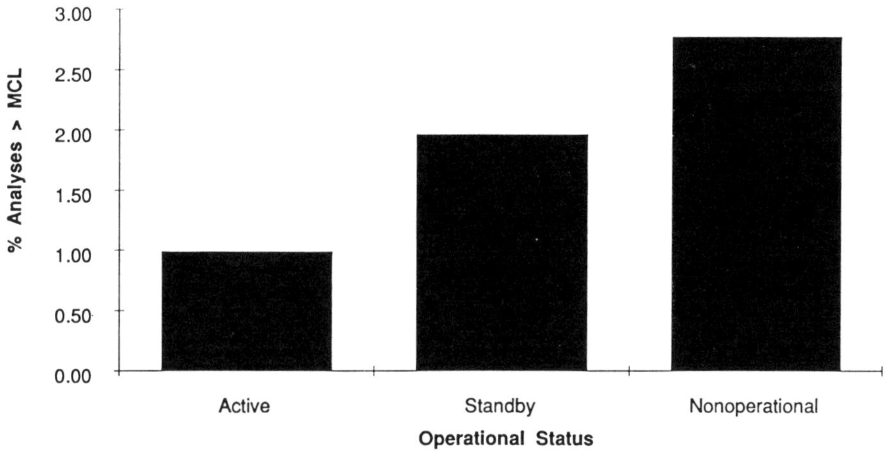

Figure 9 Percent analyses greater than MCL according to operational status of source sampled.

2029 public water sources in the area that have been tested for DBCP have exceeded the MCL, with the greatest concentration of problem wells in Fresno County. This is the highest percentage of any area in the state.

Groundwater in the San Joaquin Valley has been subjected to contamination by nitrates, both from agricultural activities and domestic waste discharges. As with other industries, agriculture tends to attract urbanization. The eastern side of the Valley has developed into several urban areas associated with agriculture and its support industries. The resulting domestic and industrial wastes have contributed to the occurrence of nitrate contamination. Almost 25% of the drinking water wells in the state with nitrates above the MCL are located in this area (2).

Selenium contamination has been an environmental problem in the San Joaquin Valley as a result of runoff from west side agricultural drainage. The runoff, contaminated with selenium, drained into the Kesterson Wildlife Refuge near Los Banos, California *via* the San Luis drain (2). Nevertheless, only two public drinking water wells have been found to exceed the selenium MCL in the San Joaquin Valley.

Industrial activities have also contributed to the presence of VOC in the San Joaquin Valley groundwater. Throughout the Valley there are localized areas of contamination caused by such industries as metal fabricators, dry cleaners, and chemical formulators. Even the food packers and processors supporting the agricultural industry have contributed to the contamination by disposal of wastes containing salts, fumigants, and fun-

Figure 10 Regions of California.

gicides (2). About 2.8% of the sources tested (and 0.27% of the total analyses) for chlorinated VOC exceeded an MCL.

Two pockets of arsenic contamination in excess of its MCL have been found, one in the city of Hanford in Kings County and another in several closely located water systems in western Kern County. The potential health impacts of arsenic are discussed later.

2. South Coast

The South Coast region, a major industrial and agricultural area, is the most extensively contaminated part of California. Comprised of the counties of Los Angeles, southwestern San Bernardino, Orange, western Riverside, and San Diego, the South Coast region is one of the most densely

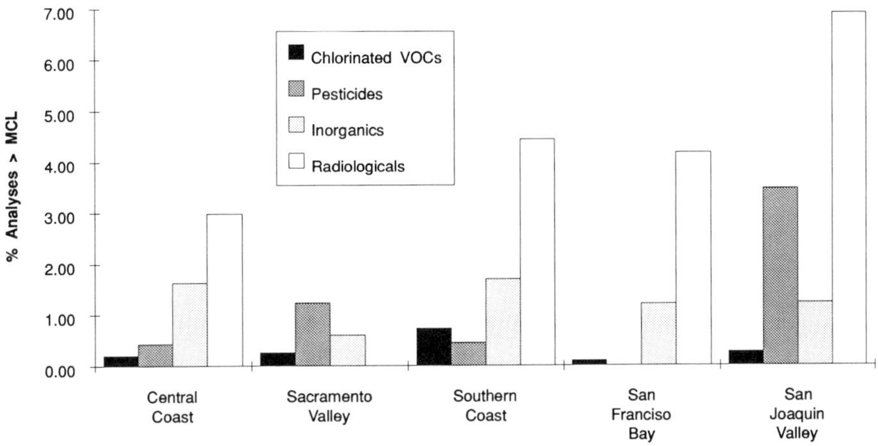

Figure 11 Percent analyses greater than MCL according to region.

populated and rapidly growing areas of the state and nation. Over half of the population of California now lives in the area. The heavy commercialization and industry that accompanied the population growth is a major factor affecting the area's water quality. As a result of this growth, much of the groundwater is contaminated to such an extent that its use as a domestic supply is restricted. The region has the largest percentage of analyses showing chlorinated VOC in excess of their MCL (Figure 11). It has a large number of wells contaminated by VOC, pesticides, total α, and nitrate in excess of their MCL. One-half or more of all the wells exceeding MCL in the state for chlorinated VOC, total α, nitrate, and fluoride have occurred in the South Coast region. The greatest number of chemically contaminated sources have been observed in an area comprising southern Los Angeles, southwestern San Bernardino, and northwestern Riverside counties (2).

The San Gabriel and San Fernando Valley groundwater basins in Los Angeles County, heavily contaminated with chlorinated VOC, have come under intensive scrutiny over the last 10 years. In southwestern San Bernardino County the Chino and Bunker Hill water basins are contaminated with chlorinated VOC and DBCP. The latter contamination is the result of extensive past use of the fumigant on citrus fruit and grape crops.

Groundwater in the South Coast region has been contaminated by nitrates from agriculturally related activities and from domestic waste discharges. Of the 451 public drinking water sources reported as exceeding the nitrate MCL statewide, over half are located in the region.

Additionally, nearly one-half of the sources exceeding the total α-particle and uranium MCL were found in the South Coast region. Gross α-particle radiation and uranium levels in excess of their MCL have also been reported in well and surface waters in Los Angeles and Orange counties by outside studies (18). In the Raymond water basin, six public drinking water sources were found to exceed the gross α MCL and seven exceeded the uranium MCL. Six sources exceeding the gross α MCL were also found in Orange County, although none are recorded in the state data base. Unlike the extensive VOC and nitrate contamination, α and uranium contamination is apparently of natural origin. It is believed to be the result of water sources originating from tunnels in the adjacent mountains of the San Gabriel Range containing uranium and rare earth deposits.

3. San Francisco Bay Area

The San Francisco Bay area includes the counties of Alameda, Contra Costa, San Francisco, and San Mateo and parts of Napa, Sonoma, and Santa Clara. Although it is a densely populated region with a wide variety of industrial and commercial activities, relatively few drinking water sources in the area have been found to exceed any MCL (2). The area has experienced less contamination by VOC than any of the five regions discussed here. Some areas of localized groundwater contamination have occurred, the most serious being contamination in the South San Francisco Bay area in Santa Clara County, where improperly stored industrial solvents have contaminated the local groundwater basin. Regardless, only two public water wells have been reported to exceed a VOC MCL in Santa Clara County despite extensive testing. Nitrate has been the most frequent contaminant found to exceed its MCL. Of the wells tested for nitrate 35 (7.8%) have exceeded its MCL in the San Francisco Bay area.

4. Sacramento Valley

The Sacramento Valley region is the origin of the two largest water transfer projects in the state, the SWP and the CVP. Comprised of Sacramento and Yolo counties and parts of Butte, Colusa, Glenn, Sutter, Yuba, and Tehema counties, it is a major agricultural area, ranked second in the nation in the production of rice. Because of the discharge of agricultural wastes from rice fields each year, the city of Sacramento is required to alter its treatment plant operation for water pumped from the Sacramento River because of the presence of herbicides (Molinate and Thiobencarb) used on rice (15). Although these contaminants are below the MCL, thiobencarb produces taste and odor problems at low concentrations. Selenium has been found in drinking water supply wells in the city of Davis, Yolo County, where 30 wells had positive results and 18 had results in excess of the MCL (2). In the city of Chico in Butte County, a cluster of six wells were found to exceed chlorinated VOC MCL during the AB 1803 study.

5. Central Coast

The Central Coast region includes Monterey, San Luis Obispo, Santa Barbara, Santa Cruz, and Ventura counties. It is less populated and industrialized than the San Francisco Bay area, but it has experienced more source contamination in excess of MCL (2). Nitrate has exceeded its MCL most frequently in the region as a result of domestic waste discharges and agricultural activities, including cattle feedlots in the Salinas Valley of Monterey County. This area had the second highest incidence of positive selenium results, 35 wells exceeding the MCL with clusters in San Luis Obispo and Ventura counties. A total of 29 sources have exceeded an MCL for chlorinated VOC.

D. Groundwater Versus Surface Water

The percentage of surface sources sampled for VOC is about one-half that of wells, and it is slightly higher for inorganics (2). Generally, the frequency of VOC being detected and exceeding MCL in surface water is lower than in groundwater. Only seven surface sources have exceeded a VOC MCL. This is not surprising because of the short residence times expected for volatile chemicals in open water bodies. The frequency of inorganic chemicals exceeding their MCL in surface waters appears to be comparable and, in some cases, to be greater than that seen for groundwater sources. Overall, 81 surface sources have exceeded an inorganic or radiological chemical MCL.

E. Trends

The amount of monitoring of California's public drinking water sources has increased considerably over the last 9 years. The most active sampling period was from 1985 through 1989, with a peak in 1989 during which the AB 1803 program and new monitoring regulations were being implemented. The number of new analyses for selected chemicals ranged from about 24,000 to 129,000 analyses per year over this time. The cumulative number of analyses recorded (number of chemicals currently having MCL that were analyzed × number of wells sampled for those chemicals) in WQM for all MCL chemicals grew from about 41,000 in 1984 to 460,000 by December 1992 (2). Over this same time, the number of new sources found to exceed MCL (Figure 12) increased concurrently (although it should be clarified that MCL for all subject chemicals had not been adopted until the late 1980s). This suggests that, contrary to early expectations, not all

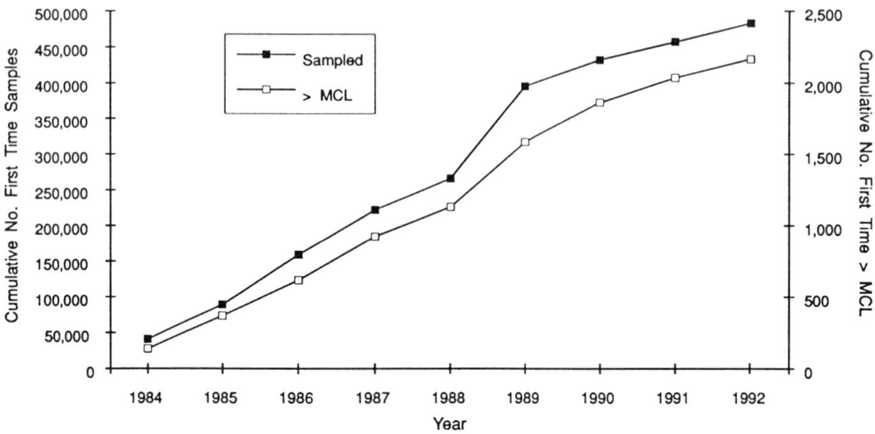

Figure 12 Comparison of the number of samples analyzed and those exceeding MCL.

problem sources were quickly identified in the first years of the intensive, targeted monitoring effort between 1985 and 1989, which focused presumably on sources vulnerable to potential contamination. For some chemicals, in fact, the percentage of sampled sources that have exceeded MCL has increased with more sampling; for others it has decreased. A noteworthy example of the former is nitrate, for which the percentage has increased from 3.3 to 17.1% (2). DBCP, on the other hand, has steadily decreased from 15.6 to 3.9%. These results can best be explained by the fact that DBCP use as an agricultural soil fumigant was discontinued in the 1970s, but overall agricultural activities and domestic waste discharges that generate nitrate continue and perhaps have increased. Additionally, DBCP monitoring has been highly targeted, concentrating on areas where past applications of the soil fumigant are known to have occurred, and is an example of effective targeted monitoring.

A projection of the number of active sources yet to be found exceeding MCL for selected chemicals on the basis of the number of unsampled active sources and a linear regression of the percentage for each chemical are illustrated in Table 12. The projections predict up to 900 sources yet to be found in excess of the nitrate MCL and 6 for DBCP. These projections should be looked upon as maximum estimates, especially for such chemicals as DBCP for which the percentages are decreasing.

Table 13 lists the statewide mean concentrations by year for the chemicals with MCL. Such broad trends need to be viewed with some reser-

Table 12 Estimate of Sources Remaining in Excess of MCL for Selected Chemicals

Chemical	% Sampled sources > MCL									Projected	Active sources	Active sources	Projected sources
	1984	1985	1986	1987	1988	1989	1990	1991	1992	1993 %	sampled	unsampled	>MCL[a]
Arsenic	0.11	0.15	1.50	0.70	0.63	0.82	1.32	1.22	0	0.93	5,934	6,766	63
Barium	0	0	0	0.10	0.12	0	0.16	0	0.80	0.42	5,940	6,760	28
Benzene	0	0.07	0	0	0.19	0.22	0	0	0	0.05	10,587	2,113	1
Cadmium	0.32	0	0.23	1.01	0.38	0.67	0.83	1.44	0.82	1.23	5,944	6,756	83
Carbon tetrachloride	0.34	1.95	0.04	0.49	0.28	0.74	1.01	0.24	0.44	0.40	10,630	2,070	8
Chromium (total)	0	0.15	0	0.20	0.13	0.60	0.49	0.73	0.41	0.70	5,940	6,760	47
Dibromochloropropane	15.63	9.77	9.40	6.25	5.85	2.00	4.53	3.45	3.87	0.09	5,254	7,446	6
1,2-Dichloroethane	0.25	0.63	0.35	0.38	0.28	0.22	0.81	0.23	0.44	0.43	10,620	2,080	9
1,1-Dichloroethylene	0	0	0.04	0.27	0.28	0.09	0.20	0	0.29	0.24	10,622	2,078	5
Cis-1,2-dichloroethylene	0	0	2.08	0.31	0.15	0	0.10	0.27	0	0.03	7,513	5,187	2
Ethylene dibromide	0	0	0.18	1.45	0	0.66	1.18	0.86	0.65	1.08	5,026	7,674	83
Fluoride	1.90	0.96	1.38	3.42	3.09	3.47	5.62	4.28	2.62	4.75	6,007	6,693	318
Lead	0	0.31	0.11	0.40	0.25	0.30	0.17	0.49	0.82	0.63	5,942	6,758	43
Mercury	0.64	1.09	0.23	0.40	0.25	0.37	0	0.24	0	0	5,935	6,765	0
Natural uranium	0	0	0	30.95	21.18	5.38	4.47	10.30	5.92	10.58	1,109	11,591	1226
Nitrate (as NO_3)	3.33	4.68	5.57	6.98	6.97	6.95	8.49	8.25	17.05	13.53	6,048	6,652	900
Selenium	0.53	0.31	0.81	0.80	2.00	1.50	2.64	2.16	0.41	2.02	5,937	6,763	137
Silver	0	0.32	0	0	0	0.15	0	0.24	0.41	0.26	5,938	6,762	17
Tetrachloroethylene	0.68	1.54	1.13	1.24	1.91	1.11	1.96	1.01	0.61	1.21	10,663	2,037	25
Total α	0.64	2.08	1.60	3.94	3.03	2.10	3.96	6.26	7.83	7.18	5,535	7,165	514
Total β	0	0	1.85	0.61	0	0.32	0	0.71	0.81	0.59	1,099	11,601	69
Total radium 226	40.00	0	0	5.88	4.17	2.33	2.78	2.13	0	0	354	12,346	0
Total trihalomethanes	0	0	0	0	0.07	0.02	0.11	0.30	0	0.15	8,336	4,364	7
Trichloroethylene	1.03	3.40	0.91	1.78	1.45	1.47	2.38	2.46	1.07	1.77	10,654	2,046	36

[a]Projected sources > MCL = number of unsampled active sources × projected 1993 percentage.

Table 13 Annual Concentration Averages for Chemicals in Public Drinking Water Sources

	Average annual concentration (μg/L)								
Chemical	1984	1985	1986	1987	1988	1989	1990	1991	1992
Aluminum	250.00	0.00	205.64	224.44	0.00	336.70	542.46	426.62	323.77
Arsenic	19.18	29.43	33.68	25.36	24.94	31.54	16.39	18.88	17.84
Atrazine	0.00	0.94	0.70	0.51	0.42	0.59	0.38	0.50	0.25
Barium	243.94	241.95	208.93	1274.97	194.42	168.18	197.28	181.80	189.34
Benzene	0.00	1.78	0.00	0.00	1.80	17.97	72.33	103.28	255.82
Cadmium	7.09	3.04	6.82	8.42	24.09	6.07	12.36	13.96	6.16
Carbon tetrachloride	5.02	4.96	3.57	4.88	2.65	2.60	2.70	3.19	2.17
Chloroform	6.77	3.53	3.26	4.33	6.42	6.33	7.24	8.26	7.08
Chromium (total)	26.59	32.68	24.56	21.30	20.06	24.64	23.79	21.58	19.35
1,1-Dichloroethane	1.30	2.08	2.37	3.21	1.60	1.94	1.14	2.09	2.00
1,2-Dichloroethane	1.55	4.40	1.51	1.68	2.73	3.43	3.32	3.65	3.86
1,1-Dichloroethylene	0.68	1.31	1.43	4.52	2.73	2.20	1.96	2.50	2.84
Cis-1,2-dichloroethylene	0.00	2.02	12.33	10.86	2.64	1.77	1.71	2.24	2.15
Trans-1,2-dichloroethylene	21.95	39.99	8.40	3.75	2.65	1.40	0.00	0.00	0.00
1,2-Dichloropropane	0.80	1.65	2.14	6.39	1.09	1.29	2.85	2.54	4.80
Dibromochloropropane	0.60	0.45	0.50	0.54	0.53	0.40	0.42	0.28	0.29
Ethylene dibromide	0.00	0.00	0.48	0.14	0.00	0.12	0.10	0.06	0.04

Fluoride[a]	1.07	0.54	0.89	0.50	0.48	0.47	0.58	0.52	0.44
Lead	22.88	33.17	26.04	26.04	27.90	33.85	14.44	13.71	13.82
Mercury	80.49	44.36	9.48	4.83	1.53	3.99	0.61	0.87	0.74
Natural uranium[b]	7.90	5.00	30.18	23.42	10.54	23.42	10.54	10.73	12.09
Nitrate (as NO_3)[a]	15.26	15.68	16.74	15.65	15.74	16.54	16.32	18.60	21.44
Selenium	16.49	26.07	12.44	14.95	15.11	11.91	10.17	11.93	7.53
Silver	23.33	41.33	23.33	25.73	25.50	33.23	19.12	40.00	205.00
Simazine	0.00	0.72	0.54	0.42	1.21	0.65	0.67	0.55	0.35
1,1,2,2-Tetrachloroethane	0.00	0.00	3.58	2.55	0.00	0.00	0.00	0.00	0.00
Tetrachloroethylene	7.99	6.76	6.66	7.11	5.25	5.13	4.53	4.70	3.89
Total α[b]	3.47	4.08	3.44	9.29	5.50	4.21	4.26	5.24	12.69
Total β[b]	7.99	4.92	12.48	10.97	8.00	12.43	7.55	9.00	8.00
Total radium 226[b]	3.04	0.00	0.70	3.38	2.23	12.90	1.31	1.32	1.00
Total trihalomethanes	0.00	0.00	12.50	16.88	13.69	12.63	19.88	21.04	21.42
1,1,1-Trichloroethane	4.24	6.02	2.23	2.94	2.78	2.03	1.49	2.17	3.36
Trichloroethylene	20.09	25.57	9.12	8.69	6.99	8.15	10.73	7.53	9.36
Trichlorofluoromethane	1.30	1.89	4.99	1.92	1.90	3.16	47.83	11.46	9.38
Trichlorotrifluoroethane	1.40	1.37	1.88	2.11	2.00	2.30	1.26	1.12	1.27
Vinyl chloride	0.00	27.67	16.50	7.50	0.00	2.10	0.00	0.00	0.00

[a]Concentration units are mg/L.
[b]Concentration units are pCi/L.

vations because the concentrations vary from one individual source to another depending on the movement of underground contaminant plumes affecting them. The concentration of a contaminant can decrease in one well of a system at the same time that it increases in a neighboring well in the same system as a contaminant plume migrates through an area. The statewide means, however, might reveal major statewide trends if they exist. For some important contaminants there are apparent trends over the 9 year period (Figures 13 through 16). For nitrate, the mean concentration has increased from 15.3 mg/L in 1984 to 21.4 mg/L in 1992 and DBCP has decreased from 0.6 to 0.3 μg/L, both trends being consistent with the observed changes in MCL percentages discussed earlier. Other increases seen were for 1,1-dichloroethylene, 1,2-dichloroethane, 1,2-dichloropropane, total THM, benzene, and total α. Decreases were observed for carbon tetrachloride, tetrachloroethylene, *trans*-1,2-dichloroethylene, trichloroethylene, and mercury. The statewide annual means for a number of contaminants have consistently exceeded their MCL. These include trichloroethylene, 1,2-dichloroethylene, carbon tetrachloride, and DBCP. Until 1992, selenium's annual mean exceeded its MCL every year. The benzene mean has risen dramatically to over 100 times the MCL since 1988.

F. Populations and Health Impacts

The known health effects of contaminants found in excess of their MCL are summarized in Table 14. Of the chemicals listed 18 have been shown to cause cancer in animals or humans through ingestion of water. For 16 of the carcinogens, the cancer risk used to establish the MCL is included in Table 14. The cancer risk represents the theoretical fraction of individuals calculated to contract cancer in a population that consumes 2 L water containing the chemical at its MCL for 70 years. It varies from 10^{-6} for many of the VOC to 2×10^{-2} for arsenic.

The populations served by individual public water systems can range from five individuals to millions of persons, and the population potentially affected by a contaminated source can vary considerably. In a small water system, only a handful of individuals may be at risk of exposure. On the other hand, if the contamination were to occur in a high production source providing water to a system serving a large population and no blending or treatment is carried out, the number of potentially exposed persons could be in the tens of thousands. Figure 17 graphically illustrates the variation between populations and the percentage of analyses for all regulated chemicals that exceeded the MCL for selected chemicals. The chart shows that the frequency for MCL being exceeded is greater in the larger systems serving greater numbers of people. Samples collected from small systems

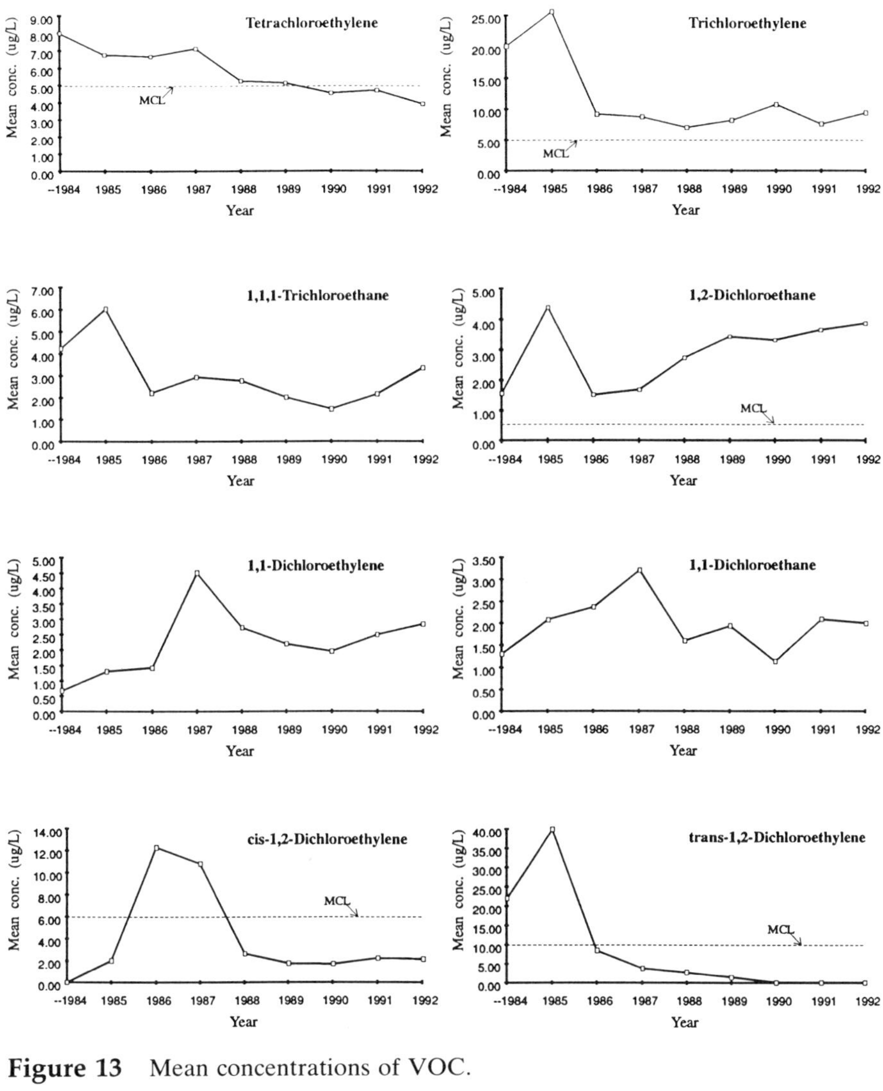

Figure 13 Mean concentrations of VOC.

serving less than 200 persons displayed only about one-seventh the frequency of the largest systems serving more than 10,000 persons. It must be cautioned that many fewer monitoring data are available for small systems, especially for inorganic and radiological chemicals.

Table 15 shows the populations potentially affected by contaminated active sources that exceed the MCL. The total populations of affected

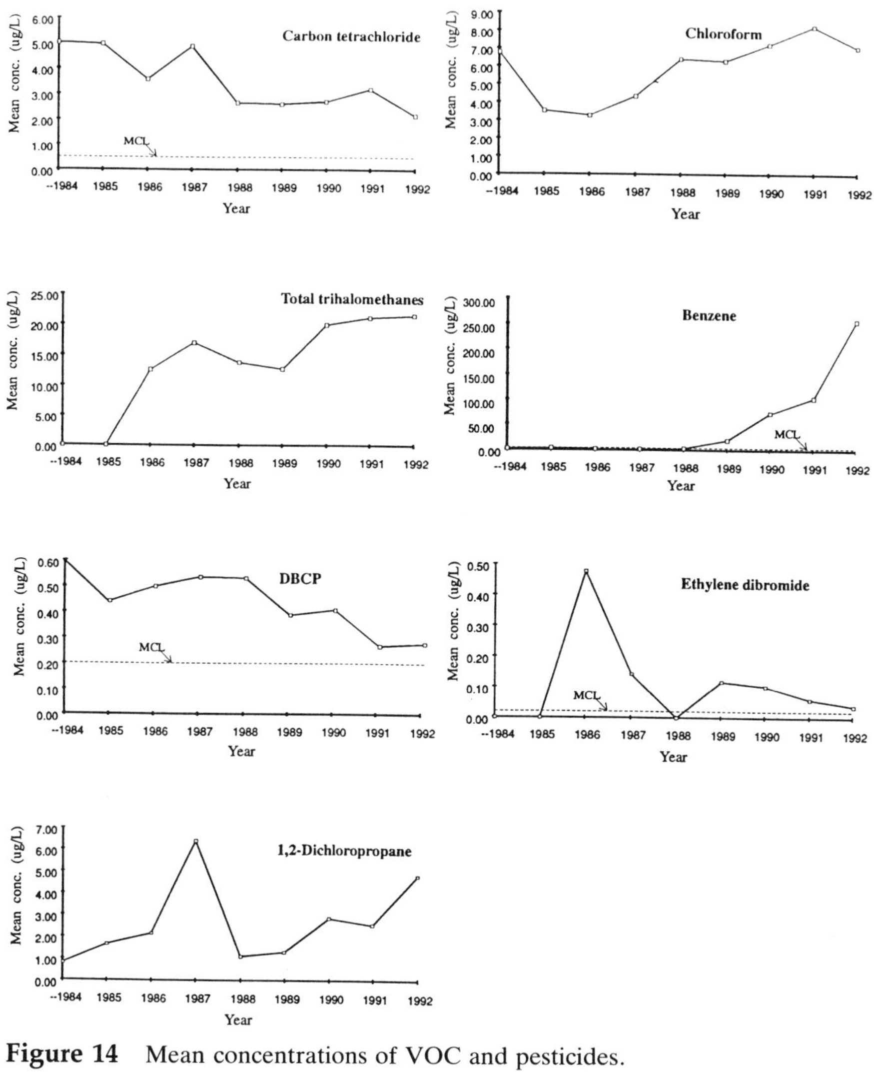

Figure 14 Mean concentrations of VOC and pesticides.

systems are the summation of the populations served by all water systems that have active sources exceeding the MCL of a given chemical. The potentially exposed population(s) (PEP) for each chemical is the total population multiplied by the ratio of the number of active sources exceeding the chemical's MCL to the total number of active sources that were sampled for the chemical in affected systems. The PEP are statistical estimates of

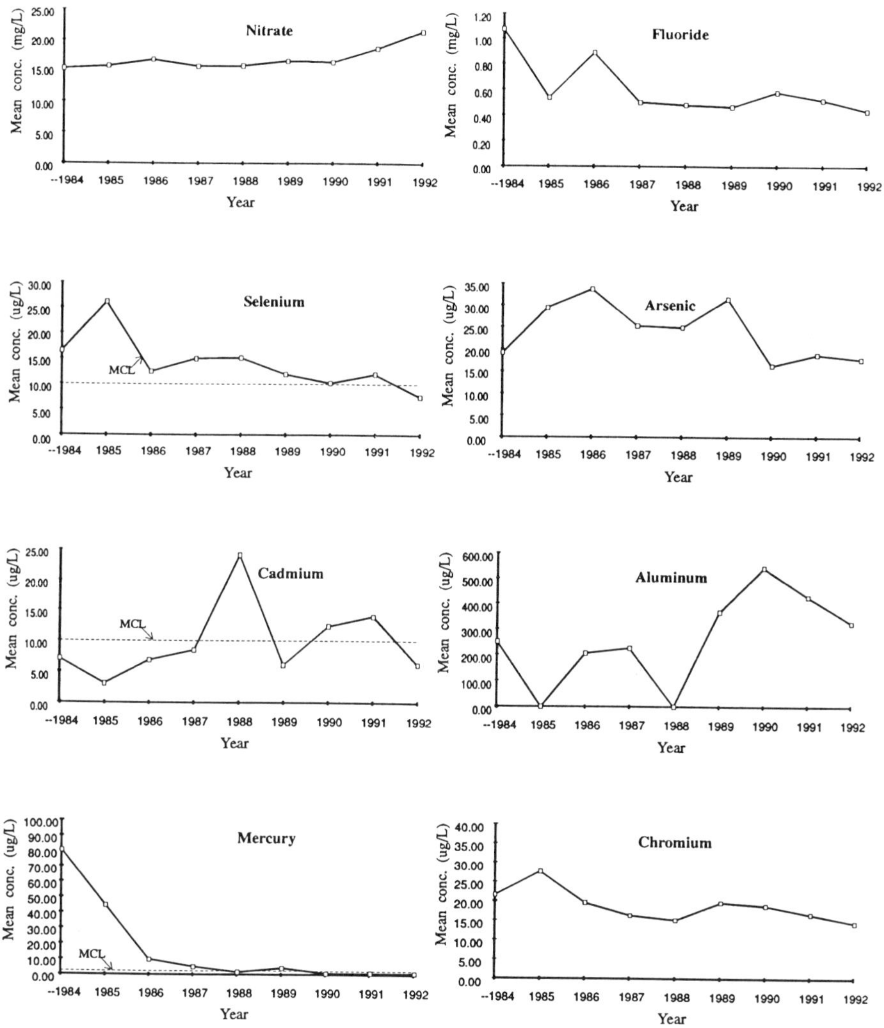

Figure 15 Mean concentrations of inorganic chemicals.

the maximum number of persons who might be exposed to the contaminant above its MCL under worst-case conditions. Those conditions might occur if the contaminated water was served directly to consumers with no treatment or blending with uncontaminated water to lower the contaminant's concentration to below the MCL. The chemicals are sorted in Table 15 according to their PEP. The last column in the table illustrates a parameter

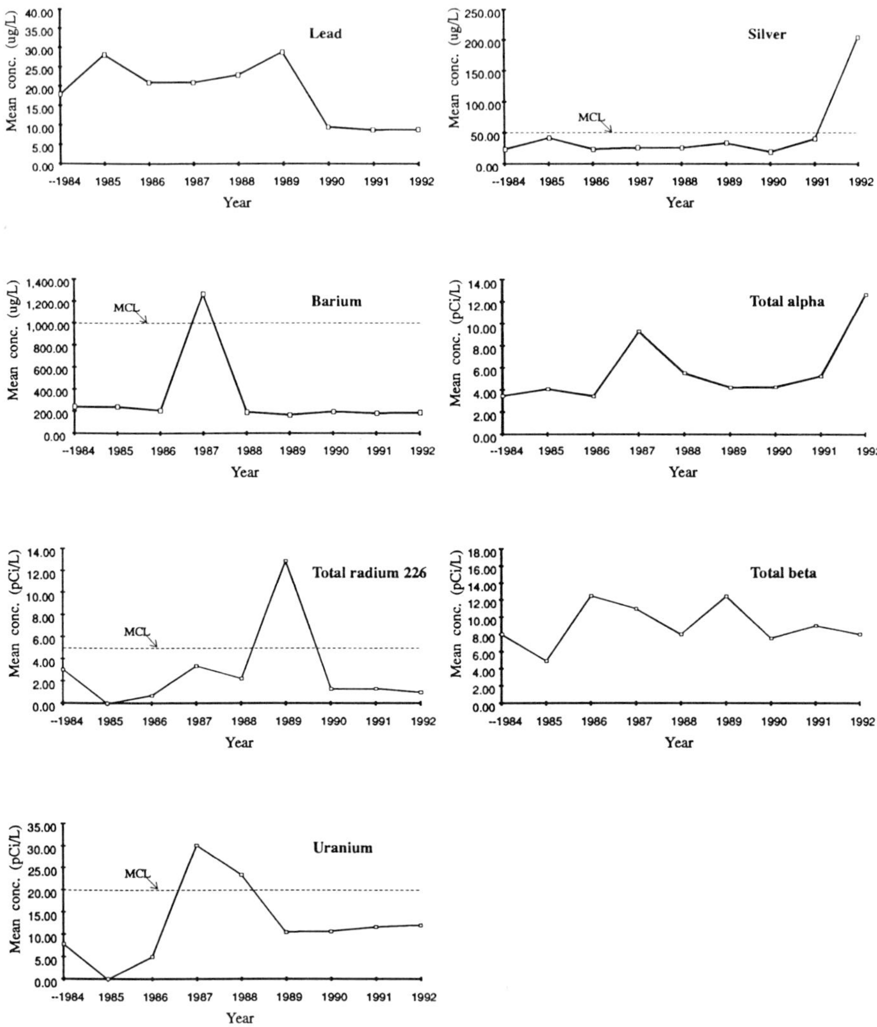

Figure 16 Mean concentrations of inorganic chemicals and radiological parameters.

Table 14 Health Effects of Chemicals Exceeding MCL in Drinking Water Sources

Chemical	Health effects[a]	Cancer risk[b]
Aluminum	*Long-term phosphate depletion and osteomalacia*; short-term gastrointestinal effects; possible neurological effects	—
Arsenic	*Skin disorders including skin cancer*, hyperpigmentation, and keratosis; tumors of the skin, lungs, genital organs, and visual organs	2×10^{-2}
Barium	*High blood pressure and nerve block*	—
Benzene	*Leukemia in humans*; blood disorders; toxic to the bone marrow	10^{-6}
Bromodichloromethane	Tumors of the large intestine and kidney in rats and tumors in kidney and liver in mice; potential human carcinogen	3×10^{-4}
Bromoform	Liver damage in male rats and kidney damage in female rats; liver damage in male mice; intestinal tumors in male and female rats	2.5×10^{-5}
Cadmium	*Impaired kidney and liver function*; high blood pressure; anemia	—
Carbon tetrachloride	*Liver tumors in three animal species*; suspected human carcinogen	10^{-6}
Chromium	(Cr^{6+} only) *Liver and kidney damage*; internal hemorrhage and respiratory disorders; ulceration of skin and dermatitis	—
Chloroform	Kidney and liver tumors in rats and mice; potential human carcinogen	1.7×10^{-5}
1,2-Dibromo-3-chloropropane (DBCP)	*Forestomach tumors in male rats*; human testicular toxin at high doses causing reduced sperm counts; probable human carcinogen	10^{-4}
Dibromochloromethane	*Liver and kidney damage in rats and mice*; *live tumors in mice*; not considered a carcinogen by the U.S. EPA	—
1,1-Dichloroethane	*Increased mortality, clinical signs of intoxication, and depressed body weight in rats*; no human or animal evidence of carcinogenicity	—

Table 14 Health Effects of Chemicals Exceeding MCL in Drinking Water Sources

Chemical	Health effects[a]	Cancer risk[b]
1,2-Dichloroethane	*Cancer at various sites in rats and mice*; potential human carcinogen	2.8×10^{-6}
1,1-Dichloroethylene	*Hepatocellular fatty degeneration in female rats*; liver and kidney effects in animals; no human or animal evidence of carcinogenicity	—
Cis-1,2-dichloroethylene	*Liver effects in rats*; no human or animal evidence of carcinogenicity	—
Trans-1,2-dichloroethylene	*Decreased immune response in body fluids and increased serum glucose levels in subchronic studies in mice*; pulmonary, central nervous system (CNS), kidney, liver, cardiac, eye, and dermal toxicant in acute, subacute, and subchronic animal studies; no chronic toxicity data available	—
1,2-Dichloropropane	*Benign breast cancer in female rats*; liver, kidney, lung, and CNS toxicant	10^{-6}
Di-(2-ethylhexyl)phthalate	*Liver tumors in rats and mice*; *liver and testes toxicant*; *liver cancer*; *reproductive and developmental effects in animal studies*	10^{-6}
Ethylene dibromide	*Stomach, nasal, and lung tumors in rats and mice*; human reproductive toxin; probable human carcinogen	10^{-6}
Fluoride	*Permanent stained spots on teeth (mottling) in children under 7*; dental and skeletal fluorosis	—
Lead	Impaired functioning of nervous system, kidneys, and red blood cell formation; reduces oxygen absorbability of blood; causes high blood pressure, low birth weights; short-term exposure can be toxic to infants and pregnant women	—

Mercury	Toxic to CNS; mental disturbances, impaired coordination of body movements, disturbances in speech, and hearing impairment; impaired kidney function; may cause birth defects	—
Nitrate (as NO_3)	Causes methemoglobinemia in infants under 3 months old and in dialysis patients by impairing ability of blood to carry oxygen to body tissues	—
Radium 226, 228	Bone and brain cancer	5×10^{-5}
Silver	Permanent blue-gray discolorization of the skin, internal organs, and eyes	—
1,1,2,2-Tetrachloroethane	*Body weight depression and changes in liver fat content in rats*; liver and CNS toxicant in humans; limited evidence of carcinogenicity	—
Tetrachloroethylene	*Liver cancer in mice and leukemia in rats*; some evidence in exposed human workers; no teratogenicity; possible fetotoxicity at high doses	2.5×10^{-6}
Total α	Cancer	Unavailable
Total β	Cancer	Unavailable
Total trihalomethanes	Refer to individual THM species (i.e., chloroform, bromodichloromethane, dibromochloromethane, and bromoform)	10^{-4}
Trichloroethylene	*Liver, lung, and kidney tumors in rats and mice*; probable human carcinogen	10^{-6}
Vinyl chloride	*Cancer in male and female Osborne Mendel rats*; known human carcinogen; results in a rare form of malignant liver tumor in animals and humans and tumors at other sites in animals	2.5×10^{-5}

[a] For many organic chemicals listed, health effect end point for the determination of the California MCL is set in italics.

[b] The theoretical risk of contracting cancer if exposed to a carcinogen through drinking water at a concentration equal to the MCL, assuming actual exposure has continued over a 70 year lifetime and 2 L water were consumed each day.

Source: From Reference 19.

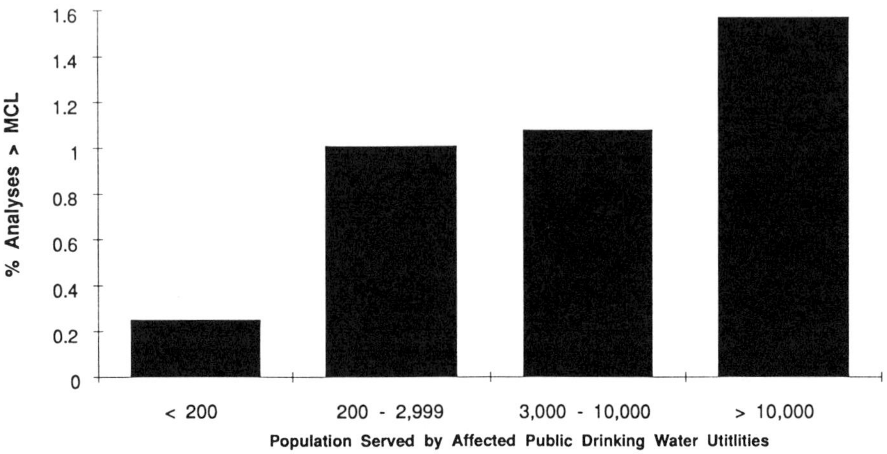

Figure 17 Percent analyses greater than MCL according to populations served.

called here the risk/exposure value (REV), which takes into account the risk associated with the carcinogens in Table 15. The REV, a product of the potentially exposed population and the cancer risk, is the theoretical number of cancer incidences if no corrective actions were taken.

VI. DISCUSSION

A. Overview

California's public drinking water sources have become seriously contaminated with chemicals in some locations. Overall, 2166 MCL violations have occurred since 1984, some being multiple for different chemicals in the same source. The areas of most concern are Southern California and the San Joaquin Valley. One of the more surprising observations is the low number of MCL wells in the highly populated San Francisco Bay area, especially in Santa Clara County, where the density of industrial activity and the hydrogeology have resulted in some areas of serious groundwater pollution. Extreme northern California, with its low populations and more limited agriculture, has had few of its public drinking water wells contaminated in excess of MCL. In view of these observations, it is not surprising that 66% of the MCL violations observed have occurred in sources of the largest water systems that serve more than 10,000 persons. These utilities are located in the more contaminated, high-population centers of the state.

The most frequently identified contaminants have been chlorinated VOC, DBCP, and nitrates, all resulting from human activities. This is consistent

Table 15 Exposure and Risk Assessment from Chemicals Exceeding MCL in Active Drinking Water Sources

Chemical	Active sources sampled in systems > MCL	Active sources > MCL	Total population of affected systems	Potentially exposed population[a]	Cancer risk	Cancer risk value[b]
Fluoride	1456	173	18,831,731	2,237,561	—	—
Mercury	279	25	17,945,528	1,608,022	—	—
Nitrate (as NO_3)	1972	304	9,480,154	1,461,444	—	—
Natural uranium	328	75	4,038,470	923,431	—	—
Trichloroethylene	1109	128	6,889,524	795,184	10^6	0.80
Total α	1345	172	5,463,819	698,719	—	—
Carbon tetrachloride	361	50	4,131,036	572,166	10^6	0.57
Cadmium	251	33	3,774,689	496,274	—	—
Total radium 226	60	8	3,483,268	464,436	5×10^6	2.32
Tetrachloroethylene	1173	85	5,166,753	374,402	2.5×10^6	0.94
Dibromochloropropane	839	195	1,508,712	350,654	10^4	35.07
Chromium (total)	190	15	3,904,527	308,252	—	—
1,2-Dichloroethane	470	32	4,523,354	307,973	2.8×10^6	0.86
Selenium	328	70	1,084,744	231,500	—	—
Ethylene dibromide	500	27	4,276,142	230,912	10^6	0.23
1,1-Dichloroethylene	209	11	3,825,964	201,367	—	—
Aluminum	235	37	962,672	151,570	—	—
Total β	152	5	3,724,950	122,531	—	—

Table 15 (continued)

Chemical	Active sources sampled in systems > MCL	Active sources > MCL	Total population of affected systems	Potentially exposed population[a]	Cancer risk	Cancer risk value[b]
Arsenic	239	40	621,041	103,940	2×10^{2}	2078.80
Total trihalomethanes	120	3	4,154,500	103,863	10^{4}	10.39
Lead	204	15	489,225	35,972	—	—
Diethylhexylphthalate	10	5	70,800	35,400	10^{6}	0.04
Silver	28	4	210,423	30,060	—	—
Cis-1,2-dichloroethylene	232	6	562,779	14,555	—	—
1,2-Dichloropropane	34	4	96,175	11,315	10^{6}	0.01
Benzene	253	6	380,001	9,012	10^{6}	0.01
1,1-Dichloroethane	109	5	100,905	4,629	—	—
Trans-1,2-dichloroethylene	4	1	6,629	1,657	—	—
Vinyl chloride	9	2	6,663	1,481	2.5×10^{6}	<0.01
Barium	13	4	4,200	1,292	—	—
Chloroform	13	1	4,500	346	1.70×10^{5}	0.01
1,1,2,2-Tetrachloroethane	3	3	<200	<200	—	—

[a]Potentially exposed population (PEP) = (active sources > MCL/active sources sampled in systems > MCL) × total population of affected systems.

[b]Risk exposure value (REV) = PEP × cancer risk.

with the observation that most wells exceeding MCL are located in the most populated and heavily agricultural areas of the state.

Contamination from chemicals of largely natural origins has also been observed. Those naturally occurring chemicals seen most frequently in excess of MCL have been fluoride, selenium, arsenic, and α emission. Fluoride and α appear to be somewhat more disperse than the contaminants of human origin, although more are seen in Southern California because of the larger number of wells in that part of the state. Selenium- and arsenic-contaminated wells that exceed MCL are lower in number; however, some occur in clusters within communities and as a result offer a potential impact on public health, especially arsenic (see later). There are few radon data in WQM; however, a special study of selected sources suggests that its occurrence in public drinking water wells might be significant.

Wells are the primary sources contaminated with chemicals and are thus the thrust of the discussion in this chapter. Surface intakes have suffered very little contamination from organic chemicals. Although nitrate was detected in 65.5% of the surface intakes sampled, none exceeded the nitrate MCL. Surface sources offer other potential hazards from microorganisms and disinfection by-products (e.g., THM) within the distribution network of water systems (2); their discussion is not within the scope of this chapter.

B. Setting Priorities

In most cases, corrective action has been taken for sources exceeding MCL (see later); however, a better procedure for assessing potential health impacts would be beneficial for setting priorities for future corrective actions, monitoring studies, and regulations. The precise public health impact resulting from the contaminated wells is not currently possible to derive because accurate concentrations of a contaminant expected at a consumer's tap cannot easily be determined on the basis of concentrations seen at a raw source serving the system. Sampling and testing a sufficient number of consumer taps for all chemicals of concern would be prohibitively expensive. A calculation approach would require a complete knowledge of the production of every contaminated well and all other wells in the parent system. The operational and design details of the distribution systems in affected water utilities would be required to know how the contaminated water is blended and diluted with other water. An elaborate computer model would have to be created to calculate the concentrations of the contaminant at every consumer water tap. This is a goal that may be attainable in the future.

In the absence of this capability at present, a broad empirical method was utilized here that uses potential exposed populations and risk/exposure values. This approach assumes worst-case scenarios in which no blending occurs within affected water systems and contaminated water is served directly to consumers. With corrective actions, the actual populations exposed to drinking water contaminants are seldom as high as the PEP. Nonetheless, PEP give a perspective about the populations potentially exposed over the MCL for each chemical. Fluoride had the highest PEP, but the MCL is based on mottled teeth rather than a serious health impact. Mercury was next. The potential health impacts of mercury can be severe; however, its occurrence in excess of its MCL and its mean annual concentration have been in decline over the last 8 years. Nitrate is third on the list; its MCL is based on methemoglobinemia, a serious health impact affecting infants under 3 months old and dialysis patients. The nitrate PEP is misleading, however, because of the specific populations affected. Infants 3 months old and younger comprise less than 1% of the population at any time in California (20).

Cancer is the health impact of most concern, and several of the contaminants discussed in this chapter have been assigned MCL on the basis of their capacity to cause cancer in animals or humans. In an attempt to take into account the potential impact of carcinogens, the PEP of carcinogens were multiplied by the theoretical cancer risk utilized to establish the MCL. The resulting REV, which take into account both the potential populations exposed and the potency of the carcinogen, are ranked as follows:

Chemical	REV
Arsenic	2078.80 (319.8)
Dibromochloropropane	35.07
Total trihalomethanes	10.39
Total radium 226	2.32
Tetrachloroethylene	0.94
1,2-Dichloroethane	0.86
Trichloroethylene	0.80
Carbon tetrachloride	0.57
Ethylene dibromide	0.23
Diethylhexylphthalate	0.04
Chloroform	0.01
Benzene	0.01
1,2-Dichloropropane	0.01
Vinyl chloride	<0.01

The chlorinated VOC, which have high frequencies of occurrence in wells over their MCL, have relatively low REV, largely as a result of the low risk of 10^{-6} utilized to establish their MCL. According to this analysis, arsenic, DBCP, and trihalomethanes rank the highest for carcinogens. Arsenic, with a REV of 2079, is much higher than any other. It is so ranked because of the very high risk of 2×10^{-2} used to establish the MCL, although more recent studies suggest a risk 6.5 times lower than previously derived (21). The arsenic REV based on the newer risk estimate is included in parentheses. Whichever is used, the arsenic REV is conservative to the extent that it considers only the 51 active sources exceeding the arsenic MCL, not the 627 that contain detectable arsenic at concentrations below the MCL. Little or no corrective action has been taken on these sources as a result of the arsenic contamination, although the theoretical risk is substantial. Additionally, the REV approach does not consider the greater potential risk of consuming water contaminated at levels much higher than the MCL. Arsenic has been observed at levels of several hundred μg/L in some wells.

The trend analysis described in Section V suggests that 63 public drinking water sources exceeding the arsenic MCL might yet remain unsampled and undetected; the annual mean arsenic levels show no apparent trend. The trend analysis suggested that 6 additional sources might be found to exceed the DBCP MCL. The percentage of sources sampled for DBCP that exceed its MCL and the overall annual concentration mean for DBCP is decreasing. The potential health impact from DBCP may therefore be expected to decrease in the future. The data suggest that nitrate offers a greater risk in the future because both the frequency of sampled wells exceeding its MCL has been increasing each year and its statewide mean concentration has been rising.

C. Corrective Actions

California regulations (Title 22, Chapter 15, CCR) require follow-up sampling and analysis to be carried out when a contaminant is found to exceed its MCL in a drinking water source. The specific requirements vary depending upon whether the contaminant is an inorganic or an organic chemical. The regulations also require that consumers be notified when a source exceeds an MCL.

The department and water systems in California respond to MCL violations in several ways to minimize actual exposures to the contaminants. They are as follows.

1. Source Closures

As shown in Section V, many sources that exceed MCL have been destroyed, abandoned, or placed on standby. Whenever feasible, this is the preferred alternative. Unfortunately, economics and practical limitations do not often allow the closure of a source and use of another in its place.

The contaminated source may be an important or sole source in the water system. An entire water basin might be contaminated, so that the expense of drilling a new well to replace one closed could not be justified.

2. Blending

Blending is the approach most often used for dealing with contaminated sources. In such cases the contaminated source water is blended within the systems with water from uncontaminated sources to lower the overall level of the contaminants to below their MCL. Blending requires the availability of uncontaminated sources and frequent distribution monitoring to ensure that the water reaching consumer taps does not exceed MCL. Blending might not be a viable alternative for smaller systems having only a few sources. Also, blending is a questionable solution for carcinogen contaminants for which it is believed no threshold limit exists.

3. Treatment

Sources are often treated at the wellhead or intake to decrease a contaminant's concentration to below its MCL. VOC are removed by aeration or granular activated carbon (GAC). Other organics and pesticides can be treated using GAC. Many contaminants, including nitrate and inorganics, can be treated using reverse osmosis. Inorganics can also be treated with ion exchange or precipitated, coagulated, and filtered. Treatment is often the most expensive option, but in many cases may be the only one available if an entire water basin is contaminated and no other water sources are at hand.

4. Other Options

Small water systems having only one or two sources cannot always shut down or destroy a contaminated source and remain in operation. Blending might not be practical, and treatment is often prohibitively expensive for a small system. Under such circumstances, the system must advise customers to consume bottled water, or it must purchase water from an uncontaminated system. Contaminated small systems have sometimes closed all of their sources and been absorbed by larger neighboring systems.

D. Future Monitoring

The maximum number of compliance analyses that can occur for the recorded public drinking water sources in California is 781,260 (number of active/standby sources × 60 chemicals with MCL). The number that has occurred to date is 408,989, or about one-half the maximum. It is not expected that all sources will be tested for all regulated chemicals because some chemicals must be tested only in sources that are expected to be vulnerable to contamination. Regulations also allow sampling of representative sources where multiple intakes and wells draw from the same water supply. Regardless, there remain sources that should be tested. This is especially the case for over 5000 small water system wells, few of which have been tested for inorganic chemicals. In view of this discussion, it is important that these sources be tested for nitrate, arsenic, mercury, selenium, radon, and total α emissions. Additionally, more monitoring requirements for drinking water sources are imminent. The U.S. EPA announced in the *Federal Register* on July 17, 1992 its intent to promulgate 23 MCL for additional inorganic and organic chemicals, VOC, and pesticides in its phase V regulations (22). The department is required to adopt conforming standards to continue administering the federal drinking water program in California. Adoption is anticipated in late 1993.

ACKNOWLEDGMENTS

The drinking water quality data presented in this chapter were generated and collected through the efforts of a large number of persons in many water utilities, water laboratories, and public agencies across the state.

Water samples were collected by persons working for water utilities, contract laboratories, the local health departments, and the Office of Drinking Water. The Office of Drinking Water 14 district offices have been responsible for receiving analyses from utilities, reviewing them, and submitting them to the Office of Drinking Water headquarters for entry into the central data base.

The author extends special thanks to his staff in the Monitoring and Evaluation Unit, who developed and maintained the drinking water data base over the last 5 years. These persons' programming and drinking water quality expertise made possible the data base as it now is.

REFERENCES

1. California Department of Water Resources, *California Water: Looking to the Future*, Bulletin 160–87, November 1987.
2. D.L. Storm and J. Windsor, *Drinking Water Quality in California*, California Department of Health Services, Office of Drinking Water, 1993 (in preparation).
3. California Department of Health Services, *Organic Chemical Contamination of Large Public Water Systems in California*, Public Water Supply Branch, April 1986.
4. California Department of Health Services, *Organic Chemical Contamination of Small Public Water Systems in California—Small Public Water System AB 1803 Final Status Report*, Public Water Supply Branch, June 1990.

5. S.M. Wong, *Monitoring and Analysis of Drinking Water in California*, California Department of Health Services, Office of Drinking Water, July 1991.
6. California Department of Water Resources, *California's Ground Water*, Bulletin 118, September 1975.
7. California Department of Water Resources, *Management of the State Water Project*, Bulletin 132–89, 1989.
8. California Department of Water Resources, *Water Conditions in California*, Bulletin 120–88, October 1988.
9. U.S. Environmental Protection Agency, *Water-Related Environmental Fate of 129 Priority Pollutants*, December 1979, EPA-440/4-79-029a.
10. T.M. Vogel and P.L. McCarty, Biotransformation of tetrachloroethylene to trichloroethylene, dichloroethylene, vinyl chloride, and carbon dioxide under methanogenic conditions, *Appl. Environ. Microbiol. 49*(5):1080–1083 (1985).
11. C. Miller Maes, M. Pepple, D. Troiano, D. Weaver, and W. Kimaru, *Sampling for Pesticide Residues in California Well Water: 1992 Well Inventory Data Base, Cumulative Report 1986–1992*, California Environmental Protection Agency, Department of Pesticide Regulation, December 1992.
12. California Department of Food and Agriculture, *Nitrate and Agriculture in California*, December 1990.
13. National Research Council, *Drinking Water and Health*, Vol. 1, National Academy Press, Washington, D.C., 1977.
14. R.H. Sakaji, *Radon in California's Ground Water: A Water Quality and Exposure Assessment*, California Department of Health Services, Office of Drinking Water, May 1990.
15. J. Culp, *Pesticide Monitoring of Surface Waters in the Northern District*, California Department of Water Resources Bulletin 120–88, April 1990.
16. California State Water Resources Control Board, *Nitrate in Drinking Water—Report to the Legislature*, Report No. 88–11 WQ, October 1988.
17. U.S. Environmental Protection Agency, *Water Pollution Control*; *National Primary Drinking Water Regulations*; *Radionuclides*, *Federal Register*, 40 CFR Part 141, September 30, 1986.
18. J. Wiegand, G. Yamamoto, and W. Gaston, Elevated levels of radioactivity in water wells in Los Angeles and Orange Counties, California, *Radon in Ground Water*, Proceedings of the NWWA Conference April 7–9, 1987, Somerset, NJ (B. Graves, ed.), Lewis Publishers, Chelsea, MI, 1987.
19. B.S. Lichti, *Health Risks Associated with California's Drinking Water*, California Department of Health Services, Office of Drinking Water, July 1991.
20. A. Oreglia and D. Canary, *Health Data Summaries for California Counties—1992*, California Department of Health Services, Health and Statistics Branch, June 1992.
21. P.B. Brown and M.A. Fan, *Arsenic—Recommended Public Health Level for Drinking Water*, California Environmental Protection Agency, Office of Environmental Toxicology Section, March 1992.
22. F. Pontius, Plan, prepare now for the impact of phase V regs, *Opflow*, American Water Works Association, *18*(11):November 1992.

5

Benefits of Deriving Drinking Water Standards Based on Comprehensive Assessments of Exposure

Robert R. Vanderslice

Rhode Island Department of Health, Providence, Rhode Island

I. INTRODUCTION

The U.S. Environmental Protection Agency (EPA) sets drinking water standards, termed maximum contaminant levels (MCL), to ensure that public water systems provide consumers with water that is safe to drink. Safe drinking water should be safe for bathing and washing clothes and dishes as well, but the U.S. EPA methodology for developing MCL does not explicitly address these other uses for drinking water. Several recent studies demonstrate that routine household use of drinking water results in both dermal exposures to drinking water contaminants via skin contact and inhalation exposures to volatilized contaminants, especially while showering (see Chapters 6, 12, and 13). This chapter describes the benefits of quantifying ingestion, dermal, and inhalation exposures as part of the drinking water standard-setting process. Recommendations for how state and local health departments can assess noningestion exposures are developed based on a review of the exposure assessment methods currently used by the U.S. EPA in deriving MCL and those used by the state of Rhode Island in handling groundwater contamination incidents.

The U.S. EPA acknowledges that exposures to drinking water contaminants occur via routes other than ingestion. The U.S. EPA Risk Assessment Council recently recommended that, "in lieu of other information, inhalation exposure to VOCs from drinking water should be assumed to be equivalent to ingestion exposures" (1). In addition, the U.S. EPA *Exposure Factors Handbook* (2) provides guidance on quantification of both inhalation and dermal exposures to contaminated drinking water. However, these methods are not used in the development of drinking water standards. To include an assessment of inhalation and dermal exposures to drinking water contaminants in its standard-setting process would represent a significant change in U.S. EPA policy.

Federal guidance for addressing noningestion exposures could be a great benefit to state and local health officials. State and local health officials routinely respond to drinking water contamination incidents and must be able to advise consumers about the appropriate uses of contaminated water. When a contamination incident warrants an advisory to use bottled water for drinking and cooking, local health officials must be prepared to answer questions on whether the water remains safe for bathing and other routine household activities.

Private well contamination presents special problems. Outside the jurisdiction of federal drinking water standards, residents using private wells must often make their own decisions about appropriate use of their water when contamination incidents occur. Local health officials face a dilemma in advising residents who rely on a contaminated well for drinking water. Although the safety of the these residents is paramount, advice to discontinue using a groundwater supply because of low-level contamination comes at a tremendous cost to residents. Typically, neither residents nor local governments have resources to finance installation of water treatment devices or new wells. If the water is classified as unfit for bathing and washing clothes and dishes, the lives of residents suffer extreme disruption.

Comprehensive exposure assessments, that is, assessments that evaluate inhalation, dermal, and ingestion exposures, are needed to determine which residents can safely use their water. It would clearly be beneficial to state and local health officials if the U.S. EPA developed drinking water standards and advisories based on such comprehensive assessments.

II. METHODS FOR ASSESSING EXPOSURES

Most MCL are derived with only cursory attention to assessment of exposures. An exposure assessment section is not included in the body of most criteria documents the U.S. EPA prepares in support of its MCL. The standard protocol for deriving MCL addresses exposures by incorporating a default assumption that ingestion of drinking water averages 1 L/day over a lifetime (3). In some cases, however, the U.S. EPA has used other methods to assess exposures to drinking water contaminants. These methods are described here, along with assessments of environmental lead exposures and protocols for addressing contaminated private wells (Table 1), both developed by the Rhode Island Department of Health (RIDOH).

A. Epidemiological Assessments

For some drinking water contaminants, such as nitrate and nitrite, arsenic, and fluoride (4), data from epidemiological studies are adequate for determining the likely health effects associated with elevated levels of these contaminants in drinking water. By comparing the disease rates for a population served by drinking water with elevated levels of a contaminant with disease rates for a similar but unexposed population, epidemiological studies can link health effects to drinking water exposures without determining how exposures occur. Exposure assessments are needed only if the exposures to the study population are not representative of the U.S. population as a whole. In its evaluation of the epidemiological studies of arsenic in drinking water in Taiwan, the U.S. EPA found a significantly greater drinking water rate among the Taiwanese than expected for U.S. populations. Accordingly, the U.S. EPA adjusted its standards to reflect these exposure assessment results (4).

Table 1 Methods for Assessing Exposures to Drinking Water Contaminants

Method	Ingestion exposures	Inhalation exposures
U.S. EPA		
Epidemiological	Not quantified	Not quantified
Default	2 L/day	Implicit in RSC of 20%
Modified default	2 L/day	Equivalent to ingestion
Radon	1.0 L/day	Calculated based on transfer coefficient
RIDOH		
Lead	Varies with concentration[a]	Assumed insignificant
Contaminated private wells	Zero–bottled water provided when MCL exceeded	Resident responsible for limiting exposures to acceptable levels

[a]First-draw samples below 5 μg/L are acceptable, and flushing is not required. Restrictions on exposure increase with increasing levels of lead found in flushed samples, as follows: <5 μg/L, unrestricted ingestion (assumed to average 1 L/day for toddlers); 5–20 μg/L, restricted tap water ingestion to 1 cup/day; 100 μg/L, no drinking or cooking; and >100 μg/L, replumbing necessary.

B. Standard U.S. EPA Methodology

For individuals familiar with the U.S. EPA terminology and acronyms, the standard methodology used to derive an MCL follows a logical progression. Guidance manuals may benefit readers who are unfamiliar with this nomenclature (3). Briefly, the methodology begins with the derivation of a dose calculated to protect the public from adverse health effects resulting from ingestion of a contaminant. This reference dose (RfD) is then divided by the estimated ingestion rate (2 L/day) to yield the drinking water equivalent level (DWEL), the level of a contaminant in water that should produce no adverse effects if all other sources of exposure are eliminated. The relative source contribution (RSC), usually assumed to be 20%, compensates for other environmental sources of exposure to this environmental contaminant. The product of the RSC and the DWEL is the maximum contaminant limit goal (MCLG). For most non-carcinogens, the MCL is set at the level of the MCLG. For carcinogens, the MCLG is set at zero and the MCL is set at the lowest technically feasible level.

This standard protocol incorporates two assumptions about exposures to drinking water contaminants: (1) ingestion of drinking water averages 2 L/day over a lifetime, and (2) ingestion of drinking water accounts for only 20% of environmental exposures.

C. Modification of the Standard Method

The U.S. EPA derived MCL for volatile organic chemicals (VOC) following the procedure just described. Drinking water ingestion rates were assumed to average 2 L/day over a lifetime, and for noncarcinogens, exposures from other sources were addressed by the RSC of 20%. In addition, the U.S. EPA estimated inhalation exposures to volatilized drinking water contaminants for the purpose of determining the health benefits of the VOC regulation. Health benefits were determined from predictions of the number of cancer cases avoided as a result of reducing the concentrations of VOC in drinking water. For this determination, the U.S. EPA assumed that inhalation exposures to volatilized contaminants were equivalent to exposures that occur by ingesting 2 L/day (5).

D. Radon Assessment

Radon is unique among regulated drinking water contaminants; the hazards associated with inhalation of radon (and its progeny) via volatilization from drinking water are believed to represent a greater hazard than drinking water ingestion (6). The U.S. EPA estimated ingestion rates to be about 1 L/day for direct tap water ingestion. The default ingestion rate of 2 L/day was not used because radon readily volatilizes from heated water used for cooking and making coffee, tea, and other beverages.

The U.S. EPA based its assessment of inhalation exposures associated with radon in drinking water on the assumption that indoor air radon concentrations increase 1 pCi/L for every 10,000 pCi/L of radon in water. To place this 1:10,000 transfer coefficient in perspective, the ratio between the default assumptions for ingestion of drinking water (2 L/day) and the default inhalation rate for light activity (20,000 L/day) is also 1:10,000 (2).

E. Rhode Island Criteria for Environmental Lead

The RIDOH is currently developing environmental lead regulations that include numerical standards for lead in drinking water, interior and exterior dust, soil, and paint (7). The RIDOH developed criteria for environmental lead in light of the pervasive lead problem that exists in the state. The majority of housing stock in Rhode Island is contaminated with lead paint, and few resources are available for remediation (8). Childhood lead exposures above background, although undesirable, are unavoidable. A childhood blood lead level of about 5 μg/dl represents background (8).

RIDOH derived environmental lead criteria that are intended to keep childhood blood lead levels below 10 μg/dl, with a maximum increase in blood lead of 5 μg/dl for any one child (7). Relying on the U.S. EPA biokinetic model for lead (9), the RIDOH developed drinking water criteria that allow an increase in blood lead content of 1 μg/dl. Criteria for ingestion of nonfood items (soil, dust, and paint) allow an increase of 4 μg/dl. Facilities that cannot meet the Rhode Island criterion for acceptable levels of lead in water of 5 μg/L must take appropriate actions to reduce tap water exposures (Table 1).

In contrast to federal drinking water regulations, which apply only to public water systems and base feasibility on the technology available to large water systems, these RIDOH lead criteria apply to individual licensed child care facilities and landlords with tenants under the age of 72 months. The RIDOH regulations seek to prevent excessive lead exposures while allowing facilities with lead hazards to continue operation. Lead is only one of many health hazards facing clients of foster homes and other oversubscribed services. Regulations that would close these facilities to prevent lead exposures would not benefit public health.

F. Rhode Island Protocol for Contaminated Private Well Protocol

In response to groundwater contamination incidents that affect private wells, the Rhode Island Department of Environmental Management, with guidance from the RIDOH, supplies bottled water for cooking and drinking purposes to affected residents. In addition, limited resources are available to provide a small number of homes with point-of-entry treatment devices (granular activated carbon filter systems) to remove organic contaminants from water used for bathing and other household purposes.

Bottled water is recommended when contamination levels either approach or exceed MCL. In response to high levels of contamination or contamination by a large number of compounds, the RIDOH conducts a case-by-case assessment on the need for installation of treatment systems. Typically, the RIDOH recommends providing treatment devices when levels of VOC exceed 10-fold the MCL.

III. RESULTS

The six methods just described share many similarities. They are all methods designed to protect the public from adverse health effects posed by drinking water contaminants. It can be argued that each of these methods implicitly or explicitly accounts for non-ingestion exposures. The evaluation given here explores whether these existing methods provide the public with an adequate margin of safety.

A. Analysis of Standard EPA Methodology

The derivation of acceptable versus expected air concentrations associated with the volatilization of a contaminant from drinking water is presented in Figure 1. *Cis*-1,2-dichloroethylene is used as an example; derivations based on other noncarcinogenic solvents, such as 1,2-*trans*-dichloroethylene or trichloroethane, yield similar conclusions. Acceptable air concentrations are derived based on an estimate of the acceptable dose an individual could receive via inhalation, given that ingestion exposures were occurring to drinking water with contaminant levels equal to the MCL. The derivation begins with the assumption that the RfD provides an estimate of the "acceptable" dose of the contaminant from all sources of exposure. Maximum acceptable concentrations for water (350 μg/L) and air (35 μg/L) represent concentrations that result in exposures equivalent to the RfD for each source. This dose is apportioned between these two sources of exposure based on the RSC of 20%, resulting in acceptable air concentrations of 28 μg/m^3. Neither ambient levels, which typically are below 1 μg/m^3 (10), nor indoor air levels from volatilized contaminants are expected to reach the estimated acceptable air concentration. The expected air concentrations due to volatilization were based on the transfer coefficient for radon of 1:10,000, a worst-case assumption. Based on this

"Acceptable" Daily Dose

$$10\ \mu g/kg/day \times 70\ kg = 700\ \mu g/day \quad (1)$$

where: 70 μg/kg/day = RfD
70 kg = assumed body weight:

"Acceptable" drinking water concentration

$$70\ \mu g/l \times 2\ l/day = 140\ \mu g/day \quad (2)$$

where: 70 μg/l = MCL
2 l/day = assumed ingestion rate

"Acceptable" exposure from air:

$$700\ \mu g/day - 140\ \mu g/day = 560\ \mu g/day \quad (3)$$

where: 700 μg/day = product of eq. 1
140 μg/day = product of eq. 2

"Acceptable" air concentration:

$$(560\ \mu g/day)\ /\ (20\ m^3/day) = 28\ \mu g/m3 \quad (4)$$

where: 560 μg/day = product of eq. 3
20 m^3 = assumed inhalation rate

Estimated Likely Concentration in Air Due to Volatilization from Drinking Water at the MCL

$$(70\ \mu g/l \times 1000\ l/m^3)\ /\ (10{,}000) = 7\ \mu g/m^3 \quad (5)$$

where: 70 μg/l = MCL
1000 l/m^3 = units conversion
10,000 = water air transfer coefficient

Figure 1 Derivation of acceptable versus expected levels of *cis*-1,2-dichloroethylene in air versus associated with volatilization from drinking water.

analysis, indoor air concentrations should remain well below levels of potential concern, provided drinking water levels remain below the MCL.

This type of unsophisticated evaluation ignores important physiological and toxicological differences that result from exposures via different routes of administration. Methods for addressing these differences are discussed in the next section.

B. Contrasting State and Federal Approaches

The RIDOH methodology differs from the U.S. EPA methods, not in how exposures are assessed as much as in how these assessments are used. The RIDOH must answer the concerns of citizens desperate for practical advice about how to cope with a contaminated water supply or a yard with unsafe levels of lead in the soil where children play. Residents want credible information on how they can reduce their exposures to environmental contaminants and on how they can protect their children. When direct assistance is not possible, the RIDOH tries to give recommendations on how these individuals can help themselves. In contrast, the U.S. EPA develops criteria without regard for an individual's ability to reduce exposures to ensure that the criteria will be protective of all individuals.

The RIDOH protocol for issuing bottled water and treatment devices was developed to prevent overexposure to private well contaminants. The RIDOH protocol is based on the assumption that the MCL is protective of public health for all routine uses of water. Bottled water is provided to residents when drinking water contaminants approach or exceed MCL. The RIDOH assumes that individuals can take precautions to reduce non-ingestion exposures while bathing and washing clothes or dishes so that water can be used safely at concentrations up to 10-fold the MCL. These precautions include shorter, cooler baths and showers, increased ventilation of bathrooms during and after bathing or showering, increased ventilation when washing clothes and dishes, and keeping toilet lids closed and flushing less often. The cutoff at 10-fold the MCL is admittedly arbitrary.

The RIDOH lead criterion uses a similar approach for allowing contaminated water to continue to be used on a restricted basis. In recent testing of school and preschool drinking water, close to 80% of Rhode Island schools were found to have drinking water contaminated by lead (11). The RIDOH cannot force replumbing throughout the state; the standard provides practical, low-cost options for how an individual or facility can limit these exposures to acceptable levels.

C. Dermal Exposures

Dermal exposures to drinking water contaminants are not directly addressed by any of the methods just described. The quantification of dermal exposures to drinking water contaminants is a challenge. The U.S. EPA originally discounted dermal exposures as irrelevant, assuming dissolved contaminants would penetrate the skin at the same slow rate as the water carrier. A later U.S. EPA analysis, although in disagreement with this assumption, again discounted most dermal exposures as insignificant (12). This conclusion was reached after a review of dermal permeation coefficients for chemicals in aqueous media indicated that the maximum likely permeation coefficient for any waterborne contaminant precluded significant dermal absorption compared with ingesting 2 L/day of that water (12).

IV. RECOMMENDATIONS AND CONCLUSIONS

As the assessment of radon exposures would indicate, the default drinking water ingestion rate of 2 L/day probably exaggerates the ingestion risks associated with volatile compounds. The methodology used to assess inhalation exposures to volatilized radon can be adapted to estimate inhalation exposures to any volatile contaminant. However, making practical recommendations based on these data is difficult for contamination incidents that involve private wells.

The U.S. EPA is creating a tool for better evaluating inhalation exposures. Reference concentrations (RfC) provide an estimate of the concentration of a contaminant in air that is not expected to result in adverse health effects. For volatile drinking water contaminants, RfC at first glance appear to provide a useful way to determine if inhalation exposures are expected to be meaningful. Given the concentration of a volatile contaminant in drinking water and a transfer coefficient as in the radon assessment, an indoor air concentration could be calculated and compared with the RfC. If the expected air concentration exceeds the RfC by a significant margin, this may imply that the water is unsafe to use for showering, even at levels below the MCL. Because of the uncertainty in the process used to develop MCL, however, the disparity may be merely an indication that the RfC provides an overly conservative estimation of acceptable air concentrations. Similarly, if the RfC exceeds the expected air concentrations, this should not necessarily be an indication that local health departments can advise residents to cease worrying about using contaminated water for washing.

The U.S. EPA will not risk changing its standard method for setting MCL, a method that has survived repeated court challenges, unless the scientific evidence is compelling. Such evidence is likely to originate with assessments based on conservative RfC values. State and local health officials must continue to offer practical advice to residents based on unsophisticated analyses of exposure and rely on common sense and good communication skills as much as state-of-the-art methodologies.

ACKNOWLEDGMENTS

Many of the ideas presented in this chapter originated during discussions with coworkers during my tenure at the U.S. Environmental Protection Agency. I owe these colleagues thanks for my initiation to the field of risk assessment. Exchanges of information with colleagues from other states during FSTRAC meetings was invaluable. I am especially grateful for the constant support and assistance from my colleagues at the RIDOH, especially Bela Matyas, Lynn Boulay, and Linda Phillips.

REFERENCES

1. E.V. Ohanian, Presentation at FASTRAC, U.S. EPA Region V, Chicago, November 1992.
2. U.S. EPA, Office of Health and Environmental Assessment, *Exposure Factor Handbook*, Final Report, EPA/600/8-89/043, U.S. EPA, Washington, D.C., 1989.
3. U.S. EPA, Workshops on Assessment and Management of Drinking Water Contamination, EPA/600/M-86/026, U.S. EPA, Washington, D.C., 1987.
4. Integrated Risk Information System (IRIS), On-Line Data Base, U.S. EPA, ECAO, Cincinnati, OH, 1992.
5. 50 FR 46880 (1986).
6. 56 FR 33050 (1991).
7. RIDOH, *Draft Rules and Regulations for Lead Poisoning Prevention (R23-24)*, Providence, RI, August 1992.
8. Environmental Lead Task Force, *A Comprehensive Plan to Eliminate Childhood Lead Poisoning in Rhode Island*, Rhode Island Department of Health, Providence, RI, March 15, 1991.
9. U.S. EPA, *Technical Support Document for Lead*, ECAO-Cin-757, U.S. EPA, Cincinnati, OH, 1990.
10. ATSDR, Toxicological Profile for Dichloroethylenes, U.S. Department of Health and Human Services, Public Health Service, 1992.
11. R.R. Vanderslice and B.T. Matyas, Rhode Island's lead in school and preschool drinking water program, *R.I. Med. 75*:222–223 (1992).
12. R.R. Vanderslice, *Dermal Exposure to Drinking Water Contaminants*, U.S. EPA, 1988.

6

Chloroform Exposure and the Body Burden Associated with Swimming in Chlorinated Pools

Clifford P. Weisel
University of Medicine and Dentistry of New Jersey—Robert Wood Johnson Medical School, Piscataway, New Jersey

Theresa A. Shepard
Rutgers—The State University of New Jersey, Piscataway, New Jersey

I. INTRODUCTION

Chloroform in swimming pools disinfected with sodium hypochlorite has been demonstrated to be at elevated levels compared with chlorinated drinking water. Typical chloroform concentrations in swimming pools exceed 100 μg/L (1–5). Chloroform and other trihalomethanes (THM) are produced in situ by the reaction of the chlorine and organic matter presumably of human origin (3). The THM concentrations have been shown to be related to the number of swimmers in the pool and the formation of the different THM to be dependent on the bromide concentration (2). Since only a small amount of water (50 ml) is expected to be swallowed by adults (1), a 40-fold higher concentration is needed in swimming pools to have an equivalent ingestion dose compared to drinking water based on the daily 2 L ingestion of tap water. This volume is used by the U.S. Environmental Protection Agency (EPA) in establishing the drinking water regulations (6). However, dermal exposure and inhalation exposure, particularly for indoor swimming pools, also result in a chloroform internal dose while swimming. In addition, children and babies may ingest quantities of water greater than 50 ml when swimming (1). It is therefore necessary to determine the chloroform exposure from swimming pools when estimating the total daily chloroform exposures to individuals who swim routinely.

Elevated concentrations of chloroform in blood plasma have been measured in individuals who swam in pools treated with sodium hypochlorite (7). The concentration in the plasma was found to be correlated with the duration of the swim, the chloroform water concentration, and the chloroform concentration in the air above the pool. Aggazzotti et al. (7) recommended the need to understand the pharmacokinetics of chloroform following exposure at swimming pools to determine any health effect association with the chloroform exposure from swimming.

Lahl et al. (2) estimated the inhalation dose for a 1 h visit (30 minute swim) by an adult to a swimming pool as 50 μg, or 0.7 μg/kg, for a 70 kg adult. This value did not include contributions from dermal adsorption. The authors suggest that the actual inhalation dose could be 10 times this value when pool conditions produced higher amounts of off-gassing of chloroform and that children would have a higher specific dose (up to 15 μg/kg) because of the lower body mass even when the lower respiration rate is included. For comparison, the estimated chloroform dose from ingesting 2 L water containing 25 μg/L of chloroform is 50 μg, or 0.7 μg/kg, and the estimated dose from a 10 minute shower, which includes both dermal and inhalation components, is 33 μg, or 0.47 μg/kg (8). The present study examines the exposure to chloroform related to swimming in a chlorinated pool.

II. METHODOLOGY

A. Exposure Scenario

An indoor pool in New Brunswick, New Jersey was visited on five occasions by a single individual, and breath samples were collected following a 30 minute exposure. During three of the visits the subject swam for the 30 minutes, during one visit the subject remained at rest in the water, and during one visit the subject remained 3 m from the pool's edge, not in the water. Following the exposure the subject left the pool area, rinsed the pool water off using tap water, and dried his body. Breath samples were collected within 5 minutes of leaving the pool area and periodically during the next several hours in a room that had low background chloroform air concentrations (<0.2 μg/m^3). A preexposure breath sample was collected before the subject entered the pool area.

B. Sampling

Water samples were collected in a clean glass bottle and capped immediately. They were analyzed within 3 h of collection by purge and trap coupled with gas chromatography/mass spectrometry (GC/MS) (9). An air sample was collected on an adsorbent trap filled with 0.5 g Tenax GC and analyzed for chloroform. The air was sampled 3 m from the pool's edge, 6 cm off the floor for 10 minutes at a flow rate of 70 ml/minute. The Tenax-filled trap was thermally desorbed into a GC/MS for analysis (10). Traps collected in series indicated that no breakthrough occurred. Quantification was done using the chloroform ion at 83 amu using a five-point external calibration curve that encompassed the amount of chloroform collected in the air and breath samples. Positive identification of the peak was based on retention time and the mass spectrum.

Breath samples were collected using a portable sampling apparatus described by Raymer et al. (11). This sampler collects primarily alveolar breath. Tenax-filled adsorbent traps were used to concentrate the organic compounds in 0.5–2 L breath. No problem because of water vapor was observed. The traps were thermally desorbed into a GC/MS, achieving a detection limit of 0.3 $\mu g/m^3$ for chloroform in a 2 minute breath sample. The breath samples were collected nominally at 5, 10, 15, 30, 60, 120, and >180 minutes after the subject left the pool area, with the exact time of collection recorded. A breath-sampling duration of 0.5, 1, or 2 minutes was used depending upon the length of time since the exposure. The geometrical mean of the postexposure time at the beginning and the end of each breath sample was used in subsequent kinetic analysis to evaluate the elimination of chloroform from the body via expiration. The breath sample concentration was corrected for the preexposure concentration by subtraction. The relative magnitude of the chloroform exposure by various modes was assessed from the area under the breath chloroform concentration-time curve. The area under the curve was calculated using the linear trapezoidal rule.

No known chloroform exposures, such as the drinking of a chlorinated beverage, occurred while the breath samples were being collected. Physical activity and food intake by the subject that could alter the metabolism and blood flow were not controlled during the sample collection period. However, no strenuous exercises were done.

III. RESULTS

The chloroform water concentration in the swimming pool ranged between 32 and 150 $\mu g/L$, with a mean value and standard deviation of 85 $\pm$ 57 $\mu g/L$. The chloroform air concentration ranged from 23 to 120 $\mu g/m^3$, with a mean value and standard deviation of 87 $\pm$ 44 $\mu g/m^3$. The preexposure breath concentrations were less than 0.5 $\mu g/m^3$ on all but 1 day, when a value of 1.6 $\mu g/m^3$ was measured. The breath concentrations measured and the corresponding time of each measurement following the exposure are given in Table 1.

The breath chloroform concentrations as a function of time after the start of the exposure via different modes (swimming, three occasions, resting in the water, and resting at poolside) are shown using linear and log-linear axes in Figure 1. Each mode of exposure lasted for 30 minutes. For comparison, average breath chloroform concentrations following a series of 10 minute showers with a water concentration range of 5–20 μg/L are also included (12). The area under the breath chloroform concentration-time curve (AUC) values calculated from the beginning of the exposure to 150 minutes later (the latest time at which data were available for all studies) are listed in Table 2. AUC values were 360 μg-minute/m^3 after resting at poolside for 30 minutes, 840 μg-min/m^3 after resting in the pool water for 30 minutes, and ranged from 720 to 1400 μg-min/m^3 after swimming for 30 minutes.

Previous studies of the chloroform breath concentration with time following environmental exposures have shown that a two-compartment model, described by the sum of two first-order exponentials, provides a better fit to the elimination rate of chloroform than a one-compartment model (12,13). These studies were solely for inhalation exposures or 10 minute mixed inhalation-dermal exposures in a shower. The sum of two declining exponentials was used to calculate the intercepts and slopes and associated half-lives of each compartment as well as the total area under the alveolar concentration-time curve for the resting by poolside inhalation exposure (Table 2), but this model did not fit the data for the 30 minute exposure within the pool while either swimming or resting.

IV. DISCUSSION

Swimming in an indoor pool resulted in a chloroform exposure and an internal dose that were measurable in exhaled breath samples collected following the exposure. For all three modes of exposure, swimming, resting in the water, and resting by poolside, the initial entry of chloroform into the body was very rapid, as evidenced by the immediate drop in concentration upon cessation of the exposure. The fall in breath chloroform concentration following dermal contact (either swimming or resting in the pool water) did not follow the pattern seen after showering. It was observed in previous studies that following a shower or an inhalation-only exposure the breath chloroform concentration was described by a sum of two declining exponentials, consistent with a two-compartment model (12,13). This was also observed for the breath samples collected after resting by the poolside (i.e., inhalation only) in this study (Figure 2), and the chlo-

Table 1 Chloroform Breath Concentrations (μg/m³) after 30 Minutes Exposure

Swimming run 1			Swimming run 2			Swimming run 3			At rest in water			At rest poolside		
Delay time (minutes)	Sample duration (minutes)	Breath concentration (μg/m³)	Delay time (minutes)	Sample duration (minutes)	Breath concentration (μg/m³)	Delay time (minutes)	Sample duration (minutes)	Breath concentration (μg/m³)	Delay time (minutes)	Sample duration (minutes)	Breath concentration (μg/m³)	Delay time (minutes)	Sample duration (minutes)	Breath concentration (μg/m³)
Pre	0.5	0.28	Pre	2	<0.3	Pre	2	<0.3	Pre	2.0	1.61	Pre	2.0	0.44
2.5	0.5	15.1	5	1	26.5	4.5	0.3	20.0	4.5	0.5	14.0	2	0.5	11.0
5	0.5	6.74	10	1	18.2	5	0.5	12.2	8	0.5	13.0	5	0.5	5.6
10	0.5	6.21	15	1	16.8	10	0.5	12.3	10	0.5	13.1	10	0.5	7.5
15	0.5	3.00	30	1	8.95	15	0.5	10.4	15	1.0	—[a]	15	0.5	4.7
30	0.5	—[a]	63	1	10.2	30	0.5	12.2	30	1.0	8.92	30	1.0	1.9
60	1	4.90	90	2	2.80	60	1	—[a]	60	1.0	—[a]	46	1.0	1.3
120	1	3.46	155	2	1.62	120	2	5.4	110	2.0	3.22	60	1.0	—[a]
180	1	0.76				180	2	2.8	160	2.0	2.69	120	2.0	0.87
						270	2	2.5	210	2.0	2.18	272	2.0	1.1
									465	2.0	2.02			
Air	—	100	Air	—	—[a]	Air	—	23	Air	—	106	Air	—	120
Water, μg/L	—	150	Water, μg/L	—	140	Water, μg/L	—	72	Water, μg/L	—	33	Water, μg/L	—	32

[a]Sample lost or invalidated.

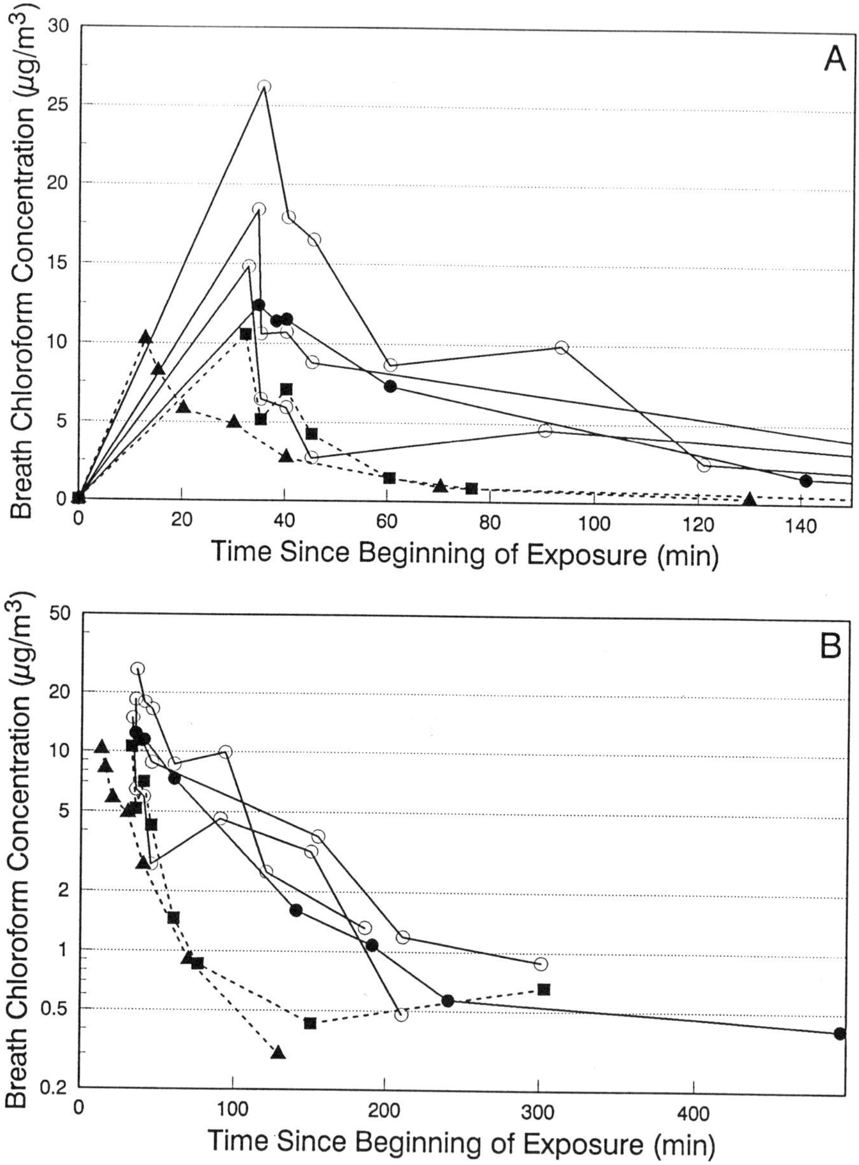

Figure 1 Breath chloroform concentration (μg/m³) as a function of time after exposure via three different modes—swimming (open circles, three occasions), resting in water (closed circles), and resting at poolside (squares)—are shown on linear (A) and log-linear axes (B). Successive points are connected by straight lines (solid where dermal contact with pool water was involved, dashed for others). For comparsion, average data following a 10 minute shower (triangles) are also included.

Table 2 Pharmacokinetic Parameters

	A ($\mu g/m^3$)	B ($\mu g/m^3$)	α ($minute^{-1}$)	β ($minute^{-1}$)	$\tau_{1/2}\alpha$ (minutes)	$\tau_{1/2}\beta$ (minutes)
At rest poolside	8.1	62	0.42	1.3	16	2.5

where $c(t) = Ae^{-\alpha t} + Be^{-\beta t}$

	Swim 1	Swim 2	Swim 3	At rest in water	At rest poolside
AUC[a] to 150 minutes, μg-minutes/m^3	720	1400	1100	840	360

[a]Area under the curve.

roform breath concentrations after swimming (or resting in the water) declined initially, but there was a secondary peak that occurred between 60 and 90 minutes. The curve resembles the shape expected from two routes of entry of chloroform into the body, one with very rapid absorption (inhalation, most likely) and the other with slower absorption (probably transdermal). A compartmental model consistent with the interpretation is shown in Figure 3. Because the data were too scanty to describe adequately the secondary peak, no attempt was made to fit the model to the observed data. The model was used, however, to demonstrate how each route of entry may contribute to the overall shape of the concentration-time curve.

Separate lines for entry of chloroform via the lungs and through the skin, as well as the sum of the two, are shown on linear and log-linear axes in Figure 4. One set of breath chloroform concentration-time data obtained after swimming is also superimposed for comparison with the simulated lines. For the simulated curve, chloroform breath concentrations are initially due to inhalation only. After a lag time of 75 minutes, the transdermal route becomes increasingly important until approximately 150 minutes, after which it is essentially the sole determinant of breath concentrations. Also, for the simulated curve, the extent of exposure via the skin is less than that via inhalation but of the same order of magnitude, as evidenced by the relative areas under the two curves (dashed versus dotted lines in Figure 4A).

The model in Figure 3 adequately describes the present data on chloroform exposure following swimming. Future studies in this area should incorporate more intensive sampling between 1 and 4 h after beginning

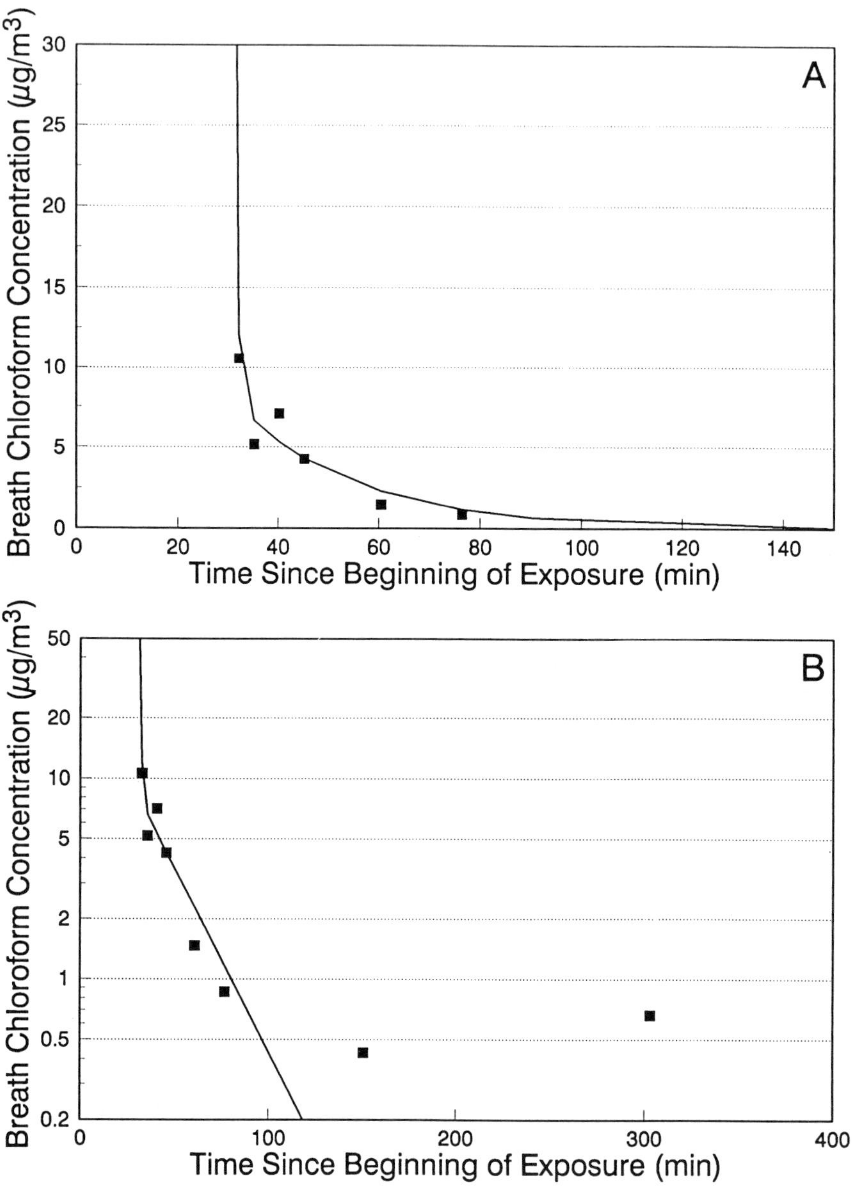

Figure 2 Breath chloroform concentration (μg/m³) as a function of time after resting at poolside for 30 minutes are plotted on linear (A) and log-linear axes (B). The fitted line is included for comparison with the observed data.

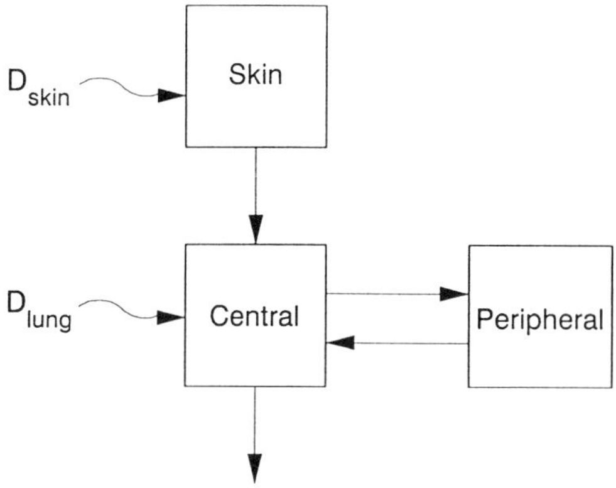

Figure 3 A compartmental model consistent with chloroform absorption and disposition following exposure via swimming or resting pool water. Two-body compartments (central and peripheral) are consistent with the biexponential decline seen following inhalation exposure. Input via inhalation is instantaneous (dose D_{lung}, directly into the central compartment). Input via the skin (dose D_{skin}) is first order with an associated time lag. Distribution to and from the peripheral compartment and elimination from the central compartment are first-order processes.

the exposure. This would allow further refinement of the model and quantitative determination of the associated parameter values.

Because the data did not warrant fitting a specific model, it was not possible to calculate a total area under the curve (i.e., extrapolated beyond the sampling time to infinity). Thus, to compare the relative magnitude of chloroform exposure by the various modes, the AUC was calculated to the same time (150 minutes after the beginning of the exposure) for each data set. Resting at poolside resulted in the smallest extent of exposure, with swimming or merely resting in pool water approximately two- to fourfold higher. There was no discernible difference in exposure between resting in pool water or swimming. It should be noted, however, that more accurate estimates of the total chloroform internal dose will only be obtained when the later portion of the breath concentration-time curve is better defined by more intensive sampling, because this will allow extrapolation of infinity to obtain the total area under the curve. Also, the total number of exposures was small and the chloroform water and air concentrations varied, and thus the actual exposures varied, compounding the difficulty in comparing the results among the different modes.

The profile of chloroform breath concentration following a shower more closely matches the profile after the inhalation exposure than the profile

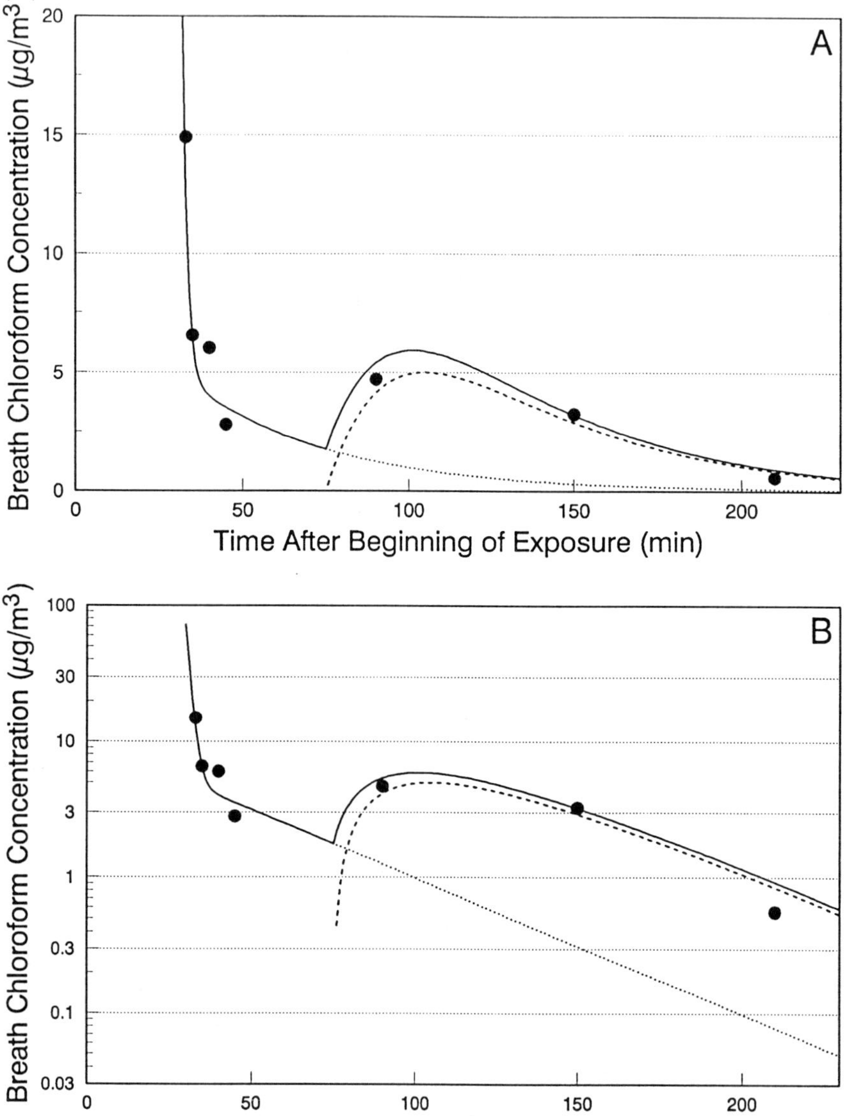

Figure 4 Simulated breath chloroform concentration-time curves resulting from inhalation (dotted line) and transdermal (dashed line) entry of chloroform into the body, as well as the sum of the two (solid line). Observed data from the set obtained following swimming (circles) are included for comparison with the simulated data.

following the swimming exposure, even though showering has a mixed inhalation and dermal exposure. If this difference is confirmed in future studies, several explanations exist as to why the dermal route was not as evident in the shower studies as during swimming. The first is that the exposure during showering was of shorter duration and at a lower intensity than exposure during swimming. This would lead to a lower overall penetration of the skin from showering and a smaller effect. A second explanation is a greater dermal exposure-inhalation exposure ratio during swimming than during showering since the air concentrations in the shower and the swimming pool area were similar but the pool water was 2–20 times the shower water concentrations. In addition, the body is completely submerged while swimming but not while showering, even though the body is constantly being sprayed with water. Therefore, the higher chloroform concentration and submergence in pool water increase the importance of the dermal route during swimming. A final explanation could be associated with the physical activity associated with swimming, which increased the circulatory flow and ventilation. Higher plasma chloroform levels were found in agonistic swimmers than nonagonistic swimmers and visitors to swimming pools (7). However, the one-breath profile of a subject who was at rest during the exposure had a shape that closely matched that of the swimmers (Figure 1), and the amount of chloroform expired was within the range seen in swimmers (Table 2).

Based on this discussion it is evident that calculations of the total exposure associated with swimming must consider the dermal contribution. The data set in Table 2 is too limited to derive the portion of the total exposure associated with inhalation exposure and the portion associated with dermal exposure, but the amount expired up to 150 minutes after the single inhalation-only exposure was between one-half and one-fourth of the amount after swimming. Some of the differences could be related to the proximity of the subject to the water since a threefold difference has been observed at a height of several meters above the water (2). However, it is also seen that the chloroform remained at elevated concentrations in the body much longer when there was a dermal exposure in addition to

---

For inhalation exposure the following equation describes the simulated line: $C(t) = 67e^{-0.71t'} + 5e^{-0.023t'}$, where C is chloroform breath concentration (μg/m^3), t is the time since beginning of exposure (minutes), and t' is the time since end of exposure (i.e., $t - 30$). For dermal exposure the following equations were used for simulation: $C(t) = 0$ for $t \leq 75$ minutes and $C(t) = 20[e^{-0.023(t'-45)} - e^{-0.046(t'-45)}]$ for $t > 75$ minutes.

an inhalation exposure. Thus, even similar exposure intensities via different exposure routes could result in different health risks.

A conservative estimate of the chloroform exposure due to swimming in a chlorinated pool was made for comparison with the ingestion exposure associated with chlorinated drinking water. The inhalation exposure to an adult who swims for 1 h/day, 3 days/week in a pool where the chloroform air concentration is 100 $\mu g/m^3$ and assuming a respiration rate of 1 m^3/h would be 300 μg. This would result in a chloroform internal dose of 210 μg, if 70% of the chloroform is absorbed across the lung barrier. If an equivalent amount enters via the dermal route, a total internal dose of 420 μg chloroform occurs each week because of swimming-related activities. These values represent a lower estimate of the internal dose since an active swimmer breathes at a faster rate than 1 m^3/h and is probably subjected to higher chloroform air concentrations than 100 $\mu g/m^3$ when swimming. In addition, a conservative estimate of the dermal-inhalation ratio was chosen based on the elimination rate data presented in Figure 1. The same individual who daily ingests 2 L water with a chloroform concentration of 25 $\mu g/L$, a standard being proposed by the U.S. EPA as the maximum allowable THM level in finished drinking water, would have a weekly exposure from ingestion of 350 μg assuming 100% absorption. The 2 L is an overestimate of the cold water ingestion since it is used by the U.S. EPA to represent the maximum exposure figure to water contaminants. In addition, a daily 10 minute shower also contributes a weekly dose of 180 μg (8). Thus, a lower estimate of the chloroform dose from the swimming is greater than the maximum estimate of the dose from ingestion of chlorinated tap water and more than twice that calculated from showering. Hence, assuming a linear dose-response relationship, any adverse health effect associated with chloroform is expected to increase by more than a factor of 2 for individuals who regularly swim in chlorinated indoor pools.

V. CONCLUSIONS

The chloroform exposure and associated health risk for an individual who swims regularly in chlorinated indoor pools are similar or greater than what is expected from the use of chlorinated tap water. Both dermal and inhalation exposure routes are important contributors to the internal dose during swimming. The elimination of chloroform from the body via exhalation following swimming is mathematically modeled by a two-compartment body model with instantaneous input by inhalation and first-order input with a time lag for the transdermal portion, which increases the duration of chloroform in the body compared with an inhalation-only route.

ACKNOWLEDGMENTS

This work was funded by the NIEHS Center Grant ES05022. The protocols for breath sampling from human subjects was approved by the UMDNJ, Institution Review Board on Human Subjects. The authors thank Sara Polgar for sample collection and analyses and Dr. James Raymer for reviewing this manuscript.

REFERENCES

1. J.A. Beech, R. Diaz, C. Ordaz, and B. Palomeque, Nitrates, chlorates and trihalomethanes in swimming pool water, *Am. J. Public Health 70*(1):79–82 (1980).
2. U. Lahl, K. Batjer, J. van Duszelin, B. Gabel, B. Stachel, and W. Thiemann, Distribution and balance of volatile halogenated hydrocarbons in the water and air of covered swimming pools using chlorine for water disinfection, *Water Res. 15*:803–814 (1981).
3. P. Chambon, M. Taveau, M. Morin, R. Chambon, and J. Vial, Survey of trihalomethane levels in Rhone-Alps water supplies. Estimates on the formation of chloroform in wastewater treatment plants and swimming pools, *Water Res. 17*:65–69 (1983).
4. G. Aggazzotti and G. Predieri, Survey of volatile halogenated organics (VHO) in Italy. Levels of VHO in drinking waters, surface waters and swimming pools, *Water Res. 20*(8):959–963 (1986).
5. D. Barnes, P.A. FitzGerald, and H.B. Swan, Catalysed formation of chlorinated organic materials in waters, *Water Sci. Technol. 24*(2):59–63 (1989).
6. U.S. EPA, *Identification and Evaluation of Waterborne Routes of Exposure from Other Than Food and Drinking Water*, 1979.
7. G. Aggazzotti, G. Fantuzzi, P.L. Tartoni, and G. Predieri, Plasma chloroform concentrations in swimmers using indoor swimming pools, *Arch. Environ. Health 45*(3):175–179 (1990).
8. W.-K. Jo, C.P. Weisel, and P.J. Lioy, Chloroform exposure and the health risk associated with multiple uses of chlorinated tap water, *Risk Anal. 10*(4):581–585 (1990).
9. U.S. EPA, *The Determination of Halogenated Chemicals in Water by the Purge and Trap Method*, Method 502.1, 1981.
10. K.J. Krost, E.D. Pellizzari, S.G. Walburn, and S.A. Jubbard, Collection and analysis of hazardous organic emissions, *Anal. Chem. 54*:810 (1982).
11. J.H. Raymer, E.D. Pellizzari, K.W. Thomas, and S.D. Cooper, Elimination of volatile organic compounds in breath after exposure to occupational and environmental microenvironments, *J. Expos. Anal. Environ. Epidemiol. 1*(4):439–451 (1991).

12. C.P. Weisel, W.K. Jo, and P.J. Lioy, Utilization of breath analysis for exposure and dose estimates of chloroform, *J. Expos. Anal. Environ. Epidemiol.* in press: (1992).
13. J.H. Raymer, K.W. Thomas, S.D. Cooper, D.A. Whitaker, and E.D. Pellizzari, A device for sampling of human alveolar breath for the measurement of expired volatile organic compounds, *J. Anal. Toxicol. 14*(6):337–344 (1990).

7

Percutaneous Absorption of Chemicals from Water Simulating Swimming and Bathing and from Vapor Exposure

Ronald C. Wester and Howard I. Maibach

University of California School of Medicine, San Francisco, California

I. INTRODUCTION

Contamination of ground and surface water and the transfer of contaminants from water into the human body are major concerns. The obvious first effect is that contaminants are ingested by drinking water. However, when the large surface area of skin is exposed to contaminated water by daily bathing and swimming, skin absorption may also be significant. Brown et al. (1) suggested that skin absorption of contaminants in drinking water has been underestimated and that ingestion may not constitute the sole, or even the primary, route of exposure.

The evolution of skin resulted in a tissue that protects precious body fluids and constituents from excessive uptake of water and contaminants in the external environment. The outermost surface of the skin that emerged for humans is the stratum corneum, which restricts but does not prevent penetration of water and other molecules. This is a complex lipid-protein structure that is exposed to contaminants during bathing and swimming.

Industrial growth has resulted in the production of organic chemicals and toxic metals whose disposal results in contamination of water supplies. As a man or woman settles into a tub or pool, the skin, the largest organ of the body (surface area approximately 18,000 cm^2), acts as a lipid sink (stratum corneum) for the lipid-soluble contaminants. Skin also serves as transfer membrane for water and whatever contaminants may be dissolved in it.

Chemicals can also transfer from vapors to skin and are subsequently systemically absorbed. The first impulse is to consider only respiratory exposure. As Johanson and Boman (2) have shown (discussed later), however, dermal exposure may be as significant.

Percutaneous absorption implies transfer of agents through the epidermis, the deeper layers of skin, and the general circulation in vivo, which includes both transport through the skin and local clearance. The first part of the process is accomplished by simple diffusion across the skin. Both in vitro and in vivo methodology can be used (3,4).

II. BINDING TO HUMAN STRATUM CORNEUM

Table 1 demonstrates the binding and partitioning of various chemical contaminants between water and powdered human stratum corneum following contact for 30 minutes (simulating a short swim or soak in a tub) (5). Lipid-soluble chemicals, such as alachlor and polychlorobiphenyls (PCB), readily partition out of water and bind to stratum corneum. Benzene partitions between air, water, and skin. Nitroaniline, a more water-soluble chemical, shows some linear binding over a 10-fold concentration range. Thus, a variety of contaminating chemicals in water are attracted or bound to skin (Figure 1). The other factor involved in bathing is the use of soap. Figure 2 shows that alachlor bound to powdered human stratum corneum partitions back into a water matrix if the water contains soap (6).

Table 1 Partitioning of Chemical Contaminants Between Water and Powdered Human Stratum Corneum

Chemical	% Dose partitioned to skin
Benzene	16.6 ± 1.4
54% PCB	95.7 ± 0.6
Nitroaniline, 0.49 μg/ml	3.9 ± 1.5
Nitroaniline, 1.6 μg/ml	3.7 ± 0.7
Nitroaniline, 4.9 μg/ml	2.5 ± 1.1
Alachlor	91.8 ± 0.1
Water wash 1	90.1 ± 0.08
Water wash 2	88.7 ± 0.05
Water wash 3	87.7 ± 0.02

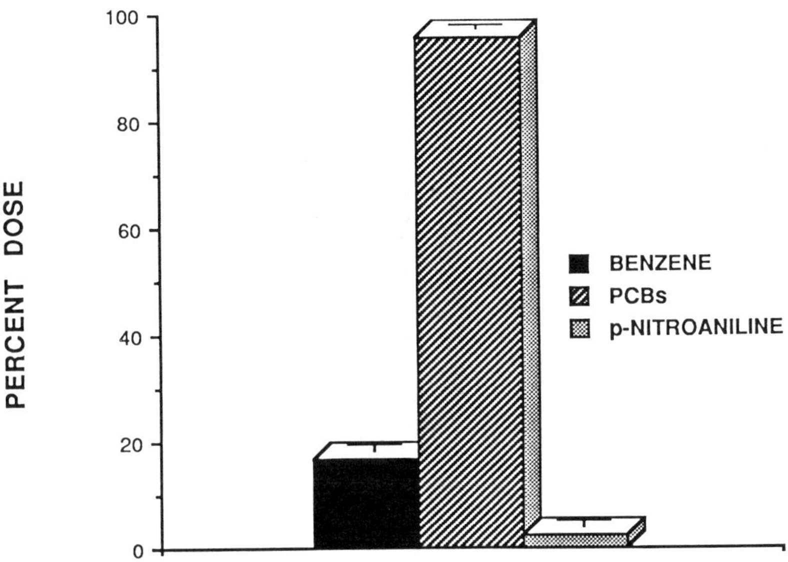

Figure 1 Chemicals partition out of water and bind to powdered human stratum corneum. This partitioning varies for different chemicals.

A. In Vitro

Table 2 shows the in vitro human skin absorption and distribution of nitroaniline (10-fold dose response) and 54% PCB. The amounts in the skin or systemic absorption for either the more water-soluble nitroaniline or the more lipid-soluble 54% PCB are considerable (5–12%) after only a 30 minute exposure. Since the percentage doses of nitroaniline in skin and absorbed through skin are the same for the 10-fold dose range, we must assume no dose dependence for absorption of nitroaniline from water over this dose range (Figure 3). Figure 4 shows the same type of in vitro data for benzene and includes comparisons for single versus three-time water exposures. Not only is benzene absorbed out of water by human skin, but multiple exposure (three times) is at least cumulative, and more likely multiple dosing results in enhanced absorption. With multiple dosing, more benzene is kept in contact with skin (evaporation continues but benzene available for absorption is replaced with the multiple dosing). Water exposure may be longer than 30 minutes, for example for children at the swimming pool who are training for the Olympics. Table 3 shows the

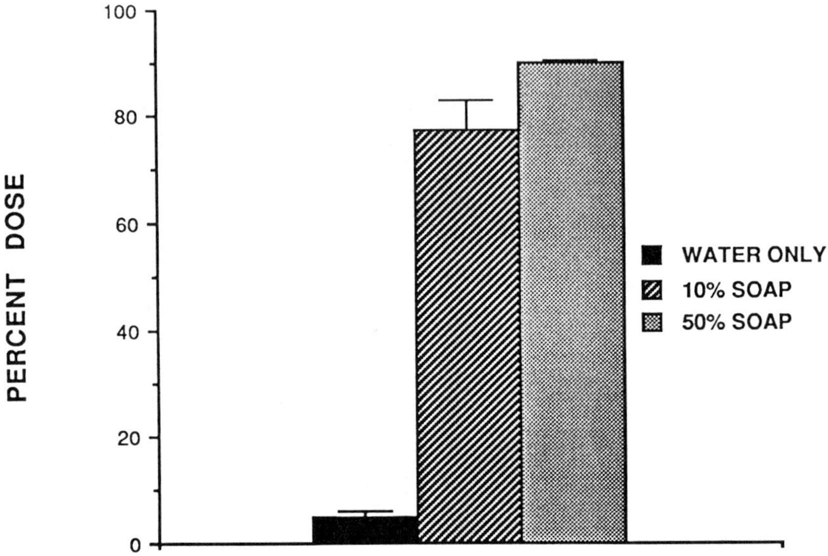

Figure 2 Alachlor binds to human powdered stratum corneum and is not washed away with water. Soap in water reverses the binding.

absorption of alachlor from water. Over time, the amount entering the skin and absorbing into plasma increases (recovery in water surface decreases: accountability remains constant).

B. In Vivo

In vitro absorption and stratum corneum binding can be confirmed by in vivo studies. Human percutaneous absorption studies in vivo are the preferred method, but some in vivo animal models, such as the rhesus monkey, are suitable substitutes (4). Table 4 shows the in vivo absorption of nitroaniline (30 minute exposure to water solution). The percentage of dose absorbed is similar to that from in vitro and binding studies. Thus, there is some validity to the surrogate systems, at least for nitroaniline. We should not overgeneralize the interpretation of this result, however; more validation is required.

III. WATER DILUTION AND PERCUTANEOUS ABSORPTION

Some marketed formulations of pesticides are diluted with water during use and diluted when they enter the environment (ground and surface

Table 2 In Vitro Percutaneous Absorption and Human Skin Distribution (%) of Nitroaniline and 54% PCB from Water

	Nitroaniline (μg/ml)			54% PCB
Parameter	4.9	1.6	0.49	(1.6 μg/ml)
Percutaneous absorption (systemic)	5.2 ± 1.6	4.4 ± 3.9	5.0 ± 0.80	0.03 ± 0.00
Surface bound/stratum corneum	0.28 ± 0.17	0.31 ± 0.04	0.21 ± 0.03	6.8 ± 1.0
Epidermis and dermis	0.61 ± 0.25	0.4 ± 0.7	2.3 ± 0.6	5.5 ± 0.7
Total (skin/systemic)	6.1	5.1	7.5	12.3
Skin wash/residual	92.4 ± 0.8	88.3 ± 2.3	85.2 ± 1.1	71.2 ± 5.5
Apparatus wash	0.47 ± 0.03	0.02 ± 0.004	0.08 ± 0.01	0.25 ± 0.2
Total (accountability)	99.0	93.4	92.8	83.3

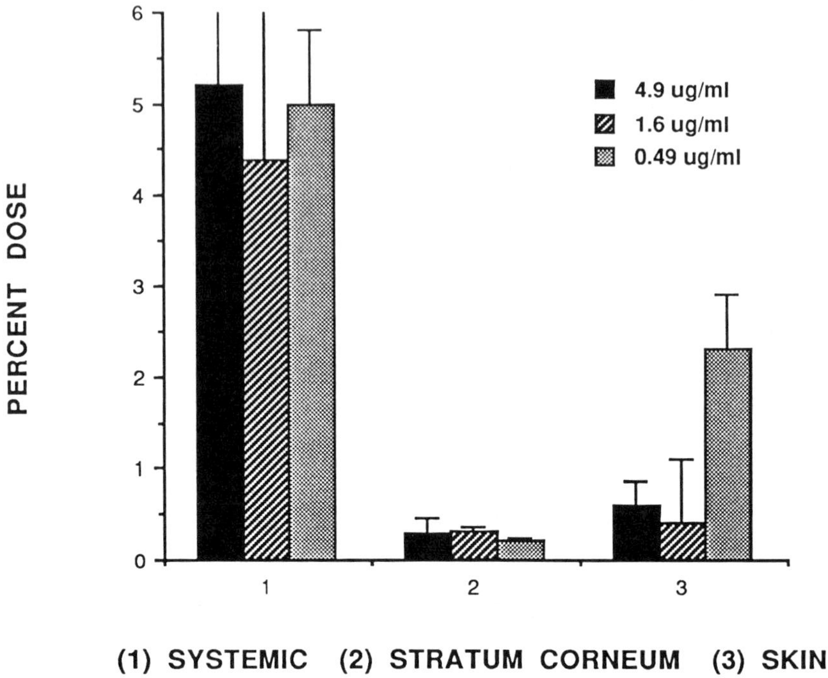

Figure 3 The in vitro human skin absorption of *p*-nitroaniline from water is dose independent over a 10-fold dose range.

water). Changes in formulation or dilution may influence percutaneous absorption.

The first example is dinoseb formulated with Premerge-3 only and with Premerge-3 diluted 1:40 with water (at two concentrations). Neither concentration change nor dilution with water changed in vivo skin absorption in the rhesus monkey (Table 5).

Table 6 shows the in vitro absorption of alachlor. Lasso emulsifiable concentrate diluted with water, 1:20, 1:40, or 1:80, was placed on human skin at two concentrations (0.5 or 1.0 ml over 5.7 cm^2 skin surface area). The data in Table 7 show that with increasing water dilution, the percentage of dose absorbed increased significantly ($p \leq 0.01$). If the mass (μg) of alachlor diffusing into the plasma receptor fluid is compared for theoretical (dilution reduces mass) and actual levels, the skin absorption increases by many times (although overall percentage efficiency remains low compared with total dose applied). It appears that the thermodynamic forces determining alachlor skin absorption change with water dilution (at least in an

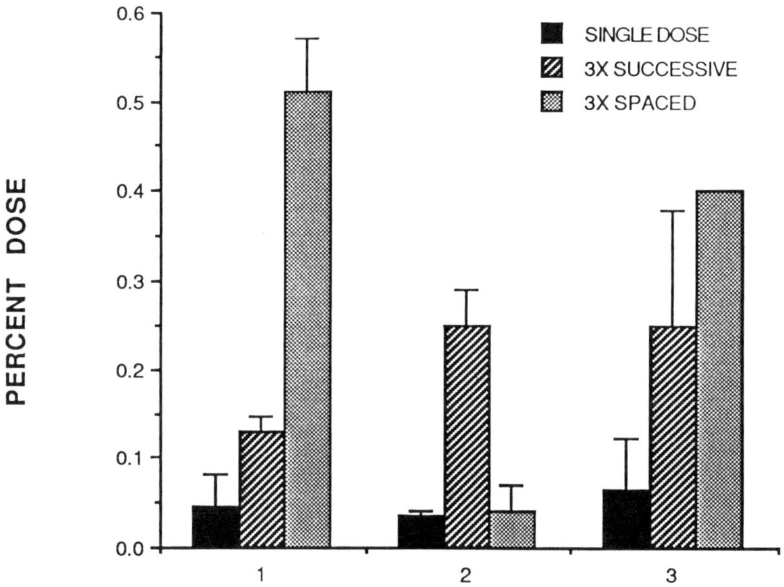

(1) SYSTEMIC (2) STRATUM CORNEUM (3) SKIN

Figure 4 Benzene absorption into and through human skin from water increases with multiple skin applications, probably because of increasing skin contact of benzene.

Table 3 In Vitro Human Skin Absorption and Distribution of Alachlor from Water over Time

	% Applied dose		
Parameter	0.5 h	4 h	8 h
Receptor fluid (plasma)	0.01	0.28	0.82
Skin digest	19.6	30.5	33.1
Skin wash	77.0	61.4	43.3
Apparatus	0.3	1.0	9.0
Total accountability	96.91	93.2	86.2

Table 4 In Vitro Percutaneous Absorption of *p*-Nitroaniline in the Rhesus Monkey Following 30 Minutes Exposure to Surface Water: Comparisons with In Vitro Binding and Absorption

Phenomenon	% Dose absorbed/bound
In vivo percutaneous absorption (rhesus monkey)	4.1 ± 2.3
In vitro percutaneous absorption (human skin)	5.2 ± 1.6
In vitro binding (powdered human stratum corneum)	2.5 ± 1.1

in vitro system using human skin)(6). However, recent in vivo studies with alachlor failed to confirm increased absorption with increased dilution (7,8). The in vitro results are thus probably due to the methodology (9).

Bogen et al. (10) determined the dermal absorption of dilute aqueous chlorform, trichlorethylene, and tetrachlorethylene in hairless guinea pigs. They were able to show dermal absorption of the compounds, and their value for chloroform was very close to one they calculated for chloroform uptake in human volunteers dermally exposed to aqueous chloroform while showering with normal tap water.

IV. PERCUTANEOUS ABSORPTION OF METALS FROM WATER

Cadmium 109 as the chloride salt was dissolved in water and applied to the surface of human skin suspended in a diffusion cell with human plasma as the receptor fluid (8). Percutaneous absorption was determined following 16 h of exposure. Table 7 gives the cadmium content of plasma, skin, surface skin wash with soap and water, and total accountability for four skin sources. Cadmium binds to human skin (Figure 5). Perhaps more disturbing is that some of the cadmium in skin partitioned into plasma

Table 5 Percutaneous Absorption of Dinoseb in the Rhesus Monkey: Formulation Diluted with Water

Dosage formulation	% Dose absorbed (n = 4; 24 h exposure)
High (3.6 mg/cm^2) Premerge-3	5 ± 3
Middle (0.2 mg/cm^2) Premerge-3 and water, 1:40	7 ± 6
Low (0.04 mg/cm^2) Premerge-3 and water, 1:40	5 ± 3

Table 6 In Vitro Percutaneous Absorption of Human Skin: Effect of Dilution on Alachlor Absorption over 8 h

Volume (ml/5.7 cm^2)	Dilution	% Dose plasma receptor fluid	Statistics		Mass (μg) plasma receptor fluid[a]	
			1:20	1:40	Actual	Theoretical
0.5	1:20	0.8 ± 0.1	—	—	90	—
	1:40	2.1 ± 0.4	NS[b]	—	130	45.0
	1:80	3.7 ± 0.7	$p < 0.01$	—	110	22.5
1.0	1:20	0.5 ± 0.1	—	—	110	—
	1:40	1.8 ± 0.1	$p < 0.01$	—	210	55.0
	1:80	3.9 ± 0.3	$p < 0.01$	$p < 0.01$	230	27.5

[a]Actual and theoretical absorption with dilution.
[b]Not significant.

Table 7 In Vitro Percutaneous Absorption of Cadmium Through Human Skin, Percent Applied Dose

Skin source	Plasma	Skin	Wash	Total
Water, 5 μl/cm²				
1	0.5 ± 0.2	8.8 ± 0.6	93.3 ± 3	103 ± 3
2	0.6 ± 0.6	12.7 ± 11.7	74 ± 1	88 ± 20
Water, 2.5 μl/cm²				
1	0.2 ± 0.2	2.4 ± 1.6	86 ± 3	89 ± 2
4	0.1 ± 0.04	3.2 ± 4.4	88 ± 13	92 ± 12

(Figure 6). Thus, the heavy metal cadmium has an affinity for human skin and the percutaneous absorption of cadmium occurs.

The previous data showed that human skin had a high capacity for cadmium. However, the exposure time was 16 h and cadmium exposure from water would be of shorter duration during a swim or when bathing.

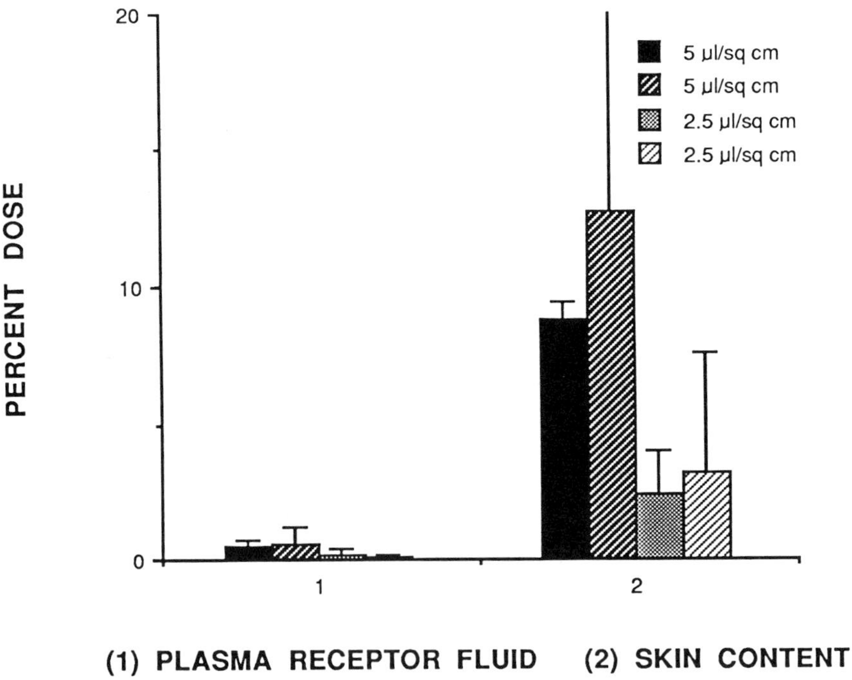

Figure 5 Cadmium absorbs into and through human skin.

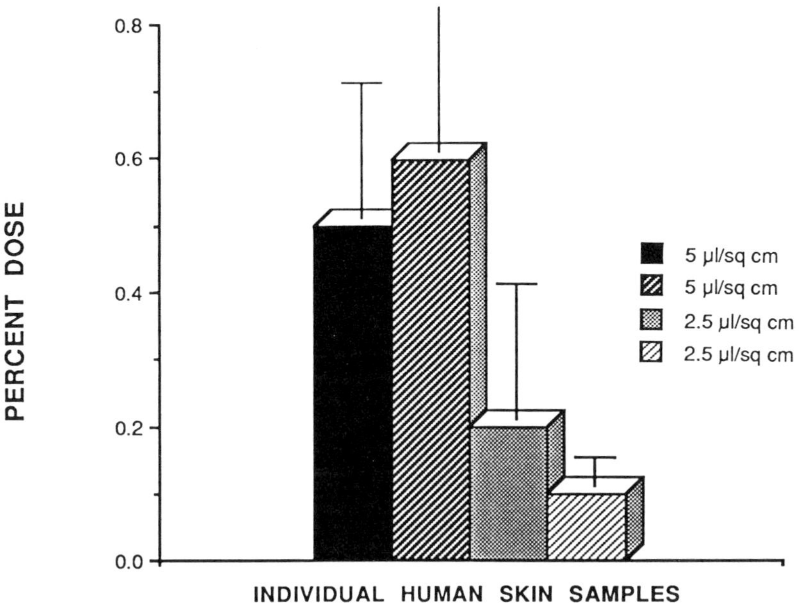

Figure 6 Cadmium in human plasma receptor fluid varies with individual human skin source.

A study was done in which human skin was exposed to cadmium in water for only 30 minutes, followed by skin surface wash with soap and water (Table 8). One-half of the replicates (9 each; 3 human skin sources × 3 replicates) were stopped at 30 minutes. Human skin content was 2.3 ± 3.3%, and no cadmium was detected in the plasma receptor fluid. The other half of the replicates were perfused for an additional 48 h. The skin content was 2.7 ± 2.2%, not significantly different ($P = 0.677$). However, some 0.6 ± 0.8% of the dose had diffused into the plasma receptor fluid ($P = 0.04$). Therefore, cadmium has the ability to bind to human skin during a short exposure in water, is not removed with soap and water wash, and subsequently is absorbed into the body.

Substantial amounts of cadmium exist in soil and water, and humans have daily contact with both these media. Cadmium partitions from both water and soil into human skin and eventually is absorbed into the body. The 0.5% absorbed into plasma receptor fluid over the exposure period is considered small. Human skin has the capacity to concentrate larger

Table 8 Exposure of Cadmium in Water to Human Skin for 30 Minutes Followed by Skin Surface Wash with Soap and Water and then 48 h Perfusion with Human Plasma[a]

Treatment	% Dose	
	Skin content	Plasma receptor fuid
30 Minutes exposure only	2.3 ± 3.3	0.0 ± 0.0
30 Minutes exposure followed by 48 h perfusion	2.7 ± 2.2	0.6 ± 0.8
Statistics	$P = 0.77$	$P = 0.04$[b]

[a]This study simulates a 30 minute exposure of human skin to cadmium in water (swim or bathe) followed by soap and water surface wash. Cadmium is able to bind to human skin in the 30 minute exposure time and then be absorbed into the body during the remainder of the day. $n = 9$ (3 human skin sources × 3 replicates each).
[b]Statistically significant difference.

amounts of cadmium, however, and this cadmium is absorbed slowly and continually over time. Because the body excretion of cadmium is so slow, half-lives of cadmium in the body are corresponding long (17–38 years). Cadmium accumulation over time becomes critical.

Arsenic is another metal found in water. Arsenic is considered a carcinogen. Percutaneous absorption of arsenic 73 from water was determined in vitro in human skin to be 1.9%. The skin absorption of arsenic was also determined in vivo in the rhesus monkey. Absorption from water was 6.4 and 2.0% for low and high doses (Table 9) (11).

Table 9 Percutaneous Absorption of Arsenic 73 from Water into Skin

	Content (% dose)		
	Receptor fluid	Skin	RF + skin
In vitro			
(human skin source)	0.93 ± 1.1	0.98 ± 0.25	1.91

each value is mean ± SD for 24 h exposure
$n = 9$ (3 skin sources × 3 replicates each)

In vivo: in rhesus monkey the percutaneous absorption of arsenic 73 from water was 6.4 ± 3.9% for a low dose and 2.0 ± 1.2% for a high dose, both for a 24 h exposure period

The U.S. Environmental Protection Agency (EPA) publication (12) assumes a default absorption for metals of $K_p^w = 1 \times 10^{-3}$ cm/h. For cadmium the default absorption would be 1.8 μg/cm² and the actual measured value was 0.6 μg/cm². This is a 3-fold difference. With arsenic the default value is 10.1 μg/cm² and the actual absorbed value is 0.04 μg/cm². This is a 250-fold difference.

V. SOAP AND WATER WASH

Bathing (as opposed to swimming) with soap may affect contaminant binding and skin absorption. Figure 2 showed the partitioning of alachlor with powdered human stratum corneum. Alachlor binds to stratum corneum and does not come off with water-only wash. However, 10% Ivory liquid soap in water wash removed 77% of the alachlor, and 50% soap in water removed 90% of the alachlor. This may be useful news for the pesticide applicator or farmer, because washing with soap and water probably removes a large portion of this and other pesticides bound to skin.

Figure 7 shows the in vivo skin decontamination of aroclor 1242 (42% PCB) with soap and water wash. PCB in mineral oil and trichlorobenzene (vehicle constituents of transformers) were applied to skin. Washing the skin with soap and water removed 70% of the surface-bound PCB (0.0

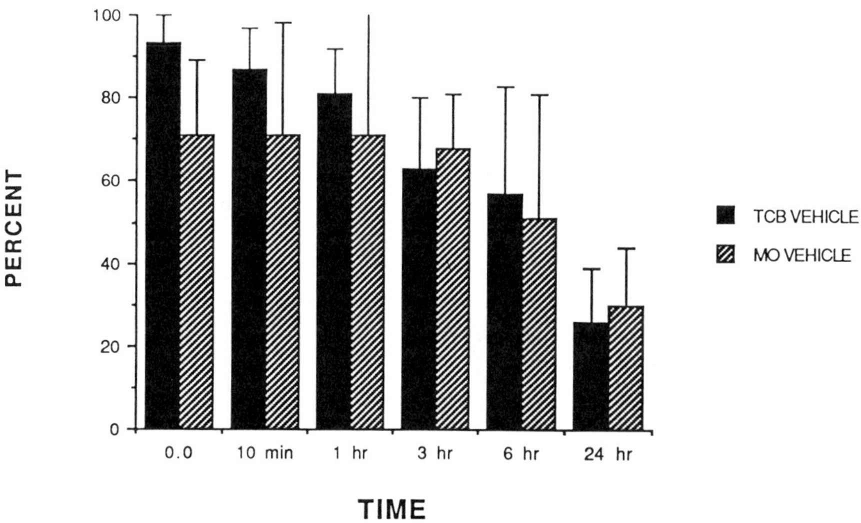

Figure 7 Aroclor 1242 (42% PCB) removal from skin with soap and water wash is time dependent.

time), and this efficiency of removal lasted for 3 h. At 6 h, removal decreased to 50%, and at 24 h only 30% was removed. With time, skin absorption of PCB limited the amount that could be removed by washing the skin with soap and water (13).

VI. PERCUTANEOUS ABSORPTION FROM VAPORS

Vapor absorption through the skin requires that the chemical first be capable of achieving a sufficient concentration in the medium to provide a significant driving force for skin penetration. Many chemicals, because of their low vapor pressure, cannot achieve adequate vapor concentration to pose a dermal exposure hazard. For chemicals that can achieve adequate vapor concentrations, it has been assumed that they are primarily absorbed by the respiratory tract (in the absence of respiratory protection). However, vapor absorption of chemicals through the skin has been demonstrated by a number of researchers (12). Because of the large surface area of the skin, compounds with even small flux values may be extensively absorbed across the skin and result in a systemic chemical burden. The potential for significant contribution to the absorbed dose by dermal exposure to chemical vapors is important to consider, particularly when respiratory protection is provided.

McDougal et al. studied the absorption of dihalomethanes in vivo in rats (14). A dose response of methylene chloride blood concentrations during 4 h dermal vapor exposure was shown. Dibromomethane penetrated fastest, followed by bromochlormethane and methylene chloride. A pharmacokinetic model was developed to predict percutaneous absorption from vapor exposure. McDougal et al. investigated the whole-body dermal penetration of styrene, xylene, toluene, perchlorethylene, benzene, halothane, hexane, and isoflurane in rats and compared the permeability constants with available human studies on vapor penetration (15). Rats with closely clipped fur were exposed to organic chemical vapors (3000–60,000 ppm) while breathing fresh air through a latex mask. Blood concentrations taken during the 4 h exposures were determined by sampling through indwelling jugular cannulas. A physiologically based pharmacokinetic model was used to predict permeability constants consistent with the experimental blood concentrations. Permeability constants (cm/h) were estimated by a least-squares optimization and ranged from 1.75 cm/h for styrene to 0.03 cm/h for isoflurane. Rat permeability constants were uniformly two to four times greater than the human constants, which were calculated from the literature. These results indicate that organic vapor permeability constants in rats are a conservative estimate of organic vapor permeability constants in humans and that the consistent differences in permeability constants between these two species may be due to physiologic differences in the skin.

Johanson and Boman (2) determined the percutaneous absorption of 2-butoxyethanol vapor in human subjects. Four male volunteers were exposed at rest for two periods of 2 h, separated by a 1 h exposure-free interval, to 50 ppm 2-butoxyethanol (BE) vapor generated in an exposure chamber. During the first 2 h period the men were exposed by mouth only via respiratory valve connected by tubes to the exposure chamber. During the second exposure period the men were exposed by skin only while sitting inside the exposure chamber, naked except for shorts and wearing a respiratory protection mask supplied with compressed air. Capillary blood samples were collected at regular intervals and analyzed for BE by a gas chromatographic method. Two experiments separated by at least 2 weeks were carried out with each volunteer, one at "normal" (23°C, 29% relative humidity) and one at raised (33°C, 71% relative humidity) air temperature and humidity in the chamber. The average concentration in blood and the calculated rate of uptake of BE were about three to four times higher during dermal exposure than during inhalation exposure. These experiments suggest that dermal uptake of BE accounts for about 75% (45–85% in individual experiments) of the total uptake during whole-body exposure to BE vapor. Thus it appears that the use of a respiratory protection mask does not protect efficiently against exposure to BE vapors. A tendency toward increased percutaneous absorption rate was seen under the raised temperature and humidity conditions.

Most studies of vapor absorption have addressed the amount of chemical actually absorbed through the skin with potential to cause systemic effects. It is also important to consider whether the presence of a chemical in the skin can damage the barrier function of the skin, cause irritant effects, or cause skin sensitization. Most methods for determining fluxes or permeability constants focus only on the amount of chemical absorbed through the skin. These methods may underestimate the amount of chemical that is actually absorbed and may give no information about the concentration of the chemical in the skin that may cause local effects.

The other major point, as demonstrated by Johanson and Boman (2), is that percutaneous absorption by vapor exposure may be as great or greater than that by respiratory exposure. Health hazard predictions must consider the skin as a port of entry for vapor exposure.

VII. DISCUSSION

We are only beginning to determine the potential health hazards associated with the skin absorption of contaminants from ground and surface water during swimming and bathing and from vapors. Human skin surface areas are large, and human skin has the potential to attract, bind, and absorb significant amounts of contaminants into the body. We have proposed methods (binding to human stratum corneum and in vitro and in vivo percutaneous absorption), and we have reported some findings (linearity of cumulative potential and the influence of soap). Additional information is needed for risk assessment.

REFERENCES

1. H.S. Brown, D.R. Bishop, and C.A. Rowan, The role of skin absorption as a route of exposure for volatile organic compounds (VOCs) in drinking water, *Am. J. Public Health 74*:479–484 (1984).
2. G. Johanson and A. Boman, Percutaneous absorption of 2-butoxyethanol vapor in human subjects, *Br. J. Ind. Med. 48*:788–792 (1991).
3. J.P. Skelly, V.P. Shah, H.I. Maibach, R.H. Guy, R.C. Wester, G. Flynn, and A. Yacobi, FDA and AAPS reports of the workshop on principles and practices of in vitro percutaneous penetration studies: Relevance to bioavailability and bioequivalence, *Pharmacol. Res. 4*:265–267 (1987).
4. R.C. Wester and H.I. Maibach, In vivo animal models for percutaneous absorption, *Percutaneous Absorption*, (R. Bronaugh and H. Maibach, eds.), Marcel Dekker, New York, 1985, pp. 251–266.
5. R.C. Wester, M. Mobayen, and H.I. Maibach, In vivo and in vitro absorption and binding to powdered stratum corneum as methods to evaluate skin absorption of environmental chemical contaminants from ground and surface water, *J. Toxicol. Environ. Health 21*:367–374 (1987).
6. D.A.W. Bucks, R.C. Wester, M.M. Mobayen, D. Yange, H.I. Maibach, and D.L. Coleman, In vitro percutaneous absorption and stratum corneum binding of alachlor: Effect of formulation dilution with water, *Toxicol. Appl. Pharmacol. 100*:417–421 (1987).
7. R.C. Wester, J. Melendres, and H.I. Maibach, In vivo percutaneous absorption and skin decontamination of alachlor in rhesus monkey, *J. Toxicol. Environ. Health 36*:1–13 (1992).
8. R.C. Wester, H.I. Maibach, L. Sedik, J. Melendres, S. DiZio, and M. Wade, In vitro percutaneous absorption of cadmium from water and soil into human skin, *Fund. Appl. Toxicol. 19*:1–5 (1992).
9. R.C. Wester and H.I. Maibach, Human skin binding and absorption of contaminants from ground and surface water during swimming and bathing, *J. Am. Coll. Toxicol. 8*:77–83 (1989).
10. K.T. Bogen, B.W. Colston, Jr., and L.K. Machicao, Dermal absorption of dilute aqueous chloroform, trichlorethylene, and tetra-chloroethylene in hairless guinea pigs, *Fund. Appl. Toxicol. 18*:30–39 (1992).
11. R.C. Wester, H.I. Maibach, L. Sedik, J. Melendres, and M. Wade, In vivo and in vitro percutaneous absorption and skin decontamination of arsenic from water and soil, *Fund. Appl. Toxicol.* (in press) (1993).

12. U.S. EPA, *Dermal Exposure Assessment: Principles and Applications*, EPA/600/8-91/011B, (1992).
13. R.C. Wester, H.I. Maibach, D.A.W. Bucks, M. McMaster, M. Mobayen, R. Sarason, and A. Moore, Percutaneous absorption and skin decontamination of PCBs. In vitro studies with human skin and in vitro studies in the rhesus monkey, *J. Toxicol. Environ. Health 31*:235–246 (1990).
14. J.N. McDougal, G.W. Jepson, H.G. Clewell, III, M.G. MacNaugton, and M.E. Andersen, A physiological pharmacokinetic model for dermal absorption of vapors in the rat. *Toxicol. Appl. Pharmacol. 85*:286–294 (1986).
15. J.N. McDougal, G.W. Jepson, H.G. Clewell, III, M.L. Gargas, and M.E. Andersen, Dermal absorption of organic vapors in rats and humans, *Fund. Appl. Toxicol. 14*:299–308, (1990).

8

A Comparison of Two Pathways to Human for Radon from Drinking Water: Inhalation and Ingestion

C. T. Hess and L. E. Haskell

University of Maine, Orono, Maine

I. INTRODUCTION

Radon 222 is a natural radioisotope created as part of the uranium decay series. As an inert gas, radon can travel freely through porous rock to escape directly into the atmosphere or to dissolve in groundwater supplies. Either route provides numerous paths into private dwellings (see Figure 1). The research discussed here focuses on radon entering through potable water supplies.

Exposure to radon and its short-lived progeny accounts for as much as 55% of the average American's annual radiation dose (1). There is an obvious division of exposure from the water source of radon: ingestion of radon-rich drinking water and inhalation of radon released during household use of such water (see Figure 2). Exposure via the ingestion route involves the stomach and gastrointestinal (GI) tract as well as blood, then the whole body, and ends with exhalation through the lungs. The inhalation exposure follows the liberation of radon from water used during cleaning, dish washing, showering, bathing, and similar activities. The liberated radon mixes with house air, releasing its short-lived progeny. The radon and progeny are inhaled into the lungs on dust particles and colloids. Persons in close proximity to the water use activities are exposed by both the highly concentrated radon and its aerosol progeny and by the average radon level in the house after the radon mixes with the house air.

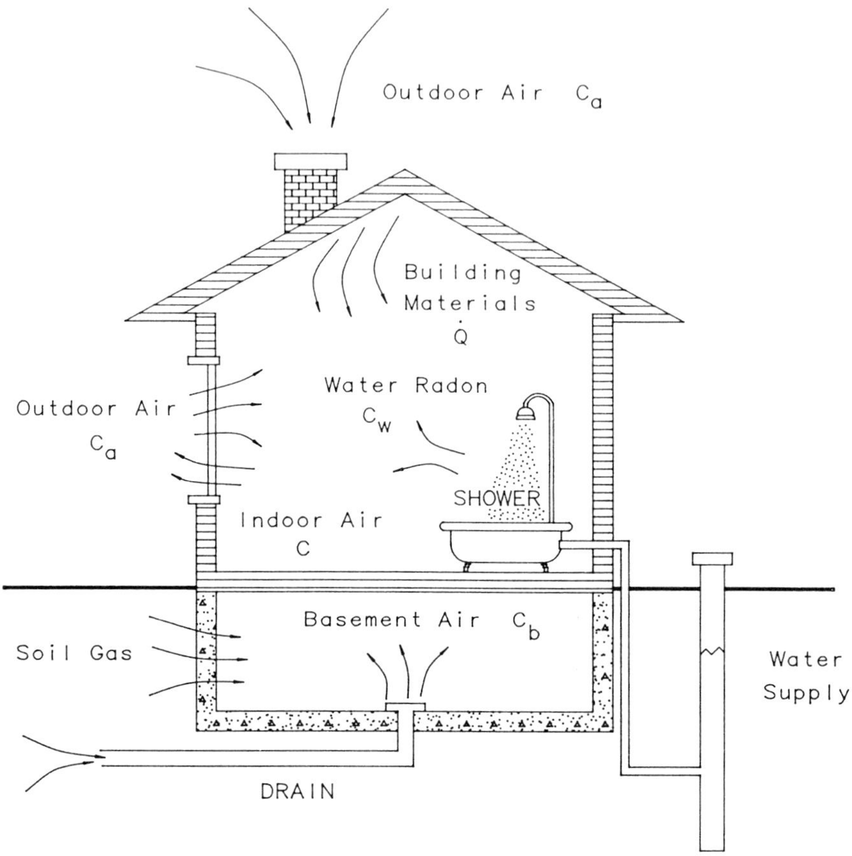

$$\dot{Q} + C_w \dot{V}_w + \dot{V}_a C_a + \dot{V}_b C_b - \dot{V}_a C - \dot{V}_b C = V\, dC/dt + VC\lambda_r$$

Figure 1 House with radon entry points.

The relationship between exposures due to the ingestion and inhalation pathways were compared by Duncan et al. in 1976 (2), who suggested 1 pCi/L in air due to 500 pCi/L in water. An extensive review of previous work done by Cross et al. in 1985 (3) resulted in 1–2.5 pCi/L in air due to 10,000 pCi/L in water. The Dundulis group's (4) calculated ingestion doses in 1984 resulted in 170 mrem/year for 400,000 pCi/L in water. Crawford-Brown (5,6) also calculated ingestion doses, and Harley et al. (7), von Dobelin and Lindell (8), Hursh et al. (9), and Brown and Hess (10) measured ingestion doses. Further comparisons are presented later. In-

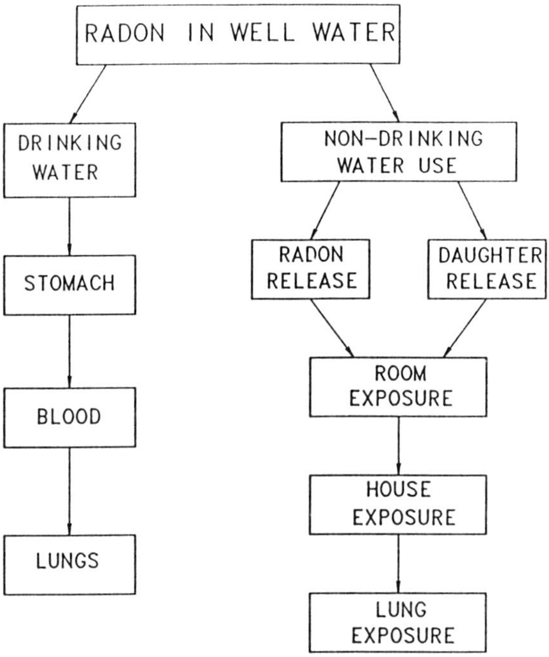

Figure 2 Radon pathway.

halation measurements undertaken by Gesell and Prichard (11) showed the transfer from water to air to be 5.2 pCi/L for water of 10,000 pCi/L. Hess et al. (12), on the other hand, measured 1.07 pCi/L in air for 10,000 pCi/L in water, and Nazaroff et al. (13) calculated an average of 0.65 pCi/L in air for 10,000 pCi/L in water.

The inhalation measurements discussed here were part of a U.S. Environmental Protection Agency (EPA) study of radon exposure from water use in 100 houses in Maine (14), which was carried out from 1987 to 1989. The study simulated 1 day's water use in a 2 h period to generate a large burst of airborne radon. A portable computerized instrument package monitored both the volume of water used and the airborne radon concentration during the 2 h waterburst and the following 22 h. In addition, the waterborne radon concentration and ventilation rate of the house were measured. Analysis of these data allowed us to estimate the portion of airborne radon contributed by the waterburst.

The second study examined the radon exposure from ingestion of radon-rich drinking water. We solicited volunteers with home water radon concentrations greater than 5000 pCi/L. We took measurements in the volunteers' homes when possible and in the laboratory when necessary. In either case, each volunteer drank approximately 500 ml water from his or her home. A flow-through radon detector monitored the radon concentration in the volunteer's breath for 35–40 minutes. We studied these data to find the fraction of the radon that escaped the body and then calculated the exposure to the stomach and the rest of the body.

Despite the different pathways involved, there are some notable similarities between the methods and the results. In part, the similarity is due to the nature of the experiments: both measure fluid flow and the radon concentration carried by that fluid. Radon enters the house during water use, diffuses throughout the building, and then exits by ventilation. Similarly, radon enters the ingesting person, flows through the digestive tract, enters the blood and lungs, and then exits as the person breathes. We measured the radon concentration in household water and the resulting radon in house air over several hours. Likewise, we measured the concentration of radon in the ingested water and calculated the amount of radon in the body over a fraction of an hour. We also measured the ventilation rates of the buildings and the breathing rates of the subjects. These data yield concentration curves showing a large increase in radon concentration followed by an exponential decline with one or more time constants. The theories of radon movement in time involve the mass flow of radon out of the building or person, with some radioactive decay.

II. RADON INHALATION

A. Theory

1. Transfer Coefficient Theory

Radon 222 enters the air in a house along four principal paths: (1) outdoor air at concentration C_a (0.25 pCi/L), which enters the house at a rate $\dot{V}_a$; (2) basement air at concentration C_b (1–150 pCi/L), which enters at rate $\dot{V}_b$; (3) building materials releasing radon at rate $\dot{Q}$ (4) water at concentration C_w (1–100,000 pCi/L), releasing radon at rate $C_w\dot{V}_w$. With these quantities and the following mass balance equation, we can determine the airborne radon concentration C for a house:

$$\dot{Q} + \dot{V}_w C_w + \dot{V}_a C_a + \dot{V}_b C_b - \dot{V}_a C - \dot{V}_b C = V\frac{dC}{dt} + VC\lambda_r \qquad (1)$$

where V is the total volume of the house and λ_r is the radioactive decay constant of radon (0.0076 h^{-1}).

The rate at which radon escapes from water to air can be characterized by a series of bursts occurring at intervals t_i, so

$$\dot{V}_w = \sum_{i=1}^{n} W(t)\varepsilon(t)\,\delta(t - t_i) \tag{2}$$

where $W(t)$ is the quantity of water released during each time interval, $\varepsilon(t)$ is the emissivity of the water radon (i.e., the fraction of radon released into the air) for each type of water use, and $\delta(t - t_i)$ describes the sudden use of water during each time interval t_i.

Integrating Equation (1), dividing by volume of the house V, and using the initial condition $C = C_0$ at $t = 0$, to $C = C(T)$ for $t = T$ and $\Delta C = C(T) - C_0$, gives

$$\Delta C = \frac{C_w W \overline{\varepsilon}}{\overline{\lambda} V T} \tag{3}$$

where the following definitions are used:

$$\overline{\lambda} = \frac{1/T \int_0^T C\lambda \, dt}{\overline{C}} \qquad \overline{C} = \frac{1}{T}\int_0^T C \, dt \qquad \text{and}$$

$$\lambda_a = \frac{\dot{V}_a}{V} \qquad \lambda_b = \frac{\dot{V}_b}{V} \qquad \lambda = \lambda_a + \lambda_b + \lambda_r \tag{4}$$

W_B is the volume of water used during the interval T_B. The average value of the emissivity $\overline{\varepsilon}$ is given by

$$\overline{\varepsilon} = \frac{\sum_i W(t_i)\varepsilon(t_i)}{\sum_i W(t_i)} \tag{5}$$

The emissivity may be calculated using λ_s, the decay constant of the house measured using sulfur hexafluoride (SF_6). Thus Equation (3) becomes

$$\varepsilon_s = \frac{\Delta C \, V \lambda_s T}{W C_w} \tag{6}$$

where W is the volume of water used during time interval T.

Now we define the transfer coefficient as the ratio of the change in airborne radon due to water use to the waterborne radon concentration. Thus the transfer coefficient is given by

$$f = \frac{\Delta C}{C_w} = \frac{W\varepsilon}{V\overline{\lambda}T} \tag{7}$$

To reflect normal water use patterns we need only consider the 22 h interval after the large waterburst. Thus Equation (7) becomes

$$f_h = \frac{(W_T - W_B)\varepsilon_s}{(T_T - T_B)V\lambda_s} \tag{8}$$

where W_T is the total water use over the 24 h experiment, W_B is the water use during the 2 h waterburst, and $T_T - T_B$ is approximately 22 h.

2. Dynamic Theory

To calculate the exposure from the liberated waterborne radon, we use the concept of the working level (WL). A *working level* is defined as any combination of the first four radon progeny in 1 L air that results in the release of 1.3×10^5 MeV of α energy. This definition assumes lung exposure is due only to α emissions from two radon progeny: polonium 218 and polonium 214. To calculate exposure we must first calculate the activity of radon and each of its first four progeny. That polonium 214 has a half-life much shorter than that of the other species in the series makes the calculation easier: we really need only calculate the activities of the first three progeny. Thus, we assume that bismuth 214 simultaneously emits a β particle and an α particle.

We calculate the amount of radon that enters a house through the water supply and decays through its short-lived progeny. In addition, we know radon escapes from the house with a decay constant λ_a. Thus we may write the following expressions for the concentration of radon:

$$\frac{dN_0}{dt} = \frac{C_w\varepsilon}{\lambda_r} \sum_{i=1}^{n} W_i\,\delta(t - t_i) - \lambda_p N_0 \tag{9}$$

If we let

$$U_i = \frac{C_w \varepsilon W_i}{\lambda_r} \tag{10}$$

then the activities of radon and the first three progeny are given by

$$\frac{dN_0}{dt} = \sum_{i=1}^{n} U_i\,\delta(t - t_i) - \lambda_p N_0 \tag{11}$$

$$\frac{dN_1}{dt} = \lambda_0 N_0 - \lambda_1 N_1 \tag{12}$$

$$\frac{dN_2}{dt} = \lambda_1 N_1 - \lambda_2 N_2 \tag{13}$$

$$\frac{dN_3}{dt} = \lambda_2 N_2 - \lambda_3 N_3 \tag{14}$$

where N_i represents the number of atoms present for species i and $\lambda_p = \lambda_r + \lambda_a$, as noted previously ($\lambda_r = 0.0076\ \text{h}^{-1}$).

Each of the four equations may be integrated in turn to solve for the respective N_i:

$$N_0(t) = \sum_{i=1}^{n} U_i\, e^{\lambda_p(t_i - t)} \tag{15}$$

$$N_1(t) = \lambda_0 \sum_{i=1}^{n} U_i \left[\frac{e^{\lambda_p(t_i - t)}}{\lambda_1 - \lambda_p} + \frac{e^{\lambda_1(t_i - t)}}{\lambda_p - \lambda_1} \right] \tag{16}$$

$$N_2(t) = \lambda_0 \lambda_1 \sum_{i=1}^{n} U_i \left[\frac{e^{\lambda_p(t_i - t)}}{(\lambda_1 - \lambda_p)(\lambda_2 - \lambda_p)} + \frac{e^{\lambda_1(t_i - t)}}{(\lambda_p - \lambda_1)(\lambda_2 - \lambda_1)} + \frac{e^{\lambda_2(t_i - t)}}{(\lambda_p - \lambda_2)(\lambda_1 - \lambda_2)} \right] \tag{17}$$

$$N_3(t) = \lambda_0 \lambda_1 \lambda_2 \sum_{i=1}^{n} U_i \left[\frac{e^{\lambda_p(t_i - t)}}{(\lambda_1 - \lambda_p)(\lambda_2 - \lambda_p)(\lambda_3 - \lambda_p)} + \frac{e^{\lambda_1(t_i - t)}}{(\lambda_p - \lambda_1)(\lambda_2 - \lambda_1)(\lambda_3 - \lambda_1)} + \frac{e^{\lambda_2(t_i - t)}}{(\lambda_p - \lambda_2)(\lambda_1 - \lambda_2)(\lambda_3 - \lambda_2)} + \frac{e^{\lambda_3(t_i - t)}}{(\lambda_p - \lambda_3)(\lambda_1 - \lambda_3)(\lambda_2 - \lambda_3)} \right] \tag{18}$$

From the instantaneous values of the N we can calculate the instantaneous exposure in working levels and thus the dose derived from the liberated radon.

B. Methods

1. House Selection

The radon inhalation project sampled 100 homes from a variety of locations around Maine and a variety of circumstances (house size, family size, water radon levels, and so on). Initially, volunteers were solicited from previous experiments on waterborne radon. Once the study began, the local media and the cooperative extension service became interested in the project and helped solicit more volunteers. Volunteers were compensated simply by the free radon testing the project offered.

Initial screening was accomplished by requiring homeowners to send water samples to our laboratory. Homeowners collected water in Teflon-capped trihalomethane bottles and mailed them to the University of Maine.

Duplicate water samples were drawn from these and measured via liquid scintillation. Only sites with more than 5000 pCi/L were accepted since concentrations below this do not permit accurate measurements. This process kept the cost low and removed low radon concentration sites before an expensive field trip was necessary. Ultimately, 101 houses ranging from small summer cottages and house trailers to modern energy-efficient homes from 9 of Maine's 16 counties were tested.

2. Measurement Techniques

Once we selected a site, we made a field trip to install the instruments and perform the experiment. The field measurements followed the procedures described by Hess et al. (12) and Ronca-Batista (15). We made airborne radon and water use measurements during two phases: a 2 h intensive use of water ("waterburst") phase and a 22 h monitoring period phase during which the residents used water according to their normal daily routines. During the waterburst, experimenters simulated activities, such as showering, laundering clothes, or washing dishes. We attempted to use 1 day's supply of water (approximately 0.2 m^3 per resident) during the waterburst. In addition we used the initial 2 h to measure the volume of the house and to collect samples of air and water for later analysis.

A computer recording system monitored the airborne radon concentration and water use volume at 10 minute intervals throughout the entire 24 h period. The system (16) consists of a Ludlum radon 222 detector (Ludlum Instrument Co., 501 Oak St., Sweetwater, TX 79556) interfaced to a Hewlett-Packard HP85 computer (Hewlett-Packard Company, Intercontinental Operations, 3495 Deer Creek Rd., Palo Alto, CA 94304) and associated components (power supply, amplifier, single-channel analyzer, and counter/timer). Radon was measured by radioassay of the daughter products collected by electrostatic precipitation on a Mylar film covering a phosphor. The light scintillations were detected and recorded by a photomultiplier tube and associated electronics.

Use of water during the radon 222 monitoring periods was measured with a Neptune water meter (Neptune Measurement Company, Emerald Rd., Greenwood, SC 29646), which was installed at each residence for this study. The meter produced one pulse per ounce of water while the data acquisition system recorded the signals.

In addition to monitoring air and water continuously, experimenters sampled and recorded various house parameters, including volume, air ventilation rate, and waterborne radon concentration. The volume of each dwelling was determined simply by measuring the dimensions of each room with a steel tape. Ventilation rate and waterborne radon were measured in the laboratory using samples gathered during the waterburst.

The air-exchange rate of each dwelling was determined by the SF_6 tracer technique described by Grot (17,18). Before the simulated water use, SF_6 was released from a small tank and allowed to mix with the house air. During the waterburst, air samples were taken using Tedlar air collection bags (SKC, Inc., 334 Valley View Rd., Eighty-Four, PA 15330-9614). Five samples were collected by pumping air into bags at 30 minute intervals. The concentration of SF_6 in the samples was determined by infrared absorption using a Myran chromatograph (Foxboro Instrument Company, 33 Commercial Rd., Foxboro, MA 02035).

We made two measurements of radon in water for each dwelling: one by the homeowners as part of the screening process and another by our researchers during the waterburst. All measurements were made using liquid scintillation techniques (19,20). We collected tap water samples in 20 ml vials that already contained 5 ml mineral oil. The aerator on a water tap was submersed in a bowl of water. Using a syringe, duplicate samples of 10 ml water were withdrawn from the bowl and then gently injected under the layer of mineral oil in the vial. The samples were transported to the University of Maine for analysis using a Packard liquid scintillation counter (Packard Instrument Co., One State St., Meriden, CT 06450).

C. Results

Figures 3, 4, and 5 show representative examples of the changes in airborne radon concentration over the course of the 24 h experiment. Superimposed on these graphs are bar graphs representing the water use (liters) during each 10 minute interval. The initial intensive water use period is evident in the first 2 h. The miscellaneous readings after the first 2 h represent the normal water use patterns of the homeowners. The apparent raggedness of the data represents the statistical variation inherent in any counting experiment.

In each figure a large peak in radon concentration occurs 1–2 h after the waterburst. The peaks in water use and radon concentration do not coincide because the liberated radon does not instantaneously mix with the house air. Once the water use has returned to its normal rate, the radon concentration slowly decreases as the house is ventilated with outside air. The radon levels decrease at a similar rate because of the similarity in ventilation rates.

The results of all house air measurement experiments are given in Tables 1 and 2. The house code is shown in column 1, the house volume is listed in column 2, the measured ventilation rate by SF_6 is in column 3, the total water used in the 2 h waterburst is in column 4, the elapsed time is in column 5, and the final and initial radon concentrations are given in picocuries/liter in columns 6 and 7. In Table 2, the house code is in column 1, water radon concentration is in column 2, the transfer of radon into air

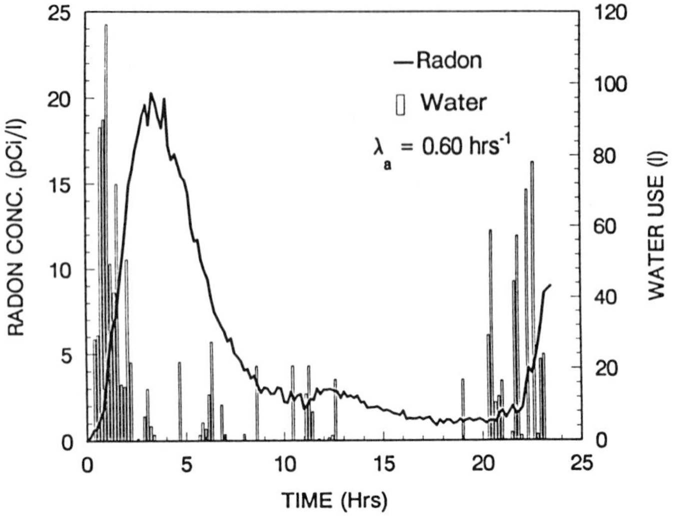

Figure 3 Water use and airborne radon over 1 day in house 500.

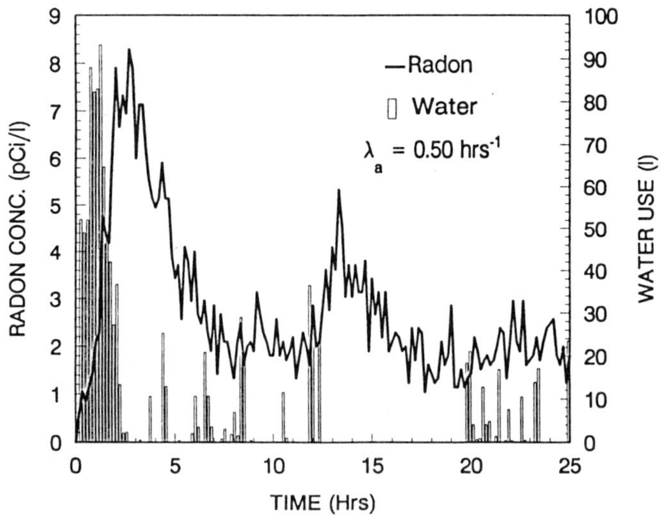

Figure 4 Water use and airborne radon over 1 day in house 780.

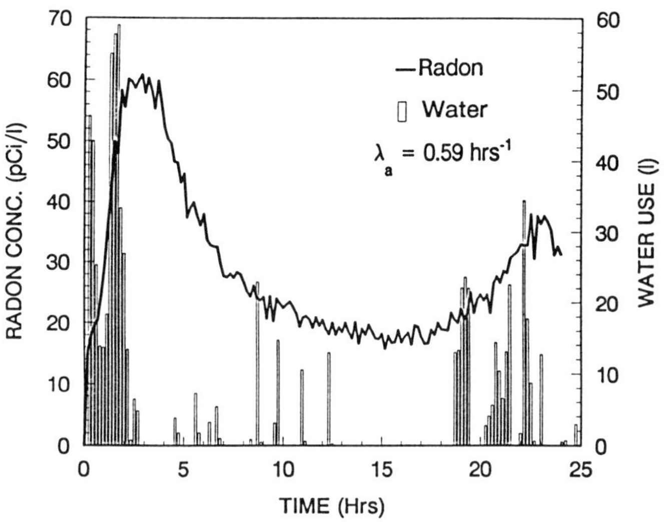

Figure 5 Water use and airborne radon over 1 day in house 1020.

ε_s is in column 3, and the calculated transfer of radon over 24 h is in column 4. Averages for these values are given in Table 3. These results are used to calculate the average dose to an individual from these houses.

From the values of the N_i [Equations (15) through (18)], it is possible to study the dynamics of radon and its progeny. Figure 6 shows the typical situation: radon and progeny start out in equilibrium; then, following a water-burst, each nuclide builds up to a maximum and then decays away. Note that the time delay in the peaks is a consequence of the radioactive decay series: each nuclide begins building up only after its parent has been generated.

The electrostatic collector used to monitor the activity does not respond to radon; rather, it responds to the first and fourth radon progeny. We can model the detector response by summing the values for N_1 and N_3 given by Equations (16) and (18). Figure 7 shows the predicted detector response compared with the measured response. The detector response does not return to zero because the model assumes that a constant concentration of radon enters the house through pathways other than the water supply (i.e., soil gas). In house 120, for example, the constant level was 1.45 pCi/L, resulting in five counts per minute between hours 15 and 25.

By properly weighting sums of N_1, N_2, and N_3 we should be able to calculate the instantaneous exposure in working levels, as shown in Figure 8; however, care must be used in applying this information. Figure 8 ac-

Table 1 House Characteristics

House code	House volume (L)	Waterburst sum (L)	Time (h)	C (T)[a]	C (0)[b]
100	6.17×10^5	817.3	2.33	4.8	0.9
110	1.91×10^5	485.6	2.16	11.4	3.8
120	4.34×10^5	907.9	2.00	16.1	1.5
130	5.23×10^5	927.6	1.83	14.4	4.0
140	9.75×10^5	864.2	2.16	3.6	1.9
150	5.35×10^5	262.5	2.50	2.0	1.0
160	4.72×10^5	749.9	2.00	2.6	1.1
200	2.65×10^5	675.6	2.50	5.1	0.9
210	4.03×10^5	482.5	2.00	5.5	1.4
220	3.58×10^5	612.7	2.00	3.0	1.1
240	3.32×10^5	368.4	2.16	7.4	2.6
250	5.36×10^5	447.6	1.66	3.3	2.0
280	2.17×10^6	414.9	1.66	4.3	3.4
290	4.16×10^5	501.8	2.00	1.5	0.1
300	1.66×10^6	609.8	1.66	0.8	0.2
320	4.75×10^5	404.9	2.50	1.3	0.7
330	5.29×10^5	536.2	2.33	20.0	3.4
340	2.71×10^5	808.7	2.33	0.9	0.1
360	1.03×10^6	571.3	2.33	4.0	1.5
370	3.32×10^5	739.0	2.33	17.4	1.2
380	5.78×10^5	514.5	2.16	1.1	0.3
390	8.90×10^5	555.5	2.66	0.1	0.0
400	5.83×10^5	652.9	1.83	3.6	1.5
410	7.71×10^5	476.7	2.33	9.2	2.0
420	2.94×10^5	803.7	3.00	12.1	2.9
440	2.27×10^5	739.5	2.33	0.8	0.2
450	5.24×10^5	516.9	2.00	2.3	0.7
460	5.15×10^5	160.2	2.16	8.4	1.5
480	5.70×10^5	628.8	2.00	14.7	2.2
490	2.27×10^5	698.6	3.16	12.7	0.5
500	4.36×10^5	618.3	2.00	20.0	1.3
520	3.71×10^5	827.8	3.00	12.4	5.4
525	3.97×10^5	459.0	1.66	12.6	1.6
530	1.09×10^6	844.8	2.00	23.4	1.1
540	5.20×10^5	920.9	2.16	8.6	1.4
550	3.43×10^5	631.5	2.16	2.0	1.1
560	2.97×10^5	453.0	2.33	3.1	0.5
570	4.34×10^5	727.9	2.00	9.9	3.0
580	3.31×10^5	291.3	1.67	0.3	0.1
600	5.51×10^5	295.4	2.00	21.7	1.6
610	3.06×10^5	835.9	2.30	5.4	1.1

Table 1 Continued

House code	House volume (L)	Waterburst sum (L)	Time (h)	C (T)[a]	C (0)[b]
620	4.94×10^5	394.9	2.30	2.4	0.5
625	2.83×10^5	837.0	2.30	1.9	0.8
630	3.62×10^5	567.0	1.84	0.5	0.3
635	4.94×10^5	396.0	2.16	3.4	1.2
640	4.44×10^5	232.0	1.84	3.4	2.3
650	8.03×10^5	635.09	2.16	5.7	2.1
680	7.30×10^5	590.0	1.84	11.6	6.5
690	4.21×10^5	810.0	1.84	6.0	0.5
700	3.10×10^5	601.0	2.00	49.6	32.0
710	1.00×10^5	385.0	2.30	8.7	2.8
720	6.63×10^5	682.0	1.84	7.2	2.2
730	7.20×10^5	528.0	2.16	5.1	2.5
740	6.72×10^5	786.0	1.84	14.3	6.7
750	7.82×10^5	411.0	2.84	9.8	4.0
760	7.82×10^5	284.0	2.60	38.8	12.0
770	8.39×10^5	110.0	2.30	6.5	3.3
780	5.42×10^5	737.0	2.60	8.3	1.7
800	4.19×10^5	467.0	2.50	3.6	1.3
810	5.64×10^5	369.0	1.16	5.3	0.7
830	6.80×10^5	331.0	1.50	23.3	4.5
870	1.71×10^5	517.0	2.16	5.8	2.0
880	5.05×10^5	593.0	1.60	2.3	1.7
900	2.14×10^5	421.0	1.16	9.8	2.2
910	6.33×10^5	612.0	1.84	16.5	5.0
920	1.93×10^5	476.0	2.00	0.5	0.3
930	2.10×10^5	342.0	1.30	1.7	0.3
940	2.81×10^5	451.0	2.00	1.9	1.5
950	4.32×10^5	615.0	2.00	5.0	1.8
960	3.09×10^5	581.0	1.84	6.8	3.0
970	4.31×10^5	567.0	2.00	12.6	3.5
980	3.15×10^5	694.0	2.30	2.9	0.3
990	3.64×10^5	527.0	2.16	3.8	0.8
1000	4.03×10^5	484.0	2.00	6.5	1.3
1010	4.28×10^5	388.0	2.00	20.5	12.5
1020	4.48×10^5	417.0	2.30	60.8	17.0
1030	9.39×10^5	392.0	2.50	36.4	15.0
1040	4.28×10^5	420.0	1.50	21.9	15.0
1050	7.08×10^5	353.0	1.50	32.0	18.5
1060	8.69×10^5	435.0	1.84	17.2	9.6
1065	3.64×10^5	429.0	1.84	19.4	12.6
1070	2.72×10^5	470.0	3.00	6.1	1.3

Table 1 Continued

House code	House volume (L)	Waterburst sum (L)	Time (h)	C (T)[a]	C (0)[b]
1080	4.11×10^5	256.0	3.16	9.6	1.4
1090	6.91×10^5	329.0	1.84	5.3	2.2
2000	3.94×10^5	397.0	1.66	12.2	10.1
2010	4.64×10^5	355.0	1.84	22.0	5.8
2020	1.00×10^5	436.0	2.33	19.0	9.6
2030	6.18×10^5	423.0	1.84	10.9	7.7
2040	5.70×10^5	296.0	1.50	21.9	8.7
2050	5.70×10^5	329.0	2.00	28.7	12.7
2060	5.70×10^5	345.0	1.66	12.7	7.9
2070	4.82×10^5	393.0	1.66	17.0	10.9
2090	2.55×10^5	258.0	1.66	10.3	8.5
3000	6.39×10^5	292.0	2.00	3.7	2.7
3020	2.55×10^5	329.0	1.33	10.2	7.0
3030	6.62×10^5	324.0	2.00	11.6	10.2
3060	3.16×10^5	207.0	1.33	5.0	3.0
3070	4.64×10^5	229.0	1.50	12.0	10.6
3080	4.64×10^5	227.0	1.50	11.1	3.4

[a]Concentration of radon at maximum.
[b]Concentration of radon at time 0.

tually represents an upper bound for the exposure. The model as outlined does not account for the removal of progeny from the respiration pathway. The progeny may attach to large aerosols or other nonrespirable surfaces (i.e., walls or furniture). Assuming that water-induced radon and constant soil gas radon are removed at about the same rate, it is clear from Figure 8 that the water-induced radon accounts for a significant exposure. In house 120 the average exposure from water-induced radon is 0.021 WL and the average exposure from the soil gas is 0.014 WL.

III. RADON INGESTION

A. Theory

Radon retention in humans is frequently modeled by a compartment theory, as described by von Dobelin and Lindel (8)

$$R(t) = \frac{1}{q_0} \sum_{i=1}^{n} A_i e^{-\lambda_i t} \tag{19}$$

Table 2 House Characteristics[a]

House code	Water	ε_S	f_h
100	30,200	0.05	6.8×10^{-6}
110	30,500	0.11	2.3×10^{-5}
120	91,600	0.10	2.3×10^{-6}
130	62,200	0.12	4.1×10^{-6}
140	24,700	0.07	5.7×10^{-6}
150	900	2.63	3.8×10^{-4}
160	3,200	0.59	2.3×10^{-5}
200	15,700	0.18	5.2×10^{-5}
210	33,900	0.08	2.1×10^{-5}
220	7,270	0.09	1.0×10^{-5}
240	13,700	0.19	2.7×10^{-5}
250	12,600	0.12	1.5×10^{-5}
280	16,900	0.04	4.3×10^{-7}
290	4,850	0.29	3.0×10^{-8}
300	4,700	0.07	1.1×10^{-6}
320	4,800	0.48	4.8×10^{-6}
330	160,000	0.12	5.3×10^{-6}
340	1,680	0.21	2.8×10^{-5}
360	25,900	0.13	8.0×10^{-6}
370	57,000	0.21	8.7×10^{-6}
380	6,230	0.14	1.1×10^{-5}
390	4,430	0.07	4.2×10^{-6}
400	5,710	0.74	8.4×10^{-6}
410	16,800	0.55	3.2×10^{-5}
420	32,600	0.15	1.5×10^{-5}
440	28,500	0.01	1.9×10^{-6}
450	1,570	1.33	2.6×10^{-5}
460	111,000	0.17	3.9×10^{-5}
480	51,900	0.08	1.0×10^{-5}
490	48,600	0.11	1.3×10^{-5}
500	74,300	0.13	2.5×10^{-5}
520	34,900	0.07	2.9×10^{-5}
525	114,000	0.04	4.2×10^{-6}
530	72,100	0.30	1.3×10^{-5}
540	27,100	0.14	2.5×10^{-5}
550	4,850	0.16	1.1×10^{-5}
560	39,500	0.07	6.0×10^{-6}
570	20,400	0.05	1.5×10^{-5}
580	194	0.85	1.4×10^{-4}
600	193,000	0.20	1.0×10^{-5}
610	5,350	0.20	8.2×10^{-5}
620	4,240	0.13	7.6×10^{-5}

Table 2 Continued

House code	Water	ε_S	f_h
625	344	1.04	1.9×10^{-4}
630	435	0.19	2.4×10^{-5}
635	7,720	0.64	2.2×10^{-5}
640	7,760	0.12	2.8×10^{-6}
650	15,500	0.17	9.4×10^{-6}
680	21,100	0.29	2.0×10^{-5}
690	34,200	0.06	9.7×10^{-8}
700	37,600	0.05	2.8×10^{-5}
710	102,650	0.01	4.5×10^{-6}
720	26,400	0.03	1.1×10^{-5}
730	7,870	0.10	5.6×10^{-5}
740	17,800	0.05	1.6×10^{-5}
750	8,840	1.61	8.4×10^{-5}
760	25,900	1.11	1.1×10^{-4}
770	23,300	0.45	5.5×10^{-5}
780	24,900	0.18	1.8×10^{-5}
800	10,900	0.16	2.6×10^{-5}
810	15,300	0.09	1.7×10^{-5}
830	14,240	1.64	3.3×10^{-5}
870	20,040	0.01	6.5×10^{-6}
880	5,880	0.02	6.9×10^{-6}
900	9,180	0.05	1.9×10^{-5}
910	7,200	0.32	1.5×10^{-4}
920	72,400	0.00	4.6×10^{-7}
930	3,390	0.11	0.0
840	10,500	0.00	4.2×10^{-6}
950	5,040	0.19	1.3×10^{-5}
960	6,170	0.07	3.8×10^{-5}
970	8,160	0.48	5.8×10^{-5}
980	10,700	0.12	1.2×10^{-5}
990	30,500	0.03	4.5×10^{-6}
1010	23,830	0.07	2.2×10^{-5}
1020	48,270	0.29	7.9×10^{-5}
1030	31,950	0.28	3.9×10^{-6}
1040	23,830	0.01	4.1×10^{-6}
1050	38,730	0.03	5.1×10^{-6}
1060	17,580	0.15	2.1×10^{-5}
1065	19,100	0.03	1.2×10^{-5}
1070	3,440	0.83	1.2×10^{-4}
1080	17,330	1.70	0.0
1090	41,980	0.11	4.3×10^{-6}
2000	4,200	0.09	3.8×10^{-5}

Table 2 Continued

House code	Water	ε_S	f_h
2010	32,060	0.23	2.8×10^{-5}
2020	15,060	0.54	6.0×10^{-5}
2030	10,300	0.08	5.6×10^{-8}
2040	27,950	0.10	2.1×10^{-5}
2050	27,950	0.23	2.8×10^{-5}
2060	6,740	0.28	1.0×10^{-5}
2070	35,600	0.05	4.6×10^{-6}
2090	5,437	0.04	5.4×10^{-5}
3000	5,621	0.11	1.3×10^{-5}
3020	5,437	0.04	2.3×10^{-6}
3030	20,577	0.02	4.4×10^{-6}
3060	5,697	0.17	2.6×10^{-5}
3070	30,497	0.02	1.4×10^{-8}
3080	30,497	0.43	1.2×10^{-6}

[a]Including water radon concentration (pCi/L), emissivity, and transfer coefficient.

Table 3 Average Values for 101 Houses

Water concentration	27,493 pCi/L
House volume	4.90×10^5 L
Waterburst volume	509 L
Ventilation rate SF_6	0.740 h^{-1}
Time of waterburst	2.05 h
Time of home water use	24.1 h
Air radon concentration $t = 0$	3.87 pCi/L
Air radon concentration $t = T$	10.58 pCi/L
Emissivity SF_6	0.640
Transfer coefficient 24 hr	$2.85 \times 10^{-5} \pm 0.57$

Although tempting, it is not correct to correlate the individual time constants λ_i to the elimination rates of the various compartments. Rather, the time constants are functions of all the elimination rates. Following this compartment model, Crawford-Brown modeled the transfer of xenon 133 through human organs as a normalized two-term exponential function (5).

In the research carried out by this laboratory, the dose was fitted to a four-parameter mathematical function similar to the function used by Crawford-Brown. The expression used is

$$C(t) = C_0(a_1 e^{-a_2 t} + a_3 e^{-a_4 t}) \tag{20}$$

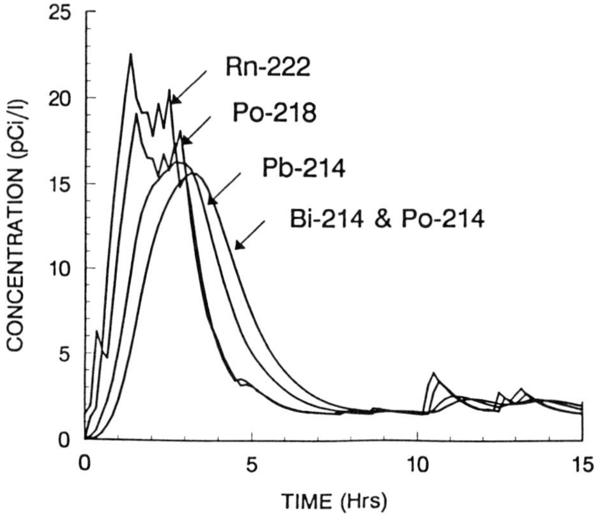

Figure 6 Theoretical radon and progeny buildup in house 120.

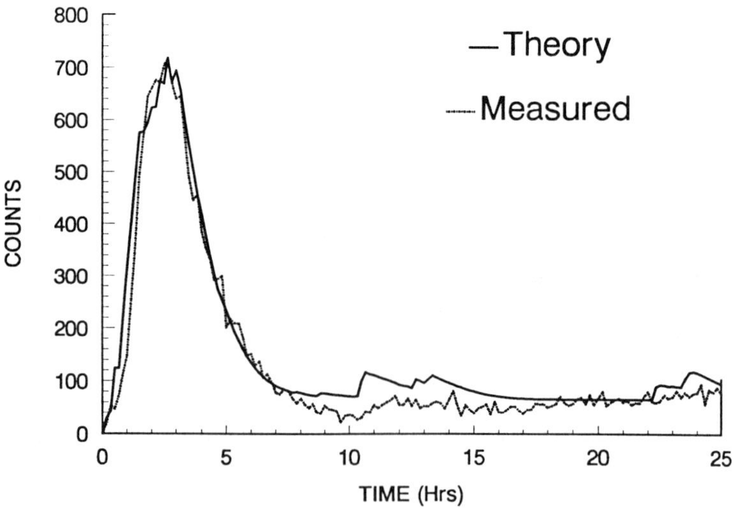

Figure 7 Predicted and measured detector responses over time for house 120.

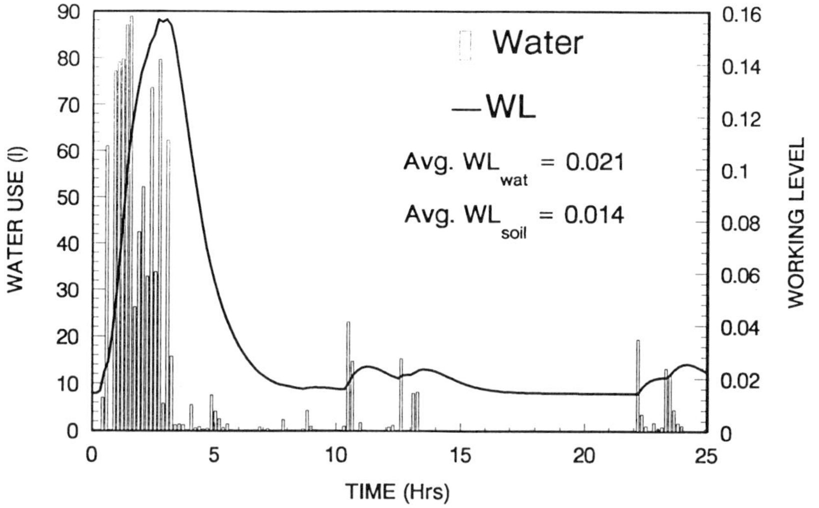

Figure 8 Water use and calculated exposure in house 120.

where C_0 is the ingested amount of radon, a_1 is the amplitude of the first time-constant term, a_3 is the amplitude of the second time-constant term, and a_2 and a_4 are two time constants for the observed radon levels in breath. These parameters were fitted by a mathematical fitting program called Passages (Passages Software Company, P.O. Box 8824, Fort Collins, CO 80525). The two terms in Equation (8) represent a buildup of radioactivity in the body due to absorption through the stomach and a loss from the body through the lungs. The buildup time constant a_2 is used to describe the stomach dose and the loss time constant, and a_4 is used to calculate the whole-body dose. The dose calculation for the stomach uses ingested radon.

B. Methods

1. Volunteer Selection

As with the inhalation study, the University of Maine Cooperative Extension Service provided assistance in locating volunteers. In all, more than 50 volunteers whose homes had waterborne radon concentrations greater than 5000 pCi/L were found. Of these, 41 subjects were tested, resulting in 29 data sets.

Each subject completed a questionnaire detailing such factors as his or her age, height, weight, level of physical activity and smoking history. The

16 men and 22 women ranged in age from 9 to 85 years. Activity levels varied from very inactive to very athletic. Following guidelines established by the University of Maine, parental and adult consent forms were used for all subjects. Again, the only reimbursement was free measurement of airborne and waterborne radon.

2. Measurement Techniques

Breath measurements of volunteers were performed using a modified flow-through scintillation cell designed by Faucher (Radon detection system, Senior Project in Physics, May 1985, University of Maine, Orono, ME 04469), and a ventilation monitoring system. A small component rack (minibin) was fitted with a power supply, high-voltage power supply, preamplifier, amplifier/discriminator, and counter/timer. A Fleisch pneumotach, pressure transducer, and analog-to-digital converter (Hewlett-Packard Pneumotach flow transducer 4730A and A/D converter 47310A, Hewlett-Packard Medical Division, 300 Minuteman Rd., Andover, MA 01810) were used to measure respiratory airflow. An HP-85 computer controlled the scaler and the analog-to-digital converter, which recorded the data. The equipment arrangement appears in Figure 9.

The home water source was run for 10 minutes, after which a measured volume (250–750 ml) of water was collected for ingestion. Subjects ingested this water and then breathed through the breath measurement system, which recorded breath volume and radon concentration for 30–40 minutes. Calibration was done using a weather balloon, as shown in Figure 10. Using the method previously described, duplicate samples of water were collected for later analysis by liquid scintillation counting. The volunteer was monitored for compliance with test procedures. With completion of permitted time, as specified by the human subjects committee or by volunteer discomfort, the experiment was completed. Every effort was made to prevent radon loss during ingestion.

C. Results

Typical results of the radon in breath study are given in Figures 11, 12, and 13 for volunteers 2, 31, and 11. The upper graphs show the radon loss rate in pCi/minute and represent the product of the radon concentration in breath times the volume of air breathed during the same minute. The integral of the radon breath loss rate gives the total picocuries of radon lost. This result is given in the lower graphs. In Figure 12 (top), the double peak is related to the presence of food in volunteer 31 and the transfer of the food along the GI tract. The data for volunteer 11 have a longer time duration (as seen in Figure 13), allowing the longer times for the bioretention curve as well. The projections done after 40 minutes of data in

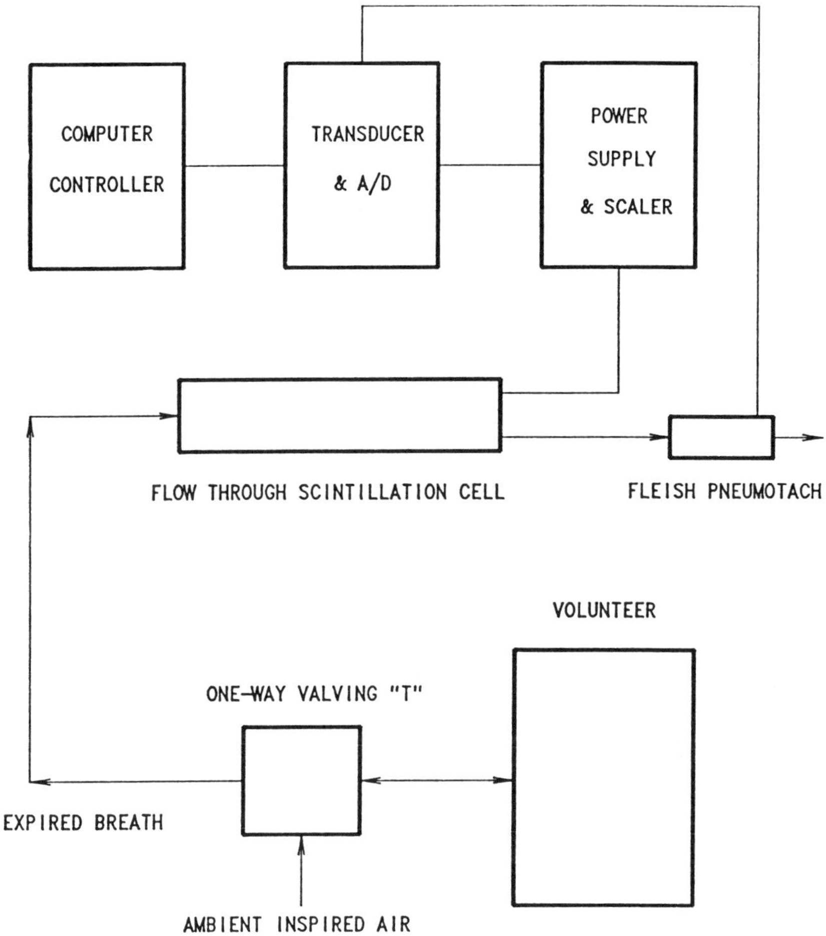

Figure 9 Radon inhalation measurement system.

Figures 11 and 12 represent constant loss of radon after the measurements were stopped by human subjects requirements. This assumption leads to a low estimate of the total dose. The results for the entire set of volunteers are given in Tables 4, 5, and 6. Table 4 shows the volunteer questionnaire information for activity level in column 2, time since eating in column 3, and time since drinking in column 4. The water temperature estimate is in column 5 (room temperature is 20°C and tap water in Maine is at 11°C). Predicted FVC (forced vital capacity) is in column 6, and predicted FEV_1

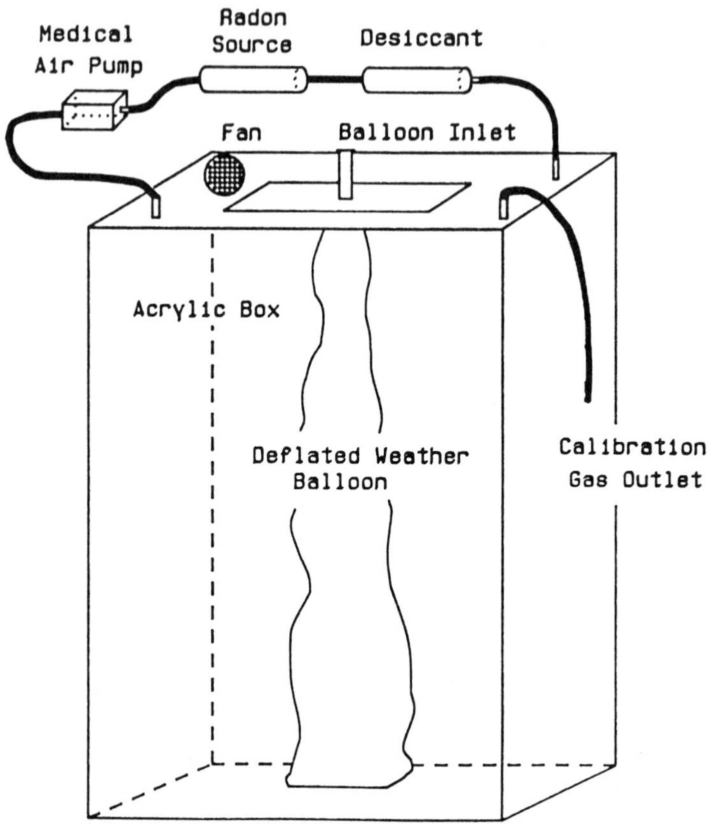

Figure 10 Radon inhalation calibration box.

(forced expiratory volume in one second) is in column 7. The average minute ventilation rate in liters per minute is in column 8, and the amount of water ingested is in column 9.

The results of the volunteer questionnaire are given in Table 5, with volunteer number in column 1, water concentration in Bq/m^3 in column 2, house or laboratory air concentration in column 3, total ingested radon in column 4, volunteer age in column 5, sex of volunteer in column 6, weight of volunteer in kg in column 7, height of volunteer in column 8, and smoking versus non-smoking in column 9.

Table 6 shows the results of the data analysis. The percentage elimination in 30 minutes is in column 2, the fitted half-life is in column 3, and the half-point time observed in the data is in column 4. Whole-body doses

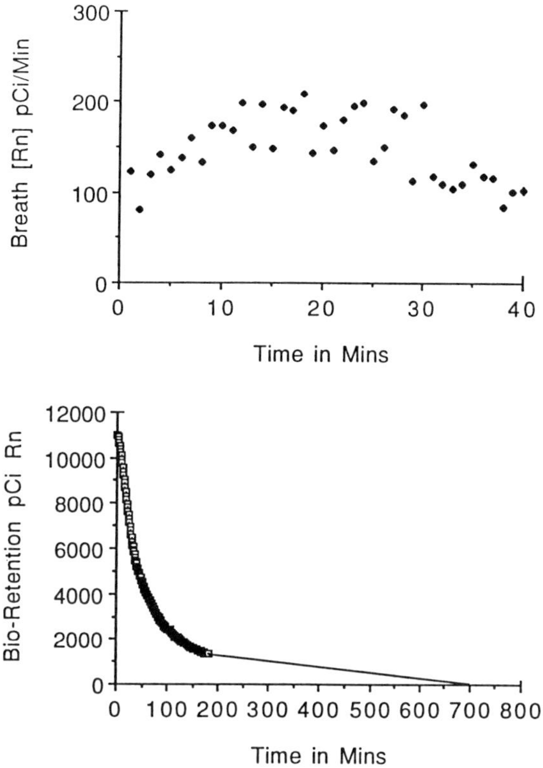

Figure 11 Radon in breath versus time for volunteer 2.

are given in SI units, and the older units are in columns 5 and 6. Standard doses are given in SI units and older units in columns 7 and 8. Averages are shown in the bottom row of Table 6.

A comparison of modeling assumptions for dose calculation is shown in Table 7 and is close to the assumptions of Crawford-Brown.

IV. CONCLUSION

We have seen a remarkable parallel between the ingestion and inhalation pathways: in each case, the measured radon concentration increases rapidly and then slowly decreases to some equilibrium level. Such behavior suggests two or more simultaneous decay processes modeled by sums of exponential decay terms. In each case the amplitude of the radon "pulse" is

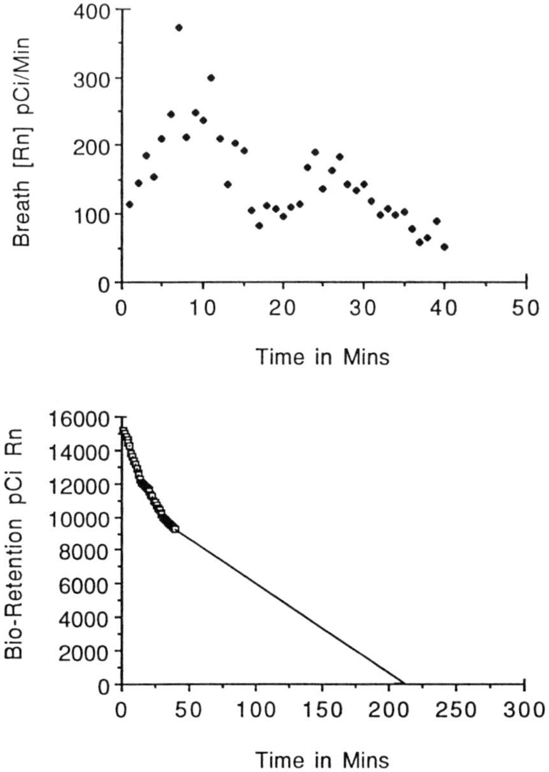

Figure 12 Radon in breath versus time for volunteer 31.

determined by the amount of radon introduced to the system and the ventilation rate of the system.

For airborne radon in houses the decay is dominated by the ventilation rate of the house, and radon is introduced by use of radon-laden water. The model agrees closely with experiment, so clearly the water is a significant contributor to indoor airborne radon. From our measurements the average transfer coefficient for radon from water to air is 2.85×10^{-5}. Our calculations predict an annual lung dose of 0.368 rem/year, assuming a waterborne radon concentration 10,000 pCi/L.

In contrast, the ingestion pathway leads to a somewhat larger dose to the stomach. Here, the decay process is dominated by the breath rate but we are interested in the stomach dose. The "pulse" of breathborne radon lasted between 30 and 180 minutes after ingestion of radon-laden water. Clearly, there are internal rates of transfer from stomach to lung that are

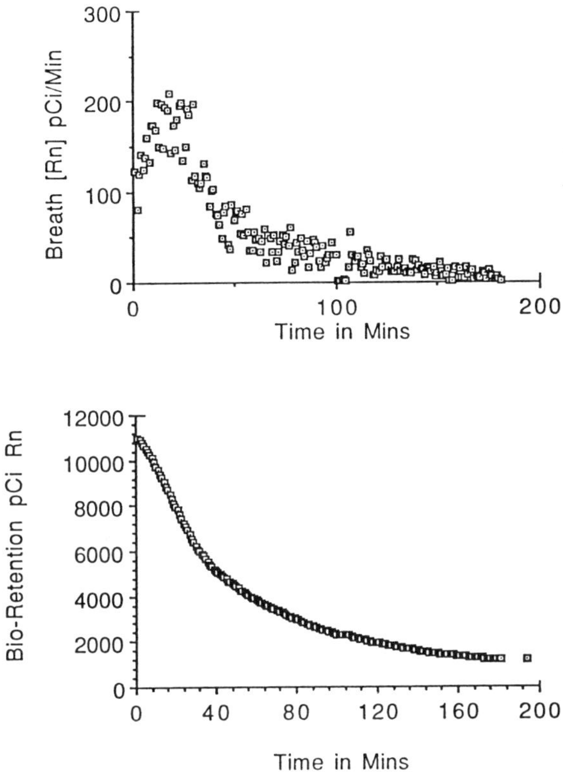

Figure 13 Radon in breath versus time for volunteer 11.

not fully encompassed by our experiments; however, we have determined a stomach dose of 1022 mR/μCi of radon ingested. Using the same 10,000 pCi/L assumption, we arrive at an annual stomach dose of 1.87 rem/year. Although the dose from ingestion is larger than that from inhalation, the cancer risk from stomach exposure is 10^{-5} per rem and it is 2×10^{-5} per rem for lung exposure (21). Thus, the cancer risk is higher from the inhalation pathway than from the ingestion pathway. This finding is agreement with past risk estimates.

Within this work we have examined the fundamentals of radon's pathways from water to humans, and as far as they go, our theories and measurements seem to agree; however, questions remain. This work provides only an upper limit on lung dose. It does not account for progeny depletion through aerosol attachment. Further work is needed to find the actual size

Table 4 Volunteer Questionnaire Information

Volume no.	Activity level	Time since food	Time since drink	Water temperature	Predicted FVC (L)	Predicted FEV_1 (L)	Minute ventilation (L/minute)	Amount water (mm^3)
1	Very	0.00	2.00	Room	4.725	3.688	8.9	500
2	Moderate	3.00	2.00	Room	5.261	3.692	8.4	250
3	Moderate	3.00	3.00	Room	5.045	3.854	8.4	500
4	Very	7.00	4.50	Room	5.067	3.736	23.2	500
6	Very	2.25	2.25	Room	5.067	3.736	21.2	500
7	Moderate	20.00	1.00	Room	6.084	4.648	16.6	500
8	Moderate	1.50	1.50	Tap	3.241	2.475	6.5	250
9	Moderate	12.00	1.00	Tap	5.415	4.084	7.5	500
10	Moderate	3.00	1.00	Tap	3.775	2.472	11.1	250
11	Moderate	12.00	2.00	Room	5.811	4.396	11.2	750
12	Moderate	3.00	4.00	Room	3.534	2.811	6.5	500
15	Moderate	3.00	2.00	Tap	2.708	1.889	5.7	250
16	Moderate	2.00	2.00	Tap	4.080	3.195	13.5	500
17	Inactive	0.00	0.00	Tap	3.534	2.811	10.4	750
19	Very	2.50	2.50	Room	2.600	2.240	6.2	500
21	Moderate	0.00	0.00	Tap	4.191	3.403	11.6	750
22	Moderate	1.50	1.50	Room	5.319	4.156	10.9	750
23	Moderate	3.00	10.00	Room	4.071	3.278	6.2	750
25	Moderate	4.00	4.00	Tap	3.462	2.736	6.4	500
29	Inactive	2.00	2.00	Tap	4.886	3.973	8.6	500
30	Moderate	2.50	1.25	Tap	3.347	2.570	9.2	500
31	Moderate	2.00	2.00	Tap	3.275	2.572	6.8	500
32	Moderate	3.00	3.00	Tap	1.403	1.146	9.1	250
33	Moderate	3.00	6.00	Tap	4.172	3.414	6.5	500
36	Moderate	6.00	0.50	Tap	4.875	3.880	5.8	500
37	Inactive	0.00	0.50	Tap	4.109	3.256	10.6	750
38	Moderate	12.00	1.00	Tap	5.715	4.468	7.4	500
41a	Moderate	5.00	3.00	Room	4.363	3.459	5.2	500
41b	Moderate	0.50	0.50	Room	4.363	3.459	3.3	500

distribution of aerosols in homes. Detailed measurements of the movement of airborne radon within a house would allow more precise determination of exposure for the occupants. Likewise, more detailed measurements and theories of the radon exchange rates from stomach to lung would lead to more precise estimates of stomach and GI tract doses. Doses to other organs could then predicted, giving a more complete picture of the radiation exposures from radon-rich water.

Table 5 Volunteer Questionnaire Information

Volume no.	Water (Bq/m³ × 10⁵)	House (Bq/m³ × 10²)	Amount ingested (Bq)	Age (years)	Sex[a]	Weight (kg)	Height	Smoker[b]
1	1.70	0.629	85.1	38	M	61.4	1.70	N
2	9.71	0.518	243	58	M	88.6	1.88	Y
3	12.7	0.925	635	40	M	75.0	1.75	N
4	5.40	0.481	270	48	M	68.6	1.80	N
	Omitted	0.666						
6	4.82	0.555	241	48	M	62.9	1.80	N
7	10.5	1.04	524	31	M	79.5	1.91	N
8	11.7	1.22	293	48	F	60.0	1.60	N
9	9.73	0.777	487	40	M	79.5	1.83	N
10	4.92	1.15	123	76	M	61.4	1.70	Y
11	5.49	0.592	411	36	M	86.4	1.88	N
12	6.83	0.555	345	31	F	56.8	1.57	N
	Omitted	3.44						
	Omitted	3.70						
15	6.79	2.37	170	75	F	65.9	1.63	N
16	11.0	3.59	552	37	F	90.0	1.73	Y
17	2.36	0.296	177	31	F	64.1	1.57	Y
	Omitted	0.740						
19	3.20	0.888	160	9	M	34.1	1.52	N
	Omitted	0.370						
21	11.5	2.07	865	18	F	65.0	1.65	N
22	13.8	1.33	1035	32	M	76.4	1.78	N
23	7.60	1.22	570	23	F	57.3	1.65	N
	Omitted	0.518						
25	6.64	0.962	332	34	F	53.6	1.57	Y
	Omitted	0.481						
	Omitted	0.370						
	Omitted	1.37						
29	10.1	0.888	503	13	F	56.8	1.77	N
30	8.83	0.888	492	46	F	59.1	1.60	N
31	11.3	1.29	565	37	F	61.4	1.55	N
32	11.3	0.629	282	9	F	22.7	1.25	N
33	11.3	1.29	565	14	F	53.6	1.63	N
	Omitted	0.740						
	Omitted	0.592						
36	13.0	0.444	649	32	M	68.2	1.70	Y
37	13.0	0.481	977	31	F	57.3	1.70	Y
38	11.4	0.555	568	28	M	79.5	1.83	Y
	Omitted	0.222						
	Omitted	4.07						
41a	11.8	3.18	590	30	F	63.6	1.75	N
41b	11.3	3.18	564	30	F	63.6	1.75	N

[a]M, male; F, female.
[b]N, no; Y, yes.

Table 6 Whole-Body and Stomach Doses

Volume no.	% Elimination in 30 minutes	Fit $t_{1/2}$	Half-point (minutes)	Whole-body dose nGy/Bq	Whole-body dose mR/μCi	Stomach dose nGy/Bq	Stomach dose mR/μCi
1	49.4	7.69	30.7	0.986	3.65	102	378
2	64.5	45.3	19.8	0.519	1.92	602	2223
3	43.0	13.0	35.5	0.897	3.32	173	640
4	47.6	7.3	30.7	1.351	5.00	97	359
6	45.1	10.1	34.9	0.419	1.55	134	496
7	68.3	6.1	17.1	0.511	1.89	81	298
8	22.9	11.5	91.9	2.622	9.70	153	565
9	18.0	25.7	104.5	2.195	8.12	34	1267
10	38.9	35.4	38.2	1.059	3.92	737	2727
11	43.5	—	36.3	1.562	5.78	231	857
12	30.8	13.3	73.5	2.300	8.51	176	653
15	35.8	—	40.8	1.070	3.96	260	963
16	25.4	39.0	92.4	1.770	6.55	517	1915
17	49.9	—	37.6	0.995	3.68	240	888
19	12.1	12.3	397.7	19.57	72.40	164	1219
21	27.1	6.6	125.0	3.432	12.70	87	325
22	15.0	17.1	142.4	3.135	11.60	227	841
23	30.9	33.9	65.2	2.049	7.58	451	1671
25	31.4	28.1	74.3	2.487	9.20	374	1385
29	28.3	41.5	75.8	2.487	9.20	661	2449
30	24.2	—	82.8	2.365	8.74	529	1956
31	33.8	18.3	70.4	2.249	8.32	250	926
32	53.2	18.3	27.0	2.522	9.33	412	1527
33	34.2	8.3	104.6	4.216	15.60	140	519
34	47.0	25.1	46.1	3.135	11.60	333	1235
35	24.1	5.3	102.4	3.108	11.50	73	272
36	50.9	5.7	28.7	1.054	3.90	74	281
41a	21.4	8.5	142.9	3.892	14.40	113	419
41b	21.8	7.8	177.4	4.865	18.00	104	387
Average	35.6	22.4	80.9	2.703	10.00	276	1022
± SD[a]		17.3		3.432	17.7	186	691

[a]Standard deviation.

Table 7 Dose Comparisons

	Whole body		Stomach (empty)		Stomach (full)	
Investigator	nGy/Bq	mR/μCi	nGy/Bq	mR/μCi	nGy/Bq	mR/μCi
Andersson	3.78	14				
Crawford-Brown	0.541	2	300	1110[a]		
Hursh			111.4	412	118.9	440
Suomela			129.7	480	205.4	760
von Dobelin	1.081	4	108.1	400[a]		
This work	2.703	10	276	1022[a]		

[a]No distinction was made between an empty and a full stomach.

REFERENCES

1. National Council of Radiation Protection and Measurements, Report 93, 1987, p. 55.
2. D.C. Duncan, T.F. Gesell, and R.H. Johnson, Jr., Radon-222 in potable water, Proc. Health Physics Soc. Tenth Mid-Year Symposium Natural Radioactivity in Mans Environment CONF-761031, Renssaelaer Polytechnic Institute Press, Troy, NY, 1976, p. 340.
3. F.T. Cross, N.H. Harley, and W. Hofmann, Health effects and risks from ^{222}Rn in drinking water, *Health Phys. 48*:649–670 (1985).
4. W.P. Dundulis, Jr., W.J. Bell, B.E. Keene, and P.J. Dostie, Radon-222 in the gastrointestinal tract: A proposed modification of the ICRP Publication 30 model, *Health Phys. 47*:243–252 (1984).
5. D.J. Crawford-Brown, The biokinetics and dosimetry of radon-222 in the human body following ingestion of groundwater, *Environ. Geochem. Health 11*:10–17 (1989).
6. D.J. Crawford-Brown, Cancer fatalities from waterborne radon (Rn-222), *Risk Anal. 11*(1):135–143 (1991).
7. J.H. Harley, E. Jetter, and N. Nelson, *Elimination of Radon from the Body*, Report No. 3, U.S. Atomic Energy Commission, New York Operations Office, 1951.
8. W. von Dobelin and B. Lindel, Some aspects of Rn-222 contamination following ingestion, *Ark. Fys. 27*:531–572 (1964).
9. J. Hursh, D. Morken, T.P. Davis, and A. Loovas, The rate of Rn-222 ingested by man, *Health Phys. 11*:465–476 (1965).
10. W.L. Brown and C.T. Hess, Measurement of the biotransfer and time constant of radon from ingested water by human breath analysis, *Health Phys. 62*:162–170 (1992).
11. T.F. Gesell and H.M. Prichard, The contribution of radon in tap water to indoor radon concentrations, *Natural Radiation Environment*, III.2 (T.F. Gesell and W.M. Lowder, eds.), NTIS, Springfield, VA, 1980, pp. 1347–1363.

12. C.T. Hess, R.E. Casparius, S.A. Norton, and W.F. Brutsaert, *Investigation of Natural Levels of ^{222}Rn in Groundwater in Maine for Assessment of Related Health Effects*, Natural Radiation Environment 111, Houston, TX, U.S. DOE CONF. 780422, *1*:529–547 (1978).
13. W.W. Nazaroff, S.M. Doyle, A.V. Nero, and R.G. Sextro, Potable water as a source of airborne radon-222 in U.S. dwellings: A review and assessment, LBL-18154, 1985.
14. C.T. Hess, M.A. Vietti, E.N. Lachappelle, and J.F. Guillemette, Radon transferred from drinking water into house air, Report to G.F. Craun, Health Effects Research Division, Environmental Protection Agency, Cincinnati, OH 45268, C.R. 813554-01-0, 1989.
15. M. Ronco-Battista, *Interim Indoor Radon and Radon Decay Product Measurement Protocols*, EPA 520/1-86-04, February 1986.
16. C.T. Hess, J.K. Korsah, and C.J. Einloth, *Radon in Houses Due to Radon in Potable Water*, American Chemical Society, New York, 1986.
17. R.A. Grot, A low cost method for measuring air infiltration rates in a large sample of dwelling, Proceedings of ASTM Symposium on Air Infiltration and Air Exchange Measurements, 1978.
18. R.A. Grot, Private communication, National Bureau of Standards, Gaithersburg, MD, *301*:921–3501 (1986).
19. H.M. Prichard and T.F. Gesell, Rapid measurements of radon-222 concentrations in water with a commercial liquid scintillation counter, *Health Phys. 33*:577–581 (1977).
20. T. Horton, *Methods and Results of Study on Radon in Drinking Water*, EPA/EERF Manual 520/5-83-021, 1983.
21. International Commission on Radiological Protection, *Recommendations of the International Commission on Radiological Protection*, Pergamon Press, New York, ICRP Publication 26, 1977.

SUGGESTED READING

Inhalation Measurements

Hess, C.T., J.K. Korsah, and C.J. Einloth. *Radon in Houses Due to Radon in Potable Water*, American Chemical Society, New York, 1986.

Hess, C.T., M.A. Vietti, E.N. Lachappelle, and J.F. Guillemette. *Radon Transferred from Drinking Water into House Air*, Report to G.F. Crann, Health Effects Research Division, Environmental Protection Agency, Cincinnati, OH 45268, C.R. 813554-01-0, 1989.

Hess, C.T., M.A. Vietti, E.N. Lachappelle, and J.F. Guillemette. Radon transferred from drinking water into house air, *Radon, Radium, and Uranium in Drinking Water* (C.R. Cothern and P.A. Rebers, eds.), Lewis Publishers, Chelsea, MI, 1990, pp. 51–67.

Hess, C.T., and W.L. Brown. *Measurement of the Biotransferral Time Constant of Radon from Ingestion Water by Human Breath Analysis*, U.S. EPA Health Effects Research Division, Cincinnati, OH 45268, C.R. 815136-01, 1991.

General

Horton, T. *Methods and Research of EPA in Drinking Water Programs*, EPA/EERF Manual, 78-1, 1978.
International Commission on Radiological Protection *Reference Manual*, Pergamon Press, Oxford, ICRP Publication 23, 1979.
National Commission of Radiological Protection, NCRP Report 93, 1987.
Wrenn, M.E., H. Spitz, and N. Cohen. Design of a continuous digital output environmental radon monitor, *IEEE Trans. Nuclear Sci. NS-22*:645–646 (1975).

9

Unified Approach to Dose and Risk Estimation for All Routes of Exposure to Radon in Water

Douglas J. Crawford-Brown

University of North Carolina, Chapel Hill, North Carolina

I. INTRODUCTION

The U.S. Environmental Protection Agency (EPA) currently is developing a regulatory standard (maximum contaminant limit, MCL) to be applied to radon in public drinking water supplies. The decision to explore regulation arose in part from earlier calculations suggesting that radon represents the single greatest source of risk of cancer in U.S. water supplies (1). These calculations, coupled with a concern for radon in general because of its presence in the air of homes (2), initiated the risk analysis process necessary before consideration of regulatory action. The previous analyses indicated lifetime risk coefficients of between 1×10^{-7} and 1×10^{-6} per pCi/L of radon in drinking water (3–5), with annual U.S. fatalities in the range of 100–200 (4–6).

The earliest risk analyses were based primarily on consideration of exposures resulting from the emanation of radon from the water and into the home air. This route was chosen both because of the existence of epidemiological data on miners suggesting a link between inhalation exposures and the risk of lung cancer and because models for risk from exposure via ingestion were not fully developed. In 1989, data concerning the relationship between water concentration and organ doses became available (7), and mathematical models of the biokinetics and dosimetry of radon in the human body following ingestion were developed (6,8).

These models indicated that the risk from the ingestion pathway for waterborne radon was in the neighborhood of 25–100% of the risk from emanation and subsequent inhalation.

These analyses employed information on average radon concentrations in the home air following emanation. In recent years, however, the regulatory community has begun to focus attention on the dermal absorption of selected environmental pollutants. This has been most true for volatile organic compounds. These compounds have been shown to represent a significant risk through dermal absorption. Given the propensity of radon to diffuse from water, there is a possibility that the dermal absorption of emanated radon represents a source of significant risk. This chapter details calculations of risk from each pathway of exposure to waterborne radon, supplementing previous analyses by developing (1) the mathematical details for the dosimetry of radon absorbed through skin, (2) the risk coefficient (lifetime risk per pCi/L of radon in water) for the absorption pathway, and (3) the contribution of dermal absorption to the total risk from waterborne radon as calculated considering all pathways of exposure.

II. RISK ANALYSIS COMPONENTS

As mentioned, risks from radon arise via several pathways of exposure. Water ingested directly without heating or agitation retains approximately 80–90% of the original concentration of radon in the tap water (9,10). Upon ingestion, the radon is transported throughout the gastrointestinal tract and from the tract into the bloodstream. Once in the bloodstream, the ingested radon is translocated to the various organs and tissues of the body, where subsequent decays lead to the emission of primarily α radiation. The fraction of radon not decaying in the body (as much as 98% of that ingested) enters the airspaces of the lungs and is exhaled with each breath. The risk of cancer is spread throughout the irradiated organs, although not evenly.

Waterborne radon may also be emanated into home air during use of the water. The concentration of radon in home air as a result of emanation from water averages approximately 0.0001 pCi/L in air per pCi/L in water (11–14), although the most recent measurements of this "transfer ratio" indicate that the average may be below 0.0001 by a factor of as much as 2–5 (15). Since the transfer ratio averaged 0.0001 in most past measurements and since this value has been used by essentially all risk analyses for waterborne radon, it is employed in the present discussion.

The risk from this emanation pathway has two components. The first is from direct inhalation of the radon itself, leading to a buildup of radon in the tissues and organs of the body. Mathematical modeling of this exposure pathway requires information on the relative solubility of radon in the various tissues and organs, all irradiated organs being at risk. The second component arises from the decay of radon to its radioactive progeny before inhalation. Airborne radon generally produces a risk significantly lower than that produced by its airborne progeny. The progeny, which are charged solids, attach readily to aerosol particles entrained in the air of a home. When inhalation occurs, the aerosols and their attached progeny are deposited in the passageways of the lung. Here, they either are removed to the stomach through movement on the mucociliary blanket or decay in situ and irradiate the surrounding lung tissue. The resulting risk is one primarily of lung cancer.

Finally, one must consider the absorption of radon through the skin of exposed individuals. Upon absorption of the radon through the skin, translocation of the radon throughout the body should be similar to the biokinetic properties of ingested radon. As in direct ingestion or inhalation of radon itself, dermal absorption should introduce risks to all irradiated organs and tissues of the body.

The total risk from waterborne radon arises from the contributions of these four components: (1) direct ingestion, (2) emanation and inhalation of radon, (3) emanation of radon and inhalation of progeny, and (4) dermal absorption. The following sections provide relatively brief discussions of the mathematical details underlying estimates of the risk posed by each of these separate pathways.

A. Direct Ingestion of Radon

Calculations of the risk from ingested radon require specification of several key factors:

1. Fraction of radon retained in the water during ingestion
2. Concentration of radon in the water at the tap (taken here to be 1 pCi/L)
3. Daily rate of ingestion of water
4. Concentration of radon and progeny in each organ or tissue of the body as a function of time following ingestion (i.e., the biokinetics of radon in the body)
5. Radiation dose rate (rad/day or Gy/day) delivered to each tissue or organ as a result of radioactive decay of the radon and/or progeny
6. Quality factor or relative biological effectiveness of the radiation at inducing cancers
7. Risk coefficient (risk/rad or risk/Gy) for the various forms and sites of cancer

As mentioned previously, approximately 20% of the radon entrained in water is emanated during the act of drawing water from the tap and ingestion. The remainder enters the stomach, from which it is translocated to either the epithelial walls or the small intestines. If translocation is into the intestinal walls, the ingested radon either decays in the epithelial tissue or is translocated further into the bloodstream and other body fluids. The same comment applies to radon entering the small intestines.

Correia et al. (7) measured the translocation of ingested inert gases from the stomach into a selection of organs and tissues of the body. Their measurements depict the concentration of radioactive xenon (an analog for radon) in the stomach, small intestines, ascending colon, descending colon, liver, lung, fat, and "general tissue" (vaguely defined) as a function of time following ingestion. Using this information, a biokinetics model of the movement of ingested radon throughout the adult body was developed (8), as depicted in Figure 1. The translocation rate constants for this model are displayed in Table 1. Table 2 displays the equations relating concentration and time since ingestion as obtained by fitting the model predictions

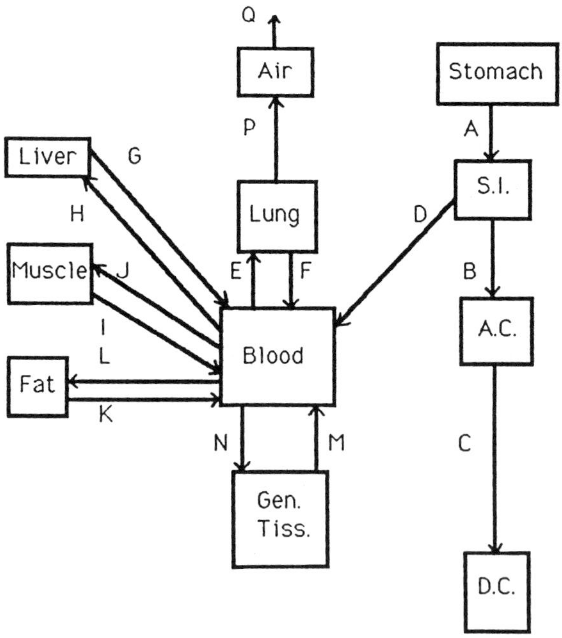

Figure 1 The biokinetics model for radon in the human body as used in this analysis.

Table 1 Transfer Rate Constants Between the Compartments of the Biokinetics Model for Radon in the Human Body

Pathway in Figure 1	Rate constant (per minute)
A	0.04
B	0.005
C	0.005
D	1.5
E	0.16
F	0.4
G	0.4
H	0.13
I	0.0073
J	0.057
K	0.0019
L	0.083
M	0.4
N	1.8
P	0.49
Q	1.74

Table 2 Summary of Concentration Functions for the Radon Biokinetics Model[a]

Organ	Concentration function
Stomach	$C(t) = 0.002(0.998e^{-0.029t} + 0.002e^{-0.0026t})$
Small intestines	$C(t) = 0.0002(0.9e^{-0.02t} + 0.1e^{-0.0045t})(1 - e^{-0.009t})$
Ascending colon	$C(t) = 0.0001(0.85^{-0.015t} + 0.15e^{-0.006t})(1 - e^{-0.099t})$
Descending colon	$C(t) = 0.00003(e^{-0.0048t})(1 - e^{-0.099t})$
Liver	$C(t) = 0.00003(e^{-0.022t} + 0.13e^{-0.0039t})$
General tissue	$C(t) = 0.000005(e^{-0.005t})(1 - e^{-0.099t})$

[a]Concentrations are pCi/ml following ingestion of 1 pCi radon. Times are minutes following ingestion.

with a multiexponential function. In some of the equations, there is an additional term that rises initially and reflects the initial buildup of radon in body tissues during the time when the stomach is the major source of radon entering the bloodstream.

For systemic organs and tissues (i.e., excluding the gastrointestinal tract), the decay of radon is assumed to occur uniformly throughout the affected organ or tissue. Since the progeny possess very short half-lives (16), it also is assumed that all progeny in the radon decay chain through ^{210}Pb decay at the site of the radon decay. The ^{210}Pb atoms are assumed to be translocated to the bone. The dose rate $\dot{D}(t)$ in an organ or tissue at any point in time following ingestion may then be calculated from the equation

$$\dot{D}(t) = C(t) \times 19.9 \times 2.22 \times 1.6 \times 10^{-8} \quad (1)$$

where $C(t)$ is the concentration of radon progeny at time t in pCi/g tissue; 19.9 is the total α energy emitted from the decay of radon and the progeny atoms in MeV; 2.22 is a conversion factor from pCi to decays per minute; and 1.6×10^{-8} is a conversion factor from MeV/g tissue to dose (rad). The units of $\dot{D}(t)$ are then rads per minute.

For the segments of the stomach and small intestines, a slight modification of this equation is necessary. Radon in these two organs diffuses from the intestinal contents (in the lumen) into the epithelial wall and, from there, into the bloodstream via the epithelial capillaries. The concentration of radon in the wall (which contains the radiosensitive cells) falls from the measured concentration in the lumen, as displayed in Table 2, to the much smaller concentration in the bloodstream at the capillary interface. Assumption of a linear concentration gradient with distance from the wall reduces the dose calculated using Equation (1) by a factor of 3. This reduction is assumed to apply in both the stomach and small intestines, at least for the component of dose produced during translocation from the intestinal tract into the bloodstream.

The remaining complication concerns lung tissue. As demonstrated previously (17), doses to the lung from ingested radon have two components. The first, formalized by Equation (1), arises from radon decaying within the cells of the lung tissue during transport through the body tissues. This in situ component is assumed to be spread uniformly throughout the lung, yielding equal dose rates to all parts of the lung tissue. The second component arises from radon translocated from the lung tissue and blood into the alveolar airspaces. From here, the radon is exhaled with each breath.

Before exhalation, however, the radon may decay. The probability of a given radon atom decaying in the airspaces of the alveolar region is the ratio of the mean residence time of the radon in the airspace to the mean lifetime resulting from radioactive decay (approximately 0.26 day, given the radiological half-life of 3.83 days). The mean residence time in the alveolar region is then given by the ratio V_a/V, where V_a is the volume of the alveolar region and V is the alveolar ventilation rate equal to the product of the tidal volume times the breathing frequency. The resulting decay of radon yields direct irradiation of the surrounding tissue by (1) the α particles from radon and/or (2) production of radon progeny that deposit in the airspaces and irradiate the lung. It is assumed that 95% of the ingested radon ultimately enters the airspaces of the lung. Of this quantity, the fraction of radon atoms decaying in the airspaces and producing progeny that also decay in the airspaces is given by the previous ratio V_a/V. The dose D_e delivered to lung tissue from the "exhaled fraction" following ingestion of 1 pCi of waterborne radon is then given by the equation

$$D_e = \frac{0.95 \times V_a \times 2.22 \times 19.9 \times 0.693 \times 1.6 \times 10^{-8}}{3.83 \times 24 \times 60 \times V} \tag{2}$$

where 0.95, V_a, V, 2.22, 19.9, and 1.6×10^{-8} are as defined previously, and 3.83 is the radiological half-life of radon in days. The factors 24 and 60 are required to convert the half-life to units of minutes. This component of the dose from ingested radon is included in all calculations reported in the present discussion.

The doses delivered to organs and tissues of the human body following ingestion of water containing 1 pCi/L at a rate of 0.66 L/day (6) over the course of 73 years (the average human life span) are displayed in Table 3. They incorporate elevations in dose rate at very young ages, as detailed elsewhere (8). The quality factor for α irradiation is taken as 8, which is a value intermediate between the value of 20 recommended by the International Commission on Radiological Protection (ICRP) for radiation protection purposes (18) and the much smaller value found in the radiobiological literature (3). The value of 8 is also employed by the U.S. EPA (19) in their analysis of radon risks. The risk factors for cancers of the various organs considered here, as developed by the U.S. EPA (19) from atomic bomb survivor data, are displayed in Table 3.

The preceding discussion may be used to determine the risk from direct ingestion of water containing radon at a concentration of 1 pCi/L over an expected lifetime of 73 years. This risk is obtained by (1) calculating the dose to each organ over this period of time (Table 3), (2) multiplying this

Table 3 Organ Doses and Organ Risk Coefficients for Ingested Radon[a]

Organ	Organ dose	Organ risk factor
Stomach	1.7×10^{-8}	0.000046
Small intestines	2.9×10^{-9}	0.0000046
Ascending colon	7.4×10^{-9}	0.0000091
Descending colon	4.1×10^{-9}	0.0000091
Lung	4.1×10^{-10}	0.00005
Liver	1.5×10^{-9}	0.00007

[a]Doses are rad per pCi ingested. Risk coefficients are risk per rem.

dose by the quality factor of 8 to obtain the dose equivalent to that organ, (3) multiplying this dose equivalent by the organ-specific risk factor (Table 3), and (4) summing the risk over all organs. The lifetime risk of fatal cancer for the stomach, small intestines, ascending colon, descending colon, lung, and liver is then 1.1×10^{-7}, 1×10^{-9}, 9×10^{-9}, 5×10^{-9}, 2×10^{-9}, and 6×10^{-9}, respectively. The total risk of fatal cancer is then equal to 1.3×10^{-7} per pCi/L of radon in water. It may be noted that this risk is dominated by the risk of stomach cancer.

B. Inhalation of Radon Progeny

Radon emanated into home air produces a concentration of airborne radon dependent upon the size and ventilation rate of the home (20–22). In general, the concentration of airborne radon C_a produced by emanation from water is given by the equation

$$C_a = \frac{C_w E W}{\lambda_v V_h} \tag{3}$$

where C_a is pCi/L of radon in air; C_w is the concentration of radon in the water supply, pCi/L of radon in water; E is the efficiency with which radon is transferred from water to air; W is the rate of water use, L/day; λ_v is the ventilation rate of the home day^{-1}; and V_h is the home volume in L. As mentioned previously, C_a averages approximately 0.0001 when C_w equals 1 pCi/L. This value of C_a is used throughout this discussion of the inhalation route.

Once in the air, atoms of radon either are removed from the home through ventilation [already considered in Equation (3)] or decay to yield radon progeny. These progeny then attach to available aerosols, with some fraction remaining "free" of attachment. It is the radon progeny, rather than the radon itself, that are of primary concern in estimates of the risk of lung cancer following inhalation. The concentration of these progeny in the home air is generally summarized by an equilibrium fraction F, taken here to be the ratio of the total rate of α decays from the radon decay chain in the air to the rate that would be present if all the progeny were in radioactive equilibrium with the radon. In structures, the effects of deposition of progeny onto wall, floor, ceiling, and fan surfaces lowers the equilibrium fraction to a value below unity, typically averaging 0.5 in U.S. homes.

The unit of radioactivity employed in discussions of radon progeny is the working level (WL). One working level results from the presence of 100 pCi radon per liter air, with the progeny in complete radioactive equilibrium. The actual value of WL holding true in a given home is then equal to

$$\mathrm{WL} = \frac{C_a F}{100} \tag{4}$$

where C_a and F are as defined earlier and 100 represents the concentration of radon assumed in the definition of WL. For a radon concentration of 0.0001 pCi, the radon progeny have an average concentration of 5×10^{-7} WL.

The risk from radon progeny is generally assumed to be related to the product of the WL (concentration) and the length of exposure, yielding the quantity working level months (WLM). The length of exposure is specified in "working months," 1 working month being equal to approximately 170 h. The number of working months in a year may be found by multiplying 8760 by an "occupancy factor" representing the fraction of each year spent by an individual in the home and dividing this product by 170. An occupancy factor of 0.75 is employed here (22), resulting in an assumed 39 working months per year or 2820 working months over a 73 year average lifetime. Multiplying this value by the value of 5×10^{-7} WL obtained previously yields an estimate of 0.0014 WLM contributed by the emanation of radon from water containing 1 pCi/L of radon.

The risk from this exposure may then be obtained through multiplication of the lung cancer risk factor (risk per WLM) by the estimate of WLM. Following the recommendations of the U.S. EPA, a risk factor of 350×10^{-6} per WLM for mining populations is adopted here. As indicated in a recent review by the National Academy of Sciences (NAS) (23), however, this risk factor must be adjusted to account for differences in age, breathing patterns, degree of attachment of progeny to aerosols, and aerosol size distribution between the mining and home populations. When the progeny are inhaled, they deposit in the various generations of the lung with an efficiency dependent upon the breathing characteristics and state of attachment to aerosols (24). Once deposited, the progeny are moved to the mucociliary blanket, resulting in either removal to the stomach or decay while in transit. The NAS modeled the mine and home environments and then compared the doses delivered to the epithelial layer of the upper passageways in the lung in deriving this correction factor. Their analyses demonstrated that miners received an approximately 30% larger dose due to inhalation of 1 WLM of progeny than expected in the home environment.

These differences reduce the risk produced by 1 WLM in the home to 70% of that found in the mines, suggesting a risk coefficient of 245×10^{-6} per WLM. Multiplying the previous estimate of lifetime exposure (0.0014 WLM) by this risk coefficient yields an average lifetime risk of 3.5×10^{-7}

for a concentration of 1 pCi/L of radon in the water. This risk is given entirely by the risk of lung cancer, particularly cancers of the tracheobronchial region.

C. Inhalation of Airborne Radon

The third exposure route considered here arises (as in Section II.B) from the emanation of radon from the water to the air of a home. In the present case, however, it is inhalation of the radon atoms themselves rather than the progeny atoms that is of interest. This route has received scant attention in the discussion on radon risks, in part because of an early memo by Bale (25) indicating that it is the progeny that produce most of the risk arising from the presence of radon in air. The intent of this section is to review the information available to test Bale's contention and to quantify the risk from inhalation of radon alone.

The extent and quality of data available to estimate risks from direct inhalation of radon are relatively poor. There are limited measurements of body content in humans exposed to radon in European spas (26,27). Correia et al. (7) exposed individuals to airborne radioactive xenon to measure retention characteristics in the various organs, discussed in Section II.A, but the exposures were not of sufficient length to establish equilibrium concentrations. However, the data were employed in Section II.A partially to determine the rate constants for the biokinetics model.

The only detailed studies of inhaled radon were conducted several decades ago (in the 1950s) and employed rats as the measured species (28,29). Nussbaum and Hursh (28) exposed adult Wistar rats for between 30 minutes and 48 h to an atmosphere containing elevated levels of radon. The rats were then sacrificed and the concentration of radon measured in each of several organs or tissues. The equilibrium concentration ratio, given by the ratio of the radon concentration in tissue to the concentration in air (both concentrations expressed in pCi/ml), was then determined. This ratio was 4.83 for omental fat, 0.41 for venous blood, 0.31 for brain, 0.31 for liver, 0.29 for kidney, 0.22 for heart, 0.18 for testis, and 0.15 for muscle.

Cohn et al. (29) also exposed rats to an atmosphere containing radon, comparing the uptake into the lung against a case in which similar rats were exposed to an atmosphere containing both radon and progeny. The amount of radioactivity in the nasal region, upper tracheobronchial region, and deep lung (alveolar region) was then determined by gamma counting. The ratio of activity (radon plus progeny to radon alone) for each lung region may be determined from their studies. This ratio was 7 for the tracheobronchial region, 28 for the nasal region, and 115 for the deep lung.

From this information, an estimate of the doses and risks to the lung from inhalation of radon alone may be obtained.

In performing the following calculation, it is assumed that 1 pCi/L of radon in water yields 0.0001 pCi/L of radon in air. As detailed in Section II.C, this airborne concentration yields approximately 0.0014 WLM from the radon progeny (i.e., not the radon gas itself). Assuming a conversion factor for the tracheobronchial region of 0.5 rad/WLM (30), the radon progeny are calculated to produce a dose of 0.0007 rad over the course of a 73 year expected lifetime. If the ratio of doses from radon progeny to radon gas alone is 7 in the tracheobronchial region, as noted, the lifetime dose from emanated radon gas is 0.0001 rad to the epithelium of this region.

For the alveolar region, the ratio of doses in the alveoli to doses in the tracheobronchial region is approximately 0.1 (31). Multiplying the dose to the tracheobronchial epithelium (0.0007 rad) by 0.1 yields an alveolar dose (from radon progeny) equal to 0.00007 rad. Dividing this latter dose by the factor of 115 found in the measurements by Cohn et al. (29) then yields a lifetime dose to the whole lung of approximately 6.1×10^{-7} rad from emanated radon gas. By contrast, Cross et al. (32) calculated a lifetime dose of approximately 1×10^{-6} rad and the recent ICRP report (33) indicates a dose of approximately 1.4×10^{-6} rad.

A reasonable estimate of the lifetime dose from radon gas alone is then 1×10^{-6} rad for the whole lung and 1×10^{-4} rad for the upper generations of the tracheobronchial region. As in inhalation of progeny, the dose to the tracheobronchial region should dominate considerations of risk. From the mining data, a risk coefficient of 245×10^{-6} per WLM was indicated. Dividing this value by the conversion coefficient of 0.5 rad per WLM yields 490×10^{-6} per rad. Multiplying this risk coefficient by the lifetime dose of 1×10^{-4} rad for radon gas in the tracheobronchial region then yields an estimated lifetime risk of 4.9×10^{-8}. If the whole lung is considered (averaging the dose throughout the 1000 g mass), the lifetime risk is equal to the product of the lung dose (1×10^{-6} rad), the quality factor (taken here to be 8), the weighting factor [0.12 from ICRP 26 (18)], and the risk coefficient [1×10^{-4} per rem from ICRP 26 (18)]. The resulting risk is 1×10^{-10} per pCi/L of radon in water. Note that this risk is far below that for the tracheobronchial region.

Information on the risk to the other organs of the body following inhalation of radon gas is less well developed. Equilibrium concentration ratios for various organs are presented as taken from the paper by Nussbaum and Hursh (28). Assuming an average ratio of 0.3 for body tissues (excluding the lung) and a radon air concentration of 0.0001 pCi/L, the average concentration of radon in the tissue is 3×10^{-5} pCi/L. This concentration may be converted to an estimate of lifetime dose through use of Equation (1) and an assumed occupancy factor of 0.75. The dose rate to these tissues is then 5.0×10^{-15} rad/minute or a dose of 6.1×10^{-7} rad over a 73 year expected lifetime. Multiplying this dose by a quality factor of 8 and the ICRP risk coefficient of 1×10^{-4} per rem yields a lifetime risk to "nonlung" organs and tissues of 4.9×10^{-10}. Adding this risk to the risk for the tracheobronchial region then yields a composite risk coefficient of 4.9×10^{-8} per pCi/L of radon in water.

D. Dermal Absorption of Radon

Information on the dermal route of exposure is the most limited of any of the exposure routes. The data on uptake into the bodies of humans exposed in spas containing radon-rich water is weakened by the inability to separate components arising from dermal absorption and inhalation of the radon present in air. The only data relevant to the present task are found in an early study by Lange and Evans (34), in which radon was entrained in an ointment and applied to the skin, and in a study by Pohl (27), in which humans bathed in radon-rich water but breathed compressed air. These situations might simulate exposures to radon in water present on or near the skin surface, but the relevance to airborne radon is questionable. It is therefore presumed that risks from airborne radon gas are dominated by the inhalation route and that dermal absorption is important only during immersion of the human body in radon-rich water and subsequent transport across the skin.

The data by Lange and Evans (34) suggest that approximately 0.1–4.5% of the radon present in applied ointment is exhaled. Given the previously cited finding that as much as 98% of ingested radon is exhaled, it is assumed here that the amount of radon absorbed through skin is equal to that exhaled. This suggests that intact skin absorbs approximately 1% of the radon present within the diffusion path of the surrounding medium. The fraction absorbed in abraded skin is higher but is not considered here.

Consider, then, a bath containing water with a radon concentration at the tap of 1 pCi/L. The bath itself is assumed to contain 100 L water (35). If the concentration of radon in the tap water is 1 pCi/L and if the fraction retained in the water is assumed to be 0.2, then the concentration of radon in the tap water is 0.2 pCi/L. This yields a total radon content of 20 pCi. Of this content, 1%, or 0.2 pCi, is transported through the dermal layer into the bloodstream.

Once in the bloodstream, it is assumed that the radon behaves with biokinetics identical to those displayed in Figure 1 and Table 1. An absorption of 0.2 pCi into the bloodstream then yields organ doses to the general tissue, lung, and liver of 1.3×10^{-10}, 8×10^{-11}, and 3×10^{-10} rad, respectively. Over the course of an expected lifetime of 73 years, assuming one bath per day, the doses to general tissue, lung, and liver would be 3.5×10^{-6}, 2.1×10^{-6}, and 8×10^{-6} rad, respectively. The risk coefficient for the lung is taken here to be 0.00005 per rad, the risk coefficient for liver cancer is taken to be 0.00007 per rad, and the risk coefficient for "general tissues" cancer is taken to be 0.0001 per rad. With an assumed quality factor of 8, this yields lifetime risks of 2.8×10^{-9} for general tissues cancer, 2×10^{-9} for lung cancer, and 4.4×10^{-9} for liver cancer. Summing these sources of risk yields an estimated lifetime risk of 9×10^{-9}.

This calculation may be compared with risk estimates based on the data of Pohl (27), who determined that the concentration of radon in the bloodstream of individuals bathing in radon-rich water is approximately 1% of that in the water at equilibrium. If the water concentration is 0.2 pCi/L, therefore, the concentration in the bloodstream is 0.002 pCi/L. The concentration of radon in the general tissue is then equal to approximately 75% of that in the blood, as indicated in the report by Nussbaum and Hursh (28). This yields a radon concentration of 0.0015 pCi/L in the general tissue, assuming continuous exposure to the water. If this assumption is dropped and a bathing time of 30 minutes/day adopted, the average radon concentration in blood as resulting from dermal absorption is 3×10^{-5} pCi/L. Using Equation (1), the dose rate to the general tissue is 2.2×10^{-14} rad/minute, or 8.5×10^{-7} rad over a 73 year expected lifetime. Employing a quality factor of 8 and a risk coefficient of 0.0001 per rem, the lifetime risk is estimated as 6.8×10^{-10}. This is a factor of 10 below the value calculated using the data of Lange and Evans. The difference between the two results may be attributed to several possible factors: (1) intersubject variability, (2) differences due to the ointment present in the Lange and Evans study, which ensures that the radon is kept close to the skin, or (3) use of a 1% absorbed fraction from the Lange and Evans data as indicated by their second experiment, rather than 0.1% as measured from exhalations during the first 4.5 h of exposure in the first experiment.

III. DISCUSSION AND CONCLUSIONS

The results in Section II support the conclusion that inhalation of radon progeny produced by the decay of emanated radon is the major source of risk arising from the presence of radon in tap water. From the analysis reported here, the following risk coefficients apply to the various routes of exposure:

Ingestion: 1.3×10^{-7} per pCi/L of radon in tap water

Progeny inhalation: 3.5×10^{-7} per pCi/L of radon in tap water
Radon inhalation: 4.9×10^{-8} per pCi/L of radon in tap water
Dermal absorption: 6.8×10^{-10} to 9×10^{-9} per pCi/L of radon in tap water

The total risk from all routes of exposure then may be obtained by summing these four components to yield a lifetime risk of approximately 5.3×10^{-7} per pCi/L of radon in tap water. The fractional contribution from the exposure pathways is 0.24 for ingestion, 0.65 for inhalation of progeny, 0.09 for inhalation of radon, and 0.02 for dermal absorption.

The risk coefficient of 5.3×10^{-7} may be compared with the value of 6.7×10^{-7} per pCi/L published by the U.S. EPA in their criteria document for radon in water (6) or the value of 4.5×10^{-7} per pCi/L when the U.S. EPA value is adjusted by the correction factor of 0.7 suggested in the National Academy of Sciences review (23). Using the risk coefficient developed in the present study, concentrations of radon in water yielding an average lifetime risk of 10^{-6}, 10^{-5}, 10^{-4}, and 10^{-3} are estimated as 1.9, 19, 190, and 1900 pCi/L, respectively.

The calculations reported here must be viewed with appropriate caution given the uncertainties inherent in the data and available scientific understanding. As indicated in a previous review of these sources of uncertainty (3), estimates of risk from waterborne radon may be characterized as accurate to within a factor of approximately 4–5 given uncertainties in the required parameters and the discrepancy between risk estimates based on the miner data and risk estimates based on direct epidemiological studies of airborne radon in homes. This suggests that the risk coefficient developed here may be expressed better as a 95% confidence interval of between 2.1×10^{-8} and 1.3×10^{-5}, with a "best estimate" of 5.3×10^{-7} per pCi/L. The 68% confidence interval would then extend from 1×10^{-7} to 2.7×10^{-6} per pCi/L.

It is important also to note that doses delivered to tissue from waterborne radon are quite small relative to background radiation. The risk from background radiation (excluding radon) delivered at a rate of 0.1 rem/year is approximately 7.3×10^{-4} from all sources of cancer using the risk coefficients presented in this chapter. The radon concentration in water must then exceed 1400 pCi/L to yield a lifetime risk equal to that from background radiation alone. The most difficult aspect of the analysis of risks from radon in water stems from the current inability to exclude the possibility of a threshold for radiation damage at the low dose rates produced by typical exposures to radon in drinking water. This suggests that the lower confidence limits on the risk coefficient for waterborne radon might include zero risk.

Several areas for future research are evident from this analysis. Foremost is continued epidemiological research on the effects of low levels of airborne radon in home air and the low systemic organ doses from radon entering the bloodstream. The emanation of radon from water to air, characterized here by a transfer ratio of 0.0001, is also in need of refinement through further measurements and proper weighting for time spent in each room of a structure. The biokinetics of ingested radon, although characterized by at least limited data, requires further study to improve estimates of dose to the lung and stomach wall. The quality factor for radiation-induced carcinogenesis at low dose rates is also highly uncertain at present. Finally, the entire issue of dermal absorption is poorly modeled at present, although the very low risk coefficient developed in the present study indicates that improved understanding of this pathway of exposure is unlikely significantly to alter estimates of the total risk from all pathways of exposure to waterborne radon.

REFERENCES

1. C. Cothern and W. Marcus, Estimating risk for carcinogenic environmental contaminants and its impact on regulatory decision making, *Regul. Toxicol. Pharmacol. 4*:265 (1984).
2. National Council on Radiation Protection and Measurements, *Exposures from the Uranium Series with Emphasis on Radon and Its Daughters*, NCRP Report No. 77, Bethesda, MD, 1984.
3. D. Crawford-Brown, *An Analysis of the Risk of Fatal Cancer Resulting from the Occurrence of Rn-222 in Public Drinking Water Supplies, the Uncertainty in Risk Estimates Employed for Establishing Standards, and the Impact of Mitigation on the Public Health*, report to the American Water Works Association, Denver, CO, 1991.
4. F. Cross, N. Harley, and W. Hofmann, Health effects and risk from Rn-222 in drinking water, *Health Phys. 48*:649, (1985).
5. D. Crawford-Brown and C. Cothern, A bayesian analysis or scientific judgment of uncertainties in estimating risk due to Rn-222 in U.S. public drinking water supplies, *Health Phys. 63*:11 (1987).
6. U.S. EPA, National primary drinking water regulations; radionuclides; proposed rule, *Fed. Register 56* (138) (1991).
7. J. Correia, S. Weise, R. Callahan, and H. Strauss, The kinetics of ingested Rn-222 in humans determined from measurements with Xe-133, Massachusetts General Hospital, Boston, MA, unpublished report, 1987.
8. D. Crawford-Brown, Cancer fatalities from waterborne radon, *Risk Anal. 11*:135 (1991).
9. J. Hursh, D. Morken, T. Davis, and A. Loovas, The fate of radon ingested by man, *Health Phys. 11*:465 (1965).

10. M. Suomela and H. Kahlos, Studies on the elimination rate and the radiation exposure following ingestion of Rn-222 rich water, *Health Phys. 23*:641 (1972).
11. C. Hess, C. Weiffenbach, and S. Norton, Variations of airborne and waterborne Rn-222 in houses in Maine, *Environ. Int. 8*:59 (1982).
12. R. McGregor and L. Gourgon, Radon and radon daughters in homes utilizing deep well water supplies, Halifax County, Nova Scotia, *J. Environ. Sci. Health A15*:25 (1980).
13. T. Gesell and H. Prichard, The contribution of radon in tap water to indoor radon concentrations, Proceedings of the Symposium on Natural Radiation Environment III, Vol. 2, U.S. DOE Conf. 780422, 1978, p. 1347.
14. A. Deb, *Contribution of Waterborne Radon to Home Air Quality*, American Water Works Association Research Foundation, Denver, CO, 1991.
15. C. Hess, M. Vietti, E. Lachapelle, and J. Guillemette, Radon transferred from drinking water into house air, *Radon, Radium and Uranium in Drinking Water* (C. Cothern and P. Rebers, eds.), Lewis Publishers, Chelsea, MI, 1990, p. 51.
16. National Academy of Sciences, *Health Effects of Radon and Other Internally Deposited Alpha Emitters*, BEIR IV, National Academy of Sciences Press, 1990.
17. D. Crawford-Brown, Age dependent lung doses from ingested Rn-222 in drinking water, *Health Phys. 52*:149 (1987).
18. International Commission on Radiological Protection, *Recommendations of the International Commission on Radiological Protection*, ICRP Publication 26, Pergamon Press, Oxford, 1977.
19. U.S. EPA, *Regulatory Impact Analysis of Proposed National Primary Drinking Water Regulations for Radionuclides*, internal report, 1991.
20. W. Nazaroff, S. Doyle, A. Nero, and R. Sextro, Potable water as a source of airborne Rn-222 in U.S. dwellings: A review and assessment, *Health Phys. 52*:281 (1987).
21. A. George and A. Breslin, The distribution of ambient radon and radon daughters in residential buildings in the New Jersey–New York area, *Proceedings of the Symposium on the Natural Radiation Environment III*, DOE Conf. 780422, 1980, p. 1272.
22. GEOMET, *Comparison of Indoor and Outdoor Air Quality*, EPRI EA-1733, Project 1309, GEOMET, Inc., Gaithersburg, MD, 1981.
23. National Academy of Sciences, *Comparative Dosimetry of Radon in Mines and Homes*, National Academy of Sciences Press, Washington, D.C., 1991.
24. D. Crawford-Brown, Dosimetry, *Environmental Radon* (C. Cothern and J. Smith, eds.), Plenum Press, New York, 1987, p. 173.
25. W. Bale, Memorandum to the files, March 14, 1951: Hazards associated with radon and thoron, *Health Phys. 38*:1061 (1980).
26. K. Furuno, Applications of radioactive spring water and excretion of radon in the expired air, *Okayama Daigaku Onsen Kenkyusho Hokoku 49*:1 (1979).
27. E. Pohl, *Biophysikalische untersuchungen uber die Inkorporation der Naturlich Radioaktiven Emanationen und deren Zerfallsprodukte*, Springer-Verlag, New York, 1965.

28. E. Nussbaum and J. Hursh, Radon solubility in rat tissues, *Science 125*:552 (1957).
29. S. Cohn, R. Skow, and J. Gong, Radon inhalation studies in rats, *Arch. Ind. Hyg. Occup. Med. 7*:508 (1953).
30. National Commission on Radiological Protection and Measurements, *Evaluation of Occupational and Environmental Exposure to Radon and Radon Daughters in the United States*, NCRP Publication 78, Bethesda, MD, 1984.
31. U.S. Public Health Service, *Control of Radon and Daughters in Uranium Mines and Calculations of Biologic Effects*, USPHS Publication 494, Washington, D.C., 1957.
32. F. Cross, C. Bloomster, P. Hendrickson, I. Nelson, B. Hooper, J. Merrill, and B. Stuart, *Evaluation of Methods for Setting Occupational Health Standards for Uranium Mines*, Batelle Pacific Northwest Laboratory, Richland, WA, 1974.
33. International Commission on Radiological Protection, *Limits for Inhalation of Rn-222, Rn-220 and Their Short Lived Daughters*, ICRP Publication 33, 1981.
34. K. Lange and R. Evans, Absorption of radon through the skin and its exhalation through the lungs, *Radiology 48*:514, 1947.
35. G. Tchobanogious and F. Burton, *Wastewater Engineering*, McGraw-Hill, New York, 1991, p. 23.

10

Use of Pharmacokinetic Models to Estimate Internal Doses from Exposure

Jerry N. Blancato
U.S. Environmental Protection Agency, Las Vegas, Nevada

Nancy Chiu
U.S. Environmental Protection Agency, Washington, D.C.

I. INTRODUCTION

In performing risk assessments situations are often simulated for which there are no data, especially in quantitative assessments. Even in qualitative assessments a certain amount of simulation takes place. For example, judgments are made regarding the potential toxicity to humans of a compound on the basis of animal studies. Of course, ethical considerations prevent toxicologists from generating toxicity data in human subjects when there is any uncertainty whatever regarding potential toxicity. In a sense, then, animals and their conditions of exposure simulate for humans and their conditions of exposure.

In the quantitative area one is faced with numerous assumptions regarding these and other simulations. These assumptions result in a certain amount of uncertainty in our assessments. Generally speaking, here is where modeling can improve the process. Well-constructed and validated models may be able to achieve two feats. First, they may be able actually to reduce some of the uncertainty; second, they may be able at least to aid in quantifying the uncertainty. It must be remembered that models are not a substitute for dependable data but rather serve to complement available data, reduce uncertainties in spheres for which data do not exist or cannot be easily obtained, and even focus the aim of additional laboratory measurements and research.

Pharmacokinetics is the description and quantification of the time course of the disposition of foreign substances in the body. Physiologically based pharmacokinetic (PBPK) models describe and predict, in a quantitative fashion, this disposition, taking into consideration various physiological parameters, including organ volumes, blood flows to individual tissues, metabolic and elimination rates, absorption rate constants, and the relative solubilities, or partitioning, of chemicals in the various tissues of the body. When properly formulated and validated, these models can simulate a variety of exposure conditions. The results of these models can then be valuable tools in exposure and risk assessments. For example, it might be possible to obtain data in humans regarding absorption fraction at very low doses, but the chemical's toxicity would prevent obtaining such data at higher concentrations. From the data acquired at low doses, a model could be formulated that could be used to predict absorption fraction at higher doses, or even longer times. Another illustration of the use of such modeling is to predict the disposition in the species at risk at low exposures from data attained in another species exposed at much higher concentrations in, for example, a bioassay. Animal bioassays are conducted at high doses to achieve a maximum response. To apply these results to humans in a typical exposure scenario, extrapolations are performed. Such extrapolations are predictions of response at doses lower than the conditions at which the original response was observed. In addition, they are predictions in a species other than the one in which the original evidence of toxicity was observed. Also, in many such bioassays the animals are exposed either continuously or at least on a daily basis for much of their lifetimes. In contrast, humans may be exposed in an intermittent and highly variable fashion. As a further complication, in many cases the animals are exposed to the putative toxin in a vehicle of administration and by a biological route other than those from which humans can expect exposure.

It is understandable, then, that the risk assessor is faced with many uncertainties even before actual quantitative risk assessment is begun. Physiologically based pharmacokinetic modeling can help in reducing, or at least revealing and quantifying, some of these uncertainties. Such models are also able to provide the risk assessor an exposure assessment of some delivered dose to the target tissue rather than merely an exposure concentration or applied dose.

Pharmacokinetic (PK) analyses or assessments can be used in two very general ways. First, they can be applied for forward analysis. Such PK analyses use exposure data to calculate biologically meaningful measures of dose. Second, they can be applied for reconstructive dose assessment. In this case, data on measured biomarkers or tissue concentrations of a compound and/or its metabolites are used to backcalculate the total dose of a xenobiotic received by an organism.

There are several ways to perform such analyses, but pharmacokinetic models are the emerging technique of choice. Pharmacokinetic models are a series of mass balance equations that describe the total disposition of the xenobiotic within the body. These models can be formulated as intricately or simply as the data allow or necessitate and are then implemented and solved on computers.

Physiologically based pharmacokinetic models tend to be direct and accurate descriptions of the actual anatomy and physiology of the body and include the representation of many of the body's actual organs and tissues. These models lend themselves to being predictors of target-level dose within and between species. PBPK models have parameters that are actual physiological constants, such as blood flows, organ volumes, and metabolic rates. Although curve-fitting routines are sometimes used to determine values for constants that are chemical and experiment specific, the values for many of the constants, such as blood flow, are determined from known biological data, allometric extrapolation, or independent experiments. As such, the models can be adjusted to predict for different species and for widely different exposure regimens. Frequently the models are formulated in conjunction with experimental data for the chemical of interest. Based on these data, other experiments, other biological data, and sometimes allometric scaling, the model can be reformulated to apply to humans exposed under various scenarios of interest. It is because of this flexibility that physiological models are being increasingly used in exposure and risk assessments. Many physiological processes are nonlinear, and the characteristics of these nonlinear processes may differ between dose levels and between species. Basically, the physiological models enable the risk assessor to account, in a quantitative fashion, for differences in pharmacokinetics that occur between different species, dose levels, and exposure regimens.

Chloroform is a major by-product of chlorine water disinfection and is potentially toxic to humans. In the chlorination process, hypochlorous acid is formed and reacts with the organic materials (e.g., humic acid) in the water, forming chloroform. Hypochlorous acid also oxidizes bromide ion present to form hypobromous acid, which reacts with the organic materials to form brominated trihalomethanes (THM). These include bromodichloromethane, dibromochloromethane, and bromoform. In drinking water, the brominated THM usually exist in far lesser amounts than chloroform (1). The most recent nationwide survey of total THM levels in finished treatment plant drinking water showed a mean concentration of 39 μg/L

(2). The U.S. Environmental Protection Agency (EPA) total exposure assessment methodology (TEAM) studies (3) of tap water in 800 homes, representing 800,000 persons, showed similar values. Mean levels of chloroform in four different states (California, Maryland, North Carolina, and New Jersey) ranged from 15 to 50 μg/L.

Chloroform is a colorless liquid at room temperature, with a sweet odor. It is very volatile, with a high vapor pressure (200 mm Hg at 25°C). Because of its high vapor pressure, chloroform volatilizes rapidly from the water. Chloroform is relatively soluble in water (8000 mg/L) and very soluble in organic solvents. The log octanol/water partition coefficient is 1.87. Since it is lipophilic, chloroform is easily absorbed by the skin during dermal contact. Humans can be exposed via household water used by three routes: (1) ingestion of chlorinated water or beverages made from chlorinated water (soup, coffee, tea, soft drinks, and so on); (2) inhalation of chloroform volatilized into indoor air from water used in the household, such as showering, cooking, clothes washing, and dish washing; and (3) dermal contact with chloroform in water while showering or bathing. Several monitoring and modeling efforts were undertaken by investigators to evaluate the magnitude of each of these exposure pathways (1,3).

Chloroform is absorbed easily via ingestion, inhalation, and dermal absorption. The absorbed chloroform is then metabolized by the mixed-function oxidase systems via either oxidative or reductive metabolic pathways in the liver and kidney. The predominant pathway for chloroform is oxidative and produces phosgene. The phosgene is very reactive, and some of it binds with cellular proteins, which may result in an alteration of cellular integrity and viability. The intensity of the binding of the metabolite parallels the extent of renal and hepatic necrosis in rats and mice (4,5). It may be that this binding is also related in some way to potential toxicity in humans.

II. PHARMACOKINETIC MODEL FOR CHLOROFORM

A. Basis for Model

The model used here is based on a pharmacokinetic model published by Corley and coworkers in 1990 (6). That model, consisting of a series of mass balance equations describing the absorption, transport, and transformation of chloroform to putative toxic metabolites, is briefly described here.

1. Biological Structure of Model

Figure 1 is a diagrammatic representation of the pharmacokinetic model. The body is depicted as several compartments perfused by arterial blood. Each compartment relates to actual anatomical tissues and organs in the body. Governing parameters for each of the organs are blood flow, organ volume, and the thermodynamic partition coefficient. The partition coefficient is actually the ratio of the concentrations of the chemical in the organ and in the blood at equilibrium. The values for blood flows, organ volumes, and partition coefficients were taken from the Corley paper (6). In addition, the liver and kidney are considered metabolically capable of metabolizing chloroform, and thus knowledge of the metabolic rates in these organs is necessary to complete the model. The rapidly and slowly perfused compartments represent organs that can be lumped together because of expected kinetic similarities. The kidney and liver are explicitly demarcated in the model because they are metabolically active and because they are potential targets of toxicity. Each organ is assumed to be well stirred, and thus the venous blood leaving the organ is represented as in equilibrium with the organ tissue itself.

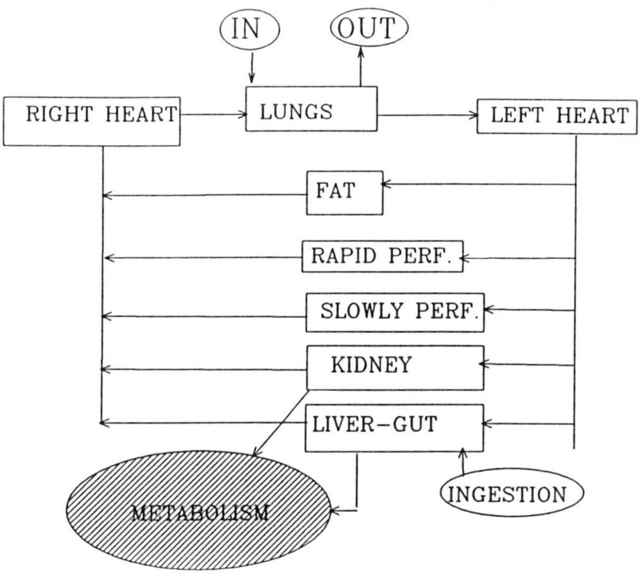

Figure 1 Physiologically based pharmacokinetic model used by Corley et al. (6).

Input of chloroform into the body is via ingestion and through inhalation. The portal circulation is not explicitly separated from the hepatic arterial circulation. Instead, absorption from the diet and drinking water is modeled as direct absorption into the liver through the gastrointestinal tract, with the hepatic and portal circulations lumped together. Clearance of chloroform is accomplished by exhalation and metabolism.

Most of the metabolism is modeled as occurring in the liver, with a small amount also occurring in the kidney. A portion of the metabolite formed in each organ is then believed to bind with endogenous molecules in that organ. The amount of metabolite bound can then be used as one measure of delivered dose. As described more fully later, the metabolite is assumed to be phosgene, which then reacts very rapidly within the tissue of formation (liver and kidney) and thus is assumed not to leave the organ of formation. Other end points, including actual chloroform concentration, the time integral of chloroform concentration, and total amount of metabolite formed, can all be used as measures of delivered dose. The choice of delivered dose depends upon the toxicological end point of interest and its mechanism(s) of action.

Figure 2 is a diagrammatic representation of the model as altered for wider applicability. The altered version enables us more easily to test the impact of various exposure patterns on the various measures of delivered dose. As previously mentioned, chloroform can be absorbed through the skin and exposure can be expected by this route. For this reason, we reconstituted the model to simulate exposure via the dermal route. In addition, our version of the model allows easier simulation of a variety of inputs by all three routes of absorption. Intermittent, continuous, repetitive, or single exposure can all be simulated with this model. Given that these models are being applied to simulate human exposure conditions realistically, input modules should be written in a consistent manner so that they may be used in more than one model.

2. Mathematical Structure of the Models

The mass balance equations in these and other pharmacokinetic models describe the rate of change of concentration or mass within the various compartments. Most often, the rate of change is described by first-order ordinary differential equations. In some cases the process of pulmonary absorption can be described as an algebraic equation, with flow, rather than volume, as one of the governing physiological parameters. Such descriptions for the lung can only be used if the chemical achieves rapid equilibrium between alveolar air and pulmonary blood. This is the case for chloroform and the models discussed in this chapter.

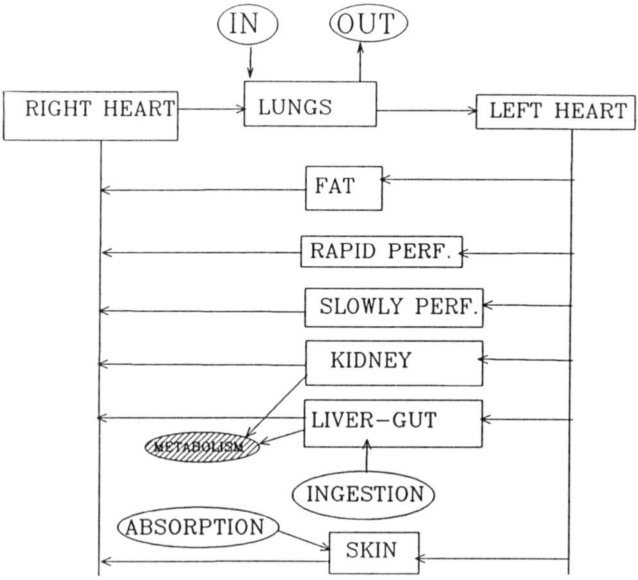

Figure 2 Physiologically based pharmacokinetic model used in this chapter.

Thus, the equation describing the concentration of chloroform in arterial blood in relation to the lung is (refer to list of abbreviations in Table 1 for definitions and nomenclature)

$$C_{\text{art}} = \frac{Q_p C_i + Q_c C_{\text{ven}}}{Q_c + Q_p / P_b} \tag{1}$$

The differential equations for the fat, rapidly perfused, and slowly perfused organs take the following form:

$$\frac{dC_i}{dt} = \frac{Q_i(C_{\text{art}} - C_i/R_i)}{V_i} \tag{2}$$

Basically, the equation indicates that the rate of change in the concentration in the tissue is the difference between the concentration entering the tissue, in this case via the arterial blood, and the concentration leaving the tissue, the venous return from organ i. This form makes certain assumptions, such as a well-stirred organ and a rapid equilibrium between the vascular, interstitial, and intracellular spaces of the organ. For details of the derivation

Table 1 Symbols, Parameters, and Parameter Values

Symbol	Parameter	Value
C_{art}	Concentration of chloroform in arterial blood	Calculated by model
Q_p	Alveolar ventilation rate	368.9 L/h
C_i	Concentration in inhaled air	Calculated by model based on environmental concentration
Q_c	Cardiac output	368.9 L/h
C_{ven}	Concentration of chloroform in venous blood	Calculated by model
P_b	Blood/air partition coefficient for chloroform	7.43
C_i	Concentration of chloroform in various organs (fat, rapidly and slowly perfused)	Calculated by model
Q_i	Blood flow to various organs	Fat: 17.4 L/h Slow: 66.1 L/h Rapid: 90.5 L/h
R_i	Tissue/blood partition coefficients for various organs	Fat: 37.69 Slow: 1.62 Rapid: 2.28
V_i	Volume of various organs	Fat: 16.17 L Slow: 42.74 L Rapid: 2.29 L
t	Time	h
C_m	Concentration of chloroform in metabolizing organs (liver and kidney)	Calculated by model
Q_m	Blood flow to metabolizing organs	Liver: 86.98 L/h Kidney: 86.98 L/h
R_m	Tissue/blood partition coefficient for metabolizing organs	Liver: 2.28 Kidney: 1.43
$V_{\max,m}$	Maximum rate of metabolism	Liver: 307.23 mg/h Kidney: 1.43 mg/h
$g(0)$	Input function for gastrointestinal absorption	Part of model code
K_m	Michaelis-Menten equilibrium constant	Liver: 0.448 mg/L Kidney: 0.448 mg/L
A	Amount of metabolite formed	Calculated by model

Table 1 Continued

Symbol	Parameter	Value
C_s	Concentration of chloroform in skin	Calculated by model
P	Permeation coefficient of chloroform across skin	0.13 cm/h
A_s	Area of skin of absorption of dose	18,500 cm^2 for whole body
C_{su}	Concentration of chloroform on skin surface (in environmental medium in contact)	Input into model by user
R_{su}	Skin/medium partition coefficient	Not applicable (see text)
Q_s	Blood flow to skin	21.0 L/h
R_s	Skin/blood partition coefficient for chloroform	1.62
V_s	Volume of skin in area of absorption	32.74 L for whole body
B	Amount of metabolite binding to endogenous organ components	Calculated by model
P_0	Total number of available binding sites	See text
EQ_b	Equilibrium binding constant	See text
a	Fraction of metabolite binding (P_0/EQ_b)	Liver: 0.002 Kidney: 0.009
C		
C_a	Concentration of chloroform in air	mg/m^3, see text
C_w	Concentration of chloroform in water	mg/L, see text
f	Fraction of chloroform volatilized from water	0.4, see text
R_w	Rate of indoor tap water use	200 L/person/day
V_h	Volume of air in house	100 m^3/person
R_v	Ventilation rate of the house	12 air exchanges/day

of this form from equations explicitly detailing the three regions of the organ, the reader is referred elsewhere (7).

The time integral of Equation (2),

$$C_i = \int_0^t \frac{dC_i}{d\tau} \qquad (3)$$

is then the instantaneous concentration in organ i. The theories and specific methods of integration are discussed in other papers (8).

The equations for metabolically active tissues take the following forms:

$$\frac{dC_m}{dt} = Q_m\left(C_{\text{art}} - \frac{C_m}{R_m}\right) - \frac{V_{\max,m}C_m}{K_m + C_m/R_m} + g(0) \tag{4}$$

$$C_m = \int_0^t \frac{dC_m}{d\tau} \tag{5}$$

In this case the change in the concentration of parent chloroform is equal to the difference in concentration between the incoming and outgoing blood minus the amount metabolized plus the amount input through an input function. For the liver, the input function results from exposure via the oral route. For other organs, such as the kidney, the input function is equal to zero.

For the models used here, the input function is related to the amount of the foreign chemical in the stomach and small intestine at any time. Absorption into the liver tissue can be by zero order, first order, or by more than one process from the various regions of the gastrointestinal tract (stomach and intestine). In this case, both first- and zero-order absorption from the stomach are considered the primary absorption pathways.

As previously discussed, the model can be expanded somewhat to have an explicit gut tissue compartment into which material passes from the lumen. From this tissue the material then diffuses into the portal circulation for delivery into the liver. It should also be noted that for compounds that exhibit significantly biliary secretion and/or enterohepatic recirculation, a more detailed model is necessary (9). For rapidly absorbed and nonrecirculated (no biliary secretion and/or enterohepatic recirculation) compounds, such as chloroform, the representation used in these models, that is, absorption directly from stomach and/or intestine into the liver, is adequate.

The amount of metabolite formed is described by

$$\frac{dA}{dt} = \frac{V_{\max,m}C_m}{K_m + C_m/R_m} \tag{6}$$

Equations (4) and (6) depict metabolism in the classic Michaelis-Menten saturable kinetic scheme. Metabolism could be occurring by other kinetic schemes, such as first- or second-order kinetics as well. It should also be mentioned that despite the appearance of Equations (4) and (6), conditions within in situ tissues are not duplicative of in vitro conditions in which the classic Michaelis-Menten parameters ($V_{\max}$ and K_m) were initially described. In the in vitro situation substrate concentration is in excess at the beginning of the analysis and its disappearance is a measure of the enzymatically controlled rate of metabolism. Under in vivo conditions the delivery of substrate (e.g., chloroform) by blood flow and absorption may not always result in its being present in excess. Other biochemical and physiological processes may be competing with or augmenting the ongoing metabolism. Hence, it might be more accurate to term the parameters the "apparent" V_{max} and K_m. In any case, inspection of Equation (6) shows that at low substrate concentrations the process shows first-order kinetics. It also follows that the actual amount (rather than concentration) of metabolite formed is described by

$$A = \int_0^t \frac{dA}{d\tau} \tag{7}$$

Dividing the amount of metabolite formed by the volume of the metabolizing organ results in the concentration of metabolite. For stable metabolites that are physiologically cleared into the blood leaving the metabolizing organs, a series of mass balance equations, similar to those for the parent compound, are written to describe the disposition of the metabolite throughout the body. For very reactive metabolites, as in the case of chloroform, which are consumed almost instantly, no physiological clearance from the organ is assumed to occur. Thus, no further mass balance equations are necessary.

The equation added to the original model (6) used to describe absorption via the skin is derived as follows:

$$\frac{dC_s}{dt} = \frac{PA_s(C_{su} - C_s/R_{su}) + Q_s(C_{art} - C_s/R_s)}{V_s} \tag{8}$$

The parameter R_{su} is the compound's partition coefficient between the skin and the medium in which it contacts the skin (e.g., water, air, or soil). For compounds that pass very quickly into the skin R_{su} is very large and Equation (8) becomes

$$\frac{dC_s}{dt} = \frac{PA_s(C_{su}) + Q_s(C_{art} - C_s/R_s)}{V_s} \tag{9}$$

Again, the time integral is the actual concentration according to

$$C_s = \int_0^t \frac{dC_s}{d\tau} \tag{10}$$

3. Model Parameters

Table 1 lists the abbreviations, the parameters, and their values as used for chloroform in this chapter. The values listed in Table 1 are mostly obtained from the Corley et al. paper (6). Some adjustments from the Corley et al. paper (6) were made to the volumes and blood flows to account for the additional skin compartment that was added for this exercise. It should be noted that the blood flows and alveolar ventilation rates shown in Table 1 are for individuals at rest. Model users should be aware that during vigorous exercise total blood flow increases. Model flows must therefore be adjusted for various levels of exercise. In addition, the distribution of blood flow may be significantly altered during exercise; that is, the muscle (the slowly perfused compartment in this model) may gain almost the entire amount of the increased cardiac output. Simply increasing the flow to each organ from the resting values proportionally without consideration for exercise-induced redistribution of flow may result in significant error when the model is applied.

For chemicals that are efficiently metabolized, such as chloroform, the metabolism parameters can greatly impact the resultant model output. Great care should be exercised in determining their values. A discussion of the biochemical in vitro and in vivo methods for their determination and interspecies extrapolation is beyond the scope of this chapter.

Partition coefficients can be determined in a variety of ways. Experimentally, they can be determined after intravenous steady-state exposure in experimental animals (10). For volatile organic chemicals in vitro techniques have been successfully applied (11).

The parameter for permeation across the skin was taken from the result of in vivo experiments with guinea pigs (12).

III. MODEL APPLICATION

A. Choice of Surrogates

The selection of the appropriate dose surrogate (or delivered dose) is somewhat complicated in that it may depend upon the specific mechanisms and target organ(s). Delivered dose may simply be the applied dose corrected for less than 100% absorption efficiency. It may be a measure of the amount of metabolites resulting from biotransformation of the compound present in the environment. In fact, it may be a concentration of a chemical or one or all of its metabolites in a particular organ, tissue, cell, or subcellular site. PBPK models like that outlined in this chapter can be formulated to describe and predict a variety of these possible surrogate dose measures. The actual delivered dose selected as the basis on which to conduct the risk assessment should depend upon the understanding of the mechanisms of toxicity, the test species, the exposure conditions, and the levels of data that are available. Effective dose may be defined as the dose causing or modulating a particular mechanism at the molecular site of action. Ideally, this effective dose should be the desired end point of any exposure assessment, but in reality it is almost never realized. The choice of extrapolation procedure between different exposure scenarios and different species depends upon the particular measure of delivered dose judged appropriate.

For chloroform several different possible surrogates can be used. Possibly, for certain effects, the peak concentration level in the tissues may be an end point of interest. Some integrated measure of chloroform concentration in the tissue is another alternative. Two other measures are concerned with biotransformation or metabolism. The total amount metabolized and the amount of metabolite bound to endogenous molecules are putative surrogate doses for chloroform. In this chapter, the amount of bound metabolites in the liver and kidney is used for illustration.

According to the published model (6), a small but significant fraction of the formed metabolite binds to endogenous components of the metabolizing tissue (liver and kidney). By this description the amount bound is linearly related to the amount of metabolite formed (parameter *a* in Table 1). Actually, binding may be described as a saturable process as in the equation

$$B = \frac{P_0 A}{EQ_b + A} \tag{11}$$

It can be realized that when the concentration of ligand (in this case the amount of metabolite formed from chloroform, A) is small compared with the equilibrium binding constant (EQ_b), the relation between binding and the metabolite present is linear and can be described according to

$$B = \frac{P_0 A}{EQ_b} \tag{12}$$

and the fraction of P_0 over EQ_b is equivalent to a linear term (a in Table 1) describing the fraction of metabolite binding.

B. Simulation

Simulation involves choosing the exposure patterns and concentrations of interest and then placing these into the model. The actual input of exposures and parameter values depends upon the exact software being used and the actual written program. We have used different software packages and programs, but currently the software that we use most frequently is SIMUSOLV. (SIMUSOLV is a trademark of the Dow Chemical Company; its mention does not in any way imply endorsement of the product by the U.S. Environmental Protection Agency.) SIMUSOLV is a simulating and optimizing program based on ACSL (ACSL, Advanced Continuous Simulation Language, is a trade mark of Mitchell and Gauthier Associates; its mention does not in any way imply endorsement by the U.S. Environmental Protection Agency) and has several numerical integration routines, as well as the capability to compare simulation results with actual data for purposes of model validation. In addition, it enables the user to estimate parameters using the developed model and actual laboratory or field-monitored data. Regardless of the software used, the modeler must be aware of both the power and limitation of numerical integration methods. Often these models exhibit a property known as *stiffness*, that is, their solutions have both rapidly and slowly changing components. Under such conditions (especially when the range of time constants differs by three or more orders of magnitude), it is best to use stiff methods of integration, such as Gear's (13) method or the LSODE (Lawrence Livermore solver of ordinary differential equations) (14,15).

C. Exposure Scenarios

The intent of this illustration is to determine the magnitude of the chloroform and metabolite dose from the normal daily use of water. Also, the doses resulting from different components (different routes) of the exposure are compared.

1. Comparison of Model with Actual Monitored Data

a. Exposure Scenarios

The utility of a model is only as great as the degree of confidence that can be placed in it. Validation is a key, and yet difficult, process for verifying all models. One difficulty with a model like this one for chloroform is that the desired end points cannot easily be monitored under in vivo conditions. The model may thus often be used to predict an end point for which no data ever exist for comparison and validation. In such a case, other, potentially less relevant end points must be used to establish confidence in the model.

In this section the model predictions of chloroform concentrations in exhaled air are compared with data from monitoring studies of human volunteers at indoor swimming pools. The data were obtained from Weisel and Shepard and are fully discussed in Chapter 6 of this volume. Although several different exposure scenarios were tested by Weisel and Shepard, here only two were used.

The first scenario was to simulate for a person at rest at poolside. It is assumed that because of no contact with the chloroform-containing pool water the only exposure would be via the inhalation route. The volunteers were monitored for chloroform concentration in the exhaled air before exposure. They then sat at poolside for 30 minutes. Their exhaled air was monitored for chloroform starting 5 minutes after leaving the pool area, that is, postexposure.

The concentrations chosen for this simulation, based on the monitoring by Weisel and Shepard, were 120 μg/m³ for air and 32 μg/L for water. The source of chloroform in the air is assumed to be from volatilization from the chlorinated swimming pool.

The second scenario was to have someone in the water, while performing no activity, for 30 minutes. The important difference from the first scenario is that here the internal dose results from both inhalation and dermal absorption.

b. Results of Comparison with Pool Data

Figures 3 and 4 show the comparison of model output and monitoring data with respect to chloroform concentration in the exhaled air. From Figure 3 it can be observed that for inhalation exposure only the model agrees with monitored data. The same can be observed in Figure 4 regarding combined inhalation and dermal exposure. The permeation coefficient was adjusted from 0.13 to 0.10 cm/h to obtain these results. The difference resulting from this adjustment was minor.

Although the model appears to be an excellent fit of actual data, caution should be exercised before concluding that such a model is validated. It

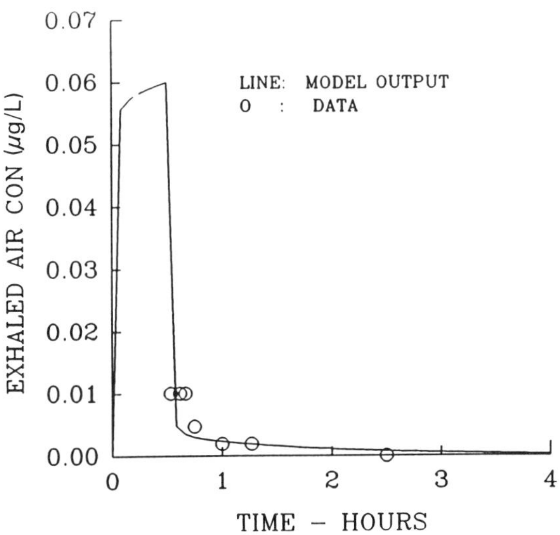

Figure 3 Comparison of model prediction of chloroform concentration in exhaled air in humans after poolside exposure. [Data from Weisel and Shepard (see Chapter 6).]

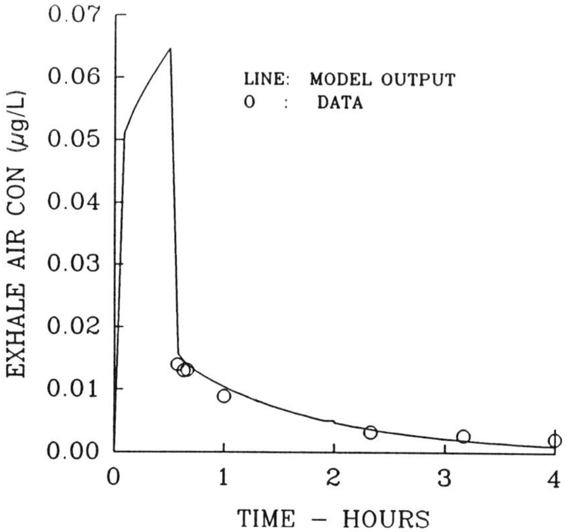

Figure 4 Comparison of model prediction of chloroform concentration in exhaled air in humans after in-pool exposure. [Data from Weisel and Shepard (see Chapter 6).]

can be concluded that for this set of experiments it adequately predicts one end point, that is, the chloroform concentrations in the exhaled breath after exposure has ceased. Nothing certain can be said about the prediction during the exposure period, because no data are available for that time period. Even more significant, it should be noted that the adequate prediction of concentrations in the exhaled air should not be taken as evidence of validation of any other end point, such as the metabolite levels. It is possible, depending upon the sensitivity of the various parameters, that changes in the metabolism constants would result in very different predicted levels of metabolite formation while still predicting indistinguishable concentrations in the breath. This is, breath concentration may be altogether insensitive to internal metabolite levels.

Other monitoring data are necessary for proper validation. Ideally, experiments should be conducted that yield data with the end point of primary interest. For example, if the amount of specific metabolite formation is the surrogate of interest for risk assessment purposes, then this is the end point that should be validated. Often, gathering such data is not possible. In such cases, the modeler and the assessor must use sound and rational scientific judgment before reaching decisions that could significantly impact the risk assessment.

2. Example of Use of Model for Total Exposure Assessment

a. Exposure Scenarios

For purposes of illustration it was decided to choose a daily exposure that is a reasonable description of chloroform exposure via contact with chlorinated water supplies. Clearly, exposure could come from drinking the water and consuming beverages made from the water, inhalation resulting from volatilization of chloroform from the water, and dermal absorption resulting from contact with the water as in showering. The actual exposure concentrations and profiles were chosen after carefully reviewing the literature and the results from various monitoring studies (3).

The potential impact of waterborne volatiles on indoor air quality has long been recognized. Chloroform levels in personal, indoor, and outdoor air were extensively monitored in four states (New Jersey, North Dakota, California, and Maryland) during 1981–1987 for the U.S. EPA TEAM studies (16). Mean personal exposures to airborne chloroform for the 800 participants generally ranged from 1 to 4 μg/m^3. Simultaneous outdoor air measurements were much lower, about 15% of the indoor levels (16). It was concluded from the TEAM study that water used in the house is the major source of chloroform in indoor air.

To calculate the background concentration of chloroform in the indoor air from a known concentration in water, a single-chamber modeling approach can be used. The equilibrium relation of air concentration to water concentration can be described according to

$$C_a = \frac{fR_w C_w}{V_h R_v} \tag{13}$$

The volatilization fraction is calculated based on the method of McKone (17–19) and yields 0.4 for chloroform. The values of the other parameters in Equation (13) are obtained from other literature sources (20,21). From these calculations an indoor air concentration of 1.6 μg/m^3 (background) is derived from water having a chloroform concentration of 24.5 μg/L. This estimate is in accordance with monitoring data from the TEAM studies (16).

Short-term inhalation exposure during showering is also examined for its contribution to internal dose. Jo et al. (22) measured the air concentration of chloroform in a full-sized shower using water levels ranging from 10 to 30 μg/L, with a mean concentration of 24.5 μg/L. Based on the various monitoring studies (3), an estimate of 200 μg/m^3 was chosen as the

Table 2 Exposure Scenarios

Exposure	Concentration	Duration and pattern
Background air	1.6 μg/m³	24 h
Shower inhalation	200 μg/m³	10 minutes
Shower dermal	14.7 μg/L	10 minutes
Drinking, ingestion	24.5 μg/L	1 L/day divided over three episodes

vapor-phase chloroform concentration within the showering compartment for purposes of this illustration. Table 2 summarizes the exposure patterns and concentrations used in this illustration.

b. Model Predictions of Dose Using Exposure Scenario

Figure 5 shows the model predictions for the amount of metabolite bound to endogenous molecules in the liver during a 1 day exposure to the chloroform-contaminated water as outlined in Table 2. It can be readily seen that exposure through drinking contributes the most to the burden of bound metabolites in the liver. The burden depicted as resulting from a 10 minute shower is via both the inhalation and dermal routes. It should also be noted

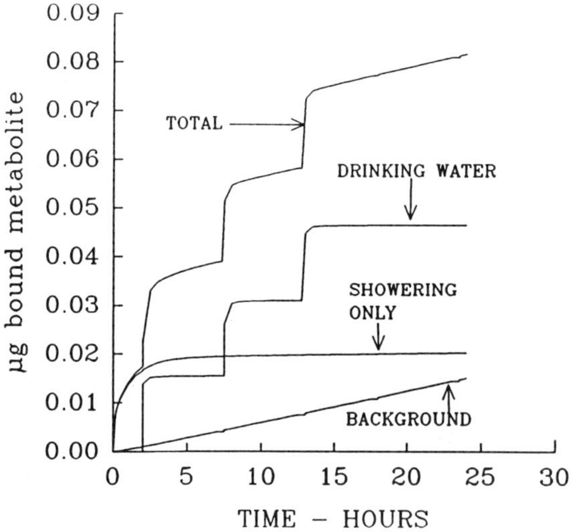

Figure 5 Model prediction of metabolite bound to endogenous components of liver resulting from typical daily exposure to chloroform-containing water.

that the burden from typical exposure to indoor air (background) contributes nearly as much as a 10 minute shower. Also, note that the continuing increase in metabolite formation observed in the plot for total exposure is due to the increasing metabolite formation resulting from the continuous background exposure.

With respect to the liver dose, the 10 minute shower contributes about 0.25 of the total dose (0.0203 μg/0.082 μg) and drinking water contributes about 0.57 of the total dose. Thus, the 10 minute shower is approximately equal to drinking about 0.436 L (0.0203 μg/0.0465 μg) of water over the course of the day. This, of course, applies only to the total amount of metabolite binding to the liver over a 24 h period, not to shorter intervals within the day. The situation for other measures of internal dose, such as the rate of metabolism or the rate of binding, would be quite different. In fact, even for metabolite bound to kidney the route-to-route comparisons show different results (see later).

Figure 6 shows the simulation in regard to this dose (metabolite bound to liver) from showering only. Under the exposure conditions used here (see Table 2), inhalation contributed about 2.1 times more than dermal absorption to the total liver dose.

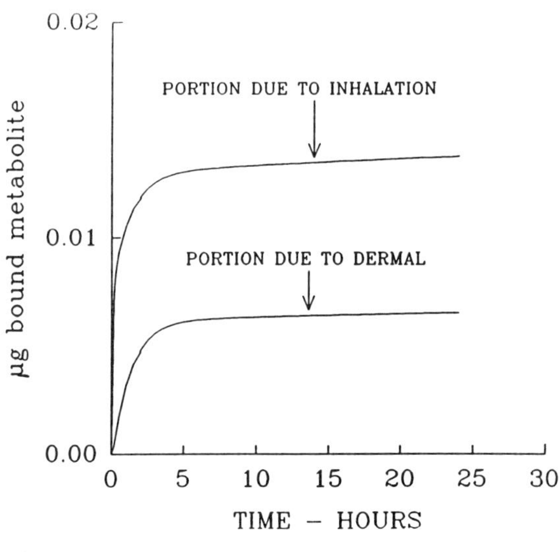

Figure 6 Model prediction of metabolite bound to endogenous components of liver resulting from the inhalation and dermal components of exposure from showering with chloroform-containing water.

Figures 7 and 8 show the results under the same exposure conditions but for the bound metabolite in the kidney rather than the liver. Compared with the liver, the total bound metabolite dose is about 12 times less. Of interest, however, is how the different components of the total exposure contribute to the total kidney dose. It is clearly evident from inspecting Figure 7 that showering contributes a relatively greater amount to the kidney dose than it does to the liver dose. This can be attributed to the absence of the liver first-pass effect that occurs with dietary ingestion. Showering accounts for more than 0.53 of the total daily bound dose to the kidney; drinking 1 L chloroform-containing water contributes only 0.07 of the total dose. Under these conditions, the 10 minute shower is equivalent to drinking almost 7.6 L (0.53/0.07) water. In other terms, drinking 1 L water contributes as much to the daily kidney dose as a shower lasting approximately 1.3 minutes. As in the case of the liver, inhaling of chloroform vapors during a 10 minute shower contributes about twice as much to the kidney dose as the dermal absorption component of showering.

IV. CONCLUSION

Properly formulated and validated, physiologically based pharmacokinetic models are valuable tools for use in the exposure and risk assessment

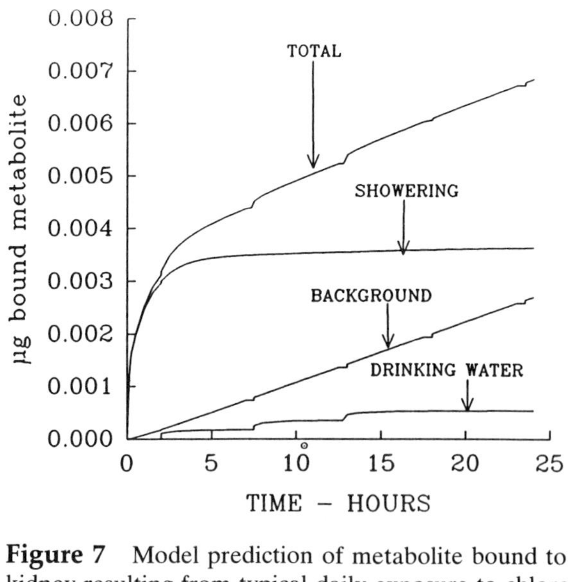

Figure 7 Model prediction of metabolite bound to endogenous components of kidney resulting from typical daily exposure to chloroform-containing water.

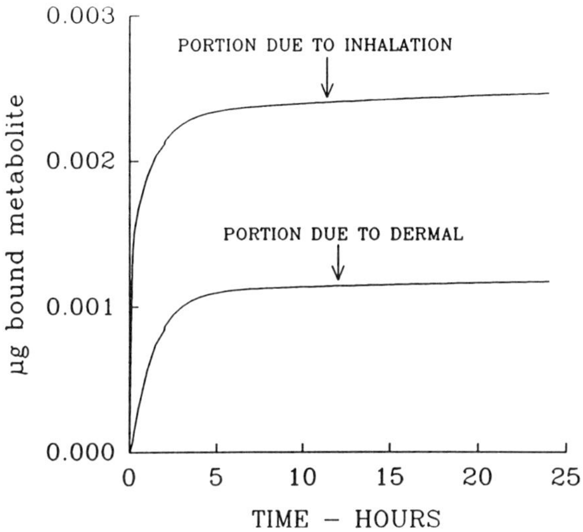

Figure 8 Model prediction of metabolite bound to endogenous components of kidney resulting from the inhalation and dermal components of exposure from showering with chloroform-containing water.

processes. As demonstrated in this chapter, they are especially helpful in determining the internal doses resulting from myriad exposure scenarios. Human exposure to potentially toxic chemicals can result from a multitude of sources. Dose results from entry of these compounds into the body by all the portals of entry. In the past, assessments have been complicated by the inability to account accurately for the differences due to these varying exposure patterns and routes of entry into the body. With these models, the kinetics of passage across the barrier membranes can be described and accurately quantified. As a result, accurate estimates of arterial blood concentrations can be achieved. From the arterial blood concentrations the models can describe the delivery of the toxin to the individual organs and tissues within the body. With this description and descriptions of metabolism, clearance, and other internal physiological and biochemical processes, the assessor now has a tool to estimate various measures of dose at or near the target site within the body. Any nonlinearities in the biochemical and physiological process that appear to add even more complexity and uncertainty to the assessment process are also accounted for within these models. Models can and have been written with much greater resolution than that presented here (8). These can be used to give estimates of dose at subcellular levels. The need for such resolution depends upon the mechanism of action and the proper selection of the metric for dose. Thus, it is clear that in the future development of this technology, a continuously iterative process between biochemists, physiologists, toxicologists, and phamacokineticists will be needed to ensure useful and accurate models for risk assessors. In such an iterative manner, rational judgments and decisions can be made, ensuring that the health of the environment and all its members, including humans, is maintained.

ACKNOWLEDGMENT

The information in this chapter was funded wholly by the U.S. Environmental Protection Agency. It was subject to agency review and approved for publication.

REFERENCES

1. U.S. EPA, Final external review draft for the drinking water criteria document on trihalomethanes, Office of Science and Technology, Office of Water, Washington, D.C., 1992.
2. M.J. McGuire, and R.G. Meadow, AWWARF trihalomethane survey, *J. Am. Water Works Ass. 80*:61 (1988).
3. L. Wallace, U.S. Environmental Protection Agency, Human exposure and body burden from chloroform and other trihalomethanes, Atmospheric Research and Exposure Research Laboratory, Research Triangle Park, NC, 1992.
4. E.L. Docks and G. Krishan, The role of glutathione on chloroform induced hepatotoxicity, *Exp. Mol. Pathol. 24*:13–22 (1976).
5. D.M. Brown, P.F. Langley, D. Smith, and D.C. Taylor, Metabolism of chloroform. I. The metabolism of ^{14}C-chloroform by different species, *Xenobiotica 4*:151–163 (1974).
6. R.A. Corley, A.L. Medrala, F.A. Smith, D.A. Staats, J.L. Gargas, R.B. Conolly, M.E. Andersen, and R.H. Reitz, Development of a physiologically based pharmacokinetic model for chloroform, *Toxicol. Appl. Pharmacol. 103*:512–527 (1990).
7. J.N. Blancato and K.B. Bischoff, The application of pharmacokinetic models to predict target dose, *Health Risk Assessment Through Dermal and Inhalation Exposure and Absorption of Toxicants*, November 1992.
8. J.N. Blancato, Physiologically based pharmacokinetic models: Examples of their use in exposure and risk assessment, in preparation for American Chemical Society Symposium Series *Biomarkers of Human Exposure to Pesticides*, 1993.
9. K.B. Bischoff, R.L. Dedrick, D.S. Zaharko, and J.A. Longstreth, Methotrexate pharmacokinetics, *J. Pharm. Sci. 60*(8):1128–1133 (1971).
10. M.J. Angelo, K.B. Bischoff, A.B. Pritchard, and M.A. Presser, A physiological model for the pharmacokinetics of methylene chloride in $B_6C_3F_1$ mice

following i.v. administrations, *J. Pharmacokinet. Biopharm. 12*(4):413–436 (1984).

11. M.L. Gargas, R.J. Burgess, D.E. Boisard, G.H. Csson, and M.E. Andersen, Partition coefficients of low-molecular weight volatile chemicals in various liquids and tissues, *Toxicol. Appl. Pharmacol. 98*:87–99 (1989).
12. K.T. Bogen, B.W. Colston, and L.K. Machicao, Dermal absorption of dilute aqueous chloroform, trichlorethylene, and tetrachloroethylene in hairless guinea pigs, *Fund. Appl. Toxicol. 17*: in press.
13. C.W. Gear, *Numerical Initial Value Problems in Ordinary Differential Equations*, Prentice-Hall, Englewood Cliffs, NJ, 1971.
14. A.S. Hindmarsh, LSODE and LSODEI, two new initial value ordinary differential equation solvers, *ACM Signum Newslett. 15*:10–11 (1980).
15. A.C. Hindmarch, ODEPACK: A systemized collection of ODE Solvers, Technical Report UCRL-88007, Lawrence Livermore National Report, 1982.
16. L. Wallace, U.S. Environmental Protection Agency, *The Total Exposure Assessment Methodology (TEAM) Study: Summary and Analysis*, 1, Office Research and Development, Washington, D.C., 1987.
17. T.E. McKone, Human exposure to volatile organic compounds in household tap water: The indoor inhalation pathway, *Environ. Sci. Technol. 21*:1194–1201 (1987).
18. T.E. McKone, Household exposure models, *Toxicol. Lett. 49*:321–329 (1989).
19. D. MacKay, Air/water exchange coefficients, *Environmental Exposure from Chemicals*, I, CRC Press Boca Raton, FL, 1989.
20. W.W. Nazaroff, S.M. Doyle, A.V. Nero, and R.G. Sextro, Potable water as a source of airborne Rn-222 in U.S. dwellings: A review and assessment, *Health Phys. 52*:281–289 (1987).
21. A.P. Becker and T.M. Lachajczyk, U.S. Environmental Protection Agency, *Evaluation of Waterborne Radon Impact on Indoor Air Quality and Assessment of Control Options*, EPA 600/7-84-093, Research Triangle Park, NC, 1984.
22. W.K. Jo, C.P. Weisel, and P.J. Lioy, Routes of chloroform exposure and body burden from showering with contaminated tap water, *Risk Anal. 10*:575–580 (1990).

11

Pharmacokinetic Interactions of Drinking Water Contaminants

Ronald Brown
International Life Sciences Institute, Risk Science Institute, Washington, D.C.

Jerry N. Blancato
U.S. Environmental Protection Agency, Las Vegas, Nevada

David Young
University of Maryland at Baltimore, Baltimore, Maryland

I. INTRODUCTION

Environmental standards for drinking water contaminants are established by regulatory agencies for individual compounds. However, drinking water may contain a complex mixture of chemical contaminants, such as trihalomethanes, pesticides, and heavy metals. Therefore, the potential exists for one or more of these contaminants to affect the toxicity or pharmacokinetic behavior of a compound of interest. Regulatory agencies are faced with the challenge of setting drinking water standards that are protective of human health in the event that such interactions occur.

Although guidelines exist for conducting a risk assessment for chemical mixtures (1), quantitative data are often unavailable to gauge the degree to which the health effects of one compound are altered by coexposure to another. For example, Vanderslice et al. (2) emphasized how the lack of data poses difficulties in accounting for interactions in the establishment of drinking water standards.

Pharmacokinetic interactions often serve as the basis for the pharmacodynamic changes produced by coexposure to two or more compounds. Pharmacokinetic models serve as tools to describe these interactions and, in the absence of experimentally derived data, allow the risk assessor to pose hypothetical questions about the potential impact of one compound on the pharmacokinetic behavior of another. Both traditional, compartmental pharmacokinetic models and physiologically based pharmacokinetic (PBPK) models can be useful for this purpose; however, PBPK models allow the risk assessor to estimate levels of the compound of interest or its metabolites in target tissues when exposure occurs in the presence of other compounds. Furthermore, PBPK models can be used to conduct route-to-route extrapolations of dose, a potentially useful feature for drinking water contaminants, since many pharmacokinetic interactions are described for volatile compounds for which the primary exposure route is inhalation. Perhaps the most important feature of PBPK models for the establishment of drinking water standards is the ability of this approach to help us understand the potential for interactions to alter significantly the pharmacokinetic behavior of the compound of interest.

Traditional pharmacokinetic models have been used for many years to describe pharmacokinetic interactions, and a large body of literature exists that addresses drug-drug interactions. Somewhat fewer data are available on the interactions occurring between environmental pollutants. Nevertheless, a growing body of literature on this topic has appeared in recent years. Despite the increasing availability of information on pharmacokinetic interactions occurring between drinking water contaminants, little guidance is available to regulatory scientists on how to evaluate the pharmacokinetic and potentially toxicological significance of these interactions. To address this need, this review attempts to provide broad guidelines for (1) determining the significance of the interaction, (2) predicting when pharmacokinetically significant interactions are likely to occur, and (3) representing these interactions in a PBPK model. Finally, this chapter explores the implications that pharmacokinetic interactions may have for the risk assessment of drinking water contaminants.

II. EXAMPLES AND SIGNIFICANCE OF PHARMACOKINETIC INTERACTIONS

A. Absorption

Interactions involving drinking water contaminants have the potential to alter the absorption of these compounds by the gastrointestinal or pulmonary tracts or across the skin. Several examples of pharmacokinetic interactions affecting absorption are presented in this section, along with a discussion of the potential significance of these interactions.

1. Gastrointestinal Absorption

The mechanisms by which interactions affect the uptake of compounds in the gut have been reviewed by a number of authors (e.g., References 3 and 4) and are summarized in Table 1.

Table 1 Mechanisms of Gastrointestinal Absorption Interactions

Physical-chemical interactions (e.g., formation of chelates)
Change in gastric emptying rate
Change in gastrointestinal motility
Damage to gut wall mucosa
Alteration in bacterial flora
Change in composition or volume of gastrointestinal fluid
Sequestration or solubilization of mixture components
Change in pH of gastrointestinal fluids

Numerous examples of these absorption interactions can be found in the literature for drugs, but relatively few studies of the effect of one environmental pollutant on the absorption of another can be identified. A notable exception is the effect of malathion on the gastrointestinal uptake of carbaryl (5). Oral administration of equal doses of malathion and carbaryl to rats in this study doubled the time required to reach peak carbaryl blood levels and decreased peak carbaryl blood levels relative to administration of carbaryl alone. These effects were probably not due to metabolic inhibition, since corresponding effects on carbaryl metabolism were not produced by coexposure to malathion. Curiously, carbaryl did not have the same effect on malathion absorption.

A number of studies report a decrease in the gastrointestinal absorption of one compound caused by another; however, facilitated uptake is also known to result from interactions occurring in the gut. Perhaps the most significant of these are the interactions between such compounds as dithiocarbamate pesticides, thiuram sulfides, or pyridinethiones and such metals as cadmium, nickel, and zinc. The organic compounds form lipophilic chelates with the metals, thereby facilitating their uptake across the gastrointestinal mucosa (6). Examples of studies showing the facilitated uptake of metals in the gut after complexation with chelating agents are provided in Table 2.

The relevance of these interactions for risk assessment is twofold. First, although the compounds that serve as chelating agents are not common drinking water contaminants, the potential for a significant percentage of the population to be exposed to these agents is quite high. For example, zinc and sodium pyridinethione are components of many antidandruff shampoos (7,8). Dithiocarbamate fungicides, such as maneb, zineb, nabam, ziram, and ferbam, are commonly used in agriculture, especially on fruit. Thiram, a related thiuram sulfide, is a widely used pesticide on a

Table 2 Facilitated Uptake of Metals in the Gastrointestinal Tract Due to the Formation of Chelates

Chelator	Metals	Reference
Zinc pyridinethione	Nickel, cadmium	7
Sodium pyridinethione	Nickel, cadmium, zinc	8
Thiram	Nickel	89
Dithiocarbamate pesticides	Nickel	89
8-Hydroxyquinolines	Nickel	6

number of crops (9). Disulfuram (tetraethylthiuram disulfide, Antabuse) is used clinically for the treatment of chronic alcoholism (10).

The second reason for concern regarding these chelator-metal interactions is the potential for clinically significant toxic effects to occur in humans. For example, use of Clioquinol (iodochlorhydroxyquin), a halogenated 8-hydroxyquinoline compound used for the treatment of diarrhea, was associated with an outbreak of subacute myelooptic neuropathy in Japan. This side effect may have been produced by an increased intestinal uptake of heavy metals.

A large body of literature exists on interactions affecting the uptake of lead in the gastrointestinal tract (11) and is not covered in this review. However, one interaction worth noting is the increased uptake of lead during coexposure to ethanol. For example, Flora and Tandon (12) demonstrated that administration of 10% vol/vol ethanol to young rats in the drinking water increased the absorption of orally administered lead relative to animals receiving lead and drinking water without ethanol.

Although many interactions result in the decreased absorption of compounds in the gut, the preceding examples demonstrate that coexposure to drinking water contaminants, especially metals, and other compounds can facilitate the absorption of the toxic compounds in the gastrointestinal tract.

When analyzing data from absorption studies, special attention should be paid to research techniques. It is especially important to consider potential vehicle effects on gastrointestinal absorption. Moderately lipophilic compounds, such as carbon tetrachloride or chloroform, are often administered to experimental animals by gavage in a corn oil vehicle. However, numerous studies have documented the differential toxicity observed when halogenated hydrocarbons are administered in corn oil compared with delivery in an aqueous vehicle. These effects may not be trivial and may impact directly on regulatory decision making. For example, Vanderslice et al. (2) pointed out the importance of considering vehicle effects on toxicity when studies are being used to derive drinking water standards. They cite two studies (13,14) in which the toxicity of the compound of interest is greater when it is delivered in corn oil versus an aqueous vehicle. Other examples can be found to support this observation. Condie et al. (15) reported an increased hepatotoxic effect of carbon tetrachloride in CD-1 mice when the compound was administered by gavage in corn oil compared with administration of an aqueous solution of the compound. In addition, Chieco et al. (16) found that 1,1-dichloroethylene (DCE) was more hepatotoxic when administered in rats in a corn oil rather than an aqueous vehicle. However, the opposite effect is also known to occur. For example, trichloroethylene (TCE) produced a dose-dependent increase in lethality in male and female B6C3F1 mice given the compound in a 20% aqueous solution of Emulphor (polyoxyethylene fatty acid esters). A similar effect was not seen in animals dosed with identical concentrations of TCE in corn oil (17).

An understanding of the pharmacokinetic behavior and mechanism of action of these compounds is key to resolving these apparently contradictory findings. Withey et al. (18) showed that corn oil delayed the rate and extent of gastrointestinal absorption and reduced the peak blood levels of four halogenated hydrocarbons in rats relative to oral gavage of the compounds in aqueous solution. Because of their moderately high oil/water partition coefficients, the halogenated hydrocarbons present as common drinking water contaminants partition more readily into lipophilic vehicles, such as corn oil, than into aqueous vehicles. As a result, these compounds are likely to be released into the gut more slowly. However, administration of the compounds in aqueous vehicle results in rapid uptake of the compound from the gastrointestinal tract, a more rapid rise in blood levels of the compounds, and increased elimination of the compound in exhaled air. Therefore, the appearance of toxicity may be related to the rate at which the compound is released from the gastrointestinal tract. Prolonged uptake may produce certain toxic effects (e.g., hepatotoxicity) that depend on the continuous slow release of the compound, whereas rapid absorption may be associated with other toxic end points, such as central nervous system (CNS) effects, seen after bolus administration of the compound. Also, toxicity may be dependent upon the transformation of the compound to toxic metabolites. The rate of appearance or the "dose rate" of metabolite formation may be a key factor in the development of toxicity. This appearance rate may depend greatly on uptake rate and may thus be affected by the vehicle of administration of the parent compound. As a result, risk assessors should carefully consider compound-vehicle interactions when evaluating studies that could serve as the basis for drinking water standards.

2. Dermal Absorption

There has recently been a great deal of interest in the potential for dermal absorption to serve as a significant route of exposure for drinking water contaminants. For example, Jo et al. (19,20) showed that the dose of chloroform dermally absorbed by adult human volunteers taking a daily 10 minute shower was similar to the dose received after ingesting 2 L water. Based on studies with hairless guinea pigs, Bogen et al. (21) also concluded that dermal absorption may be an important route of exposure to volatile organic compounds, such as chloroform, tetrachloroethylene, and TCE in drinking water.

The vehicles used in many dermal absorption studies can interact with the skin to increase or decrease drug penetration (22). For example, solvent vehicles can extract lipid components from the stratum corneum, thereby affecting the barrier function of this skin layer. Solvent vehicles also determine the thermodynamic activity and, subsequently, the permeation of penetrants (23). Therefore, it is important to consider vehicle effects if the results of dermal uptake studies involving nonaqueous vehicles are used to derive drinking water standards.

Since soap and products containing surfactants (e.g., shampoo, bath oils, and dish washing detergent) are often used for bathing, showering, or cleaning, it is also important to determine the effect that these agents could have on the dermal absorption of drinking water contaminants. Přiborsky et al. (24) recently showed that the dermal absorption of indomethacin was increased in rats over 10-fold in the presence of 1% sodium dodecyl sulfate. Other surfactants also increased the uptake of indomethacin in this study; however, two surfactants slightly reduced the absorption of this compound relative to controls. Similar contradictory results have also been reported. For example, Chowhan and Pritchard (25) reported an increase in the transdermal penetration of naproxen in the presence of surfactants, whereas Zatz (26) described a decrease in the uptake of Benzocaine (ethyl aminobenzoate) under certain conditions when the drug was applied to mouse skin in vitro with aqueous solutions of a nonionic surfactant. This apparent contradiction can be explained as follows. At low concentrations, the surfactant is likely to interact directly with the skin, thereby increasing the absorption of the penetrant molecule; however, at concentrations of surfactant that enable micelles to form, an interaction between the penetrant compound and micelle may occur, which reduces the concentration of free compound at the skin surface and in turn reduces the flux of the compound across the skin (22). Given the potential for daily exposure of the skin to both surfactants and drinking water contaminants and the ability of surfactants to increase or decrease the dermal absorption of a wide range of compounds, it is important to obtain flux data for drinking water contaminants in the presence and absence of relevant concentrations of surfactants before accurate predictions of the effect of surfactants on the dermal absorption of drinking water contaminants can be made. Until these data are available, the exposure and risk assessor should keep in mind that low concentrations of surfactants have the ability to enhance the dermal absorption of many compounds, whereas high concentrations may produce the opposite effect.

B. Distribution

Pharmacokinetic interactions can affect the distribution of a compound in the body by altering plasma protein or tissue binding or by modifying tissue uptake through changes in membrane transport. A demonstration of such an effect is presented elsewhere (27).

Enhanced uptake of compounds in the gut, as just described, may result in a corresponding increase in the concentration of the compound of interest in various tissues. Examples of studies showing increased tissue uptake of compounds that occur after coadministration of another compound are provided in Table 3.

Elevated tissue levels of one compound, when coadministered with other compounds, are often due to facilitated uptake in the gut; however, the mechanism responsible for increased tissue levels of some compounds may be more complex. Ethanol may facilitate the uptake of lead in the brain, not only by increasing blood lead levels but also by impairing the blood-brain barrier (28). Other compounds, such as mercury (29), are also known to impair the blood-brain barrier, thereby increasing the likelihood that coadministered compounds could be taken up by the CNS. Lipophilic metal chelates formed by coadministration of dithiocarbamates, thiram sulfides, or pyridinethiones and metals may persist in the body after absorption from the gut, making it easier for the chelates to cross the blood-brain barrier than for the metals themselves. Other potential mechanisms responsible for altering the level of one metal in tissues after prior or simultaneous exposure to another include competition for carriers and induction of metal binding proteins.

Most of the interactions listed in Table 3 result in increased tissue levels of drinking water contaminants; however, other interactions produce decreased tissue levels of the compounds of interest. Although one might expect these interactions to be beneficial, there are examples of toxicologically significant effects resulting from reduced tissue levels of essential minerals. For example, coadministration of TCE or DCE and radiolabeled chromium to rats reduced chromium levels in a number of organs, including

Table 3 Examples of Pharmacokinetic Interactions Affecting the Distribution of Environmental Pollutants[a]

Compound 1	Compound 2	Effect	Reference
Carbaryl	Malathion	Malathion caused a 26 and 41% increase in levels of carbaryl in kidney and fat, respectively, when coadministered with carbaryl	5
Malathion	Carbaryl	Carbaryl caused a 50% increase in tissue levels of malathion	5
Mercury	Selenium	Selenate reduced ^{203}Hg levels in the kidney and increased ^{203}Hg levels in liver and blood compared with control	90
Nickel	Halogenated 8-hydroxyquinolines	Halogenated 8-hydroxyquinolines, when administered to mice orally immediately after radiolabeled nickel, increased nickel levels in nervous tissues	6
Nickel	Thiram and dithiocarbamate pesticides	Thiram and dimethyldithiocarbamates increased nickel levels in all tissues relative to controls; ethylene- and propylenebisdithiocarbamates had no effect or lowered nickel levels in tissues	89
Nickel, cadmium, zinc	Sodium pyridinethione	Increased levels of metals in all tissues relative to controls	7
Nickel, cadmium	Zinc pyridinethione	Increased levels of metals in all tissues relative to controls	8
Lead	Ethanol	Increased levels of lead in blood, liver, and brain observed after simultaneous exposure to lead and ethanol	12
Aluminum	Ethanol	Concentration of aluminum in the blood and liver was higher after coadministration of aluminum and ethanol	28
DDT	Aldrin, dieldrin	Increased concentration of each of these compounds in blood and fat after coadministration	91

[a]Effect of compound 2 on the distribution of compound 1.

the heart, relative to controls (30). This study was conducted because TCE, DCE, and chromium were found in the drinking water in an area of Tucson, Arizona where an increased incidence of cardiac malformations was observed in newborn children. The authors of this study speculate that TCE or DCE affects the distribution or metabolism of chromate and thereby plays a role in the cardiac teratogenic effects of these compounds.

Displacement of one compound by another from binding sites on plasma proteins can also affect blood and tissue levels of the compound of interest. Although many examples of this type of interaction can be found in the literature, the clinical or toxicological significance of protein displacement interactions is probably overstated. McInnes and Brodie (31) point out that displacement of a drug from plasma proteins results in a transient increase in the free concentration of the drug in the plasma. This unbound drug distributes quickly to the tissues. A new steady-state free concentration of the drug is established, similar to the concentration that existed before exposure to the displacing agent, because of the compensatory increase in metabolism and/or excretion that usually follows the increase in unbound drug in the plasma. McInnes and Brodie further stated that protein binding displacement is likely to be pharmacologically significant only when it is accompanied by enzyme inhibition.

Plasma protein binding affects the hepatic clearance of compounds as well as their distribution in the body. These effects are discussed in Section II.C.1.f.

C. Metabolism

1. Inhibition

Inhibition of the metabolism of one compound by another is undoubtedly the most common pharmacokinetic interaction reported in the literature. Although reports of metabolic interactions involving drugs predominate, metabolic interactions occurring between drinking water contaminants have been documented. A number of these interactions are summarized in Table 4.

From Table 4, it is clear that the potential exists for one drinking water contaminant to inhibit the metabolism of another. However, the challenge facing risk assessors is to identify those interactions that are likely to be pharmacokinetically or toxicologically significant. Various factors should be considered in evaluating the significance of metabolic inhibitions. These factors include the following:

Mechanism of the enzyme inhibition
Dose dependence of the interaction
Tissue solubility of the inhibitor

Table 4 Examples of Metabolic Inhibition Reactions Affecting Potential Drinking Water Contaminants

Substrate	Inhibitor(s)	Effect	References
Acetone	Methyl ethyl ketone	No effect on the elimination of either compound	71
Aniline	Cadmium/nickel	Reduced aniline hydroxylase and other MFO activity	92
Aniline	Manganese	Inhibition of aniline hydroxylase and other MFO activity	93
Benzene	Toluene	Reduced excretion of total phenols and hippuric acid	39
Benzene	Toluene	Accumulation of benzene and its metabolites in bone marrow and other tissues	40
Benzene	Toluene	Dose-dependent inhibition of benzene metabolism	41
Benzene	Toluene	Mutual metabolic suppression	42, 94
Benzene	Toluene	Mutual metabolic suppression	87
Benzene	Gasoline components	Decreased benzene metabolism	66
Butadiene	Styrene	Inhibition of butadiene metabolism	43
Bromobenzene	Toluene	Inhibition of bromobenzene metabolism	95
Dichloroethylene	Trichloroethylene	Mutual metabolic suppression	36
Dichloroethylene	Trichloroethylene	Dose-dependent mutual metabolic suppression	96
Ethylbenzene	*m*-Xylene	Mutual metabolic suppression	97
n-Hexane	Toluene	Decreased urinary excretion of *n*-hexane metabolites	98
n-Hexane	Methyl ethyl ketone	Decreased formation and clearance of *n*-hexane metabolites	99
Methylene chloride	Toluene	Reduced metabolism of methylene chloride to CO	100
Methylene chloride	Ethanol	Reduced metabolism of methylene chloride to CO	83
Methylene chloride	Pyrazole	Reduced metabolism of methylene chloride to CO	84
Methylene chloride	Benzene, toluene, or xylene	Reduced metabolism of methylene chloride to CO	85
Styrene	Toluene	Reduced metabolism of styrene	39
Styrene	Acetone	No effect on the metabolism or elimination of styrene	101

Substrate	Inhibitor(s)	Effect	References
Styrene	Acrylonitrile	Reduced metabolism of styrene	102
Toluene	Trichloroethylene	Mutual metabolic suppression	103
Toluene	*n*-Hexane	No effect on blood levels of *n*-hexane or excretion of hippuric acid	104
Toluene	Trichloroethylene or benzene	Inhibition of toluene metabolism	105
Toluene	Xylene	Mutual metabolic suppression	86
Toluene	Xylene	No change in amount of hippuric and methyl hippuric acid excreted into the urine	106
Toluene	Xylene	Reduced urinary excretion of hippuric and methyl hippuric acid; elevated levels of toluene and xylene in the blood	107
Toluene	Xylene	Dose-dependent metabolic interaction	108
Trichloroethane	Tetrachloroethylene	Reduced urinary excretion of trichlorates	109
Trichloroethylene	Carbon tetrachloride	Mutual metabolic suppression, but carbon tetrachloride has a greater inhibitory effect on TCE metabolism than vice versa	110
Trichloroethylene	Tetrachloroethylene	Reduced urinary excretion of trichlorates	111
Trichloroethylene	Groundwater contaminants	Inhibition of TCE metabolism	112
Trichloroethylene	Vinyl chloride	Mutual metabolic suppression that occurs by competitive inhibition; GSH depletion induced by vinyl chloride	113, 114
m-Xylene	1,1,1-Trichloroethane	No effect on blood levels of *m*-xylene	115
m-Xylene	Aspirin	Mutual metabolic suppression	51
Xylene	Methyl ethyl ketone	Reduced metabolism of xylene	116

Isozyme primarily responsible for metabolism of the substrate
Ability of the substrate to undergo metabolism or elimination by multiple routes
Rate at which the substrate is cleared by the liver

The effect of each factor on determining the significance of the metabolic inhibition is discussed here.

a. Mechanism of Enzyme Inhibition

Enzyme inhibition mechanisms have been thoroughly described in many textbooks of biochemistry (e.g., Reference 32) and review articles (e.g., Reference 33). Classification of a metabolic inhibition reaction as competitive, noncompetitive, or uncompetitive may be necessary to determine the best way to represent these interactions in a PBPK model. However, this distinction may carry little relevance for risk assessment decision making. A more important distinction is whether the inhibitor acts reversibly or whether it irreversibly destroys the enzyme. Andersen et al. (34) commented that competitive inhibition is likely to occur only at the relatively high exposure concentrations seen in occupational settings. More significant for the risk assessment of drinking water contaminants may be the enzyme inactivation that occurs after exposure to "suicide substrates" or mechanism-based inhibitors of enzyme activity. Numerous mechanism-based enzyme inactivators have been identified (35), and several of these inhibitors are known to be drinking water contaminants. For example, Andersen et al. (36,37) showed that *cis*- and *trans*-1,2-dichloroethylene serve as suicide inactivators of the enzyme system responsible for trichloroethylene metabolism. Both isomers of DCE have been detected in surface water, groundwater, and drinking water samples. Carbon tetrachloride, another important drinking water contaminant, undergoes reductive metabolism in the liver, yielding trichloromethyl and lipid radicals that destroy cytochrome P_{450} (35). In addition, parathion and malathion produce the autocatalytic destruction of cytochrome P_{450} (38). Therefore, a number of commonly occurring drinking water contaminants have the potential to destroy metabolic enzyme systems, thereby inhibiting the metabolism of other substrates metabolized by the same enzymes. At the relatively low doses of compounds received from the ingestion of drinking water, enzyme destruction is likely to result in a more significant metabolic inhibition than competition of two (or more) substrates for the same enzyme. The latter mechanism is probably more relevant in scenarios involving high exposure concentrations.

b. Dose Dependence

Many of the metabolic interactions listed in Table 4 were observed only after relatively high doses of the mixture components were used in the studies. However, it is not clear whether these same interactions would occur under environmentally relevant exposure conditions. Although it seems intuitive that metabolic interactions, especially those involving competitive inhibition, occur as a dose-dependent process, few interactions have been evaluated with regard to their dose dependence.

One interaction that has been shown to be dose dependent is the mutual suppression of benzene and toluene metabolism after coadministration of these compounds. The inhibition of benzene metabolism by toluene, and vice versa, is among the most studied metabolic interactions, perhaps because benzene and toluene often occur simultaneously in occupational settings, as components of gasoline, and as groundwater contaminants. Early studies in animals established that the oxidative metabolism of benzene and toluene is mutually suppressed when the compounds are administered together and that this inhibition is most likely due to competitive inhibition of mixed-function oxidase enzymes (39,40). However, these studies used relatively high concentrations of benzene and toluene. A subsequent study by Sato and Nakajima (41) established the dose dependence of this inhibition. These investigators found that coadministration of 24.2–97.7 mg benzene/kg body weight to rats and an equimolar dose of toluene had no effect on the pharmacokinetic behavior of either compound; however, metabolic inhibition occurred at higher doses of benzene and toluene.

Because of the dose dependence of this inhibition, it is important to determine if the interaction is likely to occur in humans exposed to benzene and toluene in environmentally relevant settings. The results of Inoue et al. (42) suggest that a mutual metabolic interaction between benzene and toluene in humans occurs at levels much lower than those to which rats were exposed in the study conducted by Sato and Nakajima (41). Inoue et al. (42) measured the urinary metabolites of benzene and toluene in workers exposed to these compounds singly or in combination. Coexposure to a mean concentration of 17.9 ppm benzene and 20.5 ppm toluene over an 8 h workday (equivalent to 8.2 mg benzene and 10.8 mg toluene per kg body weight for a 70 kg worker breathing 9.8 m^3/8 h shift) resulted in suppression of the formation of benzene and toluene metabolites in coexposed workers relative to levels measured in workers exposed to similar concentrations of the individual compounds.

c. *Tissue Solubility*

The poor solubility of some compounds may limit their ability to inhibit the metabolism of other compounds because of their inability to accumulate in tissues. Laib et al. (43) investigated the potential metabolic interactions occurring between 1,3-butadiene and styrene in rats exposed to these compounds by inhalation. They found that styrene inhibited the metabolism of 1,3-butadiene, whereas 1,3-butadiene had little effect on the metabolism of styrene. Laib et al. (43) postulated that one reason for these findings was the greater tissue solubility of styrene relative to 1,3-butadiene, enabling styrene to accumulate to a greater degree in tissues.

d. Isozyme Specificity

Although we often refer to cytochrome P_{450} as a single enzyme, over 30 forms or isozymes of this enzyme exist in many species, including humans (44). Each of these forms exhibits some degree of substrate specificity. For example, cytochrome P_{450} 2E1 is largely responsible for catalyzing the oxidation of a number of important drinking water contaminants, such as benzene, chloroform, carbon tetrachloride, and methylene chloride, as well as ethanol and such drugs as acetaminophen (Table 5). The substrate specificity of other P_{450} isozymes is also well known (45). Clearly, knowledge of the isozyme that serves as the primary catalyst of a compound is important in determining whether another compound also metabolized by the same isozyme is likely to serve as a competitive inhibitor of the metabolism of the compound of interest. Many of the inhibited reactions listed in Table 4 are the result of competitive inhibition between two P_{450} 2E1 substrates. In addition, a number of preferential inhibitors of specific P_{450} isozymes are known to exist. For example, disulfuram serves as a mechanism-based inhibitor selective for P_{450} 2E1. Knowledge of the selectivity of this inhibitor not only gives investigators the ability to use disulfuram as an isozyme-specific inhibitor of P_{450} 2E1 but also enables risk assessors to predict that patients taking disulfuram clinically may be less able to metabolize the compounds listed in Table 5.

e. Presence of Alternative Pathways

The presence of alternative pathways may be one of the key factors in determining the pharmacokinetic, toxicological, or clinical significance of

Table 5 Potential Drinking Water Contaminants with Specificity Toward Cytochrome P_{450} 2E1

Acrylonitrile	Ethylene dichloride
Benzene	Methylene chloride
Carbon tetrachloride	Tetrachloroethylene
Chloroform	1,1,1-Trichloroethane
Chloromethane	Trichloroethylene
1,2-Dichloropropane	Vinyl bromide
Ethylene dibromide	Vinyl chloride

Source: From Reference 117.

an interaction. Indeed, relative to clinically significant drug interactions, McInnes and Brodie (31) commented:

> Metabolic drug interactions are most likely to be important if elimination is by a single process; they are less relevant if 2 or more routes are available, unless the alternative pathways are saturable to give rise to toxic products. This is an important reason why drug interactions due to induction or inhibition are less prominent clinically than might be expected.

Since a number of important drinking water contaminants are cometabolized by different enzyme systems (e.g., cometabolism of methylene chloride by mixed-function oxidase- and glutathione-S-transferase (GST)-mediated pathways), the presence of alternative pathways may become an important consideration in determining the potential for significant interactions between drinking water contaminants.

f. Hepatic Blood Flow and Plasma Protein Binding

The factors already described (e.g., mechanism of the enzyme inhibition, dose dependence, and tissue solubility) address direct interactional effects on metabolic enzyme systems; however, hepatic clearance Cl_H is not governed solely by enzyme activity. Such factors as blood flow Q and binding of the substrate to plasma proteins also play a role in determining the rate at which a compound is cleared by metabolically active organs. Therefore, pharmacokinetic interactions that affect blood flow or plasma protein bindings also have the potential to affect the hepatic clearance of the compounds of interest. However, changes in blood flow or plasma protein binding do not have an effect on the clearance of all compounds. Changes in hepatic blood flow have a negligible effect on the clearance of compounds with an extraction ratio $E < 0.3$ ($E = Cl_H/Q$), because clearance rate is largely determined by intrinsic enzyme activity. For compounds with high extraction ratios ($E > 0.7$), however, hepatic elimination is closely linked to blood flow rate, since the rate of substrate delivery becomes the limiting step in clearance. Although this type of interaction has not been documented for drinking water contaminants, other compounds with high exposure potential, such as caffeine (46) and the therapeutic agent cimetidine (47), are known to reduce hepatic blood flow and thereby inhibit the elimination of highly cleared compounds in the liver.

The effects of protein binding on hepatic elimination have also been well characterized. Using the well-stirred model of hepatic elimination, Wilkinson and Shand (48) predicted the effects of altered binding on hepatic extraction of substrates. When the free intrinsic clearance-blood flow ratio (and, therefore, extraction ratio) is low, the unbound fraction of substrate in the blood becomes the limiting factor for extraction. This relationship, termed restrictive elimination by Wilkinson and Shand (48), holds true for a number of poorly extracted substrates. For example, the extraction ratio of tolbutamide, a poorly extracted substrate, is directly proportional to the fraction of the unbound compound in the perfusate in an isolated, perfused liver preparation (49). In contrast, at high extraction ratios, the extent of protein binding has little effect on elimination. The state occurring when the extraction ratio is greater than the unbound fraction of substrate in blood entering the liver has been termed nonrestrictive by Wilkinson and Shand (48).

From the factors described here, it is evident that estimates of the rate at which a compound is cleared by the liver enable the risk assessor to determine whether interactions affecting hepatic blood flow and/or plasma protein bindings are likely significantly to inhibit the hepatic elimination of the compounds of interest. Thus, for highly cleared compounds, hepatic blood flow is the most important parameter, and intrinsic clearance and plasma protein binding are more influential for poorly cleared compounds.

g. Inhibition of Phase II Metabolic Reactions

A majority of the metabolic interactions listed in Table 4 involve inhibition of phase I (mixed-function oxidase-mediated) enzyme reactions; however, inhibition of phase II (conjugative) reactions by drinking water contaminants is also known to occur, either through direct enzyme inhibition or by depletion of necessary cofactors. For example, various haloacetonitriles known to occur as by-products of the drinking water disinfection process (e.g., trichloroacetonitrile) are potent inhibitors of glutathione-S-transferase in liver cytosolic fractions, albeit at relatively high concentrations (50). However, environmentally or occupationally relevant exposures can also inhibit phase II enzyme systems. Campbell et al. (51) reported that the metabolism of *m*-xylene to *m*-methylhippuric acid in human volunteers exposed to xylene at a level equivalent to the occupational exposure limit (100 ppm) was inhibited 50% by coadministration of a therapeutic dose (1500 mg over 4 h) of aspirin. The metabolism of aspirin to its glycine conjugate was also inhibited to a similar extent. According to Campbell et al. (51), the likely mechanism for this interaction is mutual inhibition of the enzymes acyl coenzyme A-synthetase or glycine *N*-acylase.

Depletion of cofactors required for phase II metabolic reactions by one compound also has the potential to effect the metabolism and subsequently the toxicity of other compounds that undergo a similar biotransformation step. Depletion of the important intracellular cofactor glutathione (GSH) is a well-known mechanism by which certain compounds induce their hepatotoxic effects (52). Drinking water contaminants, such as chloroform, reduce levels of GSH in isolated hepatocytes (53). This mechanism is thought to be responsible for the additive hepatotoxic effect of chloroform and cadmium in isolated hepatocytes relative to the effect seen on administration of either compound alone (54).

Inhibition of GST enzymes or depletion of intracellular GSH may not always result in increased toxicity of the compounds of interest. In many cases, biologically active intermediates can be generated by the GSH conjugation pathway (55,56). Therefore, reduced generation of GSH conjugates may actually exert a protective effect against toxicity.

2. Induction

The metabolic activity of xenobiotic enzymes can be increased (induced) by prior exposure to numerous compounds, many of which can occur as drinking water contaminants. Table 6 lists a number of environmental pollutants known to induce enzyme systems. In addition to the compounds listed in Table 6, a number of drugs are known to induce mixed-function oxidase (MFO) activity (reviewed in References 57 and 58), as can such agents as cigarette smoke and ethanol, to which humans can be exposed in high doses.

Similar to the situation with enzyme inhibitors, the challenge facing risk assessors is not only to identify the compounds that serve as enzyme inducers but also to determine when enzyme induction leads to pharmacokinetically and toxicologically significant effects. The stimulation of MFO enzymes is known to potentiate the toxic effects of compounds that require bioactivation to exert their toxicity. For example, exposure to methyl ethyl ketone potentiates the neurotoxic effects of *n*-hexane. A recent study (59) suggested that this potentiation may be due, in part, to the induction of microsomal enzymes responsible for the bioactivation of *n*-hexane to its

Table 6 Examples of Environmental Pollutants That Serve as Enzyme-Inducing Agents

Inducing agent	Reference	Inducing agent	Reference
Acetone	118	Isopropanol	123
Benzene	119	Methyl *n*-butyl ketone	118
Cadmium	92	Methyl ethyl ketone	59
Crude oil	120	Methyl isobutyl ketone	122
Ethylbenzene	61	Nickel	92
Gasoline	121	Toluene	124
n-Hexane	122	Xylene	125

presumed toxic metabolite, 2,5-hexanedione. Therefore, understanding the mechanism by which mixture components exert their toxic effect will aid in determining which interactions are likely to be toxicologically significant.

Knowledge of extraction ratio values for the compounds of interest is important for determining the significance of induction as well as inhibition effects. Since the elimination of highly cleared compounds in the liver is largely under the influence of hepatic blood flow, increased metabolic activity has little effect on clearance. However, metabolic stimulation can markedly increase the clearance of compounds with low extraction ratios ($E < 0.3$). Nevertheless, Miller (60) points out that enzyme induction from exposure to cigarette smoke affects only the hepatic clearance of drugs with extraction ratios greater than 0.7. In fact, it is only when the extraction ratio approaches 1.0 that enzyme induction has an appreciable effect on hepatic clearance.

The significance of isozyme-specific inhibition of cytochrome P_{450} for predicting when important interactions will occur was discussed in the previous section. Similarly, a number of environmental agents are known to induce specific isozymes of cytochrome P_{450}. For example, prior exposure of rats to styrene was shown selectively to induce P_{450} 2B1 (61). The drug phenobarbital is also a well-known selective inducer of P_{450} 2B1, and ethanol is an established inducer of P_{450} 2E1. Knowledge of isozyme-specific induction patterns after exposure to one compound and information on the isozymes responsible for metabolizing important drinking water contaminants will assist the regulatory scientist in making decisions about the significance of these interactions.

When the effects of enzyme induction on clearance or target organ toxicity are being evaluated, it is important to recognize that xenobiotic compounds can stimulate metabolic enzymes in extrahepatic tissues as well. For example, exposure of rats to 300 ppm *p*-xylene for 6 h/day for 5 days increased aryl hydrocarbon hydroxylase activity in the lung by 41% over controls (62). Increased microsomal enzyme activity in the lung could result in greater first-pass clearance of volatile drinking water contaminants by the lung, thereby potentially increasing the carcinogenicity or toxicity of compounds requiring metabolic activation.

D. Excretion

In addition to the effects described earlier, pharmacokinetic interactions can also alter the biliary or urinary excretion of drugs or environmental pollutants. Although there are few known examples of interactions that affect the biliary excretion of compounds, a number of studies document the effect of one drug on the urinary excretion of another. Interactions involving urinary excretion have been reported for certain pesticides as well. For example, Martinez Tabache and Posadas del Rio (63) observed that subchronic oral administration of parathion to female rats affected the elimination of sodium salicylate, partially through inhibition of metabolism and partially via effects on the renal excretion of the drug. Also, Waldron Lechner and Abdel-Rahman (5) observed that malathion significantly reduced the urinary excretion of carbaryl when the two compounds were coadministered to rats. This finding is especially important since urinary excretion is the primary route by which carbaryl is eliminated.

The rate of urinary excretion of compounds is determined by three factors: (1) glomerular filtration, (2) secretion, and (3) reabsorption. Pharmacokinetic interactions can affect any of these processes, thereby altering the urinary excretion of the compound of interest. In terms of significance, many of the principles described earlier as relating to metabolic inhibition also apply to urinary excretion. A compound with a high renal extraction ratio is more affected by changes in renal blood flow than by altered protein binding or changes in active transport mechanisms. Conversely, interactions involving protein binding or active secretion are more likely to affect the urinary excretion of compounds that are poorly cleared by the kidney (64). Also in terms of significance, Aarons (64) points out that interactions involving urinary excretion are important only for those compounds that undergo elimination primarily by this route. The example involving malathion and carbaryl emphasizes the potential for interactions to affect the excretion of important drinking water contaminants. It is also important to remember that although there are few known examples of environmental pollutants altering urinary excretion of other compounds, there are numerous examples of drugs affecting this process. Clearly, these drugs have the potential to affect the urinary excretion of drinking water contaminants eliminated by the same route.

III. MODELING THE INTERACTION PROCESS

A. Classic Compartmental Models

Most reports that describe the interaction between compounds have been qualitative or, at best, semiquantitative in nature. For example, alterations in the metabolic schemes of a number of binary mixture components have been investigated by determining the changes in the percentage of specific metabolites eliminated. Using a more quantitative approach, a number of interactions have been studied by investigating the alteration in specific pharmacokinetic parameters. These parameters have been estimated using classic pharmacokinetic procedures before and after administration of a second compound. The parameters that have received the most attention are the following:

Biological, distribution, and/or absorption half-life
Extent of absorption
Area under the curve
Clearance
Peak concentration
Time to peak concentration
Volume of distribution
Fraction bound by plasma proteins

Comparison of values for these parameters before and after administration of a second compound may be used by risk assessors initially to determine if an interaction between the mixture components is occurring. This approach may be useful to screen mixture components for the presence of an interaction; however, it does not provide risk assessors with estimates of the target tissue concentrations of the mixture components, nor does it mechanistically describe the interaction process. As an alternative, PBPK models can be used to describe the pharmacokinetic behavior of one compound in the presence of another.

B. Physiologically Based Pharmacokinetic Models

Physiologically based pharmacokinetic models have been developed to describe the pharmacokinetic behavior of a number of important individual drinking water contaminants, such as benzene (65,66), chloroform (Reference 67 and Chapter 10), carbon tetrachloride (68), lead (69), methylene chloride (37), and tetrachloroethylene (70). In addition, PBPK models have been developed by various researchers to describe the pharmacokinetic behavior of binary or complex mixture components. These models are listed in Table 7.

There are two basic approaches for using PBPK models to describe these interactions: (1) adjust the value of an input parameter so the model-derived simulation can be superimposed onto experimentally derived data, or (2) incorporate specific terms into the model to represent the interaction. The former approach was used by Brown et al. (71) to describe the potential effect of acetone and methyl ethyl ketone (MEK) on their reciprocal uptake and elimination. Using the prototypic PBPK model for volatile compounds described by Fiserova-Bergerova (72), these investigators successfully simulated the blood concentration of acetone and MEK in humans exposed to the individual compounds for 4 h. Coexposures to acetone and MEK at one-half of the original atmospheric concentration of each could be simulated simply by reducing the exposure concentration term in the model

Table 7 Physiologically Based Pharmacokinetic Models of Chemical Mixture Components

Mixture	Interaction modeled	References
Acetone/MEK	Metabolic inhibition	71
Benzene/gasoline	Metabolic inhibition	66
Benzene/toluene	Metabolic inhibition	87
Caffeine/β-naphthoflavore	Enzyme induction	76
Carbon tetrachloride/methanol	Enzyme induction	126
Salicylic acid/phenolsulfthalein	Renal elimination	78, 79
TCE/1,1-DCE	Metabolic inhibition	36, 96
TCE/ethanol	Metabolic inhibition/ enzyme induction	127
TCE/vinyl chloride	Metabolic inhibition	115, 116
TCE/groundwater contaminants	Metabolic inhibition	114
Toluene/xylene	Metabolic inhibition	128

by one-half. Therefore, no interaction between these two solvents could be demonstrated under these exposure conditions. Obviously, if an interaction was occurring, halving the exposure concentration value for the model would not necessarily result in an adequate simulation of the data.

A preferred approach to developing a PBPK model for chemical mixtures is to incorporate specific terms to account mechanistically for the pharmacokinetic interaction. The most common interaction represented in a pharmacokinetic model is an inhibition of some process, usually metabolism.

The approach used by Andersen and colleagues (36) to represent inhibition is attractive because of its ability to account for competitive, noncompetitive, uncompetitive, and substrate inhibition mechanisms for any saturable process, not only metabolism. The general inhibition equation in this model is

$$\frac{d\mathrm{AMT}_1}{dt} = QC_{a1} - QC_{v1} - \frac{V_{\max 1}C_{v1}}{K_m(T_1) + C_{v1}(T_2)} \tag{1}$$

$$T_1 = 1 + \frac{C_{v2}}{K_{mi2}} + \frac{(C_{v2})^2}{K_{mi2}K_{m22}}$$

$$T_2 = 1 + \frac{C_{v2}}{K_{m21}} + \frac{C_{v1}}{K_{m11}}$$

where $d\mathrm{AMT}/dt$ represents the rate of change in the amount of a compound in an organ or compartment; Q is the blood flow to that organ; C_a is the concentration of the compound in arterial blood; C_v is the concentration of the compound in venous blood; V_{max} is the maximum velocity of the reaction (usually metabolism); and K_m is the Michaelis binding constant. The subscripts 1 and 2 designate compounds 1 and 2, respectively, in the terms.

As shown here, this general equation can be reduced to components that represent the competitive, uncompetitive, noncompetitive, and substrate inhibition of saturable rate processes:

$$\frac{d\mathrm{AMT}_1}{dt} = QC_{a1} - QC_{v1} - \frac{V_{\mathrm{max}i}C_{v1}}{C_{v1} + K_{mi}(1 + C_{v2}/K_{i21})} \quad \text{competitive inhibition} \tag{2a}$$

$$\frac{d\mathrm{AMT}_1}{dt} = QC_{a1} - QC_{v1} - \frac{V_{\mathrm{max}i}C_{v1}}{C_{v1} + (1 + C_{v2}/K_{i21}) + K_{mi}} \quad \text{uncompetitive inhibition} \tag{2b}$$

$$\frac{d\mathrm{AMT}_1}{dt} = QC_{a1} - QC_{v1} - \frac{V_{\mathrm{max}i}C_{v1}}{(C_{v1} + K_{mi})(1 + C_{v2}/K_{i21})} \quad \text{noncompetitive inhibition} \tag{2c}$$

Andersen and colleagues (36) were able to describe the uptake by rats of trichloroethylene and 1,1-dichloroethylene from a closed chamber as individual compounds and as a binary mixture solely by accounting for competitive metabolic enzyme inhibition. Although interactions potentially occur between mixture components during absorption, distribution, or excretion as well, it was not necessary to account for these processes in the model.

The strength of Equation (1) is its broad utility. Although the equation was written specifically with terms to describe enzyme inhibition, the inhibition of any pharmacokinetic process in which saturable reactions occur can be represented by modifications to this equation. Exceptions to this general approach are described here.

1. Absorption

There are numerous examples of oral absorption interactions for binary mixtures of drugs and environmental pollutants. The approach to describing the interaction has generally been to estimate the absorption rate constant A and bioavailability F or maximum concentration and area under the curve with and without the interaction. However, a modification of

Equation (1) can be used to describe active transport absorption interactions. To our knowledge, competition of similar substances for a carrier system (competitive inhibition) is the only active transport interaction known to occur in the gut. If other interaction mechanisms are revealed, however, Equations (2b) and (2c) may be employed to describe these interactions.

Degradation of one substrate in the gastrointestinal lumen by a second compound (or by bacteria in the gut) is another type of interaction that can alter the absorption of ingested compounds. The following equation, developed by Aarons (64), can be used to describe the rate of appearance in the blood of a compound undergoing degradation in the gut:

$$\text{Rate of appearance} = kaDe^{-(k_a + k_d)t} \tag{3}$$

where k_a and k_d are the absorption and degradation rate constants, respectively, and D is the dose.

2. Binding

The binary interactions associated with the alteration of distribution can occur at both the plasma and tissue levels for the interacting compounds. Competition for these binding sites may result in displacement of the bound chemical from the binding site. For this to occur, the concentration of the displacing compound must approach the molar concentration of the sites on the binding protein, thereby reducing the sites available for binding (competitive inhibition). Equation (2a) can be used to represent these displacement processes. For example, in their PBPK model of tolbutamide-sulfonamide interaction in rats, Sugita et al. (73) used a modification of Equation (2a) to account for the competitive inhibition of sulfonamide for tolbutamide plasma protein binding.

In addition, noncompetitive binding inhibition is known to occur at binding sites on albumin as a result of tertiary conformational changes (74). Equation (2b) is appropriate to describe these interactions in a PBPK model.

3. Metabolism

a. *Inhibition*

Binary metabolic inhibition has been the most widely investigated pharmacokinetic interaction. Each of the mechanisms by which enzyme inhibition can occur (competitive, noncompetitive, uncompetitive, and substrate) can be represented by Equation (1).

b. *Induction*

Although numerous examples of enzyme induction have been documented, few investigators have attempted to model this process. Levy et al. (75)

developed an equation to describe the change in maximum velocity V_{max} that occurs during enzyme induction:

$$V_{max}(t) = V'_{max} - (V'_{max} - V_{max})e^{-k't} \tag{4}$$

where $V_{max} = K_2S/K$, and $V'_{max} = K_2S'/K'$. The terms S/K and S'/K' represent the basal and induced steady-state enzyme concentrations, respectively, K_2 is the rate constant for dissociation of the enzyme-substrate complex to product and free enzyme, and k' is the rate constant of degradation. This equation can be used in a PBPK model of binary mixture interactions in which enzyme induction is known to occur.

O'Flaherty (76) modeled the induction of caffeine metabolism by β-naphthoflavone and the appearance of its three first-generation metabolites by developing an individual PBPK model for the parent compound and each metabolite. The fit of the model was optimized to experimentally derived blood concentration data to obtain values for the metabolic parameters. The use of PBPK models to analyze the effect of coexposure on metabolic rate and target tissue binding was also demonstrated by Blancato (27).

c. *Cofactor Depletion*

As discussed previously, conjugation of a compound with GSH generally results in detoxification of that compound. For example, conjugation of the reactive metabolites of bromobenzene, acetaminophen, and chloroform with GSH prevents the covalent binding of these reactive intermediates to cellular constituents. A depletion of GSH results in an increase in the hepatotoxicity of these compounds. Conversely, GSH may be responsible for the bioactivation of such compounds as halogenated ethylenes. Therefore, a reduction in toxicity of these compounds results from GSH depletion. Clearly, the available levels of this cofactor significantly impact the resultant toxicity of a wide range of compounds.

D'Souza et al. (77) modeled the time-dependent depletion of hepatic GSH and incorporated this description of GSH depletion into a PBPK model for ethylene dichloride (EDC). The conjugation reaction of GSH with EDC also accounts for the reaction of the EDC metabolite, chloroacetaldehyde, with GSH. Therefore, this approach could be readily modified to account for the competition of two (or more) substrates for the available intracellular GSH pool in a PBPK model for mixtures.

4. Excretion

A physiologically based compartmental model for renal elimination interactions was developed by Russel and colleagues (78,79). A Michaelis-Menten constant to account for competitive interaction between substances with an affinity for the same carrier can be incorporated into the model. This model is ideal for describing tubular secretion interactions and can be connected to a PBPK model representing the rest of the body. In addition, noncompetitive interactions in active renal transport processes can occur as a result of the degradation of carriers (80). These interactions can be represented in a PBPK model by Equation (2b).

Interactions that result in changes in glomerular filtration rate due to hemodynamic alterations can be represented simply by using appropriate values for blood flow in the model.

5. Physiological Parameters

PBPK models require input values for physiological parameters, such as blood flow and organ volume. Values for these parameters are usually obtained from the literature or from compendia, such as the recent U.S. Environmental Protection Agency (81) document, *Reference Physiological Parameters in Pharmacokinetic Modeling*. However, since changes in regional blood flow can occur after exposure to chemical mixtures, the published blood flow values may not be accurate for use in PBPK models of chemical mixtures. Therefore, if values for blood flow changes are known, these data can be incorporated into the PBPK model. In the absence of such data, the modeler should recognize that the use of published blood flow values may introduce uncertainty into the model.

IV. IMPLICATIONS FOR RISK ASSESSMENT

Regulatory agencies are often presented with extensive pharmacokinetic data generated after exposure of experimental animals or humans to individual compounds. However, risk assessors must often estimate the equivalent internal dose of one compound in the presence of other compounds that may alter its pharmacokinetic behavior. Although a small but growing body of literature exists to document the known interactions between environmental pollutants or between drugs and pollutants, little guidance is available to the regulatory scientist on means to evaluate the significance of these interactions for risk assessment. To address this need, this review attempts to provide broad guidelines for (1) determining the significance of the interaction, (2) predicting when pharmacokinetically significant interactions are likely to occur, and (3) representing these interactions in a PBPK model. The implications of considering pharmacokinetic interactions for the risk assessment of drinking water contaminants are explored more fully here.

A. Evaluating the Significance of Pharmacokinetic Interactions

The importance of understanding the mechanism by which a pharmacokinetic interaction occurs cannot be overstated. This is especially true for metabolic interactions. For example, enzyme inhibition that occurs via a competitive mechanism is likely to be transient and short-lived; however, "suicide-substrate" inhibition that alters the structure of the enzyme is likely to have more important pharmacokinetic implications. In addition, if the mixture components undergo MFO-mediated biotransformation, knowledge of the cytochrome P_{450} isozymes primarily responsible for their metabolism is vital for predicting the likelihood that an interaction between the mixture components will occur.

McInnes and Brodie (31) point out two other factors that should be considered in assessing the potential significance of pharmacokinetic interactions:

> An appreciation of those drug interactions that really do matter can be best achieved by combining a practical knowledge of the pharmacological mechanisms involved with the awareness of the most vulnerable patients (those with little reserve capacity) and the drugs associated with the greatest risk (those with a narrow therapeutic index).

Although this guidance was developed for clinicians concerned with potential drug interactions, these principles are equally valid for the risk assessment of environmental pollutants. Populations at high risk for developing a toxic response from a pharmacokinetic interaction include not only those with diminished "reserve capacity" but also those whose intrinsic metabolic activity may be significantly higher or lower than the population mean because of enzyme polymorphism.

Although many of the interactions described in this chapter could produce a toxic effect that is greater than that produced by the individual compounds, it is important to remember that pharmacokinetic interactions could lead to reduced toxicity. Furthermore, the temporal pattern of exposure may produce increased or decreased toxicity for the same two compounds depending on the sequence in which they were administered. An example of such an effect exists for methylene chloride. This compound undergoes metabolism by both a GST- and a MFO-mediated pathway. A growing body of evidence exists to support a role for the GST pathway in producing a metabolite involved in the carcinogenicity of methylene chloride (37). However, the possibility that a metabolite produced by the MFO pathway also contributes to the formation of tumors in exposed animals cannot be excluded. Since this step is primarily catalyzed by cytochrome P_{450} 2E1 (82), exposure of individuals to inducers or inhibitors of P_{450} could result in an increase or decrease in MFO-mediated methylene chloride metabolism. In a series of studies, Pankow and colleagues (83–85) showed that simultaneous coexposure of rats to ethanol, pyrazole, benzene, toluene, or xylene and methylene chloride results in a decrease in carboxyhemoglobin (therefore carbon monoxide from the MFO pathway) production, but prior administration of these agents serves to enhance methylene chloride metabolism via the MFO pathway. The significance of these results for methylene chloride risk assessment obviously depends on the sequence and timing of the coexposures and whether the MFO pathway is involved in the production or detoxification of reactive metabolites. The nature of the inhibition was not defined in these studies, but the ability of these compounds to serve as substrates for P_{450} 2E1 suggests that competitive inhibition is occurring. If the carcinogenic mechanism for methylene chloride proposed by Andersen et al. (37) is correct, suppression of MFO activity in individuals coexposed to methylene chloride and certain industrial solvents or ethanol may result in the increased production of GST pathway metabolites with a subsequent increase in cancer risk. Conversely, enhanced MFO activity may reduce cancer risk from exposure to this compound.

This example emphasizes the difficulty in evaluating the significance of interactions even between two compounds that interact by a well-known mechanism. It is also possible that multiple pharmacokinetic processes could be impacted by the interaction, further complicating the assessment. There are numerous examples of multiple mechanisms playing a role in altering the pharmacokinetic behavior of mixture components. Waldron Lechner and Abdel-Rahman (5) reported that malathion decreased the rate of absorption of carbaryl as well as the altered tissue distribution of this pesticide. The clearance of toluene by the perfused rat liver in the presence of ethanol was reduced by changes in hepatic blood flow and intrinsic enzyme activity (86). Salicylate alters the elimination kinetics of parathion through combined effects on renal excretion and biotransformation (63).

The risk assessor should be aware that multiple pharmacokinetic interactions are possible; however, the majority of the interactions described in this chapter are the result of one compound affecting a single key determinant that mediates the pharmacokinetic behavior of another. Therefore, these principles should have broad applicability in risk assessment.

B. Pharmacokinetic Models of Chemical Mixtures

The ability of PBPK models to describe the pharmacokinetic behavior of important individual drinking water contaminants has been addressed

(Chapter 10), and the means to represent interactions in these models are described in this review. Along with data on the mechanism of toxicity and dose-response relationships, PBPK models of chemical mixtures can aid in the regulatory decision-making process by

Extrapolating the dose of parent compounds or metabolites in target tissues
Scaling up animal data to humans
Conducting route-to-route extrapolations

The ability of PBPK models to extrapolate a dose from one route to another allows risk assessors to take full advantage of the important pharmacokinetic interaction data on drinking water contaminants administered by different routes of exposure.

In addition, PBPK models can be used to ask "what if" questions and test hypotheses that are not easily testable by direct experimentation. Similarly, they can be used to devise or refine experimental studies of chemical mixtures.

Although it is possible to represent interactions in a PBPK models to obtain reasonable estimates of the pharmacokinetic behavior of mixture components, this representation may not be consistent with the presumed mechanism of the interaction. For example, Purcell et al. (87) were able to simulate the mutual inhibition of benzene and toluene metabolism in rats exposed by inhalation to this binary mixture by representing the interaction as a noncompetitive inhibition; however, in vitro data suggest that this interaction is likely to be due to competitive inhibition of one compound by another (40). Therefore, a PBPK model, like any mathematical model, should not necessarily be used to elucidate a biological mechanism of action but can be quite useful in a predictive or descriptive mode once a valid mechanism has been represented in the model.

Although PBPK models have the potential to assist in the evaluation of pharmacokinetic interactions, this methodology is not available to all risk assessors charged with setting exposure standards for drinking water contaminants, nor is this level of sophistication always necessary. As a first approximation, risk assessors may find that changes in classic pharmacokinetic parameters (e.g., clearance, area under the curve, and volume of distribution) are all that is required to draw conclusions about whether an interaction is occurring and whether the interaction is pharmacokinetically significant. The suggestions offered in this chapter for determining when an interaction is likely to be significant should be helpful to risk assessors charged with this responsibility.

C. Use of Pharmacokinetic Interaction Data in Risk Assessment

Once it has been established that a pharmacokinetic interaction is occurring, a major challenge facing regulatory scientists is to use this information in the risk assessment and regulatory decision-making process. As discussed here, such information may be useful in setting exposure standards that are protective of human health by identifying levels of the compound of concern in target tissues during coexposure to other pollutants. Knowledge of the way one compound affects the pharmacokinetic behavior of another is also valuable in the exposure assessment phase when biological monitoring data are used.

1. Set Drinking Water Standards

Vanderslice and colleagues (2) emphasized the problems faced by regulatory scientists in assessing the risks of mixtures of contaminants in drinking water. They provide an example of how data on the metabolic interaction between carbon tetrachloride and TCE were evaluated in deciding whether the existing drinking water standards for carbon tetrachloride were protective of human health. The existing data were found by these investigators to be inadequate for determining a no observed adverse effect level (NOAEL) for carbon tetrachloride in the presence of TCE. It is also not possible from the existing data to extrapolate the effects seen in rats to those expected in humans, nor is it possible to determine how the effects might differ if the compounds were administered by an alternative route of exposure. Knowledge of the mechanism by which TCE enhances carbon tetrachloride hepatotoxicity and a means to predict concentrations of these compounds in the liver at doses lower than those used in the study would serve as powerful tools in making these extrapolations.

2. Interpret Biological Monitoring Data

Exposure assessors involved in biological monitoring programs should be aware of the potential for simultaneous or previous exposure to some compounds to result in pharmacokinetic interactions that may influence the expression of an exposure biomarker of interest. Pharmacokinetic interactions may increase or decrease the levels of a parent compound in blood or exhaled air, as well as alter the levels of parent compound or metabolite appearing in the urine. Therefore, these interactions, if they occur, could pose difficulties for validation of the biomarkers, introduce uncertainties into a reconstructive exposure assessment based on biomarker data, and contribute to the interindividual variability in biomarker values.

Several examples of binary mixture interactions presented earlier in the chapter underscore the importance of considering the effect of interactions on the expression of biomarkers. For example, muconic acid serves as a common urinary biomarker for exposure to benzene; however, coexposure to benzene and toluene reduces the urinary excretion of this biomarker.

Biological monitoring is widely used to assess exposure of workers to toxic substances in occupational settings; however, it is being increasingly used to assess exposure to environmental pollutants in nonoccupational settings as well. In terms of significance, exposure of humans to environmental pollutants and to ethanol, drugs, or such agents as cigarette smoke may be more important than coexposure to environmental pollutants because of the relatively greater dose of compounds taken intentionally for therapeutic or recreational purposes. For example, cigarette smoke is a well-known inducer of enzyme activity. Enzyme induction may in turn alter the levels of the biomarker in blood or urine. As a result, it may be useful also to measure cotinine (a metabolite of nicotine) in the urine of subjects selected for a biological monitoring study of environmental pollutants as an index of their exposure to tobacco smoke. For similar reasons, knowledge of each person's drug and ethanol intake is also important for the interpretation of biomarker data.

In conclusion, pharmacokinetic interactions have the potential to alter the toxicity of environmental pollutants that exist as drinking water contaminants. Wosilait and Luecke (88) pointed out that "It has been said that drug interactions are not necessarily dangerous, but not knowing they are occurring can be hazardous." To this end, this chapter has attempted to provide risk assessors with broad guidelines to predict when interactions are likely to occur and to evaluate the significance of these interactions if they occur. Physiologically based pharmacokinetic models should serve as powerful tools to evaluate interactions between chemical mixture components. As with all models, however, they should not be used in lieu of data. Knowledge of mechanism of action and of special species and individual sensitivity must be considered when developing and implementing PBPK models. These models, when properly formulated, can take into account as much mechanistic data as the risk assessor has available. As such, PBPK models are a distinct improvement over the nondeterministic, curve-fitting models used in the past.

ACKNOWLEDGMENTS

Many of the concepts described in this chapter were discussed at an expert panel workshop held on September 9, 1989 in Bethesda, Maryland. This workshop was convened to provide guidance to the U.S. Environmental Protection Agency on issues relating to the pharmacokinetics of chemical mixture components. The authors gratefully acknowledge the input of the following workshop attendees: Linda Birnbaum, Ph.D., National Institutes of Health (NIH) (now at EPA); Kenneth Bischoff, Ph.D., University of Delaware; James Bruckner, Ph.D., University of Georgia; Robert Dedrick, Ph.D., NIH; Bruce Fowler, Ph.D., University of Maryland; Timothy Gerrity, Ph.D., EPA; Elaine Grose, Ph.D., EPA; Rogene Henderson, Ph.D., Lovelace Inhalation Toxicology Research Laboratory; Kim-Chi Hoang, EPA; Robert Lutz, Ph.D., NIH; Lorenz Rhomberg, Ph.D., EPA; and Sorrell Schwartz, Ph.D., Georgetown University. The authors also appreciate the insightful technical comments of Myron Weiner, Ph.D. of the University of Maryland and Barbara Davis, Ph.D. of the EPA on this chapter, as well as the editorial review of the chapter provided by Julie Dunlap, Ph.D. The views expressed in this paper are solely those of the authors and do not represent official policy.

REFERENCES

1. U.S. EPA, *Guidelines for Health Risk Assessment of Chemical Mixtures*, 51FR34014, September 24, 1986.
2. R. Vanderslice, J. Orme, E. Ohanian, and C. Sonich-Mullin, Problems in assessing the risks of mixtures of contaminants in drinking water, *Toxicol. Ind. Health 5*:747–755 (1989).
3. F. Murad and A. Gilman, Drug interactions, *Goodman and Gilman's Pharmacological Basis of Therapeutics* (A.G. Gilman, L.S. Goodman, T.W. Rall, and F. Murad, eds.), 7th ed., Macmillan, New York, 1985, pp. 1734–1750.
4. P.J. Neuvonen, Drug absorption interactions, *Drug Absorption* (L.F. Prescott and W.S. Nimmo, eds.), ADIS Press, New York, 1987, pp. 228–237.
5. D. Waldron Lechner and M.S. Abdel-Rahman, Kinetics of carbaryl and malathion in combination in the rat, *J. Toxicol. Environ. Health 18*:241–256 (1986).
6. K. Borg and H. Tjälve, Uptake of $^{63}Ni^{2+}$ in the central and peripheral nervous system of mice after oral administration: Effects of treatments with halogenated 8-hydroxyquinolines, *Toxicology 54*:59–68 (1989).
7. S. Jasim and H. Tjälve, Effect of zinc pyridinethione on the tissue disposition of nickel and cadmium in mice, *Acta Pharmacol. Toxicol. 59*:204–208 (1980).
8. S. Jasim and H. Tjälve, Effect of sodium pyridinethione on the uptake and distribution of nickel, cadmium and zinc in pregnant and non-pregnant mice, *Toxicology 38*:327–350 (1986).
9. R.T. Meister, *Farm Chemicals Handbook*, Meister Publishing, Willoughby, OH, 1990.
10. J.M. Ritchie, The aliphatic alcohols, *Goodman and Gilman's Pharmacological Basis of Therapeutics* (A.G. Gilman, L.S. Goodman, T.W. Rall, and F. Murad, eds.), 7th ed., Macmillan, New York, 1985, pp. 372–386.
11. ATSDR, *Toxicological Profile for Lead*, Agency for Toxic Substances and Disease Registry, 1989.

12. S.J.S. Flora and S.K. Tandon, Effect of combined exposure to lead and ethanol on some biochemical indices in the rat, *Biochem. Pharmacol. 36*:537–541 (1987).
13. T. Jorgenson, E. Meierhenry, C. Rushbrook, R. Bull, and M. Robinson, Carcinogenicity of chloroform in drinking water to male Osborne-Mendel rats and B6C3F1 mice, *Fund. Appl. Toxicol. 5*:760–769 (1985).
14. H.J. Kim, S. Odend'hal, R. Ramanathan, C.E. Dallas, S. Muralidhara, and J.V. Bruckner, Effect of oral dosing vehicles on acute hepatotoxicity of carbon tetrachloride (CCl_4) in rats, *Toxicologist 7*:265 (1986).
15. L.W. Condie, R.D. Laurie, and T. Mills, Effect of gavage vehicle on hepatotoxicity of carbon tetrachloride in CD-1 mice: Corn oil versus Tween-60 aqueous emulsion, *Fund. Appl. Toxicol. 7*:199–206 (1986).
16. P. Chieco, M.T. Moslen, and E.S. Reynolds, Effect of administrative vehicle on oral 1,1-dichloroethylene toxicity, *Toxicol. Appl. Pharmacol. 57*:146–155 (1981).
17. B.A. Merrick, M. Robinson, and L.W. Condie, Differing hepatotoxicity and lethality after subacute trichloroethylene exposure in aqueous or corn oil gavage vehicles in B6C3F1 mice, *J.Appl. Toxicol. 9*:15–21 (1989).
18. J. Withey, B. Collins, and P. Collins, Effect of vehicle on the pharmacokinetics and uptake of four halogenated hydrocarbons from the gastrointestinal tract of the rat, *J. Appl. Toxicol. 3*:249–253 (1983).
19. W. Jo, C. Weisel, and P. Lioy, Routes of chloroform exposure and body burden from showering with clorinated tap water, *Risk Anal. 10*:575–580 (1990).
20. W. Jo, C. Weisel, and P. Lioy, Chloroform exposure and the health risk associated with multiple uses of chlorinated tap water, *Risk Anal. 10*:581–585 (1990).
21. K. Bogen, B. Colston, and L. Machicao, Dermal absorption of dilute aqueous chloroform, trichloroethylene, and tetrachloroethylene in hairless guinea pigs, *Fund. Appl. Toxicol. 18*:30–39 (1992).
22. J.W. Wiechers, The barrier function of the skin in relation to percutaneous absorption of drugs, *Pharm. Weekbl.* [*Sci.*] *11*:185–198 (1989).
23. K.B. Sloan, S.A.M. Koch, K.G. Siver, and F.P. Flowers, Use of solubility parameters of drug and vehicle to predict flux through skin, *J. Invest. Dermatol. 87*:244–252 (1986).
24. J. Přiborsky, K. Takayama, Z. Přiborska, E. Mühlbachová, and N. Tsuneji, The influence of detergents on skin barrier properties, *Pharmacol. Toxicol. 70*:344–346 (1992).
25. Z.T. Chownhan and R. Prichard, Effect of surfactants on percutaneous abosrption of naproxen. I. Comparisons of rabbit, rat and human excised skin, *J. Pharm. Sci. 67*:1272–1274 (1978).
26. J.L. Zatz, Fundamentals of transdermal controlled drug administration: Physicochemical considerations, *Drug Dev. Ind. Pharm. 9*:561–577 (1983).

27. J.N. Blancato, Pharmacokinetics, chemical interactions and toxicological risk assessment in perspective. *Proceedings of the Fourth European ISSX Meeting* *1*:132 (1992).
28. S.J.S. Flora, M. Dhawan, and S.K. Tandon, Effects of combined exposure to aluminum and ethanol on aluminium body burden and some neuronal, hepatic and haematopoietic biochemical variables in rat, *Hum. Exp. Toxicol.* *10*:45–48 (1991).
29. L.W. Chang and H.A. Hartmann, Blood-brain barrier dysfunction in experimental mercury intoxication, *Acta Neuropath. (Berl.)* *21*:179–184 (1972).
30. J.B. Ulreich, B.J. Snider, E.J. Correa, B.J. Dawson, and S.J. Goldberg, Redistribution of chromium in organs of rats exposed to trichloroethylene (TCE) and dichloroethylene (DCE), *Toxicologist* *11*:217 (1991).
31. G.T. McInnes and M.J. Brodie, Drug interactions that matter: A critical reappraisal, *Drugs* *36*:83–110 (1988).
32. A.L. Lehninger, *Principles of Biochemistry*, Worth Publishers, New York, 1982.
33. B. Testa, Mechanisms of inhibition of xenobiotic-metabolizing enzymes, *Xenobiotica* *20*:1129–1137 (1990).
34. M.E. Andersen, K. Krishnan, and H.J. Clewell, Physiologically based pharmacokinetic analysis of interactions in simple mixtures, *Proceedings of the Fourth European ISSX Meeting* *1*:135 (1992).
35. P.R. Ortiz de Montellano, Suicide substrates for drug metabolizing enzymes: Mechanisms and biological consequences, *Prog. Drug Metab.* *11*:99–148 (1988).
36. M.E. Andersen, M.L. Gargas, H.J. Clewell, III, and K.M. Severyn, Quantitative evaluation of the metabolic interactions between trichloroethylene and 1,1-dichloroethylene in vivo using gas uptake methods, *Toxicol. Appl. Pharmacol.* *89*:149–157 (1987).
37. M.E. Andersen, H.J. Clewell, III, M.L. Gargas, F.A. Smith, and R.H. Reitz, Physiologically based pharmacokinetics and the risk assessment process for methylene chloride, *Toxicol. Appl. Pharmacol.* *87*:185–205 (1987).
38. M. Murray and G.F. Reidy, Selectivity in the inhibition of mammalian cytochromes P-450 by chemical agents, *Pharmacol. Rev.* *42*:85–101 (1990).
39. M. Ikeda, H. Ohtsji, and T. Imamura, In vivo suppression of benzene and styrene oxidation by co-administered toluene in rats and effects of phenobarbital, *Xenobiotica* *2*:101–106 (1972).
40. L.W. Andrews, E.W. Lee, C.M. Witmer, J.J. Kocsis, and R. Snyder, Effects of toluene on the metabolic disposition and hemopoietic toxicity of (H-3) benzene, *Biochem. Pharmacol.* *26*:293–300 (1977).
41. A. Sato and T. Nakajima, Dose-dependent metabolic interaction between benzene and toluene in vivo and in vitro, *Toxicol. Appl. Pharmacol.* *48*:249–256 (1979).
42. O. Inoue, K. Seiji, T. Watanabe, M. Kasahara, H. Nakatsuka, S. Yin, G. Li, S. Cai, C. Jin, and M. Ikeda, Mutual metabolic suppression between

benzene and toluene in man, *Int. Arch. Occup. Environ. Health 60*:15–20 (1988).

43. R.J. Laib, M. Tucholski, J.G. Filser, and G.A. Csanady, Pharmacokinetic interaction between 1,3-butadiene and styrene in Sprague-Dawley rats, *Arch. Toxicol. 66*:310–314 (1992).
44. D.W. Nebert, D.R. Nelson, M.J. Coon, R.W. Estabrook, R. Feyerseisen, Y. Fujii-Kuriyama, F.J. Gonzalez, F.P. Guengerich, I.C. Gunsalus, E.F. Johnson, J.C. Loper, R. Sato, M.R. Waterman, and D.J. Waxman, The P450 superfamily: Update on new sequences, gene mapping, and recommended nomenclature, *DNA Cell. Biol. 10*:1 (1991).
45. S.A. Wrighton and J.C. Stevens, The human hepatic cytochromes P450 involved in drug metabolism, *Crit. Rev. Toxicol. 22*:1–21 (1992).
46. J. Onrot, O. Shaheen, I. Biaggioni, M. Goldberg, J. Feely, G. Wilkinson, A. Hollister, and D. Robertson, Reduction of liver plasma flow by caffeine and theophylline, *Clin. Pharmacol. Ther. 40*:506–510 (1986).
47. G. Kochak, A. Rakhit, T. Thompson, and M. Hurley, Pentopril-cimetidine interaction caused by a reduction in hepatic blood flow, *J. Clin. Pharmacol. 28*:222–227 (1988).
48. G.R. Wilkinson and D.G. Shand, A physiologic approach to hepatic clearance. *Clin. Pharmacol. Ther. 18*:377–390 (1976).
49. Schary and Rowland 1983.
50. A.E. Ahmed, S.A. Soliman, J.P. Loh, and G.I. Hussein, Studies on the mechanism of haloacetonitriles toxicity: Inhibition of rat hepatic glutathione S-transferases in vitro, *Toxicol. Appl. Pharmacol. 100*:271–279 (1989).
51. L. Campbell, H.K. Wilson, A.M. Samuel, and D. Gompertz, Interactions of *m*-xylene and aspirin metabolism in man, *Br. J. Ind. Med. 45*:127–132 (1988).
52. D.J. Reed and M.W. Fariss, Glutathione depletion and susceptibility, *Pharmacol. Rev. 36*:25S–33S (1984).
53. T. Ekstrom and J. Hoberg, Chloroform-induced glutathione depletion and toxicity in freshly isolated hepatocytes. *Biochem. Pharmacol. 29*:3059–3065 (1980).
54. N.H. Stacey, Assessment of the toxicity of chemical mixtures and isolated rat hepatocytes: Cadmium and chloroform, *Fund. Appl. Toxicol. 9*:616–622 (1987).
55. O.J. Igwe, Biologically active intermediates generated by the reduced glutathione conjugation pathway, *Biochem. Pharmacol. 35*:2987–2994 (1986).
56. T.J. Monks, M.W. Anders, W. Dekant, J. Stevens, S.S. Lau, and P.J. van Bladeren, Glutathione conjugate mediated toxicities, *Toxicol. Appl. Pharmacol. 106*:1–19 (1990).
57. M. Barry and J. Feely, Enzyme induction and inhibition, *Pharmacol. Ther. 48*:71–94 (1986).
58. B.K. Park and N.R. Kitteringham, Assessment of enzyme induction and enzyme inhibition in humans: Toxicological implications, *Xenobiotica 20*:1171–85 (1990).

59. P. Robertson, E.L. White, and J.S. Bus, Effects of methyl ethyl ketone pretreatment on hepatic mixed-function oxidase activity and on in vivo metabolism of *n*-hexane, *Xenobiotica 19*:721–729 (1989).
60. L.G. Miller, Cigarettes and drug therapy: Pharmacokinetic and pharmacodynamic considerations, *Clin. Pharm. 9*:125–135 (1990).
61. D.J. Sequeira, C.S. Eyer, G.F. Cawley, T.G. Nick, and W.L. Backes, Ethylbenzene-mediated induction of cytochrome P450 isozymes in male and female rats, *Biochem. Pharmacol. 44*:1171–1182 (1992).
62. D.M. Silverman and R.A. Schatz, Pulmonary microsomal alterations following short-term low level inhalation of *p*-xylene in rats, *Toxicology 65*:271–281 (1991).
63. L.M. Martinez Tabche and F.A. Posadas del Rio, Effects of subchronic parathion administration on sodium salicylate excretion kinetics in female rats, *J. Appl. Toxicol. 9*:5–8 (1989).
64. L. Aarons, Kinetics of drug-drug interactions, *Pharmacol. Ther. 14*:321–34 (1981).
65. M.A. Medinsky, P.J. Sabourin, G. Lucier, L.S. Birnbaum, and R.F. Henderson, A physiological model for simulation of benzene metabolism by rats and mice, *Toxicol. Appl. Pharmacol. 99*:193–206 (1989).
66. C.C. Travis, M.T. Fox, W.M. Simmons, and B.F. Lyon, Co-exposure to gasoline vapor decreases benzene metabolism in Fischer-344 rats, *Toxicol. Lett. 62*:231–240 (1992).
67. R.A. Corley, A.L. Mendrala, F.A. Smith, D.A. Staats, M.L. Gargas, R.B. Conolly, U.E. Andersen, and R.H. Reitz, Development of a physiologically-based pharmacokinetic model for chloroform, *Toxicol. Appl. Pharmacol. 103*:512–527 (1990).
68. D.J. Paustenbach, H.J. Clewell, III, M.L. Gargas, and M.E. Andersen, A physiologically based pharmacokinetic model for inhaled carbon tetrachloride, *Toxicol. Appl. Pharmacol. 96*:191–211 (1988).
69. J.W. Dalley, P.K. Gupta, and C.T. Hung, A physiological pharmacokinetic model describing the disposition of lead in the absence and presence of L-ascorbic acid in rats, *Toxicol. Lett. 50*:337–348 (1990).
70. R.C. Ward, C.C. Travis, D.M. Hetrick, M.E. Andersen, and M.L. Gargas, Pharmacokinetics of tetrachloroethylene. *Toxicol Appl. Pharmacol. 93*: 108–117 (1988).
71. W.D. Brown, J.V. Setzer, R.B. Dick, F.C. Phipps, and L.K. Lowry, Body burden profiles of single and mixed solvent exposure, *J. Occup. Med. 29*:877–883 (1987).
72. V. Fiserova-Bergerova, Physiological models for pulmonary administration and elimination of inert vapors and gases, *Modeling of Inhalation Exposure to Vapors: Uptake, Distribution, and Elimination* (V. Fiserova-Bergerova, ed.), CRC Press, Boca Raton, FL, 1983, pp. 73–100.
73. O. Sugita, Y. Sawada, Y. Sugiyama, T. Iga, and M. Hanano, Physiologically based pharmacokinetics of drug-drug interaction: a study of tolbutamide-

sulfonamide interaction in rats. *J. Pharmacokinet. Biopharm.* *(10)3*:297–316 (1982).
74. McElnay and D'Arcy 1983.
75. R.H. Levy, A.A. Lai, and M.S. Dumain, Time-dependent kinetics. IV. Pharmacokinetic theory of enzyme induction, *J. Pharm. Sci.* *68*:398–399 (1979).
76. E.J. O'Flaherty, Application of pharmacokinetic principles to exposure to chemical mixtures, *Toxicol. Ind. Health* *5*:667–680 (1989).
77. R.W. D'Souza, W.R. Francis, and M.E. Andersen, Physiological model for tissue glutathione depletion and increased resynthesis after ethylene dichloride exposure, *J. Pharmacol. Exp. Ther.* *245*:563–568 (1987).
78. F.G.M. Russel, A.C. Wouterse, and C.A.M. van Ginneken, Physiologically based pharmacokinetic model for the renal clearance of phenolsulfophthalein and the interaction with probenecid and salicyluric acid, *J. Pharmacokinet. Biopharm.* *15*:349–368 (1987).
79. F.G.M. Russel, A.C. Wouterse, and C.A.M. van Ginneken, Physiologically based pharmacokinetic model for the renal clearance and the interaction with phenolsulfophthalein in the dog, *Drug Metab. Dispos.* *15*:695–701 (1987).
80. C.A. Van Ginneken and F.G. Russel, Saturable pharmacokinetics in the renal excretion of drugs. *Clin. Pharmacokinet.* *(16)1*:38–54 (1989).
81. U.S. EPA, *Reference Physiological Parameters in Pharmacokinetic Modeling*, Office of Health and Environmental Assessment, Washington, D.C., EPA/600/6-88/004, 1988.
82. D. Koop, Oxidative and reductive metabolism by cytochrome P450 2E1, *FASEB J.* *6*:724–730 (1992).
83. D. Pankow, S. Kretchmer, and P. Hoffmann, Inhibition and stimulation of dichloromethane metabolism to carbon monoxide in the rat following administration of ethanol, *Arch. Toxicol. Suppl.* 566 (1991).
84. D. Pankow, S. Kretschmer, and M. Weise, Effect of pyrazole on dichloromethane metabolism to carbon monoxide, *Arch. Toxicol. Suppl.* *14*:246–248 (1991).
85. D. Pankow, F. Matschiner, and H.J. Wiegmann, Influence of aromatic hydrocarbons on the metabolism of dichloromethane to carbon monoxide in rats, *Toxicology* 68:89–100 (1991).
86. M. Wallen, S. Holm, and M.B. Nordquist, Coexposure to toluene and *p*-xylene in man: Uptake and elimination, *Br. J. Ind. Med.* *42*:111–116 (1985).
87. K.J. Purcell, G.H. Cason, M.L. Gargas, M.E. Andersen, and C.C. Travis, in vivo metabolic interactions of benzene and toluene, *Toxicol. Lett.* *52*:141–152 (1990).
88. W.D. Wosilait and R.H. Luecke, Displacement interactions resulting from competition for binding sites on proteins, *Toxic Interactions* (R.S. Goldstein, W.R. Hewitt, and J.B. Hook, eds.), Academic Press, San Diego, 1990, pp. 116–147.

89. K. Borg, and H. Tjälve, Effect of thiram and dithiocarbamate pesticides on the gastrointestinal absorption and distribution of nickel and mice, *Toxicol. Lett. 42*:87–98 (1988).
90. M. Cikrt and V. Bencko, Mercury-selenium interaction: Distribution and excretion of $^{203}Hg^{2+}$ in rats after simultaneous administration of selenite or selenate, *Toxicol. Lett. 48*:159–164 (1989).
91. W.B. Diechman and M.L. Kepling, Dieldrin and DDT retention in dogs fed aldrin and DDT individually and as a mixture. *Toxicol. Appl. Pharmacol. 12(2)*:287 (1968).
92. M. Iscan, T. Coban, and M. Iscan, Combined effect of cadmium and nickel on rat hepatic monooxygenases: Possible stimulation of certain cytochrome P-450 isozymes, *Toxicol. Lett. 62*:191–199 (1992).
93. M.J. Deimling and R.C. Schnell, Effect of manganese on the hepatic microsomal mixed function oxidase enzyme system in the rat, *Fund. Appl. Toxicol. 4*:1009–1018 (1984).
94. O. Inoue, K. Seiji, H. Nakatsura, T. Watanabe, S.N. Yin, G.L. Li, S.X. Cai, C. Jin, and M. Ikeda, Urinary *t,t*-muconic acid as an indicator of exposure to benzene, *Br. J. Ind. Med. 46*:122–127 (1989).
95. A. Koizumi, T. Sadamoto, H. Naganuma, and M. Ikeda, Mechanisms for modification of bromobenzene hepatotoxicity by co-administered toluene and chlorobenzene. *Am. J. Ind. Med. 6*:241–250 (1984).
96. H.E. El-Masri, J.D. Tessari, and R.S.H. Yang, Physiologically based pharmacokinetic (PBPK) modeling of the toxicologic interaction between trichloroethylene (TCE) and 1,1-dichloroethane, *Toxicologist 13*:438 (1993).
97. K. Engstrom, V. Riihimaki, and A. Laine, Urinary disposition of ethylbenzene and *m*-xylene in man following separate and combined exposure, *Int. Arch. Occup. Environ. Health 54*:355–363 (1984).
98. L. Perbellini, R. Leone, M.E. Fracasso, F. Brugnone, and M.S. Venturini, Metabolic interaction between *n*-hexane and toluene in vivo and in vitro, *Int. Arch. Occup. Environ. Health 50*:351–358 (1982).
99. E. Shibata, J. Huang, Y. Ono, N. Hisanaga, M. Iwata, I. Saito, Y. Takeuchi, Changes in urinary *n*-hexane metabolites by co-exposure to various concentrations of methyl ethyl ketone and fixed *n*-hexane levels. *Arch. Toxicol. 64(2)*:165–8 (1990).
100. H.P. Ciuchta, G.M. Savell, and R.D. Spiker, Effect of alcohols and toluene upon methylene chloride-induced carboxyhemoglobin in the rat and monkey. *Toxicol. Appl. Pharmacol. 49*:347–354 (1979).
101 E. Wigaeus, S. Holm, and I. Astrand, Exposure to acetone: uptake and elimination in man. *Scand. J. Work Environ. Health 7*:84–94 (1981).
102. J. Normandeau, S. Chakrabarti, and J. Brodeur, Influence of simultaneous exposure to acrylonitrile and styrene on the toxicity and metabolism of styrene in rats. *Toxicol. Appl. Pharmacol. 75*:346–349 (1984).
103. M. Ikeda, Reciprocal metabolic inhibition of toluene and trichloroethylene in vivo and in vitro, *Int. Arch. Arbeitzmed. 33*:125–130 (1976).

104. T. Suzuki, S. Shimbo, and H. Nishitani, Muscular atrophy due to glue sniffing. *Int. Arch. Arbeitsmed. 33(2)*:115–123 (1974).
105. E. Wigaeus, A. Lof, and M. Norqvist, Uptake, distribution, metabolism, and elimination of styrene in man. A comparison between single exposure and co-exposure with acetone. *Br. J. Ind. Med. 41*:539–546 (1984).
106. M. Ogata and T. Fujii, Urinary excretion of hippuric acid and *m*-methylhippuric acid after administration of toluene and *m*-xylene mixture to rats. *Int. Arch. Occup. Environ. Health 43*:45–51 (1979).
107. R. Tardif, G.L. Plaa, and J. Brodeur, Evidence of metabolic interactions between toluene and xylene in rats after inhalation, *Toxicologist 9*:250 (1989).
108. R. Tardif, R. Goyal, and J. Brodeur, Assessment of occupational health risk from multiple exposure: Review of industrial solvent interaction and implication for biological monitoring of exposure, *Toxicol. Ind. Health 8*:37–52 (1992).
109. A. Koizuni, M. Kumai, and M. Ikeda, *In vivo* suppression of 1,1,1-trichloroethane metabolism by coadministered tetrachloroethylene: an inhalation study. *Bull. Environ. Contam. Toxicol. 29*:196–199 (1982).
110. J.M. Gallo, V. Srivatsan, and J.V. Bruckner, Pharmacokinetic (PK) evaluation of the interaction between carbon tetrachloride (CCl_4) and trichloroethylene (TCE), *Toxicologist 11*:218 (1991).
111. K. Seiji, O. Inoue, C. Jin, Y.T. Liu, S.X. Cai, M. Ohashi, T. Watanabe, H. Nakatsuka, T. Kawai, and M. Ikeda, Dose-excretion relationship in tetrachloroethylene exposed workers and the effect of tetrachloroethylene co-exposure and trichloroethylene metabolism, *Am. J. Ind. Med. 16*:675–684 (1989).
112. A.A. Constan, J.D. Tessari, and R.S.H. Yang, Effects of repeated exposures to a mixture of groundwater contaminants in F344 rats: A pilot study to PB-PK modeling of mixtures, *Toxicologist 13*:438 (1993).
113. H.A. Barton, J.A. Creech, C.S. Seckel, and C.S. Godin, Mixtures of chlorinated ethylenes: Physiologically based pharmacokinetic (PBPK) modeling, *Toxicologist 13*:357 (1993).
114. J.R. Creech and H. Barton, Inhalation uptake and metabolism of vinyl chloride (VC) and trichloroethylene (TCE) mixtures, *Toxicologist 13*:436 (1993).
115. K. Savolainen, V. Riihimäki, A. Laine, and J. Ketoni, Short-term exposure of human subjects to *m*-xylene and 1,1,1-trichloroethane, *Int. Arch. Occup. Environ. Health 49*:89–98 (1981).
116. J. Liira, V. Riihimäki, K. Engström, and P. Pfäffli, Coexposure of man to *m*-xylene and methyl-ethyl-ketone. *Scan. J. Work Environ. Health 14*:322–327 (1988).
117. F.P. Guengerich, D.H. Kim, and M. Iwasaki, Role of human cytochrome P-450 IIE1 in the oxidation of many low molecular weight cancer suspects, *Chem. Res. Toxicol. 4*:168–179 (1991).
118. A. Kobusch, B. Bailey, and P. du Souich, Enzyme induction by environmental agents: Effect on drug kinetics, *Interactions Between Drugs and*

Chemicals in Industrial Societies (G. Plaa, P. du Souich, and S. Erill, eds.), Elsevier, Amsterdam, 1987, pp. 29–46.

119. F.U. Saito, J.J. Kocsis, and R. Snyder, Effect of benzene on hepatic drug metabolism and ultrastructure. *Toxicol. Appl. Pharmacol. 26*:209–217 (1973).
120. S. Khan, M. Martin, A.D. Rahimtula, and J.F. Payne, Effect of a Prudhoe Bay crude oil on hepatic and placental drug metabolism in rats. *Can. J. Physiol. Pharmacol. 65(12)*:2400–2408 (1987).
121. J.F. Brady, F. Xiao, J.M. Gapac, S.M. Ning, and C.S. Yang, Alteration of rat liver microsomal monooxygenase activities by gasoline treatment. *Arch. Toxicol. 64(8)*:677–679 (1990).
122. D.M. Lapadula, C. Habig, R.P. Gupta, and M.B. Abou-Donia, Induction of cytochrome P450 isozymes by simulataneous inhalation exposure of hens to *n*-hexane and methyl *iso*-butyl ketone (M*i*BK), *Biochem. Pharmacol. 41*:877–883 (1991).
123. K. Zahlsen, K. Aarstad, and O.G. Nilsen, Inhalation of isopropanol: Induction of activating and deactivating enzymes in rat kidney and liver. Increased microsomal metabolism of *n*-hexane, *Toxicology 34*:57–66 (1985).
124. K. Pyykko, Time-course of effects of toluene on microsomal enzymes in rat liver, kidney and lung during and after inhalation exposure, *Chem. Biol. Interact. 44*:299–310 (1983).
125. R. Toftgård, O.G. Nilsen, and J. Gustafsson, Changes in rat liver microsomal cytochrome P-450 and enzymatic activities after the inhalation of *n*-hexane, xylene, methyl ethyl ketone and methylchloroform for four weeks, *Scand. J. Work Environ. Health 7*:31–37 (1981).
126. M.V. Evans and J.E. Simmons, Enhanced bioactivation and the hepatotoxicity of carbon tetrachloride in methanol pretreated rats, *Toxicologist 13*:178 (1993).
127. A. Sato, K. Endoh, T. Kaneko, and G. Johanson, Effects of consumption of ethanol on the biological monitoring of exposure to organic solvent vapours: A simulation study with trichloroethylene, *Br. J. Ind. Med. 48*:548–556 (1991).
128. S. Laparé, K. Krishnan, J. Brodeur, and R. Tardif, Physiologically-based modeling of the toxicokinetic interaction between toluene and xylene in humans, *Toxicologist 13*:356 (1993).

12

Need for Physiologically Based Toxicokinetic Models in Estimating Target Organ Dosage Following Oral Ingestion of Cadmium

John M. Frazier

The Johns Hopkins University,
Baltimore, Maryland

I. INTRODUCTION

The major goal of toxicological risk assessment is to estimate the quantitative risk to human populations of exposure to chemicals in the food, drinking water, and air. The general process by which the quantitative risk is estimated has been described previously (1,2). The role of physiologically based toxicokinetic (PBTK) modeling in this process is to provide better estimates of the concentration of the parent chemical and/or its metabolites at the target tissue. In the example to be discussed here, the need for a PBTK model for cadmium, the objective is to define the specifications of a mathematical model capable of describing the systemic kinetics of cadmium following exposure via the drinking water supply. Since renal toxicity is a major concern with cadmium, any useful model must focus on estimating the concentration of the reactive form of cadmium in the target cells, the renal tubular epithelial cells.

Cadmium is a particularly valuable example since it illustrates several general issues that arise in the risk assessment of chemicals in the environment. One issue is how to evaluate chemicals that have a long biological half-life. In such cases, the toxic chemical can accumulate in target organs over long periods of time, even at low levels of exposure, ultimately reaching the critical organ concentration X_c. The critical organ concentration is the average concentration in the organ when toxicity is first observed in any sensitive cell type within that organ. In the example discussed here, the critical organ concentration is the average renal cortical concentration when proximal tubular toxicity is first observed.

A second issue is the nature of the toxicity threshold observed in the target tissue (for cadmium, the kidney cortex). Threshold phenomena are a general concern in risk assessment, having a significant influence on the quantitation of dose-response relationships at low doses. When coupled to the slow rate of accumulation of a toxicant over a major portion of the lifetime, the expression of target organ toxicity becomes an extremely complicated issue. Overt signs of toxicity are late to develop, and once toxicity occurs, the effects are usually irreversible since it is difficult to reduce the body burden.

The mechanistic basis of threshold phenomena fall into several categories: (1) saturation of metabolic detoxication pathways; (2) saturation of binding to nontoxic sites on proteins in general or specific detoxication proteins (e.g., metallothionein, a low-molecular-weight, cysteine-rich protein capable of binding several metals, including cadmium, where binding is thought to serve as a detoxication mechanism (3–5); (3) depletion of a detoxication factor (e.g., glutathione); and (4) overwhelming of repair and/or compensation mechanisms, to name a few. In risk assessment, an empirical approach to account for threshold effects arising from these processes is to estimate the population distribution of the critical organ concentration. This distribution gives the probability that an individual has a critical organ concentration between X_c and $X_c + \Delta X_c$. In general, this distribution depends on age, sex, physical condition, existence of diseases, and many other host factors. For chemicals with long biological half-lives that therefore accumulate to varying degrees over the lifetime of the individual, the effect of age on the distribution of critical organ concentrations is particularly important.

Within the context of the critical organ concentration approach to risk assessment, the fundamental measure of risk is the probability that an individual has an organ concentration of the toxicant that exceeds the critical organ concentration. Since both the risk determining factors, actual organ concentration and critical organ concentration, vary from individual to individual, it is necessary for quantitation of the associated risk to develop models to estimate the probability distributions for each of these factors.

To apply these concepts to the renal toxicity resulting from cadmium exposure via drinking water, it is necessary to develop models for both (1) the long-term systemic toxicokinetics of cadmium (which must include an explicit description of reactive metal concentrations in the target tissue, the kidney proximal tubules), and (2) the critical organ concentration of cadmium in the kidney. The total lifetime risk for renal toxicity can then be computed as the integral of the overlap (product) of these two probability distributions. The risk can be evaluated for various possible exposure scenarios to determine the minimum regulatory requirements for cadmium in drinking water to assure an acceptable lifetime risk. Although this procedure has not been fully implemented, the requirements of a PBTK model for cadmium kinetics that can serve this purpose are described here.

II. GENERAL DESCRIPTION OF CADMIUM KINETICS IN HUMAN POPULATIONS

Many autopsy studies of cadmium concentrations in human tissues have been conducted (6–12). The major features of the toxicokinetics of cadmium in human liver and kidney cortex are illustrated in Figure 1. Cadmium concentrations in all organs are extremely low at birth because of the limited ability of cadmium to cross the placenta and deposit in the fetus (7). Over the lifetime of the individual, cadmium tends to accumulate preferentially in the liver and kidney, particularly the kidney cortex. The actual levels observed in these tissues depend on several factors. The geographical location is important, probably as a consequence of exposure in food and drinking water. The highest levels in nonoccupationally exposed individuals occur in Japan and the lowest levels in Sweden. A second factor that affects cadmium levels in general populations is cigarette smoking. Cadmium is present in cigarette smoke (13), and thus the inhalation route of exposure is important in these individuals. Cigarette smokers have almost twice the body burden of cadmium as nonsmokers from the same location (14).

A particularly interesting aspect of cadmium accumulation in human populations is the peaking of cadmium concentrations in the kidney cortex at between 50 and 60 years of age and the subsequent decrease in older individuals. This effect is thought to be a consequence of renal aging in which renal tubular tissue is preferentially lost (15). It is well documented that a similar effect is observed following occupational exposure to high concentrations of cadmium (16). Such exposure results in renal tubular toxicity and a concurrent decrease in renal cortical cadmium concentrations. One outstanding toxicological question is whether cadmium concentrations in the kidney tubules below the critical organ concentration for overt renal toxicity could contribute to an accelerated renal tubular aging effect. The development of a realistic PBTK model for cadmium kinetics in human beings will contribute to a better evaluation of the hazard posed by long-term accumulation of cadmium in the renal cortex.

Although the ultimate goal is to develop a PBTK model to describe the long-term behavior of cadmium in human beings, the data base derived

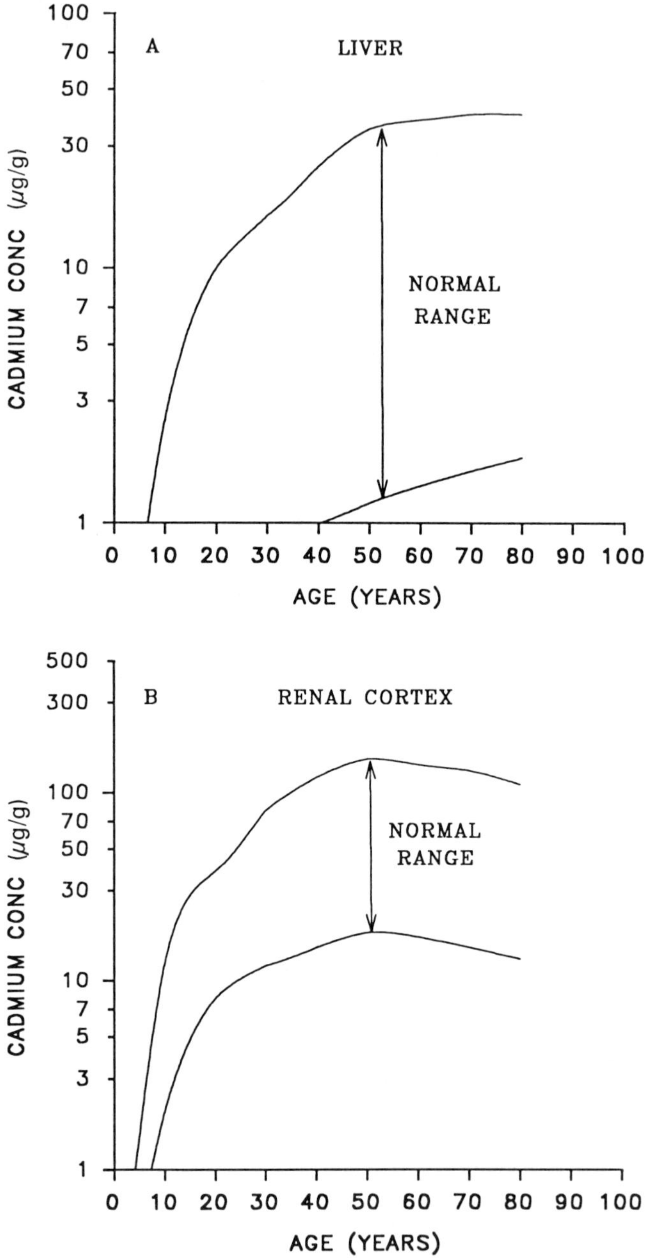

Figure 1 Cadmium concentrations in human liver (A) and kidney cortex (B) as a function of age. Data summarize the range of "normal" tissue concentrations obtained from several epidemiological studies of autopsy cases. (From Refs. 6–12.)

from the epidemiological studies provides only a framework in which human cadmium kinetics can be described. Essential experimental studies to elucidate the mechanisms of kinetic phenomena, such as the inducibility of metallothionein in the liver or kidney cortex and mechanisms of membrane transport, cannot be undertaken in human beings for ethical reasons. Therefore, it is necessary to investigate the basic mechanisms that control cadmium kinetics in mammalian systems by studying animal models. Based on the mechanistic understanding obtained from such experimental studies in animal models, a PBTK model for cadmium can be constructed and extrapolated to humans. This strategy is used as a framework for the discussion here.

III. EXPERIMENTAL STUDIES IN ANIMAL MODELS

A. Short-Term Studies

Basic knowledge concerning the biochemical and physiological mechanisms that control the toxicokinetics of cadmium in mammalian systems has been developed over the last 20 years by an international group of researchers (17,18). This discussion is not an attempt to review the literature but highlights some of the most important aspects relevant to the modeling issue. The major conclusions are summarized in Table 1, and brief comments are provided here.

One conspicuous observation following intravenous (IV) injections of ionic cadmium is the rapid clearance of cadmium from the systemic plasma (19). Within the first 15 minutes following an IV injection, greater than 90% of the dose is cleared from the plasma into soft tissues. The majority of the dose (greater than 60%) is accumulated in the liver, and less than 5% is accumulated in the kidneys. Over the next 24 h, little cadmium is excreted in the urine and less than 20% is excreted in the bile (20,21). These observations suggest three important conclusions. First, since IV injections result in widespread systemic distribution of the metal via the vasculature, that such a large fraction of the dose accumulates in the liver suggests that liver-specific mechanisms of accumulation are involved. Second, that clearance is so rapid suggests that transport mechanisms other than simple diffusion of the ionic species contribute to tissue uptake. Third, that little cadmium is excreted in the urine, coupled with limited accumulation of cadmium in the kidney, suggests that glomerular filtration of cadmium in plasma is limited, possibly as a consequence of cadmium binding to plasma proteins.

Immediately following hepatic accumulation of cadmium, the major portion of the intracellular metal is associated with high-molecular-weight

Table 1 Major Kinetic Observation in Animal Experiments That Must Be Described by a Mathematical Model for Cadmium Kinetics

Cadmium kinetics
At birth, the cadmium body burden is very small and accumulates over the lifetime of the individual
Renal cadmium concentrations continually increase until late middle age and then decrease
As cadmium dosage increases, qualitative differences in kinetics appear (e.g., hepatic accumulation of cadmium occurs at higher doses), indicating nonlinearities in system responses
Rapid selective transport of cadmium into the liver is an important early event in systemic processing
Cadmium is strongly bound to plasma albumin and high-molecular-weight intracellular ligands
Renal processing of cadmium and cadmium-protein complexes limits urinary excretion of cadmium
Metallothionein kinetics
The major portion of intracellular cadmium is associated with metallothionein
Under conditions of chronic exposure, the induction of metallothionein is a continuing process to balance protein degradation
Systemic processing of metal-metallothionein complexes is a controlling factor in renal accumulation of cadmium
Essential metal kinetics
Zinc concentrations in the liver and kidney are elevated concurrently with cadmium exposure, and most of the increased tissue burden is associated with metallothionein
Copper concentrations in the kidney cortex, particularly in the renal tubules, is elevated by cadmium exposure; renal tubular copper concentrations decrease earlier and more rapidly than cadmium concentrations

macromolecules in the hepatic cytosol (22,23). Over a period of the next 6 h, intracellular redistribution of cadmium occurs, resulting in the sequestering of cadmium in a newly synthesized metallothionein pool. The total binding capacity of the newly synthesized metallothionein generally exceeds the total hepatic cadmium burden, resulting in the accumulation of excess metal binding capacity in the cystosol. This excess capacity is occupied by endogenous zinc, resulting in mixed metal-metallothionein complexes, such as cadmium/zinc–metallothionein.

During the first few hours after injection, a small but measurable rate of biliary excretion of cadmium is observed (21). However, biliary excretion of cadmium shuts down when the majority of intrahepatic cadmium is bound to metallothionein. It appears that metallothionein-bound cadmium is not available for biliary excretion.

Cadmium is effectively trapped in hepatocytes as the metallothionein complex despite the fact that metallothionein has a finite lifetime (24–26). This trapping of cadmium is a consequence of excess metallothionein present in the cell as the zinc-metallothionein complex just mentioned. Since cadmium has a higher affinity for metallothionein than zinc, any cadmium released when cadmium metallothionein complexes are degraded immediately displaces zinc from existing zinc metallothionein. Thus, cadmium is trapped in the hepatocytes by a cyclic process of binding, release, and rebinding.

Cadmium, as the cadmium-metallothionein complex, is ultimately released from hepatocytes as the cells turn over during the normal life cycle of the cell. Hepatocytes have a lifetime of between 90 and 120 days. When the cell dies, the cellular membranes lyse releasing the intracellular contents into the hepatic circulation. Most intracellular cadmium is released into the plasma as the metallothionein complex. This complex is rapidly filtered at the renal glomerulus since metallothionein is a small protein (27–29). As long as the kidney proximal tubular epithelial cells function properly, metallothionein is efficiently reabsorbed from the tubular lumen by a receptor-mediated endocytotic mechanism that is a component of a peptide conservation pathway (30). Metallothionein, along with the associated metals, is processed by the endocytotic vesicle system and is thought to be degraded by lysosomal enzymes. The cadmium released is subsequently sequestered by renal metallothionein and trapped in the renal tubular cells. Current thinking suggests that renal metallothionein acts as a detoxication mechanism with a finite capacity. When this capacity is exceeded, that is, when renal cortical concentrations exceed 200 μg/g wet weight, the detoxication capacity of metallothionein is saturated and overt renal proximal tubular toxicity occurs. This is the basis of the threshold effect observed for cadmium-induced renal toxicity.

Based on these animal model studies, many components of cadmium toxicokinetics have been described. From the risk assessment point of view, the main objectives are to develop (1) a toxicokinetic model to describe the systemic processing of cadmium and (2) a description of the reactive form of cadmium, presumed to be the free cadmium pool, in the renal proximal tubules since this component determines the critical organ concentration.

B. 2 Year Cadmium Feeding Study

Several long-term studies of cadmium kinetics have been conducted in animal models. Of particular relevance to the issue at hand is a study conducted at Johns Hopkins University (31). This 2 year study involved exposing male Wistar rats to cadmium in drinking water at several dose levels (control and 0.5, 5.0, and 50 μg Cd/ml). Groups of animals from each dose level were sacrificed at 3 month intervals, and tissue concentrations of cadmium, zinc, and copper were determined by atomic absorption spectroscopy. Several useful data sets were obtained.

The concentration of cadmium in renal cortical tissue and isolated renal tubules was observed over the 2 year study. Figure 2 illustrates the cadmium concentrations in the renal cortex for the two highest dose groups. The pattern of cadmium accumulation over the lifetime of the animals is similar to that observed in human beings, that is, the concentration in renal cortex increases over a significant portion of the animal's lifetime, and then peaks and drops off in the aged animals. In Figure 3 the cadmium concentration in the renal tubules is compared with the renal cortical concentration for the highest dose group. It is interesting to note that the cadmium concentration in the renal tubules exceeds the average cortical concentration up to the peak and then drops off more rapidly in the last 6 months, ultimately ending lower than the average cortical concentration. In this animal model, the rapid decrease in tubular cadmium concentration could be accounted for by the renal aging process, since renal tubules degenerated at a faster rate than other renal tissues.

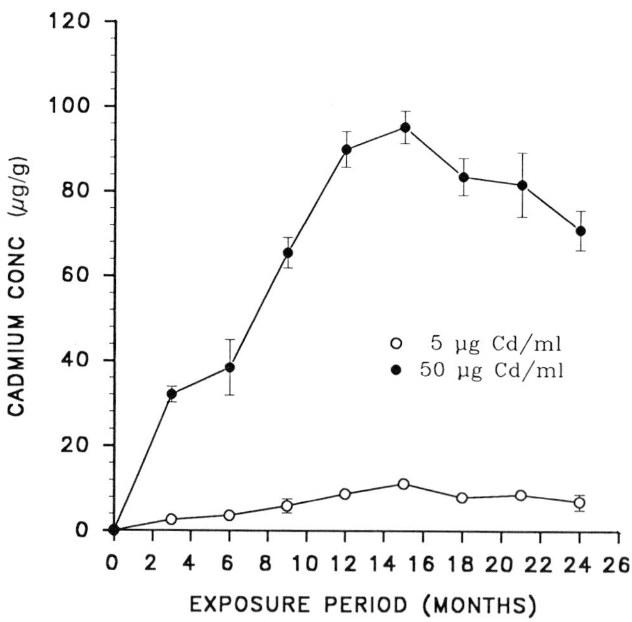

Figure 2 Renal cortical cadmium concentrations in rats during a 2 year cadmium feeding study. The tissue concentrations for the two highest dose groups (5 and 50 μg Cd/ml in drinking water) are illustrated. Data are from Ref. (31).

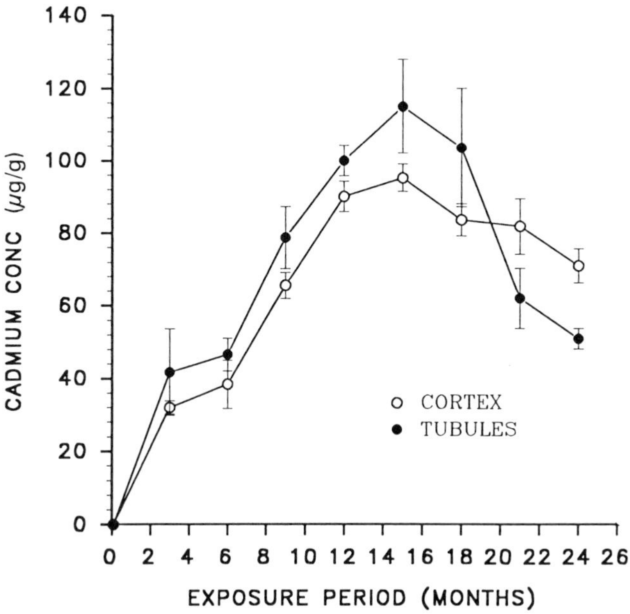

Figure 3 Comparison of cadmium concentrations in renal cortex and renal tubules of rats during a 2 year cadmium feeding study. Renal cortex and tubular cadmium concentrations were determined during exposure to 50 μg Cd/ml in drinking water. Data are from Ref. (31).

Cadmium concentrations were determined in other tissues: liver (Figure 4) and intestine (Figure 5). In the liver, the cadmium concentration increased continuously throughout the study at the 50 μg Cd/ml dose level, reaching the highest concentration at the termination of the study. At lower doses, the hepatic cadmium concentrations were much lower and there was no apparent tendency to accumulate cadmium between months 3 and 24. The loss of regulation of hepatic cadmium concentration at doses between 5.0 and 50 μg Cd/ml implies that hepatotoxic manifestations can be expected to develop at higher doses.

Further investigation of the distribution of cadmium in liver tissues indicated that the majority of the metal was bound to metallothionein in the hepatic cytosol. A Sephadex G-75 gel permeation chromatography profile for the highest dose group is given in Figure 6. As observed in experimental animals after large acute parenteral doses of cadmium, induction of metallothionein and the subsequent sequestration of cadmium is a major factor controlling hepatic intracellular kinetics of the metal following oral dosing.

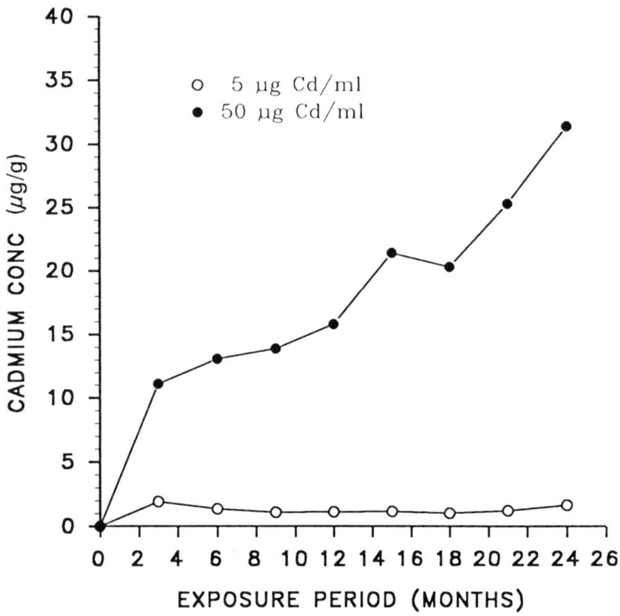

Figure 4 Hepatic cadmium concentrations in rats during a 2 year cadmium feeding study. Hepatic concentrations for the two highest doses (5 and 50 μg Cd/ml in drinking water) are illustrated. Data are from Ref. (31).

In the intestine (Figure 6), the cadmium concentration tended to reach a steady state at all dose levels between 3 and 6 months and to maintain those concentrations for the duration of the experiment. That the intestinal cadmium concentrations did not continue to increase even at the highest dose is probably a consequence of the relatively short lifetime of the epithelial cells lining the gastrointestinal (GI) tract. Although these cells are exposed to high concentrations of cadmium in the lumen of the GI tract and probably accumulate cadmium at a significant rate, the constant sloughing off of these cells limits the overall accumulation of cadmium in this tissue. Any attempts to model cadmium kinetics in the GI tract must take these processes into consideration.

The total body burden of cadmium gives some indication of the net balance of cadmium uptake from drinking water and elimination via all routes. Unfortunately, the total body burden was not directly determined in this study. However, since the cadmium content of the liver and kidney represent the major portion of the body burden, the "apparent body burden" (defined as the sum of the total cadmium content of the liver and

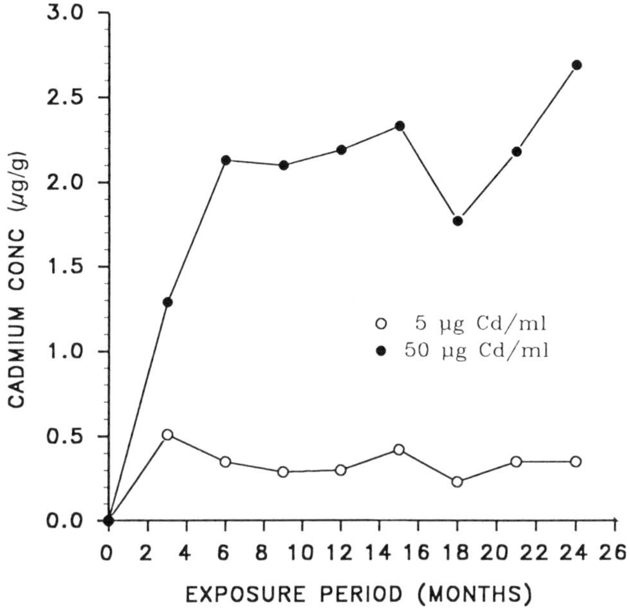

Figure 5 Intestinal cadmium concentrations in rats during a 2 year cadmium feeding study. Intestinal concentrations for the two highest doses (5 and 50 μg/ml in drinking water) are illustrated. Data are from Ref. (31).

both kidneys) could be determined. These data for the highest dose group are given in Figure 7. Even though the kidney has the highest cadmium concentrations, because of the greater mass of the liver, the liver contains the largest portion of the body burden. As a consequence, at the highest dose the apparent body burden increases throughout the lifetime of the rats, reflecting the steady increase in hepatic cadmium content despite the decrease in renal cadmium content.

An important aspect of cadmium exposure is the concurrent effect on essential metal metabolism. Both zinc and copper kinetics were altered in this study. The effect on essential metals is tissue specific. Cadmium exposure significantly increased hepatic zinc concentration at the highest dose, with little effect at lower doses (Figure 8). Renal zinc concentrations were slightly elevated, but the effect was not as dramatic as in the liver. The accumulation of zinc in the liver was accounted for by the increase in zinc associated with metallothionein (Figure 6). Any model to describe the toxicokinetics of cadmium must include a description of the effects on zinc metabolism, in particular the accumulation of zinc in the metallothionein pool.

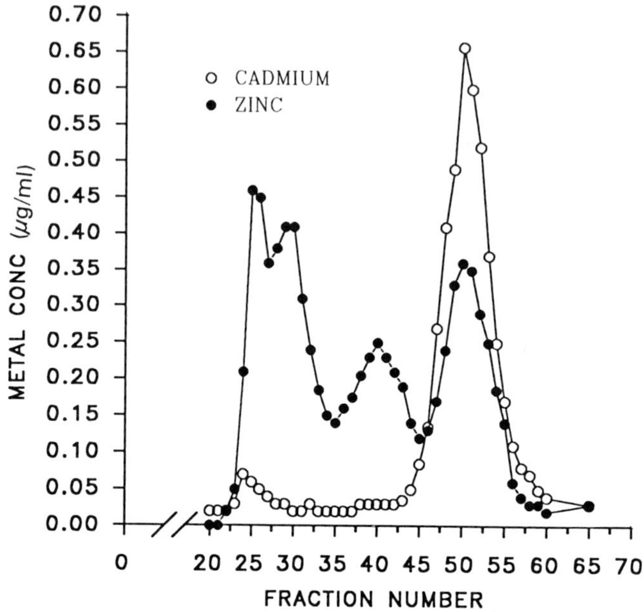

Figure 6 Distribution of cadmium and zinc among hepatic cytosolic ligands following cadmium exposure in drinking water for 12 months. Gel permeation chromatograph on Sephadex G-75 of hepatic cytosol obtained from male Wistar rats exposed to 50 μg Cd/ml.

Hepatic copper concentrations were not affected by cadmium exposure. However, renal tubular copper concentrations were increased (Figure 9). At the highest cadmium dose, this effect was apparent at 3 months and persisted throughout the study; at the 5.0 μg Cd/ml dose, the effect was significant only during the 12–18 month period. When the cortical concentration is compared to the tubular concentration (Figure 10), a pattern similar, but not identical, to that exhibited by cadmium concentrations is observed. Since copper is an essential metal, a measurable concentration of copper is present at the beginning of the study (at $t = 0$ corresponding to 28-day-old weanling rats). A more important effect from a toxicological point of view is the observation that the renal tubular copper concentration appears to decrease before tubular cadmium concentrations and the decrease is more dramatic. One hypothesis that could account for this effect is that tubular cells with high copper burdens have more metallothionein binding capacity occupied with copper and thus have less detoxification capacity for cadmium. Therefore, renal tubular cells with high copper

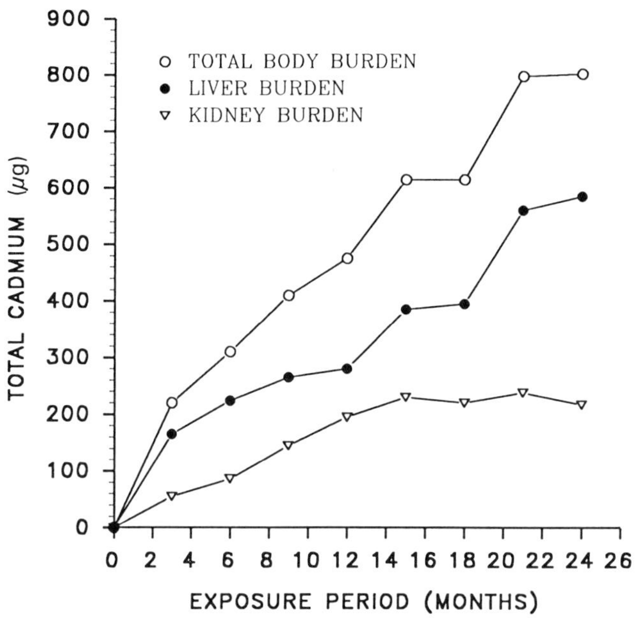

Figure 7 Cadmium "body burden" in rats during a 2 year cadmium feeding study. The body burden was computed as the sum of hepatic and renal cadmium for the two highest doses (5 and 50 μg Cd/ml). Data are from Ref. (31).

burdens are more susceptible to cadmium toxicity. This hypothesis should be more thoroughly tested since it implies an important toxicological interaction between an essential metal and a toxic metal.

The important features of cadmium kinetics that must be described by any mathematical model are summarized in Table 1. The major issues can be classified into three categories: (1) the basic kinetics of cadmium itself, (2) the behavior of metallothionein as a critical detoxication mechanism, and (3) interactions with essential metals.

IV. EXISTING MODELS FOR CADMIUM KINETICS

Several mathematical models have been developed to describe cadmium kinetics in mammalian systems (15,32–35). Of these models, the Nordberg-Kjellstrom model has been used most frequently for risk assessment purposes. This model is a linear multicompartment model that includes many of the known features of cadmium kinetics. The compartments are treated phenomenologically, and no attempt is made to identify associated organ

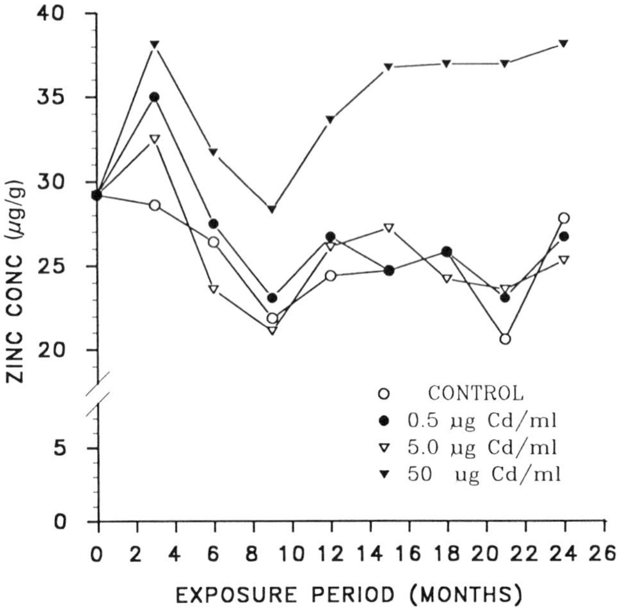

Figure 8 Hepatic zinc concentrations in rats during a 2 year cadmium feeding study. Hepatic zinc concentrations for all dose groups are illustrated. Data are from Ref. (31).

volumes or blood flows. Mass-transfer coefficients are constant with respect to cadmium concentrations (linear model), with the renal reabsorption coefficient (C_{17}) and the renal elimination coefficient (C_{19}) time dependent to account for the decrease in renal cadmium concentrations after 50 years of age. This model has served risk assessment needs well in the past; however, there are two major limitations. First, the linear nature of the model does not allow an adequate description of known nonlinearities in biological responses to cadmium dosing. Second, the phenomenological approach does not provide a basis for incorporating biological variability into the model to predict population distributions of responses. These limitations can be addressed by a PBTK model.

V. REQUIREMENTS FOR A PHYSIOLOGICALLY BASED TOXICOKINETIC MODEL FOR CADMIUM

The development of a PBTK model for cadmium in mammalian systems must take into consideration the major issues just outlined. A schematic

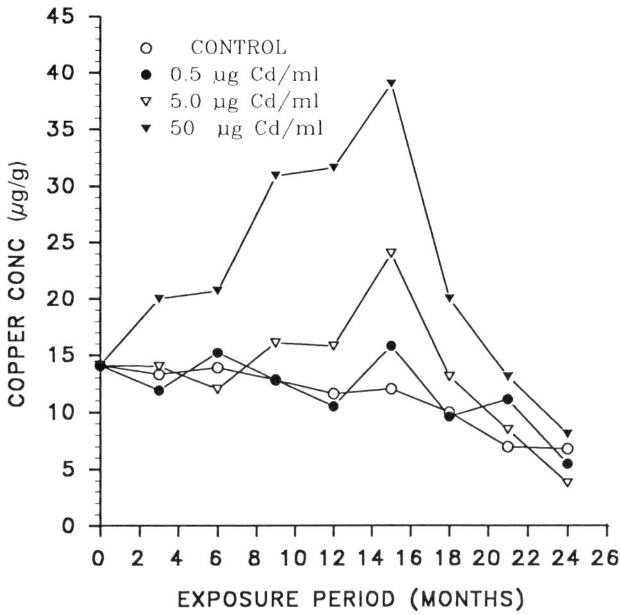

Figure 9 Renal tubular copper concentration in rats during a 2 year cadmium feeding study. Concentrations for all dose groups are illustrated. Data are from Ref. (31).

diagram of a general model for cadmium is given in Figure 11. The organs on which most of our attention is focused are the liver and kidney. In addition, it is essential to include the GI tract as the organ of exposure and uptake when dosing is via the drinking water. The "other tissue" is included to account for the mass balance of cadmium, since 20–30% of the body burden is nonspecifically distributed in muscle and other tissues. The lung is included in the model for two purposes: (1) to separate the venous and arterial blood compartments, and (2) to provide an exposure organ for inhalation of cadmium fumes, as in the case of cigarette smokers. In the model to be discussed, the inhalation route is not a primary source of exposure, but the model can be expanded if necessary.

The compartmental structure of each organ in a PBTK model depends on the degree of complexity required for an adequate description of the effects observed. The three organs, liver, other, and lung, can be described by a relatively simple compartmental structure (Figure 12). In this structure, a tissue consists of three connected compartments: the extracellular compartment (a combination of the plasma and interstitial spaces), the red

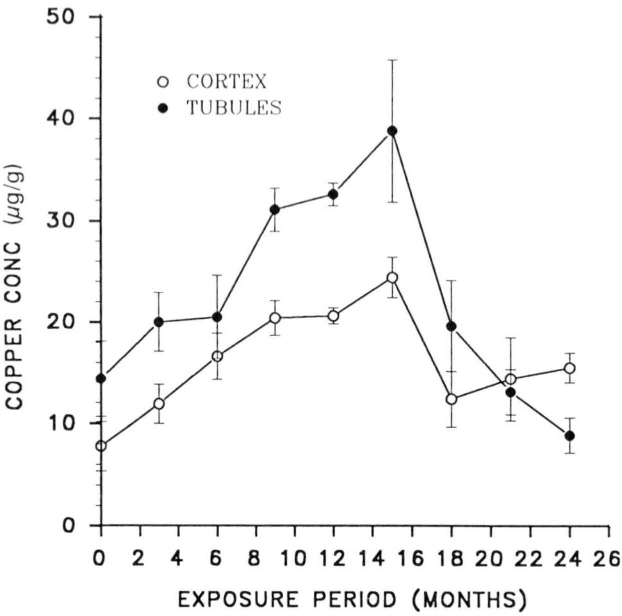

Figure 10 Comparison of renal cortical and renal tubular copper concentrations in rats during a 2 year cadmium feeding study. Concentrations for the 50 μg Cd/ml dose group are illustrated. Data are from Ref. (31).

blood cells (RBC; which are segregated to model RBC kinetics as a possible marker for longer term exposure), and the intracellular compartment. Each tissue also contains a lipid phase (the tissue fat) and a solid phase (which is relatively small for soft tissues) that are not connected to the model since cadmium does not accumulate in these compartments but must be included to account properly for compartmental volumes and tissue weight.

For more complicated organs (from the point of view of cadmium kinetics), such as the GI tract and the kidney, additional structural details must be included. Although the GI tract is described as a single compartment in this simple model, the lumen is modeled as a separate subcompartment. Thus, input to the system from the oral route is modeled as input to the GI lumen from which metals can be absorbed into the intracellular compartment of the GI tract and subsequently enter the systemic circulation. Metals that are not absorbed are eliminated via the fecal route. Metals that are absorbed into the systemic circulation pass directly to the hepatic circulation before mixing into the venous return, thus allowing "first-pass clearance" by the liver.

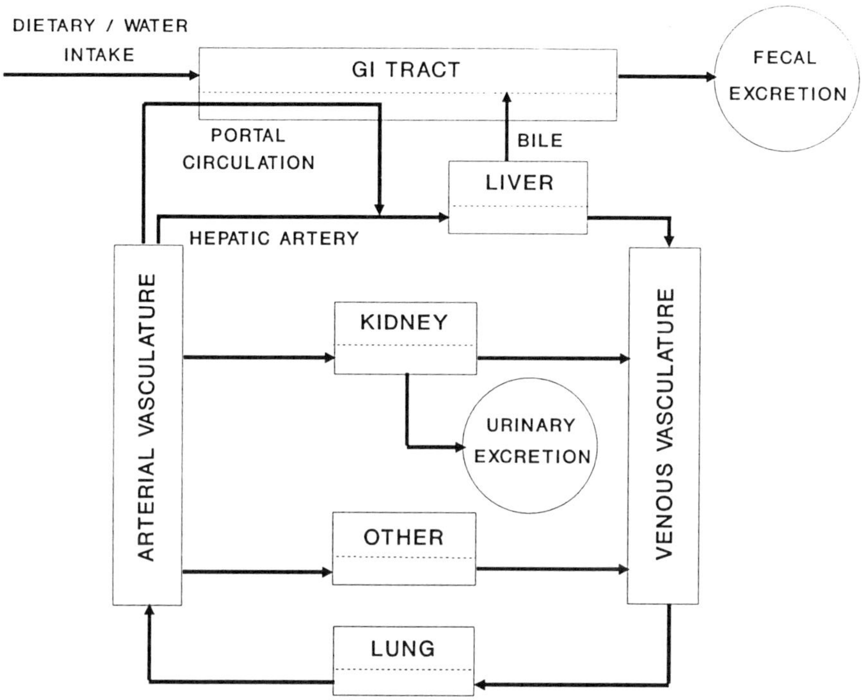

Figure 11 PBTK model for cadmium.

In the kidney (Figure 13), two aspects are particularly important: (1) glomerular filtration, which allows small molecules in the plasma to pass directly into the tubular lumen, and (2) the separation of tubular tissues, which have direct contact with the contents of both the vasculature and the lumen, from other renal tissues, which communicate only with the vasculature. This structure allows the explicit modeling of the tubular compartment where renal toxicity occurs. In addition, urinary excretion is expressed as a balance between three processes—glomerular filtration, reabsorption by tubular cells, and sloughing off of tubular cells—which are the primary mechanisms controlling this route of excretion.

Protein binding is an important aspect of metal kinetics in general and cadmium kinetics in particular. In any model of cadmium kinetics several ligands must be taken into consideration. In the plasma, binding to albumin is a dominant factor. Albumin binding prevents plasma cadmium from being filtered at the glomerulus unless it is bound to metallothionein. Metallothionein binding in the intracellular cytosolic compartment of var-

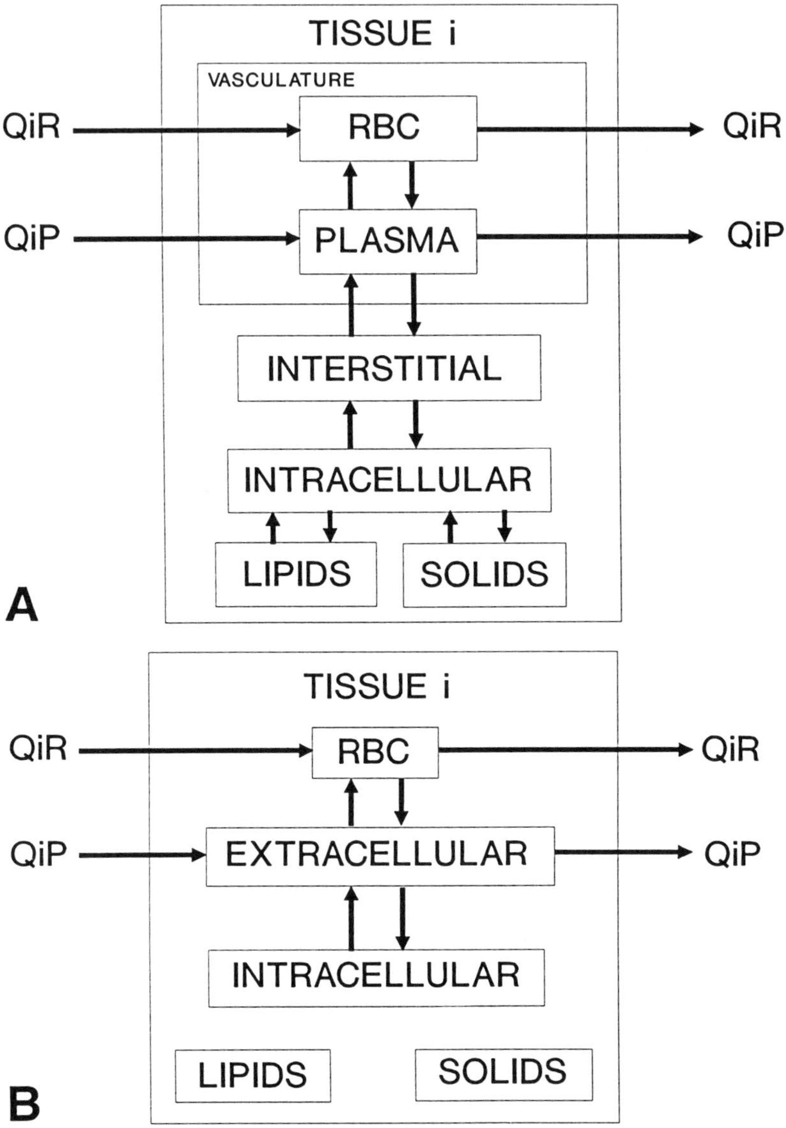

Figure 12 Compartmental structure of a general organ for the PBTK model. (A) In general, each organ is initially divided into six compartments: two vascular compartments consisting of the plasma and red blood cells, the interstitial aqueous space, the intracellular aqueous space, the lipid phase, and the solid phase. (B) Since cadmium does not accumulate to a significant level in either the lipid or solid

ious tissues acts as a detoxication mechanism and also traps the metal, reducing its availability for excretion by biliary and urinary routes. In addition, intracellular binding to cytosolic proteins (both enzymes and regulatory proteins) must be taken into consideration and is thought to contribute to cellular toxicity. For these reasons, protein binding must be included in various compartments of the model, and the distribution of cadmium between free and bound forms must be computed in each compartment.

Another interesting problem to be accounted for is the induction of metallothionein in both the liver and the kidney. Not only does existing metallothionein bind intracellular cadmium, but the binding capacity of this pool can expand in response to cadmium exposure. The issue of metallothionein induction has been actively investigated over the last 15 years (36). The generally accepted view of the induction process following cadmium loading assumes that there is a metal regulatory factor (MRF, presumably a protein) present in cellular cytosol that is activated by the presence of free cadmium (and other metals, including zinc and mercury), probably through a metal binding reaction. The activated metal-MRF complex migrates into the nucleus and binds to a metal regulatory element (a DNA sequence) that controls the expression of the metallothionein gene. Once the metal-MRF-DNA complex is formed, the transcription of the metallothionein gene is increased, ultimately increasing the metallothionein mRNA pool in the cytosol and the rate of metallothionein synthesis. As long as the metal-MRF complex exists, the rate of metallothionein mRNA synthesis is elevated and the metallothionein pool continues to increase above basal levels. Increasing the metallothionein pool provides a larger metal binding capacity that scavenges free cadmium as well as cadmium bound to intracellular ligands, which have a lower affinity for cadmium. Since the affinity of cadmium for metallothionein is extremely high ($K_A > 10^{15}$), cadmium is preferentially bound to metallothionein relative to other intracellular ligands. At some point sufficient metallothionein is synthesized by the cell to sequester essentially all the intracellular cadmium, including the metal that was bound to the MRF. At this time, little cadmium-MRF complex exists and the enhanced transcription of the

---

phase, in the cadmium PBTK model these two compartments are not connected kinetically to the model but merely account for physical space in organs. Furthermore, it is assumed that equilibrium is rapidly established between the plasma and interstitial spaces, and thus these two compartments are combined to form the extracellular compartment.

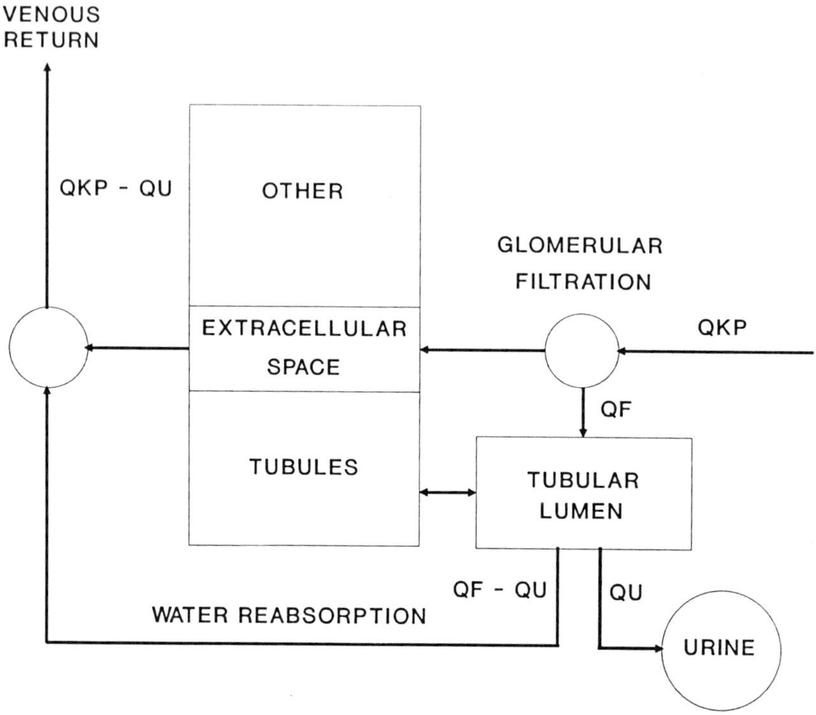

Figure 13 The kidney compartmental model. The kidney is divided into two major tissues: the tubular tissue and the other tissue, which represents all other cell types in the kidney. Plasma flow to the kidney is split by glomerular filtration, and a portion (the filtrate) flows into the tubular lumen. The remaining plasma perfuses the renal vasculature. Free metals and metals bound to metallothionein are filtered into the lumen at a rate dependent on the glomerular filtration rate. The tubular tissues are exposed to metals at both the vasculature and the luminal interfaces. Other tissues are exposed via the vasculature only. Metals not reabsorbed from the lumen by the tubular tissue are excreted via the urine. Metallothionein in the tubular lumen is reabsorbed into a lysosomal compartment and enters the tubular intracellular aqueous compartment following degradation. Free metal in the tissues can bind to high-molecular-weight macromolecules, bind to preexisting metallothionein, or induce new metallothionein synthesis. It should be noted that the concentration of metal species is altered by glomerular filtration, and this effect must be taken into account in the renal vasculature.

metallothionein gene dies out. This does not immediately halt metallothionein synthesis in the cytosol since the metallothionein mRNA has a finite lifetime. Thus, excess metallothionein is synthesized, "excess" in the sense that more metallothionein is synthesized than is needed to sequester the pool of intracellular cadmium. Zinc binds to this excess metallothionein, and at later times both cadmium and zinc metallothionein are present in the cell cytosol. When zinc metallothionein is degraded intracellularly, the zinc released mixes with the relatively large intracellular zinc pool, which rapidly readjusts to the homeostatically controlled level. On the other hand, when cadmium metallothionein degrades, the cadmium released can immediately displace zinc from preexisting zinc metallothionein and thus cadmium is trapped in the cell. These processes must be modeled to describe accurately cadmium kinetics in mammalian systems.

One final issue to be addressed is the kinetic interactions between cadmium and the essential metal zinc, which were just described. These interactions occur at several levels of the kinetic model. The three main aspects that must be taken into consideration are (1) competition for binding to proteins (albumin in plasma, high-molecular-weight macromolecules, and metallothionein intracellularly), (2) interactions at the level of metallothionein induction, and (3) competition for membrane transport. Competition for protein binding can be taken into account by including a competition factor in the binding relationship similar to the term that appears in the Michaelis-Menten relationship for substrate competition. For metallothionein binding, it is assumed that cadmium has an absolute binding preference over zinc. Thus, all available binding sites on metallothionein are filled with cadmium until either the intracellular free cadmium concentration is reduced to essentially zero or the metallothionein binding capacity is saturated. If any binding sites on metallothionein remained unoccupied after cadmium binding is satisfied, these excess sites are stoichiometrically filled with zinc. Transport competition is described by a competitive factor in the transport relationship for mediated metal transport. It can be assyumed that there is no competition for diffusional transport. With these kinetic interactions included in the model, it is clear that the zinc status of the individual can have a significant influence on the toxicokinetic behavior of cadmium. These cadmium-zinc interactions are not well studied at the critical sites, GI tract, liver, and kidney, that control the long-term accumulation of cadmium in the target organ.

Any PBTK model to describe cadmium kinetics for risk assessment must address these issues in a mechanistically realistic manner. Without the inclusion of specific mechanisms for protein binding, nonlinearities in transport processes, induction of metallothionein, and interactions with essential metals, it will be impossible to address quantitatively the fundamental problem of describing the population distribution of renal cadmium accumulation as a function of exposure and age. In addition, a mechanistically based PBTK model will also provide useful insights into the critical organ concentration issue since the threshold effect in the kidney is thought to be related to the detoxication properties of metallothionein. Thus, an adequate PBTK model for cadmium should not only account for the systemic distribution of cadmium but should also describe the free cadmium concentrations in the renal proximal tubular compartment, which controls the toxicodynamic responses. In this case, a good PBTK model will have multiple applications in the risk assessment process.

ACKNOWLEDGMENTS

The author thanks Ms. Dynita Parker for her secretarial assistance in the preparation of this manuscript. This project was made possible by support from the Johns Hopkins Center for Alternatives to Animal Testing and a grant from the International Life Sciences Institute–Risk Science Institute.

REFERENCES

1. N. Grossblatt (ed.), *Risk Assessment in the Federal Government: Managing the Process*, National Academic Press, 1983.
2. C.C. Travis (ed.), *Carcinogen Risk Assessment*, Plenum Press, New York, 1988.
3. D.H. Hamer, Metallothionein, *Annu. Rev. Biochem. 55*:913–951 (1986).
4. M.A. Dunn, T.L. Blalock, and R.J. Cousins, Metallothionein (42525A), *Proc. Soc. Exp. Biol. Med. 185*:107–119 (1987).
5. M.P. Waalkes and P.L. Goering, Metallothionein and other cadmium-binding proteins: Recent developments, *Chem. Res. Toxicol. 3*:281–287 (1990).
6. C.G. Elinder, T. Kjellström, L. Friberg, B. Lind, and L. Linnman, Cadmium in kidney cortex, liver, and pancreas from Swedish autopsies. Estimation of biological half time in kidney cortex, considering calorie intake and smoking habits, *Arch. Environ. Health 31*:292–302 (1976).
7. S.B. Gross, D.W. Yeager, and M.S. Middendorf, Cadmium in liver, kidney, and hair of humans, fetal through old age, *J. Toxicol. Environ. Health 2*:153–167 (1976).
8. T.L.M. Syversen, T.K. Stray, G.B. Syversen, and J. Ofstad, Cadmium and zinc in human liver and kidney, *Scand. J. Clin. Lab. Invest. 36*:251–256 (1976).
9. T. Kjellström, Exposure and accumulation of cadmium in populations from Japan, the United States, and Sweden, *Environ. Health Perspect. 28*:169–197 (1979).
10. N.E. Kowal, D.E. Johnson, D.F. Kraemer, and H.R. Pahren, Normal levels of cadmium in diet, urine, blood, and tissues of inhabitants of the United States, *J. Toxicol. Environ. Health 5*:995–1014 (1979).

11. E. Vuori, Huunan-Seppälä, J.O. Kilpiö, and S.S. Salmela, Biologically active metals in human tissues. II. The effect of age on the concentration of cadmium in aorta, heart, kidney, liver, lung, pancreas and skeletal muscle, *Scand. J. Work Environ. Health 5*:16–22 (1979).
12. R. Scott, E. Aughey, G.S. Fell, and M.J. Quinn, Cadmium concentrations in human kidneys from the UK, *Hum. Toxicol. 6*:111–119 (1987).
13. M. Nandi, H. Jick, D. Slone, and S. Shapiro, Cadmium content of cigarettes, *Lancet 1*:1329–1330 (1969).
14. G.P. Lewis, W.J. Jusko, L.L. Coughlin, and S. Hartz, Contribution of cigarette smoking to cadmium accumulation in man, *Lancet 5*:291–292 (1972).
15. C.C. Travis and A.G. Haddock, Interpretation of the observed age-dependency of cadmium body burdens in man, *Environ. Res. 22*:46–60 (1980).
16. K.J. Ellis and S.H. Cohn, Cadmium inhalation exposure estimates: Their significance with respect to kidney and liver cadmium burden, *J. Toxicol. Environ. Health 15*:173–187 (1985).
17. G.F. Nordberg, Cadmium metabolism and toxicity. Experimental studies on mice with special reference to the use of biological materials as indices of retention and the possible role of metallothionein in transport and detoxification of cadmium, *Environ. Physiol. Biochem. 2*:7–36 (1972).
18. I. Bremner, Mammalian absorption, transport and excretion of cadmium, *The Chemistry, Biochemistry and Biology of Cadmium* (Webb, ed.), Elsevier/North-Holland Biomedical Press, Amsterdam, 1979, pp.
19. J.M. Frazier, Cadmium and zinc kinetics in rat plasma following intravenous injection, *J. Toxicol. Environ. Health 6*:503–518.
20. M.G. Cherian and J.J. Vostal, Biliary excretion of cadmium in rat. I. Dose-dependent biliary excretion and the form of cadmium in the bile, *J. Toxicol. Environ. Health 2*:945–954 (1977).
21. C.D. Klaassen and F.N. Kotsonis, Biliary excretion of cadmium in the rat, rabbit, and dog, *Toxicol. Appl. Pharmacol. 41*:101–112 (1977).
22. M. Cempel and M. Webb, The time-course of cadmium-thionein synthesis in the rat, *Biochem. Pharmacol. 25*:2067–2071 (1976).
23. J.M. Frazier and J. Puglese, Dose dependence of cadmium kinetics in the rat liver following intravenous injection, *Toxicol. Appl. Pharmacol. 43*:461–474 (1978).
24. S.L. Feldman, K.S. Squibb, and R.J. Cousins, Degradation of cadmium-thionein in rat liver and kidney, *J. Toxicol. Environ. Health 4*:805–813 (1978).
25. S.H. Oh, J.T. Deagen, P.D. Whanger, and P.H. Weswig, Biological function of metallothionein. IV. Biosynthesis and degradation of liver and kidney metallothionein in rats fed diets containing zinc or cadmium, *Bioinorg. Chem. 8*:245–254 (1978).
26. K. Cain and D.E. Holt, Metallothionein degradation: Metal composition as a controlling factor, *Chem.-Biol. Interact. 28*:91–106 (1979).
27. M.G. Cherian and Z.A. Shaikh, Metabolism of intravenously injected cadmium-binding protein, *Biochem. Biophys. Res. Commun. 65*:863–869 (1975).

28. K. Tanaka, K. Sueda, S. Onosaka, and K. Okahara, Fate of ^{109}Cd-labeled metallothionein in rats, *Toxicol. Appl. Pharmacol. 33*:258–266 (1975).
29. M. Webb and A.T. Etienne, Studies on the toxicity and metabolism of cadmium-thionein, *Biochem. Pharmacol. 26*:25–30 (1977).
30. E.C. Foulks, Renal tubular transport of cadmium-metallothionein, *Toxicol. Appl. Pharmacol. 45*:505–512 (1978).
31. S.A. Perlin, K. Kawata, and J.M. Frazier, The effects of cadmium on renal aging—A chronic cadmium feeding study in rats, EPA Project Summary Report No. EPA-600/S1-84-007, 1984.
32. T. Kjellström and G.F. Nordberg, A kinetic model of cadmium metabolism in the human being, *Environ. Res. 16*:248–269 (1978).
33. G.F. Nordberg and T. Kjellström, Metabolic model for cadmium in man, *Environ. Health Perspect. 28*:211–217 (1979).
34. J. Matsubara-Kahn, Compartmental analysis for the evaluation of biological half-lives of cadmium and mercury in mouse organs, *Environ. Res. 7*:54–67 (1974).
35. K.E. Shank, R.J. Vetter, and P.L. Ziemer, A mathematical model of cadmium transport in a biological system, *Environ. Res. 13*:209–214 (1977).
36. D.J. Thiele, Metal-regulated transcription in eukaryotes, *Nucleic Acids Res. 20*:1183–1189 (1992).

13

Estimating Dermal and Inhalation Exposure to Volatile Chemicals in Domestic Water

John Schaum, Kim Hoang, Russell Kinerson, Jackie Moya, and Rhoda G. M. Wang*

U.S. Environmental Protection Agency, Washington, D.C.

I. INTRODUCTION

This chapter presents a methodology for estimating exposure to volatile compounds in domestic water supplies. The method encompasses two exposure pathways: inhalation and dermal contact. The importance of these exposures are evaluated by comparing them to direct ingestion of drinking water.

II. ESTIMATING EXPOSURE TO VOLATILE CHEMICALS

A. Inhalation Exposure

The following discussion was developed to provide a screening method for estimating the indoor air concentrations of volatile chemicals from indoor water uses and the resulting human inhalation exposures, with an emphasis on showers. This procedure evolved from research done by Andelman at the University of Pittsburgh under funding from the Exposure Assessment Group at the U.S. Environmental Protection Agency (EPA) in Washington, D.C. This research also involved the development of a more sophisticated model, MAVRIQ (model for analysis of volatiles and residential

---

**Present affiliation*: California Environmental Protection Agency, Sacramento, California

indoor air quality). This model should be available soon and will allow refinement of exposure estimates derived from the screening procedure since it more accurately accounts for human behavioral and water use patterns.

1. Showering Exposure

As a first step in a screening procedure, the assessor can use a worst-case type of analysis to evaluate the possible exposure during showering. The maximum possible concentration of a contaminant in air (as a result of volatilization from water) can be estimated using the Henry's law constant. This approach predicts the equilibrium partitioning between the water and air phases. In reality, such conditions are expected to occur only in a closed system or at the water/air interface in an open system. This equilibrium level in air is easily estimated as

$$C_a = C_w H \tag{1}$$

where:
C_a = concentration of contaminant in air, mg/m^3
C_w = concentration of contaminant in water, mg/L × 1000 L/m^3
H = Henry's law constant (unitless)

The unitless Henry's law constant can be calculated from the more traditional form of the Henry's law constant by using the equation

$$H = \frac{H'}{RT} \tag{2}$$

where:
H' = Henry's law constant, atm-m^3/mol
R = gas constant, atm-m^3/mol-K
T = temperature, K

The health risk associated with this concentration can then be evaluated. For noncarcinogens, this concentration is compared to the reference concentration (RfC, as reported in the EPA integrated risk information system, IRIS). For carcinogens, this concentration is multiplied by the inhalation unit risk (also as reported in IRIS) to yield a risk estimate. Both these approaches assume chronic continuous exposures, which is clearly not the case for showering exposures. As shown later, the exposure can be easily averaged over time to make the exposure estimate more realistic, but the intent in this first screening step is to keep the approach as simple as possible. Also, this approach is consistent with the philosophy of a worst-case analysis, that is, if the risks estimated via this approach are low, then it is safe to conclude that this pathway is not a concern. If the risks are not low, however, then more sophisticated approaches are needed to draw conclusions.

As a second step, a somewhat more realistic yet simple model can be used to predict possible air concentrations. This model treats the bathroom as one compartment and yields an air concentration averaged over the time of the actual shower and the time spent in the bathroom after the shower. The model was derived by assuming that the contaminant volatilizes at a constant rate and instantly mixed uniformly with the bathroom air and that ventilation with clean air does not occur. This implies that the contaminant in the air increases linearly from zero to a maximum at the end of the shower and then remains constant during the time an individual spends in the bathroom immediately after the shower.

$$C_a = \frac{(C_{a\max}/2)t_1 + C_{a\max}t_2}{t_1 + t_2} \tag{3}$$

where:

C_a = concentration of contaminant in air, mg/m^3
$C_{a\max}$ = maximum concentration of contaminant air, mg/m^3
t_1 = time of shower, h
t_2 = time after shower, h

$C_{a\max}$ is estimated as

$$C_{a\max} = \frac{C_w f F_w t_1}{V_a} \tag{4}$$

where:

$C_{a\max}$ = maximum air concentration in bathroom, mg/m^3
C_w = water concentration, mg/L
f = fraction volatilized (unitless)
F_W = water flow rate, L/h
V_a = bathroom volume, m^3

Water concentration is the concentration of the contaminant in the water at as it enters the shower. It must be determined on a site-specific basis.

Fraction volatilized (or *transfer efficiency*) is the mass fraction of contaminant in water that volatilizes over the course of the shower. It is a chemical-specific value that is not easily predicted. The volatilization rates depend on such properties as Henry's law constant and molecular weight, but the relationship has not been well established. Andelman (1) reports volatilization fractions ranging from 0.5 to 0.9 based on experiments with trichloroethylene and chloroform. These chemicals are assumed to be representative of other volatiles (i.e., chemicals with Henry's constants that are similar or greater). However, this range of values for fraction volatilized would not be applicable to less volatile chemicals. McKone (2) has suggested an approach in which the volatilization fraction for an untested chemical can be predicted from a tested chemical using a ratio of their overall mass-transfer coefficients:

$$f_i = f_j \frac{(2.5/D_w^{0.67} + RT/D_a^{0.67}H)_j}{(2.5/D_w^{0.67} + RT/D_a^{0.67}H)_i} \tag{5}$$

where:

f_i = volatilization fraction for chemical i
f_j = volatilization fraction for chemical j
D_a = diffusion coefficient in air, m^2/s
D_w = diffusion coefficient in water, m^2/s
R = gas constant, atm-m^3/mol-K = 8.21×10^{-5}
H = Henry's law constant, atm-m^3/mol-K
T = temperature, K

Water flow rate is the rate at which water flows into the shower. Bond et al. (3) report an average per capita water use of 75 L/day for bathing. Assuming that this represents one 0.1 h shower, then the actual flow rate during the shower would be 750 L/h. Such flow rates vary depending on a variety of site-specific factors, such as available water pressure, pipe size, shower head design, and personal preference. For default purposes, a range of 500–1000 L/h is assumed.

Bathroom volume is volume of the bathroom including the shower stall. No relevant data could be found. Assuming an 8 foot ceiling and floor dimensions ranging from 5×5 to 7×10 feet, the volumes range from 200 to 560 ft^3, or 6 to 16 m^3.

Shower time is the actual time of the shower. A survey in Australia of 2500 households found that shower times average about 7 minutes and have a 90th percentile of about 12 minutes (4).

Time after shower is the time an individual spends in the bathroom after the shower is turned off. No relevant data could be found. Obviously this could vary significantly depending on individual behavior patterns. Based on judgment, it was assumed to vary from 10 to 30 minutes, or 0.2 to 0.5 h.

The default values to use when data are not available are summarized in Table 1.

The assumptions of constant volatilization and no ventilation tend to make this model very conservative; that is, it is likely to overestimate actual levels. Additional assumptions made by this model include the following:

The concentration of the chemical in the air is assumed to be zero at the beginning of the shower, that is, any residual contaminant concentra-

Table 1 Recommended Default Values for Shower Model

Variable	Value or range	Reference
Fraction volatilized, f	0.5–0.9[a]	1
Water flow rate F_W, L/h	500–1000	See text
Shower period t_1, h	0.08–0.3	4
After shower period t_2, h	0.2–0.5	See text
Bathroom volume V_a, m^3	6–16	See text
Exposure duration ED, years	9–30	5

[a]This range applies to volatiles.

tions (remaining from previous showers taken by other members of the family or other water use activities) are assumed negligible.

The exchange between the air in the shower chamber and that in the bathroom is so rapid that the combined volume of these two compartments can be treated as a single chamber with a single concentration of volatilized chemical.

The model does not account for the exchange rate that occurs when an exhaust fan is turned on. Modeling results using MAVRIQ indicate that exposure could be reduced by about 20% if an average-sized exhaust fan is used.

Once the concentration has been estimated, the risk is then evaluated using the RfC or inhalation unit risk for the chemical of concern. However, an adjustment should be made to reflect that the exposure occurs intermittently rather than continuously:

$$\mathrm{HI} = \frac{C_a}{RfC}\left(\frac{t_1 + t_2}{24\ \mathrm{h}}\right) \qquad \text{noncarcinogens} \tag{6}$$

where:
HI = hazard index (unitless)

$$\mathrm{Risk} = 1 - e^{-X} \qquad \text{carcinogens} \tag{7}$$

where:

$$X = \mathrm{UR}C_a\left(\frac{t_1 + t_2}{24\ \mathrm{h}}\right)\left(\frac{\mathrm{ED}}{70\ \text{years}}\right)$$

and
UR = unit risk, m^3/mg
ED = exposure duration, years

The exposure duration term in these equations represents the overall time of exposure. This period could be determined by the time that contamination of the water supply persists or by the time an individual uses the contaminated water supply. Lacking site-specific data, a default range of 9–30 years is recommended, which corresponds to the average and 90th percentile time that a person is likely to live at the same residence (5).

Since the risk factors in Equations (6) and (7) were derived from continuous exposure experiments, application to an intermittent exposure (even with the time adjustments) introduces considerable uncertainty. Ideally, pharmacokinetic models are used to estimate how the dose delivered to the target organ changes relative to a continuous exposure situation. Alternatively, the assessment approach can be shifted to the exposure associated with inhaling contaminants throughout the house.

Although this approach does not account for the short-term peak exposures occurring during showers, it has two advantages. First, it better represents the total exposure to volatiles that may occur. Second, it involves more continuous exposures, which better match the conditions for which the risk factors were designed and therefore should provide more valid risk estimates (unless threshold effects occur during the short-term peak exposures). This approach is presented here.

2. Whole-House Exposure

A similar model may be used to describe the average indoor air concentrations that occur as a result of volatilization from water use throughout the house. Water uses that can contribute volatiles to the air include dish washing, bathing, clothes washing, and cooking. The model does not address the time and space variations encountered throughout the day in the home and thus does not provide air concentrations at the point of water use, such as during showering. Rather, it produces an estimate assumed to represent a spatial average over the house.

The air concentration can be estimated by using the equation

$$C_a = \frac{\mathrm{WHF} \times C_w \times f}{\mathrm{HV} \times \mathrm{ER} \times \mathrm{MC}} \tag{8}$$

where:

C_a = concentration in air, mg/m^3
C_w = concentration in water, mg/L
WHF = water flow rate in whole house, L/day
HV = house volume, m^3
ER = exchange rate, air changes/day
MC = mixing coefficient (unitless)
f = fraction of contaminant that volatilizes (unitless)

Like the shower model presented earlier, this is a simple dilution model. However, this model includes the exchange rate term to account for ventilation with clean air and also includes a mixing coefficient to account for the fact that perfect mixing is unlikely to occur over the large volume of a house. Each term in the model is discussed here.

Water concentration is the concentration of the contaminant in the water as it enters the house. It must be determined on a site-specific basis.

Water flow rate in whole house is the flow rate for water entering the house for all indoor uses. Total water use on a per capita basis is commonly assumed to be 600 L/day, or 160 gallons/day. However, indoor water use only per family has been estimated as 890 L/day. This estimate was derived by subtracting water use for lawn watering and car washing (3).

House volume is the total enclosed volume of a house. It does not include garages or screened porches. Assuming most houses have floor areas ranging from 1000 to 3000 feet^2 and 8 foot ceilings, then volumes range roughly from 200 to 700 m^3. Prichard and Gessel (6) observed a range of 150–680 m^3 in the Houston area.

Exchange rate is the rate at which the household air is replaced with outdoor air. It can vary widely depending on house design. Newer houses use tighter construction methods, reducing infiltration. Climate and seasonal changes also affect variability. ASHRAE (7) reports that typical exchange rates for houses in North America range from 0.2 to 2.0 changes per hour. Figure 1 shows the distribution for exchange rates in older, lower income houses based on a survey of 266 houses from 16 different cities in the United States. This histogram indicates a 50th percentile of about 0.5 changes per hour and a 90th percentile of about 2 air changes per hour. Figure 2 shows the distribution for exchange rates in newly constructed houses based on a survey of 312 houses from various locations in North America. This histogram indicates a 50th percentile of about 0.5 changes per hour and a 90th percentile of about 1.2 air changes per hour. On the basis of these surveys, a range (from middle to high values) of 10–50 air changes per day is recommended for default purposes.

The *mixing coefficient* represents how well mixed the contaminant is in the household air, where 1.0 represents perfect mixing and zero represents a complete lack of mixing. It is difficult to estimate because it is highly site specific, depending on ventilation patterns and room geometry. Values near the center of the range have been measured in homes (8) and are judged to be most representative of normal home conditions. Thus, a central range of 0.3–0.7 is recommended for default purposes. Ideally, these values should be refined on the basis of site-specific measurements.

Fraction volatilized is the mass fraction of contaminant in water that volatilizes as a result of water uses throughout the house. The same default

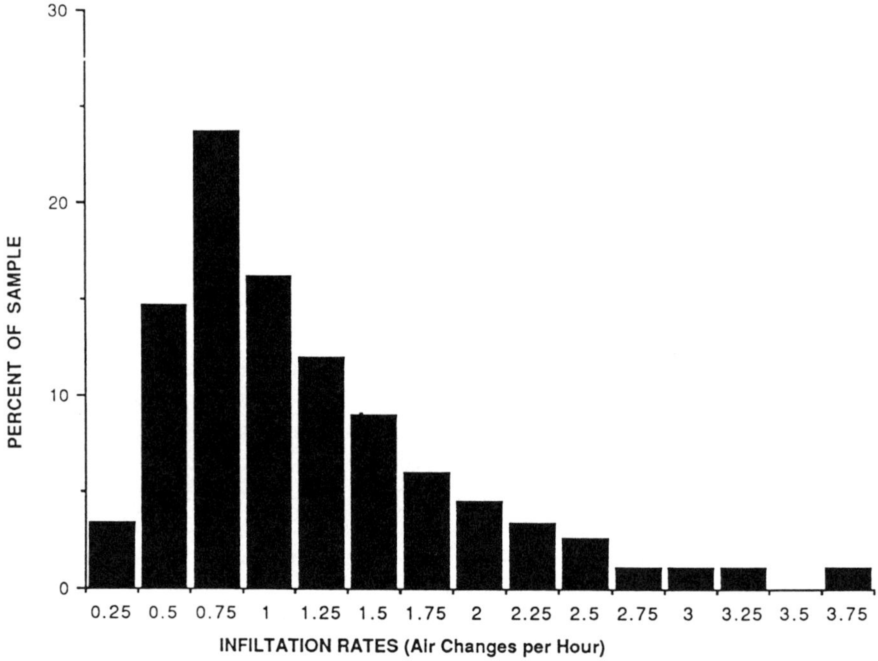

Figure 1 Infiltration rates in low-income houses.

values as used for the shower model are recommended, based on the assumption that the volatilization fraction observed for showers represents an average of all water uses throughout the home. Actually, this fraction is likely to vary depending on the particular source. Prichard and Gesell (6) measured transfer efficiencies for radon that varied from a high of 90% for dishwashers and washing machines to a low of 30% for toilets and sinks. Showers and baths were found to have values of 63 and 48%, respectively. As with the shower default, this value applies to volatiles only and should not be applied to chemicals with Henry's constants less than those of chloroform (see Table 2).

The risks associated with the whole-house air concentration can be evaluated using equations identical to Equations (6) and (7) except that the quantity $t_1 + t_2$ is replaced by the time (h/day) a person spends in the house. The U.S. EPA (5) reviewed a number of behavior pattern surveys and reported that individuals spend about 90% of the time indoors and an average of 115 h/week (or about 16 h/day) inside the residence. This value is recommended for default purposes. The default values for the exposure

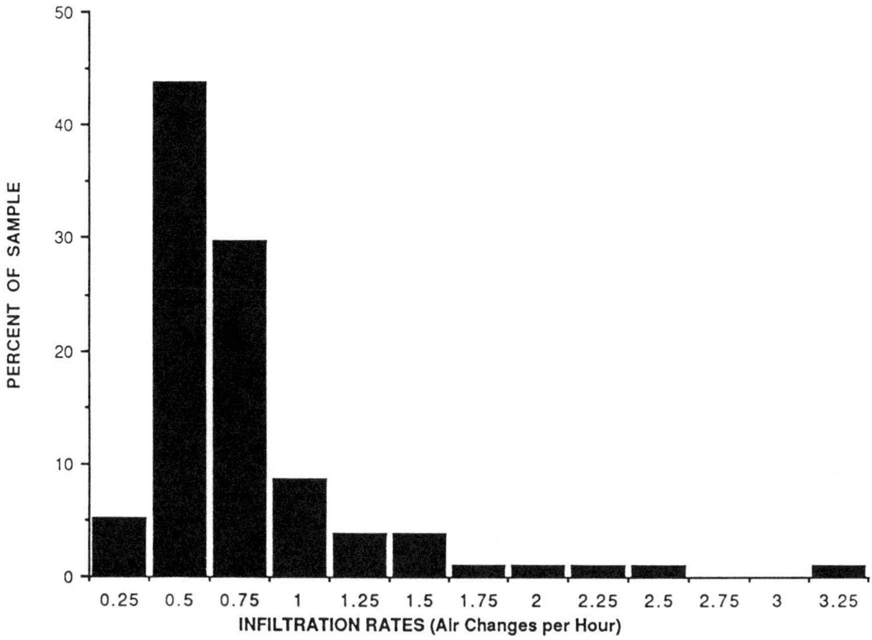

Figure 2 Infiltration rate in new construction.

Table 2 Experimental Data on Fraction Volatilized

Chemical	Temperature (°C)	*H* (unitless)	% Volatilized
Trichloroethylene	46	1.14	81.8
Chloroform	42	0.35	56
Dibromochlorpropane	42	0.03	22.8

Source: From Giardino and Andelman, 1991 (15).

duration term in Equations (6) and (7) are the same as indicated previously, that is, 9–30 years.

The default values to use for estimating whole-house exposures when data are not available are summarized in Table 3.

B. Estimating Dermal Exposure

The following discussion presents a method for estimating the water concentrations encountered during showering and the resulting dermal ex-

Table 3 Recommended Default Values for Whole-House Model

Variable	Value or range	Reference
Water flow rate WHF, L/day	890	3
House volume HV, m^3	200–700	See text
Exchange rate ER, air changes/day	10–50	7
Mixing coefficient MC (unitless)	0.3–0.7	See text
Fraction volatilized f (unitless)	0.5–0.9[a]	1
Time in residence, h/day	16	5
Exposure duration t, years	9–30	5

[a]This range applies to volatiles.

posure. The procedures are derived largely from the EPA document *Dermal Exposure: Principles and Applications* (9).

1. Contaminant Levels in Shower Water

Table 1 indicates that about 75% of the volatile contaminants in shower water volatilize. As a first approximation, it is assumed that this occurs linearly over time, so that the average concentration in the water can be estimated as

$$\begin{aligned} C_{wav} &= \frac{C_{wi} + C_{wf}}{2} \\ &= \frac{C_{wi} + 0.25C_{wi}}{2} \\ &\approx 0.6C_{wi} \end{aligned} \tag{9}$$

where:

C_{wav} = average contaminant concentration in water, mg/L
C_{wi} = initial contaminant concentration in water, mg/L
C_{wf} = final contaminant concentration in water, mg/L

As more sophisticated volatilization models are developed, mass balance procedures can be used to estimate more accurately how contaminant concentration varies over time.

2. Dermal Exposure

Given the permeability coefficient K_p of any chemical, the total mass per unit area (M/A) entering through the exposed surface during the exposure period, as evaluated by Fick's first law at steady state, can be evaluated as

$$DA_{\text{event}} = \frac{M}{A} = K_p C_w t_{\text{event}} \tag{10}$$

where:

DA_{event} = dose per exposure event, mg
K_p = permeability coefficient, cm/h
t_{event} = time of exposure event, h
C_w = contaminant concentration in water, mg/cm^3

This equation has been used traditionally to estimate the absorbed dose from dermal exposure to environmental contaminants. However, the short contact times associated with typical water exposure scenarios (i.e., bathing and swimming) may prevent attainment of steady state. Also, after the contact event chemicals stored in the skin lipids may continue to leach into the body. Therefore, the total amount absorbed in actual exposure scenarios is always underestimated using this equation. Cleek and Bunge (10) developed the following general scheme [as adapted by the U.S. EPA (9)] to estimate (M/A) in actual exposure scenarios, depending on whether the exposure period is shorter or longer than the non–steady-state period of the flux of chemicals through the skin.

The mass absorbed per unit area during the non–steady-state time period can be estimated with the expression

$$DA_{\text{event}} = \frac{M}{A} = 2K_p C_w \sqrt{\frac{6\tau t_{\text{event}}}{\pi}} \tag{11}$$

where τ = lag time (h).

The total mass absorbed per unit area for exposure events longer than the non–steady-state period can be estimated by the equation

$$DA_{\text{event}} = \frac{M}{A} = K_p C_w \frac{t_{\text{event}} + 2(1 + 3B)\tau}{1 + B} \tag{12}$$

where B = constant (unitless).

Whether to use Equation (11) or (12) depends on the duration of the exposure event and the value of B. B describes the relative contribution of the permeability coefficients of the chemical in the stratum corneum and the viable epidermis. This parameter B is defined as (10)

$$B = K_{ve/v} \frac{K_{p,sc} K_{sc/ve}}{K_{p,ve} K_{sc/v}} \tag{13}$$

where:

$K_{p,sc}$ = permeability coefficient of the chemical in the stratum corneum, cm/h

$K_{p,ve}$ = permeability coefficient of the chemical in the viable epidermis, cm/h

$K_{sc/v}$ = partition coefficient of the chemical between the stratum corneum and the vehicle

$K_{ve/v}$ = partition coefficient of the chemical between the viable epidermis and the vehicle

$K_{sc/ve}$ = partition coefficient between the stratum corneum and the viable epidermis

Cleek and Bunge (10) suggest that B can be estimated as

$$B = \frac{K_{ow}}{10{,}000} \tag{14}$$

where K_{ow} = octanol-water partition coefficient

Based on calculations by Cleek and Bunge (10), the time it takes to reach steady state (t^*) can be evaluated as a function of B:

$$t^* = 2.4\tau \qquad \text{for } B \leq 0.1 \tag{15}$$

$$t^* = (2.4 + 6 \log B)\tau \qquad \text{for } 0.1 \leq B \leq 1.17 \tag{16}$$

$$t^* = 6(b - \sqrt{b^2 - c^2})\tau \qquad \text{for } B \geq 1.17 \tag{17}$$

where b and c are defined as

$$b = \frac{2}{\pi}(1 + B)^2 - c \tag{18}$$

$$c = \frac{1 + 3B}{3} \tag{19}$$

This approach also requires estimates of τ, which is defined by Scheuplein and Blank (11) as

$$\tau = \frac{l_{sc}^2}{6D_{sc}} \tag{20}$$

where:

l_{sc} = thickness of the stratum corneum, μm

D_{sc} = diffusivity of contaminant in stratum corneum, cm^2/h

Finally, Cleek and Bunge (10) derived the following expression to estimate D_{sc} for organics:

$$\log \frac{D_{sc}}{l_{sc}} = -2.72 - 0.0061\text{MW} \qquad (21)$$

where MW = molecular weight. The stratum corneum thickness in humans has been found to be in the range of 10–20 μm for most areas of the body (11).

The key assumption in this approach is that all the chemical that entered the skin during the exposure period eventually becomes available to the body. This further implies that no loss of the chemical present in the skin occurs by metabolism, irreversible binding, evaporation, or desquamation.

III. EVALUATION OF IMPORTANCE OF EXPOSURE TO VOLATILES

The importance of exposures (via inhalation or dermal contact) to contaminants volatilized from the water supply can be estimated by comparing them to direct ingestion of the water. This approach has some uncertainties but provides an initial basis for evaluating the relative importance of these pathways. For example, the same contaminant concentration is assumed to be present in the shower water as in ingested water. However, the elevated temperatures associated with showers probably result in more volatilization (and consequently lower contaminant concentrations) than would occur from tap water during the preparation of cold beverages. Alternatively, the preparation of heated beverages made from tap water (i.e., coffee and tea) may result in more volatilization than occurs during showering. Additional uncertainties associated with this approach are discussed here.

A. Vapor Inhalation

Using the whole-house model, the ratio of the daily amount of contaminant inhaled to amount ingested is calculated as

$$\frac{\text{Inhaled dose}}{\text{Ingested dose}} = \frac{\text{WHF} \times C_w \times f \times \text{INH}(t/24)/\text{HV} \times \text{ER} \times \text{MC}}{\text{IR} \times C_w} \qquad (22)$$

where:

INH = inhalation rate, m^3/d
IR = ingestion rate, L/d
t = time indoors, h/d

This ratio can vary depending on site and chemical-specific factors. Figure 3 shows how the ratio varies with the fraction volatilized for three different scenarios. Scenario 1 corresponds to an older house with a high air exchange rate and relatively high mixing coefficient, scenario 3 corresponds to a newer house with low air exchange rate and low mixing coefficient, and scenario 2 represents an intermediate situation between scenarios 1 and 3. For scenarios 1 and 2, the inhalation exposure approaches the ingestion exposure levels only for very volatile compounds. For scenario 3, even moderately volatile compounds create higher inhalation exposures than ingestion.

This approach assumes that the health effects associated with these pathways are proportional to the potential doses. This assumption limits the validity of the approach in two ways. First, many compounds are more readily absorbed via the lungs than via the GI tract, suggesting that this approach may underestimate the importance of the inhalation route for such compounds. Second, some compounds cause portal of entry effects, that is, direct effects on lungs or the gastrointestinal (GI) tract. Obviously this approach does not apply to such compounds.

B. Dermal Contact

When the same water supply is used for drinking and bathing, the importance of dermal contact with water can be evaluated by comparing the possible absorbed dose occurring during bathing relative to that occurring

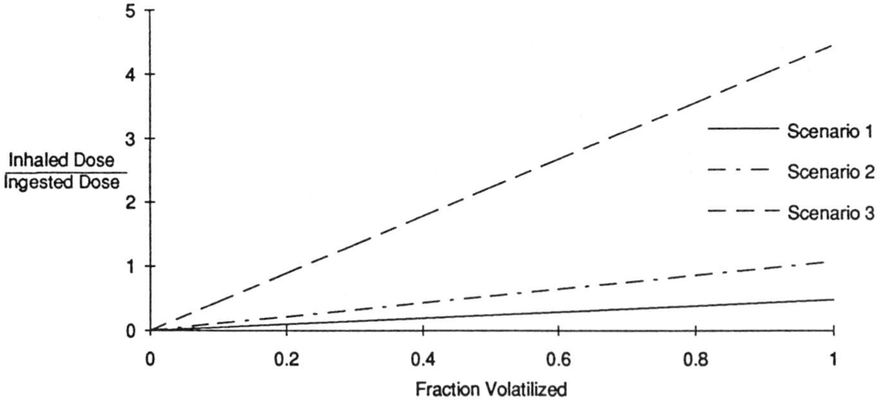

Figure 3 Ratio of inhaled to ingested dose versus fraction volatilized from household wastewater supply.

as a result of ingestion:

$$\frac{\text{Dermal dose}}{\text{Ingested dose}} = \frac{2C_w K_p (6\tau t/\pi)^{0.5} A}{C_w \times \text{IR} \times \text{AF}_{\text{GI}}} \tag{23}$$

where:

C_w = contaminant concentration in water, mg/cm^3
K_p = permeability coefficient, cm/h
t = exposure time, minutes/day
A = exposed skin area, cm^2
IR = ingestion rate, L/day
AF_{GI} = GI tract absorption fraction

Assuming an average adult ingestion rate (IR) of 2 L/day, GI tract absorption fraction of 1 (i.e., the fraction of the administered dose that becomes absorbed), shower time of 10 minutes, and skin area of 20,000 cm^2, this ratio becomes

$$\frac{\text{Dermal dose}}{\text{Ingested dose}} = 10K_p\sqrt{\tau} \tag{24}$$

So the dermal dose exceeds the ingested dose when

$$K_p\sqrt{\tau} > 0.1 \tag{25}$$

The ratio of dermal dose to ingested dose was computed for about 200 compounds of environmental concern and plotted as a function of K_p in Figure 4. This plot suggests that the dermal dose exceeds the ingested dose when K_p is greater than about 0.1 cm/h. The theoretical maximum K_p is approximately 1 cm/h, and it appears that most compounds have K_p less than 0.1 (only about 20 of the 200 in this data set had K_p above 0.1 cm/h).

This analysis suggests that when the same water supply is used for drinking and bathing, dermal exposure while showering or bathing is not important to consider for most contaminants but may be important for the small percentage of compounds that penetrate fastest. Several uncertainties are associated with this conclusion. First, the degree to which the 200 tested chemicals are representative of all environmental contaminants is unknown. Second, multiple uncertainties are introduced in the measurement or estimation of K_p. Third, most dermal experiments are conducted at ambient temperatures, and the elevated temperatures during showering and bathing likely increase volatilization, reducing the amount available for dermal absorption. Finally, this conclusion could be affected to some extent by the assumption for exposure duration. If time spent showering or bathing differs from the assumed 10 minutes/day, this will affect the exposure proportionally.

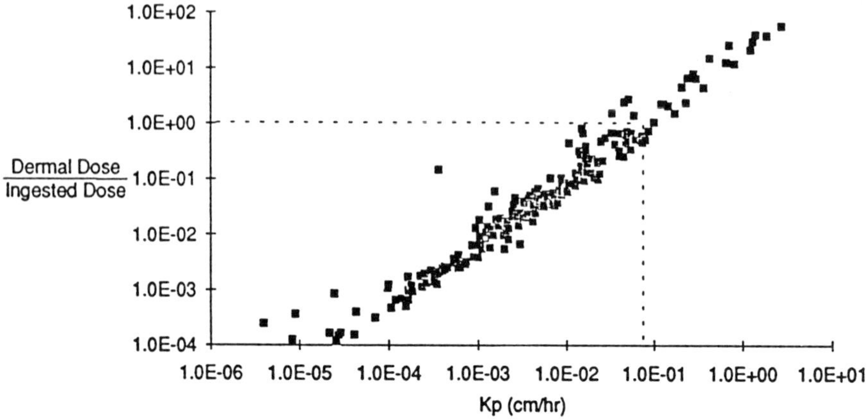

Figure 4 Ratio of dermal dose to ingested dose versus dermal permeability coefficient.

Similar conclusions can be drawn for swimming. The U.S. EPA (9) suggests default assumptions for time swimming, implying a central estimate of about 3 h/year to an upper estimate of 150 h/year. Bathing time totals 60 h/year as the cumulative total of daily 10 minute baths. Thus, when the same water source is used for swimming, bathing, and drinking, the absorbed dose from dermal water contact may be two to three times as much as would occur from bathing alone but would still be much less than direct ingestion for most contaminants.

Some experimental support for these conclusions can be found in the recent work by Jo et al. (12,13). Chloroform levels in breath after 10 min showers were measured when the subjects first wore no clothing and then wore protective rubber suits. The breath levels dropped by about half when wearing the rubber suits, leading Jo et al. (12) to conclude that the chloroform dose from inhalation and dermal contact were about equal during normal showering. Jo et al. (13) also evaluated direct consumption of the water. The dose from ingestion of 2 L/day was estimated as about three times greater than the dermal (or inhalation) dose alone. K_p values for chloroform have been measured in the range of 10^{-1} cm/h (14). Thus, even a relatively rapid permeant was found to cause a lower dose by dermal contact than by direct ingestion.

REFERENCES

1. J.B. Andelman, Total exposure to volatile organic compounds in potable water. In: Cantor, K.P.; Christman, R.F.; Ram, M.M., eds. *Significance and*

Treatment of Volatile Organic Compounds in Water Supplies. Lewis Publishers, Chelsea, MI, 1990, pp. 485–504.

2. T.E. McKone, Household exposure models. *Toxicol. Lett. 49*:321–339 (1989).
3. R.G. Bond, C.B. Straub, and R. Prober (eds.), *Handbook of Environmental Control*, Vol. III, *Water Supply and Treatment*, CRC Press, Cleveland, OH, 1973.
4. I.R. James and M.W. Knuiman, An application of Bayes methodology to the analysis of diary records from a water use study. *J. Am. Stat. Assoc. 82*(399):705–711 (1987).
5. U.S. EPA, *Exposure Factors Handbook*, EPA 600/8-89/043, (1989).
6. G.M. Prichard and T.F. Gessel, An estimate of population exposures due to radon in public water supplies in area of Houston, TX, *Health Phys. 41*:599–606 (1981).
7. ASHRAE, Natural ventilation and infiltration, *ASHRAE Fundamentals Handbook*, ASHRAE, Inc., Atlanta, GA, 1985.
8. R.A. Wadden and P.A. Scheff, *Indoor Air Pollution*, Wiley Publishers, New York, 1983.
9. U.S. EPA, *Dermal Exposure Assessment: Principles and Applications*, Draft Report, Office of Health and Environmental Assessment, Washington, D.C., EPA/600/8-91/011 (1991).
10. R.L. Cleek and A.L. Bunge, A new method for estimating dermal absorption from chemical exposure. Annual Meeting of the Society for Risk Analysis, December 1991, Baltimore, MD, 1991.
11. R.J. Scheuplein and I.H. Blank, Permeability of the skin, *Physiol. Rev. 51*:702–747 (1971).
12. W.K. Jo, C. D. Weisel, and P.J. Lioy, Chloroform exposure and the health risk associated with multiple uses of chlorinated tap water, *Risk Anal. 10*(4):581–585 (1990).
13. W.K. Jo, C.D. Weisel, and P.J. Lioy, Routes of chloroform exposure and body burden from showering with chlorinated tap water, *Risk Anal. 10*(4):575–580 (1990).
14. K.T. Bogen, B.W. Colston, and L.K. Machicao, Dermal absorption of dilute aqueous chloroform, trichloroethylene and tetrachloroethylene in hairless guinea pigs, *Fund. Appl. Toxicol. 17*:1–13 (1991).
15. N.J. Giardino and J.B. Andelman, Modeling volatilization of chemicals from showers, Annual Conference of American Water Works Association, June 1991, Philadelphia, PA, 1991.

14

Percutaneous Absorption of Dilute Aqueous Chlorinated Organic Solvents in the Hairless Guinea Pig

Kenneth T. Bogen, Bill W. Colston, Jr., and Lesia K. Machicao

University of California,
Livermore, California

I. INTRODUCTION

Dermal absorption of extremely dilute aqueous compounds, particularly volatile organic chlorinated compounds (VCOC), has not been well studied. Dermal uptake of VCOC has been studied in humans exposed to neat liquid solvents by thumb immersion (1,2) and similar studies have been conducted using live shaved guinea pigs, shaved mice, and hairless mice (3–6). In vivo dermal absorption of aqueous solutions of the pesticide dinoseb was measured recently using the rhesus monkey (7). Jo et al. (8,9) recently detected dermal absorption of dilute aqueous chloroform (CF) in humans indirectly by monitoring CF concentration in exhaled breath of volunteers following showers with and without rubber wet suits using normal tap water at approximately 40°C containing CF at approximately 5–35 ppb (parts per billion). Elevated levels of chloroform in blood have also been measured in samples taken from swimmers immersed in continuously chlorinated indoor pools containing water with chloroform ranging from 210 to 2580 ppb compared with nonswimmers; chloroform concentrations in the pool water as well as in air directly over the water were found to be significantly correlated with chloroform concentrations in swimmers' blood (10).

Studies have been performed using dilute aqueous solutions of nonchlorinated aromatic compounds, such as toluene, styrene, xylene, ethylbenzene, and benzene (11–13). In these studies, a volunteer's entire hand was immersed for 1 h in a 1 L beaker containing the test compound in dilute aqueous solution ranging from 66.5 to 600 mg/L at 23–25°C; uptake was determined by measuring the difference between initial and final concentrations. From these data, Brown et al. (14) estimated dermal permeability constants for styrene, toluene, and ethylbenzene and used them to extrapolate dermal uptake of these compounds in humans bathing in aqueous concentrations as low as four orders of magnitude below those at which Dutkiewicz and Tyras made their empirical determinations of dermal permeability.

The present study was designed specifically to measure dermal uptake of very dilute aqueous ^{14}C-radiolabeled chloroform, trichloroethylene (TCE), and tetrachloroethylene (PCE) in hairless guinea pigs dermally exposed over a majority of their surface area. These compounds are all suspected human carcinogens found in groundwater throughout the United States in the ppb range (15–21). We used hairless guinea pigs because this animal's skin has been shown to provide a fair approximation of human skin for the purpose of quantifying epicutaneous permeability, particularly in comparison with that of other laboratory rodents (22–27).

II. METHODS

A. Test Chemicals

Chemicals used were [^{14}C]CF (2.0 mCi/mmol at >98% purity; ICN Biomedicals, Inc., Irvine, CA), TCE (>99.99% purity; Burdick & Jackson Labs., Muskegon, MI); [1,2-^{14}C]TCE (4.1 mCi/mmol at >99% purity; Sigma Chemical Co., St. Louis, MO), and [1,2-^{14}C]PCE (2.7 mCi/mmol at >98% purity; Sigma Chemical Co., St. Louis, MO). Each ampule was crushed into a 5 ml beaker of methanol held in ice to yield an initial, continuously refrigerated stock of labeled sample (LS) with activities of 21.1 μCi/ml (CF), 27.6 μCi/ml (TCE), and 63.7 μCi/ml (PCE).

B. Animals

A total of 16 female Crl:IAF(HA)BR hairless (but euthymic) guinea pigs (Charles River Laboratories, Chicago, IL, approximately 10–40 weeks old (~400–700 g), were used in a total of 36 dermal exposure (DE) and/or positive control (PC) experiments. In no case was an animal reused before radioactivity measured in urine and feces remained at background levels (see Section II.I), which in all cases was at least 3 weeks after use. Animals were reused because the experimental conditions, involving very small, infrequent doses of compounds known to be fairly rapidly metabolized by rodents to mostly excreted products (15–19), were not considered to affect significantly the dermal absorption properties studied. Food and water were available ad libitum except as noted.

C. Anesthesia

Sterile sodium pentobarbital (PB) at 50 mg/ml (Nembutal; Abbott Laboratories, N. Chicago, IL), diluted in an equal volume of sterile 0.9% NaCl in H_2O, was used to achieve sedation in all experimental animals (from which food and water were withheld 24 h before sedation), using doses of 30 mg/kg body weight administered intraperitoneally (20 minutes before preparation for exposure) plus 15 mg/kg administered subcutaneously (35 minutes into the exposure period, to achieve an even level of sedation lasting 70 minutes). Animals generally recovered from deep sedation within 30–60 minutes after dosing but generally remained immobile in light sleep for an additional 30–120 minutes. The relatively low dose levels and the ≤3 week interdose period used were selected to minimize the recovery time after sedation while minimizing the chance for PB-induced hepatic metabolism in light of the relatively small and nonchronic nature of the PB dose used (see Reference 28).

D. Dermal Exposure Chamber

The glass dermal exposure chamber (DEC; see Figure 1) was used (height = 15.0 cm and inner diameter = 8.4 cm) to which a water- and airtight diaphragm could be affixed by means of a flexible rubber O ring. The diaphragms used (one per experiment) consisted of a 12-cm square of latex rubber sheeting (thickness = 0.64 ± 0.08 mm; McMaster-Carr, Los Angeles, CA) into the center of which was cut a circular hole (diameter = 2.9 cm). The diaphragms were backed by one to two layers of aluminum/polyester adhesive tape sections (8.9-μm aluminum foil plus 51-μm polyester film backed with pressure-activated acrylic adhesive; stock no. A25, Lamart Corporation, Clifton, NJ) affixed to the circular area of latex to be enclosed by the O ring. This formed an aluminum tape undercoating (~9-cm diameter) with a central circular hole that, when stretched, fit snugly around the guinea pig chest yet allowed regular breathing under sedation.

E. Dermal Exposure Experiments

A group of 16 "low-concentration" DE experiments were conducted using animals dermally exposed to extremely dilute aqueous concentrations ranging from ~20 to ~110 ppb CF, TCE, and PCE (hereafter referred to as

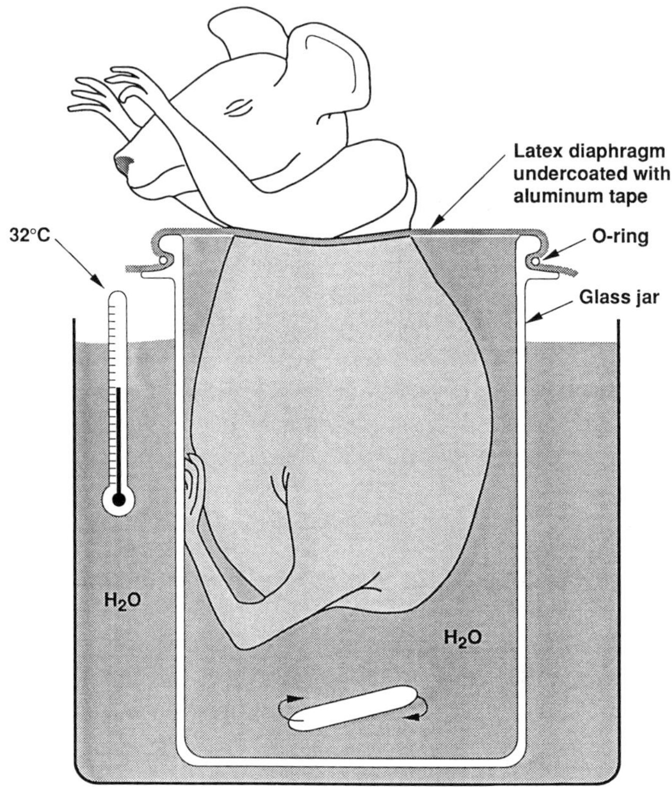

Figure 1 Dermal exposure chamber for guinea pigs (capacity 0.86 L with no animal present). The sedated animal is placed through a precut hole in the latex diaphragm, which is then undercoated with aluminum-faced tape. Once attached to the chamber, the diaphragm maintains an air- and watertight seal around the animal, allowing aqueous dermal exposure with no airspace.

experiments CF:D1–6, TCE:D1–5, and PCE:D1–5, respectively). Five additional "high-concentration" experiments (hereafter referred to as experiments TCE:D6–D10) involved dermal exposure to 100,000 ppb aqueous TCE prepared beforehand. Before each DE experiment, from 40 to 100 μl LS was diluted in 1.0 ml double-distilled water (DDW); this "input solution" (IS) was injected into the DEC before each experimental exposure.

All DEC and diaphragm materials and the stir bar were weighed dry before each experiment. In DE experiments, the latex diaphragm was placed around the upper chest of the animal just below the front legs and shoulders, with the edge of the latex curving upward at the point of dermal contact. Sections of aluminum tape were then attached to the underside of the diaphragm as close to the chest of the animal as possible, exposing the minimum possible latex surface area to the fluid placed subsequently into the DEC. A Teflon-coated magnetic stir bar was placed into the DEC, which was then filled to two-thirds capacity with DDW (or, in experiment TCE:D6–10, with a 100,000 ppb solution of TCE in DDW) preheated to 32°C. The animal was then quickly lowered into the DEC, the diaphragm was affixed to the chamber with one or two rubber O rings, and any residual air bubble in the DEC was removed by syringe through the diaphragm. The DEC was weighed at this time to allow determination of its initial fluid capacity V, taking into account a DDW density of 0.995 g/ml at 32°C. Chamber water was continuously stirred and maintained at an average temperature of 32°C (± 2°C range) throughout each test period.

Before the beginning of a test period (hereafter referred to as t_0), samples were taken of DDW (0.5 ml, to serve as "blank controls"), of the fluid contained in the sealed DEC (0.5 ml, through the diaphragm to serve as "chamber controls"), and of IS (20 μl in experiments TCE:D1–5 and corresponding control experiments and 50–100 μl in all other DE and corresponding control experiments). At t_0, 0.5 ml IS was injected by syringe into the DEC through the diaphragm, so that at t_0 the DEC fluid volume V_0 was approximately equal to V. Samples of DEC fluid were then withdrawn by syringe through a distant location in the diaphragm at 10, 15, 20, 30, 40, 50, 60, and 70 minutes after t_0. (Hereafter, t_n denotes n minutes after t_0.) All transdiaphragm samples were taken using a 500-μl gas-sampling syringe with a Teflon-tipped plunger (Series C; Dynatech Precision Sampling Corp., Baton Rouge, LA), which was always triple-rinsed with DDW before each use. A separate 500-μl syringe was used for injecting IS into the DEC at t_0. After the last fluid sample was taken at t_0 + 70 minutes, the guinea pig was removed from the DEC and placed in a metabolism cage in a negative airflow unit. The total exposure duration T_e (minutes) in each DE experiment was actually 72 minutes (or 75 and 85 minutes in TCE:D2 and TCE:D1, respectively) because of time required to complete the final DEC fluid sample and the removal of the animal from the DEC. A single DEC was used for all experiments. After each experiment, the DEC was rinsed with methanol, triple rinsed with DDW, and air dried (resulting in background radioactivity levels upon subsequent sampling; see Section II.I).

F. Chamber Control Experiments

A total of 20 chamber control (CC) experiments were conducted corresponding to the low- and high-concentration experiments with CF, TCE, and PCE. These CC experiments are referred to here as experiments CF: C1–5 (a set of CC experiments corresponding to CF:D1–6), TCE:C1–5 (corresponding to TCE:D1–5), TCE:C6–10 (corresponding to TCE:D6–10), and PCE:C1–5 (corresponding to PCE:D1–5). In all CC experiments, in place of an animal in the DEC, a glass beaker 4–5 cm in diameter was inserted into the diaphragm hole to a depth of approximately 0.5 cm (the minimum depth required to attain a secure seal). Thus, in CC experiments the animal surface area was replaced by 60–65 cm^2 of glass surface similar to that of the DEC. All CC experiment durations were extended to 190 minutes (except TCE:C1–3, which lasted 70 minutes) to enable the detection of potentially significant loss rates, but all other procedures used (i.e., for sampling and temperature maintenance, among others) were identical to the corresponding DE procedures just described.

G. Positive Control Experiments

A total of 15 positive control experiments (experiments CF:P1–5, TCE: P1–5, and PCE:P1–5) were conducted using very small doses of the test compounds. In all PC experiments, animals were sedated twice as in the DEC experiments and, within 15 minutes of the second sedative administration, were administered LS dissolved in corn oil (0.5 ml total volume) as a single subcutaneous (SC) dose in the animal's midback region slightly to the right of center [except in experiments TCE:P1 and P2, in which a single intramuscular (IM) dose dissolved in methanol (0.2 ml total volume) in the back of animal's left thigh was used]. In each of the PC experiments, an estimate of injected dose was obtained from separate samples of the dosing solution used taken before and after animal injection.

H. Collection of Urine and Feces

All urine and feces produced after exposure in all PC and low-concentration DE experiments were collected for 2–4 weeks; samples were obtained for analysis at approximately 3, 15, and 24 h postexposure and thereafter approximately five times per week until measured activity did not differ significantly from background, at which time cage rinses were performed. Total urine volumes and fecal output mass were recorded for each animal during postexposure observation.

I. Radiolabel Measurement

All samples taken of "blank" DDW, IS, t_0 chamber control fluid, DEC fluid, urine (two 1-ml volumes/sample), diaphragm material (six 1-cm^2 sections each of latex and of aluminum tape per experiment), and PC dosing vehicle and vehicle + dose were placed directly in 15 ml Universol

(ICN Biomedicals, Inc., Irvine, CA) liquid scintillation cocktail for quantification of ^{14}C in counts per minute (cpm) in a Tri-Carb 4530 scintillation counter (Packard Instrument Co., Downers Grove, IL). All radiolabel measures in cpm were converted to equivalent disintegrations per minute (dpm) using a calibration curve for ^{14}C based on external standards, which were supplied by Packard Instrument Co. Measured dpm were then converted to net dpm (ndpm) by subtracting dpm measured in appropriate control blanks. For each CC and DE experiment, measured ndpm was used (after scaling to account for the ratio between tested and total amounts) to estimate total ndpm injected into the DEC at t_0, referred to here as I_0; total ndpm present in DEC water at t_{10}, referred to here as L_{10}; and total ndpm present at T_e in the DEC diaphragm (total in latex plus aluminum tape), referred to here as D_e.

Fecal radioactivity was assayed as follows: 10 fecal pellets were selected from each weighed fecal collection, weighed, and then crushed by mortar and pestle; the fecal powder was then rinsed with 20 ml methanol. Two 1.0-ml aliquots of the rinsate were used to measure radiolabel as described, and the measured ndpm was appropriately scaled to convert to estimated equivalent total ndpm for each entire fecal collection. Fecal samples from nonexposed animals were used to establish background dpm levels by this procedure. Similarly, measured ndpm in urine samples was normalized to each collected total volume, and urine collected from nonexposed animals was used to establish background for urine. Urinary and fecal ndpm were added to obtain total postexposure ndpm in urine and feces. Total recovered net cage rinse activities were added to corresponding excreted radiolabel totals.

J. Estimation of Dermal Surface Areas

The animals used in the DE experiments were placed directly into metabolism cages postexposure. Exposed dermal surface area A (cm^2) was therefore estimated by interpolation from the allometric relation $A = aW^b$, where W = body weight in grams and a and b were parameters estimated by log-log least-squares regression of experimentally determined surface areas of 10 female hairless guinea pigs weighing between 360 and 760 g. The measured areas were selected to correspond as closely as possible to areas exposed in the DE experiments. Each surface area measurement was performed immediately after sacrifice by weighing a cutout tracing of the dissected skin on preweighed paper.

K. Data Analysis

Radiolabel loss during all CC and DE experiments was expressed as percentage of label remaining in sampled DEC fluid compared with that first measured (at t_{10}), where the latter amount was assigned the value of 100%. Observed rates R (%/h) of relative net radiolabel loss over time from sampled DEC fluid in DE experiments were anticipated, following Scheuplein and Blank (22,29), to reflect first-order loss in accordance with Fick's law, which states that as a limiting approximation for dilute solutions, the flux or permeation rate J_s (mg/cm^2-h) of solute passively diffusing across a given area equals the concentration difference ΔC_s (mg/L) across that tissue multiplied by a permeability constant k_p (ml/cm^2-h = cm/h), assuming ideal mixing far from equilibrium. Data on relative radiolabel loss over time from sampled test chamber water were thus analyzed using simple linear regression to model the initial, approximately linear portion of anticipated exponential decay (observed loss was generally 30% or less in all DE experiments; see Section III). Approximate linearity of radiolabel loss was also assumed for all CC experiments (in which observed loss was much less than that in the DE experiments; see Section III). Thus, each set of CC or DE data on relative radiolabel loss from DEC fluid over the duration of the experiment was analyzed to obtain a corresponding value of R (% loss/h), the negative of the slope obtained from linear regression on those data. Corresponding k_p values were estimated under the assumption that linear net radiolabel loss from the DEC was due entirely to corresponding linear uptake into exposed animals during dermal exposure; that is, as

$$k_p = \frac{(V_0 - 2.0\text{ ml})(R - R_0)}{100\%A} \tag{1}$$

where $V_0 - 2.0$ ml represents the approximate average volume of DEC water during exposure (accounting for withdrawn sample volumes) and where R_0 is the average rate of relative radiolabel loss (i.e., the negative of the linear slope fit to) for the corresponding set of pooled CC data. For each low-concentration DE experiment, absorbed dermal dose was then estimated [under the same assumptions used for Equation (1)] as

$$\text{Dose }(\mu\text{g}) = \frac{k_p T_e C_{10} A}{60\text{ minutes/h}} \cdot \frac{9.95 \times 10^{-4}\ \mu\text{g}}{\text{ml-ppb at 32°C}} \times \left(1 + \frac{R\ 10\text{ minutes}}{100\%\ 60\text{ minutes/h}}\right) \tag{2}$$

in which C_{10} is the calculated aqueous concentration of test compound (ppb) in DEC fluid sampled at t_{10} (based on the corresponding measured value of L_{10}). Such estimated doses, in turn, were used with corresponding radiolabel excretion data to obtain the average fraction of estimated dermal dose excreted as urinary and fecal metabolites. The latter fraction was then compared with the average fraction obtained in corresponding PC experiments to test the assumptions that underlay the application of Equations (1) and (2).

For each CC and DE experiment, the recovery of input I_0 in DEC fluid was examined by estimating a hypothetical number L_0 of ndpm in DEC fluid at t_0 corresponding to the measured value L_{10} under the assumptions of instantaneous and ideal mixing at t_0 and of linear radiolabel loss between t_0 and t_{10}, as

$$L_0 = L_{10}\left(1 + \frac{R \text{ 10 minutes}}{100\% \text{ 60 minutes/h}}\right) \tag{3}$$

The value of L_0 obtained was then expressed as a percentage of I_0. Similarly, for each of these experiments, the number (hereafter denoted D_{72}) of ndpm in the DEC diaphragm (total in aluminum tape and latex) at t_{72} (a time picked to allow a standardized comparison of uptake into diaphragm material in CC and DE experiments) was estimated as

$$D_{72} = D_e \frac{\text{72 minutes}}{T_e} \tag{4}$$

and expressed as a percentage of L_0.

Unless otherwise specified, reported means are all given ± the coefficient of variation (CV = standard deviation as a percentage of mean). Reported significance levels (p values) are all for two-tailed t-tests, except for means comparisons involving samples of unequal variance (based on an F test with $p \leq 0.05$), in which case the reported p values (denoted by p^*) refer to two-tailed Wilcoxon rank-sum tests. Reference to statistical nonsignificance without a reported value of p or p^* means that a value greater than 0.05 pertains.

III. RESULTS

The 10 measured dermal surface areas (cm^2), fit to the function aW^b (where W = body weight, grams), yielded best fit parameter values (95% confidence limits) of $a = 24.5$ (8.31–72.3) and $b = 0.394$ (0.221–0.568); $r^2 = 0.77$, $p = 0.00078$, Figure 2).

Data on relative radiolabel losses from the DEC observed during CC and corresponding DE experiments with linear fits are shown in Figures 3 through 5. Corresponding data on animal weights, estimated surface areas, DEC water volumes, measured values of C_{10} and R, and estimated values of R_0 and k_p (calculated as explained in Section II), are also given in Table 1. As indicated in Table 1, the values R_0 obtained for pooled CC data are all greater than zero, but not significantly so in the case of the pooled CC data for CF. When compared with these R_0 values, the loss

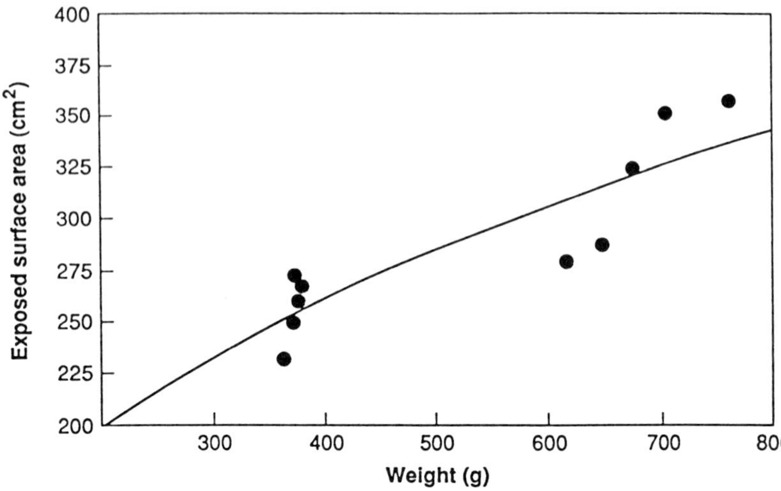

Figure 2 Measured dermal areas of 10 female hairless guinea pigs versus body weight (data points), with areas corresponding to those used in dermal exposure experiments, fit the allometric regression equation: area (cm^2) = 24.5(body weight, g)$^{0.394}$.

rates R from corresponding DE experiments are all significantly greater, except in the case of two values obtained from DE data for CF. Table 1 shows that the mean k_p values obtained from the four DE data sets are in the following compound-related order CF < TCE < PCE. Notably, the mean k_p value for the low-concentration TCE experiments (TCE:D1–5) does not differ significantly ($p^* = 0.22$) from that based on the high-concentration experiments (TCE:D6–10). The mean k_p value for CF is significantly less than that for TCE:D1–5 ($p = 0.0019$) and for TCE:D1–10 ($p^* = 0.011$); that for PCE is significantly greater than that for TCE:D1–5 ($p = 0.0039$) and for TCE:D1–10 ($p = 0.0042$). The k_p values obtained from the three low-concentration experiments are most consistent (CV range 17–29%); the k_p values obtained from the high-concentration TCE experiments are the most variable (CV = 58%).

Table 2 presents a comparison of radiolabel excretion data from low-concentration DE and PC experiments. Table 2 shows that, for each of the three compounds tested, the average excreted percentage of estimated dermal dose in the DE experiments (based on information from Table 1, as explained in Section II) does not differ significantly from the average excreted percentage of the injected dose in the corresponding PC experiments. The time taken to excrete 95% of the measured metabolized dose,

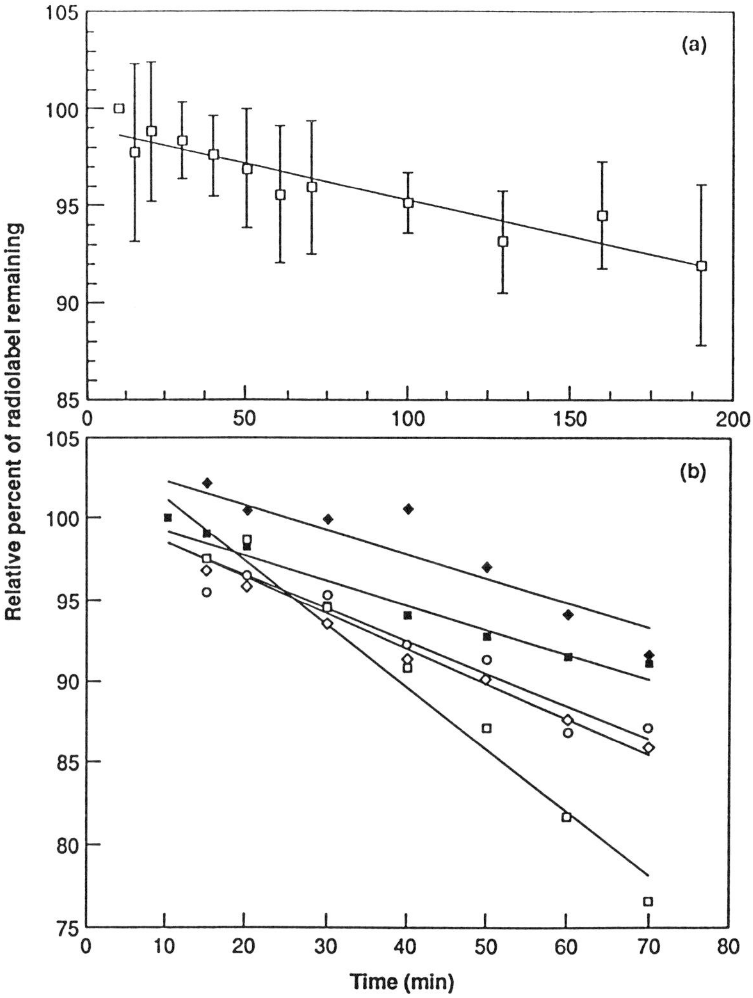

Figure 3 Losses of radiolabeled trichloroethylene (TCE) from chamber water compared with that measured 10 minutes after input (TCE concentrations: ≈100,000 ppb). (a) Five control experiments (no animal): data points from each experiment at each sampling time are shown with a linear fit to the combined data. (b) Five dermal exposure experiments (animal present): data points from each experiment at each sampling time are shown with corresponding linear fits (all slopes significantly less than that of pooled control data; see Table 1).

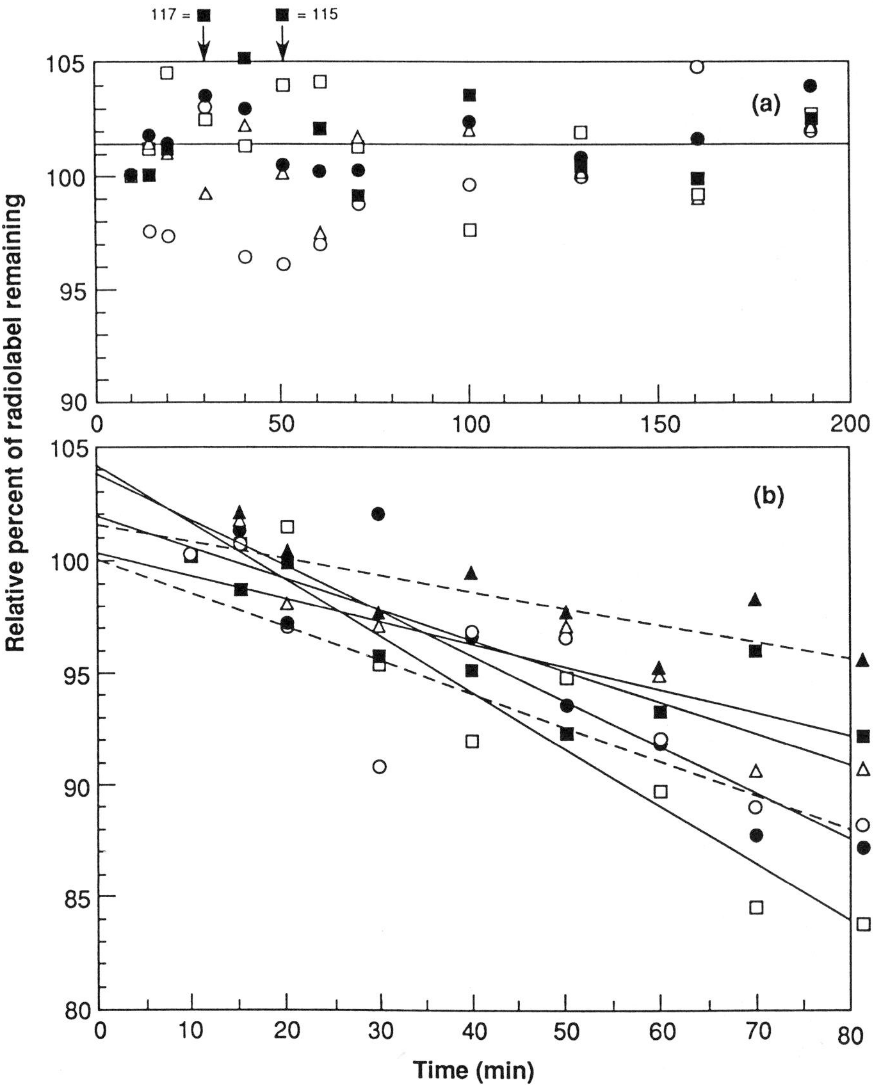

Figure 4 Loss of radiolabeled chloroform (CF) from chamber water compared with that measured 10 minutes after input (CF concentrations 22–52 ppb). (a) Five control experiments (no animal): data points from each experiment at each sampling time are shown with a linear fit to the combined data. (b) Six dermal exposure experiments (animal present): data points from each experiment at each sampling time are shown with corresponding linear fits (all slopes significantly less than that of pooled control data, except those corresponding to the dashed lines; see Table 1).

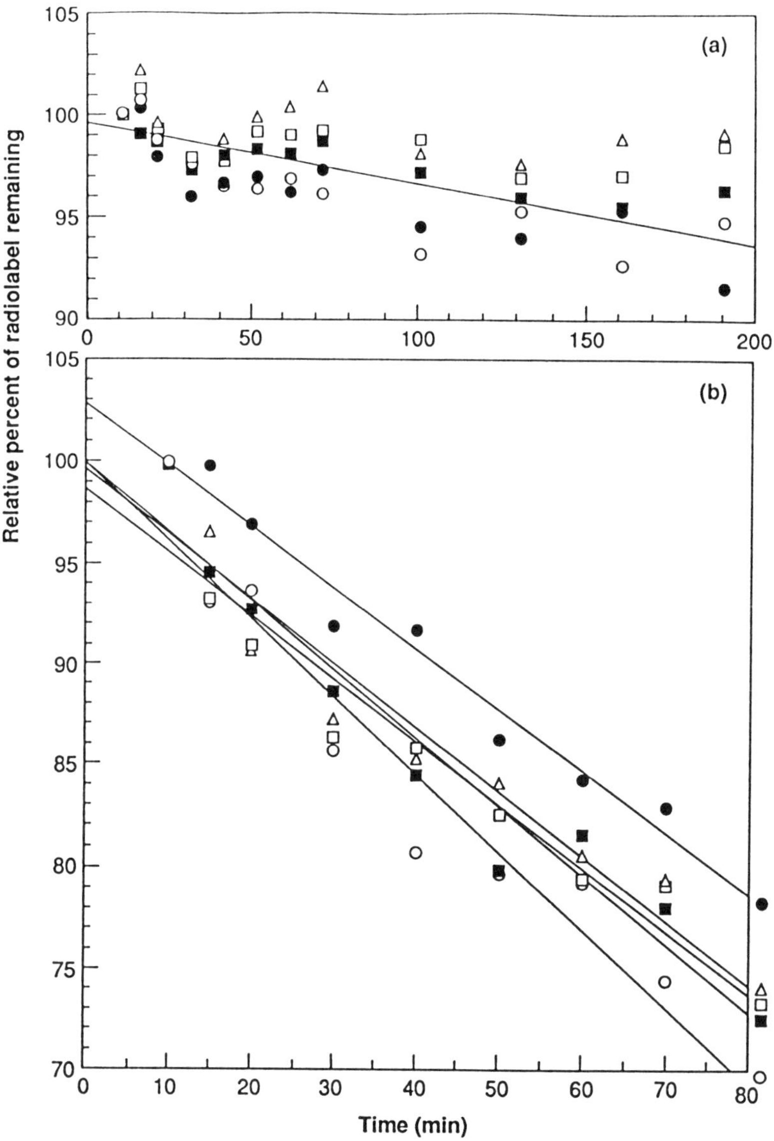

Figure 5 Loss of radiolabeled tetrachloroethylene (PCE) from chamber water compared with that measured 10 minutes after input (PCE concentrations 23–64 ppb). (a) Five control experiments (no animal): data points from each experiment at each sampling time are shown with a linear fit to the combined data. (b) Five dermal exposure experiments (animal present): data points from each experiment at each sampling time are shown with corresponding linear fits (all slopes significantly less than that of pooled control data; see Table 1).

Table 1 Percutaneous Absorption of Aqueous TCE, CF, and PCE in Hairless Guinea Pigs[a]

Experiment	Animal weight W (g)	Exposed surface area A (cm^2)	Initial water volume V_0 (ml)	Concentration at time = 10 minutes C_{10} (ppb)	Radiolabel loss rate[b] R (%/h)	Permeability constant[c] k_p (ml/cm^2-h)
TCE:C1–5	—	—	794.9–812.9	15–84	1.6 ± 18	—
TCE:D1	473	278	449.9	87	17[d]	0.25
TCE:D2	492	283	478.6	110	17[d]	0.26
TCE:D3	423	266	509.4	58	16[d]	0.27
TCE:D4	535	292	488.0	20	13[d]	0.19
TCE:D5	488	282	520.7	19	12[d]	0.19
						0.23 ± 17
TCE:C6–10	—	—	778.8–844.4	≈100,000	2.3 ± 17	—
TCE:D6	386	257	449.9	≈100,000	13[e]	0.24
TCE:D7	451	273	478.6	≈100,000	9.2[e]	0.16
TCE:D8	368	252	509.4	≈100,000	9.1[e]	0.21
TCE:D9	411	263	488.0	≈100,000	12[e]	0.23
TCE:D10	385	257	520.7	≈100,000	23[e]	0.47
						0.21 ± 58
CF:C1–5	—	—	797.9–836.4	22–44	0.078 ± 590	—
CF:D1	604	306	428.1	52	9.1[f]	0.13
CF:D2	623	310	426.5	24	12[f]	0.16
CF:D3	654	316	387.5	44	15[f]	0.18
CF:D4	576	301	470.7	34	6.1[g]	0.094
CF:D5	514	288	482.4	19	8.3[f]	0.14
CF:D6	481	280	500.3	19	4.5[g]	0.079
						0.13 ± 29
PCE:C1–5	—	—	768.2–815.2	23–150	1.3 ± 18	—
PCE:D1	375	254	577.7	27	23[h]	0.49
PCE:D2	420	266	543.8	51	19[h]	0.36
PCE:D3	489	282	495.3	64	19[h]	0.31
PCE:D4	490	282	525.1	56	21[h]	0.37
PCE:D5	458	275	530.4	56	19[h]	0.34
						0.37 ± 18

[a]TCE = trichloroethylene, CF = chloroform, PCE = tetrachloroethylene. See Section II for description of notation for column variables and (if applicable) their calculation.
[b]For each set of chamber control experiments, the value listed is the estimated loss rate R_0 (± CV, %), for the pooled control data (see Section II).
[c]Below each set of related values is the corresponding arithmetic mean (± CV, %).
[d]Significantly greater than corresponding loss rate for pooled control data (by analyses of covariance, $p < 10^{-6}$).
[e]Significantly greater than corresponding loss rate for pooled control data (by analyses of covariance, $10^{-9} < p < 0.022$).
[f]Significantly greater than corresponding loss rate for pooled control data (by analyses of covariance, $10^{-4} < p < 0.020$).
[g]Not significantly different from corresponding loss rate for pooled control data (by analysis of covariance; $p = 0.085$ for CF:D4, $p = 0.21$ for CF:D6).
[h]Significantly greater than corresponding loss rate for pooled control data (by analyses of covariance, $p < 10^{-12}$).

Table 2 Excretion of Radiolabel in Urine and Feces by Hairless Guinea Pigs Exposed to TCE, CF, or PCE[a]

Experiment	Animal weight (g)	Exposure regimen	Administered dose (μg)	Excreted % administered dose (% ± CV, %)	Time to excrete 95% of metabolized dose (days ± CV, %)
TCE:D1		Dermal	8.7	64	11
TCE:D2		Dermal	10	50	6
TCE:D3		Dermal	5.2	58	7
TCE:D4		Dermal	1.4	86	14
TCE:D5		Dermal	1.2	40	—
				59 ± 30[b]	8.6 ± 44[c]
TCE:P1	467	IM	41	49	8
TCE:P2	550	IM	20	47	17
TCE:P3	469	SC	2.6	76	22
TCE:P4	436	SC	2.7	76	21
TCE:P5	561	SC	2.7	77	22
				65 ± 24[b]	18 ± 33[c]
CF:D1		Dermal	2.4	12	2.5
CF:D2		Dermal	1.5	15	11
CF:D3		Dermal	3.1	1.4	1.5
CF:D4		Dermal	1.2	9.3	6
CF:D5		Dermal	0.91	0	—
CF:D6		Dermal	0.50	0	—
				6.3 ± 110[d]	5.2 ± 82[e]
CF:P1	646	SC	3.3	6.1	8
CF:P2	618	SC	3.3	5.5	8
CF:P3	653	SC	3.3	3.4	1
CF:P4	582	SC	3.3	9.7	4
CF:P5	532	SC	3.3	25	5.5
				9.9 ± 88[d]	5.3 ± 56[e]
PCE:D1		Dermal	4.2	22	8.5
PCE:D2		Dermal	6.0	17	8
PCE:D3		Dermal	6.9	8.9	4.5
PCE:D4		Dermal	7.0	18	7
PCE:D5		Dermal	6.5	4.8	2
				14 ± 50[f]	6.0 ± 50[g]
PCE:P1	716	SC	3.7	4.5	1
PCE:P2	686	SC	3.7	17	1
PCE:P3	613	SC	3.7	23	1
PCE:P4	648	SC	3.7	15	1
PCE:P5	532	SC	3.6	8.5	2
				14 ± 53[f]	1 ± 40[g]

[a]TCE = trichloroethylene, CF = chloroform, PCE = tetrachloroethylene. Below each set of related values listed is the corresponding arithmetic mean (± CV, %). See Section II for explanation of experiment designations, positive control and dermal exposure protocols, procedure used to estimate administered dose to animals in dermal exposure experiments based on information from Table 1, and CV, %. Time until 95% excretion was estimated to the nearest half-day. Weights of animals used in dermal exposure experiments are listed in Table 1.
[b]Not significantly different ($p = 0.60$).
[c]Significantly different ($p = 0.018$).
Not significantly different ($p \geq 0.45$).
[g]Significantly different ($p^* = 0.012$).

however, was significantly different in DE and corresponding PC experiments for TCE (about twofold less) and PCE (about sixfold greater).

Data on recovery of radiolabel in DEC water and diaphragm materials in all CC and DE experiments are given in Table 3. As shown in Table 3, there is no statistically significant difference between any CC and corresponding DE experiment in the average recovery of radiolabel injected into the DEC based on the first withdrawn DEC fluid sample taken at t_{10}. These recoveries, except for experiments TCE:D1–5 and pooled experiments TCE:C1–5 + TCE:D1–5, are not significantly less than 100%. Table 3 also shows that measured or estimated recovery of radiolabel in the DEC diaphragm at t_{72} (as a percentage of L_0—see Section II) is significantly different in all CC and corresponding DE experiments except those involving low-concentration TCE. There is no consistent pattern among these differences, however, and they reflect mean levels of radiolabel uptake into the DEC diaphragm that, when significantly greater than zero, are quite small ($\leq 1\%$) in relation to L_0.

Table 3 Recovery of Radiolabel in DEC Water and Diaphragm[a]

Experiments	n	% Input present in DEC water at t_0 = 100%L_0/I	Radiolabel in diaphragm at t_{72} as % in DEC water at t_0 = 100%D_{72}/L_0
TCE:C1–5	5	89 ± 9.8	0.41 ± 0.34
TCE:D1–5	5	93 ± 4.5[b,c]	0.71 ± 1.0
TCE:C6–10	5	100 ± 9.0	0.74 ± 0.36[d,e]
TCE:D6–10	5	100 ± 12	0.19 ± 0.076[d,e]
CF:C1–5	5	98 ± 3.4	0.15 ± 0.22[f]
CF:D1–6	6	103 ± 6.0	1.1 ± 0.97[f]
PCE:C1–5	5	102 ± 14	0.28 ± 0.29[e,g]
PCE:D1–5	5	102 ± 17	0.96 ± 0.35[g]

[a]DEC = dermal exposure chamber. See Section II for details on measurement of input net dpm I and on estimation of net dpm in DEC water at t_0 (L_0) and in DEC diaphragm materials at t_{72} (D_{72}). Percentage values are given as mean ± standard deviation.
[b]Significantly less than 100% ($p = 0.023$); other values in this column do not differ significantly from 100% ($p > 0.05$).
[c]Significantly less than 100% ($p = 0.023$), which is also true of the mean (91 ± 7.5%) for pooled TCE:C1–5 + TCE:D1–5 data ($p = 0.0035$).
[d]Difference significant ($p^* = 0.0079$).
[e]Significantly greater than zero ($p < 0.01$); other values in this column are not significantly greater than zero ($p > 0.05$).
[f]Difference significant ($p^* = 0.0067$).
[g]Difference significant ($p = 0.010$).

IV. DISCUSSION

Our analysis of surface area measurements provides a reasonable basis for interpolation of the exposed surface areas of the medium-sized hairless guinea pigs used in our DE experiments. The best fit empirical allometric exponent of body weight obtained (0.39 ± 19%) is significantly lower ($p < 0.05$) than values of 0.67 or 0.75 typically used for *interspecies* extrapolation of surface areas (30).

The data summarized in Figures 3 through 5 and Table 1 support the conclusions that relative radiolabel losses from the DEC in DE and CC experiments were approximately linear over time, that the rates of such loss in DE experiments were greater than those observed in corresponding CC experiments, and that the k_p values estimated from the difference between these DE and corresponding CC loss rates are reasonably consistent for each of the three VCOC tested. The results of our PC experiments (Table 2) address the question of whether k_p values we obtained actually reflect radiolabel losses from DEC as a result of percutaneous absorption, rather than loss by some other process(es). Multicompartment, physiologically based pharmacokinetic models for VCOC predict that the metabolized fraction of a very small administered VCOC dose primarily metabolized in liver, such as the dose received in any of our PC and low-concentration DE experiments, is likely to be similar whether that dose is administered by SC or IM injection or by dermal absorption (31). Therefore, the consistencies between DE and corresponding PC data on excreted radiolabel recovery for all three VCOC tested (Table 2) and between DE and corresponding CC data on recovery of radiolabel injected into DEC (Table 3) jointly provide clear evidence that dermal uptake accounted for the decline of radiolabel in DEC fluid over time in the DE experiments over that observed in corresponding CC experiments.

Although differences were observed in the *kinetics* of radiolabel excretion between DE and corresponding PC experiments using TCE and PCE (Table 2), they are understandable in light of the different exposure routes involved and do not detract from our general conclusion from these DE/PC comparisons that our DE and CC results were used successfully to measure in vivo uptake of the tested compounds through guinea pig skin. The data obtained on recovery of radiolabel in DEC diaphragms (Table 3) indicate that such uptake, when it occurred in significant amounts, could be at most only a very small source of error in these measurements.

The average k_p value obtained for TCE at a relatively high aqueous concentration (~100,000 ppb) was not found to differ significantly from that obtained at concentrations three to four orders of magnitude lower. This invariance indicates that dermal uptake of TCE is strictly proportional to its concentration over a very large range, in accordance with Fick's law (see Section II), and that similar "linearity" in dermal uptake over time is likely for closely related VCOC such as CF and PCE.

The average k_p values obtained for dilute aqueous CF, TCE, and PCE (Table 1) are compared in Table 4 with others obtained or calculated (under Fick's law assumptions) for selected compounds tested in permeability studies in vitro or in vivo involving human skin or tissue samples. As shown in Table 4, the mean k_p values we obtained are far greater (by roughly two to three orders of magnitude) than those obtained for water (using human tissue) or neat CF and PCE (using live shaved mouse); they are greater than those obtained for aliphatic alcohols as large as decanol; they are similar to those obtained for 2-ethoxyethanol and a series of phenolic compounds; and they are roughly two- to threefold less than those obtained

Table 4 Average Dermal Permeability Constants Obtained for Selected Compounds[a]

Compound	Dermal tissue[b]	Study type[c]	k_p (ml/cm^2-h)	Reference
Water	HAE, HASC	R	0.002–0.0073	29, 47
Decanol	HAE	R	0.080	29
Phenol	HAE	R	0.082	34
CF (neat)	SM	V	0.0016	3
CF	HGPS	V	0.13	This study
Cresols	HAE	R	0.15–0.18	34
TCE[d]	HGPS	V	0.23	This study
2-Ethoxyethanol	HAE	R	0.30	22
Chlorophenols	HAE	R	0.33–0.36	34
PCE (neat)	SM	V	0.00015	3
PCE	HGPS	V	0.37	This study
Benzene	HH	V	0.41	13
Chlorocresol	HAE	R	0.55	34
Styrene	HH	V	0.64	12
Toluene	HH	V	0.94	12
Ethylbenzene	HH	V	1.2	12

[a]Compounds listed are dilute aqueous (unless noted otherwise).
[b]HAE = human abdominal epidermis; HASC = human abdominal stratum corneum, HGPS = hairless guinea pig skin; HH = human hand; SM = shaved mouse skin.
[c]R = in vitro, V = in vivo.
[d]Based on low-concentration studies (TCE:C1–5 and TCE:D1–5).

for some simple aromatic hydrocarbons in studies in vivo (discussed in Section I).

Although permeability constants were not calculated in the study of Jo et al. (8,9) on dermal absorption of aqueous CF by showering volunteers, their data may be used to estimate an average effective value of k_p for skin exposed to CF in shower water, and this value may be compared with that we obtained using hairless guinea pigs. Consistently significant differences measured in CF concentration in exhaled breath following 10 minute showers by persons with and without wet suits (intended to prevent dermal but not respiratory uptake) were used by Jo et al. as the basis of an estimate that dermal uptake from showering constituted approximately 90% of that due to inhalation alone. At present, we assume this figure of 90% is correct for the purpose of calculating a value of k_p reflecting the results of the Jo et al. study. From Table IV of that study, we calculate that the mean air concentration of CF (μg/L) in the breathing zone of shower stalls used was 0.66% of the corresponding concentration (μg/L) of CF in the water used for each shower. Respiratory uptake of CF is thought to be driven by the alveolar ventilation rate, which we assume for a reference 70 kg adult is 378 L/h (32). This reference adult has a dermal surface area of approximately 18,000 cm^2 (33), about 80% of which we assume [following, e.g., Brown et al. (14)] was effectively immersed continuously during showering. From these stated assumptions, the data of Jo et al. (8,9) imply the average, effective k_p value of

$$k_p = \frac{(0.90)(378{,}000 \text{ ml/h})(0.0066)}{(0.80)(18{,}000 \text{ cm}^2)} \approx 0.16 \text{ ml/cm}^2\text{-h}$$

which is very close to the mean k_p value of 0.13 ml/cm^2-h we obtained for CF using hairless guinea pigs (Table 1). It may be that the total dermal CF dose was underestimated in the Joe et al. study because its design may not have properly accounted for potential dermal compartment-induced delay in the release of dermally absorbed CF into blood subject to equilibration with exhaled air. Roberts et al. (34), for example, showed that linear diffusion of phenolic compounds through human epidermis in vitro at 25°C, yielding k_p values close to those obtained here for CF (Table 4), occurred only after significant lag periods ranging from 30 to 80 minutes. Such lag periods observed in in vitro studies may overestimate corresponding in vivo lag periods, however, particularly at elevated temperatures typical of water used in showering that may facilitate dermal penetration by, for example, stimulating blood perfusion or opening pores (35). Thus, some of the CF dermally absorbed by showerers in the Jo et al. study may not have been reflected in the first breath samples taken (5 minutes) after showering ended, and these were the only samples used in that study to calculate the relative contribution of dermal dose.

Health risks associated with exposure to water containing potentially carcinogenic VCOC has been a continuing regulatory concern (20,36–39). When first considered quantitatively, dermal VCOC absorption from domestic water appeared to constitute a significant uptake pathway compared with ingestion or respiration (14,40). Integrated assessments considering multiple exposure pathways subsequently led to the conclusion that dermal absorption of particular VCOC from domestic water supplies may constitute a substantial (21,41–44) or minor (20,37,44) fraction of corresponding total uptake from all routes.

In vitro methods have been used to study the penetration through human skin of a range of dilute aqueous compounds, including steroids, alcohols, phenols, glycol ethers, and pesticides (e.g., References 6, 7, 22, 29, 34, and 45). However, because of the limitations associated with extrapolating in vitro data to estimate absorbed dose for regulatory purposes, in vivo data are currently the preferred basis for human dose estimation, particularly for lipophilic compounds (25,46). As discussed, the hairless guinea pig may provide a reasonable model for human percutaneous absorption of dilute aqueous VCOC, such as CF, TCE, and PCE. To the extent that this is true, the results we obtained using hairless guinea pigs may be useful in assessing human exposure to these compounds. Assuming that the k_p values we obtained for these compounds for hairless guinea pigs are similar to corresponding values for humans, we estimated dermal uptake of these compounds by humans exposed to them through bathing or swimming (Table 5). These estimates, given in terms of the equivalent dose that would be received by ingesting a given volume of the water used (assuming 100% ingestive absorption), show that dermal absorption may be an important (and perhaps, for regular swimmers, the primary) route of human exposure to these compounds from all water-related sources. These exposure estimates are therefore in general agreement with the studies cited earlier suggesting that dermal absorption is a significant pathway for human exposure to aqueous VCOC.

ACKNOWLEDGMENTS

We are very grateful to Dr. J.P. Knezovich, Ms. D.J. Bishop, Mr. B.F. Brunckhorst, and Mr. S.M. Rose for their helpful suggestions and assistance. This work was performed under the auspices of the U.S. Department of Energy at Lawrence Livermore National Laboratory under Contract W-7405-ENG-48, with funds from the U.S. Air Force, H.G. Aerospace Medical Research Laboratory, Toxic Hazards Division (AF/DOE AML/86-20),

Table 5 Extrapolation of Human Dermal Uptake of Dilute Aqueous Compounds Tested in This Study

Compound	Permeability constant based on data from hairless guinea pigs k_p (ml/cm²-h) Mean	95% UCL[a]	Equivalent ingested volume[b] for bathing[c] (L)	Equivalent ingested volume[b] for swimming[d] (L)
Chloroform	0.13		0.62	2.3
		0.17	0.82	3.1
Trichloroethylene[e]	0.23		1.1	4.1
		0.28	1.3	5.0
Tetrachloroethylene	0.37		1.8	6.7
		0.46	2.2	8.3

[a]UCL = upper two-tailed confidence limit.
[b]Estimated dermal dose is that received if the water volume listed were ingested, assuming 100% ingestive absorption.
[c]Assumed bathing scenario: 18,000 cm² total dermal surface area (for a reference 70 kg adult), with 80% effective dermal immersion for 20 minutes.
[d]Assumed swimming scenario: same as bathing scenario, except 100% immersion for 1 h is assumed.
[e]Based on data from low-concentration experiments (TCE:C1–5 and TCE:D1–5).

the California Department of Health Services (MOU-3, 87-T0102), and the U.S. Environmental Protection Agency Office of Research and Development (DW89934205).

REFERENCES

1. R.D. Stewart and H.C. Dodd, Absorption of carbon tetrachloride, trichloroethylene, tetrachloroethylene, methylene chloride, and 1,1,1-trichloroethane through human skin, *J. Am. Ind. Hyg. Assoc. 25*:439–446 (1964).
2. C.L. Hake and R.D. Stewart, Human exposure to tetrachloroethylene: Inhalation and skin contact, *Environ. Health Perspect. 21*:231–238 (1977).
3. H. Tsurata, Percutaneous absorption of organic solvents. 1. Comparative study of the in vivo percutaneous absorption of chlorinated solvents in mice, *Ind. Health 13*:227–236 (1975).
4. J.E. Wahlberg, Percutaneous toxicity of solvents: A comparative investigation in the guinea pig with benzene, toluene and 1,1,2-trichloroethane, *Ann. Occup. Hyg. 19*:115–119 (1976).

5. T. Kronevi, J. Wahlberg, and B. Holmberg, Histopathology of skin, liver, and kidney after epicutaneous administration of five industrial solvents to guinea pigs, *Environ. Res. 19*:56–69 (1979).
6. A.S. Susten, B.L. Dames, and R.W. Niemeier, In vivo percutaneous absorption studies of volatile solvents in hairless mice. I. Description of a skin-depot, *J. Appl. Toxicol. 6*:43–46 (1986).
7. R.C. Wester and H.I. Maibach, Human skin binding and absorption of contaminants from ground and surface water during swimming and bathing, *J. Am. Coll. Toxicol. 8*:853–859 (1989).
8. W.K. Jo, C.P. Weisel, and P.J. Lioy, Routes of chloroform exposure and body burden from showering with chlorinated tap water, *Risk Anal. 10*:575–580 (1990).
9. W.K. Jo, C.P. Weisel, and P.J. Lioy, Chloroform exposure and the health risk associated with multiple uses of chlorinated tap water, *Risk Anal. 10*:581–585 (1990).
10. G. Aggazzotti, G. Fantuzzi, P.L. Tartoni, and G. Predieri, Plasma chloroform concentrations in swimmers using indoor swimming pools, *Arch. Environ. Health 45*:175–179 (1990).
11. T. Dutkiewicz and H. Tyras, A study of the skin absorption of ethylbenzene in man, *Br. J. Ind. Med. 24*:330–332 (1967).
12. T. Dutkiewicz and H. Tyras, Skin absorption of toluene, styrene, and xylene by man, *Br. J. Ind. Med. 25*:243 (1968).
13. T. Dutkiewicz and H. Tyras, The absorption of benzene through the skin of man (Polish), *Zes. Nauk. Bromat. Chem. Toksykol. 1*:159 [reported in B. Baranowska-Dutkiewicz, Skin absorption of aniline from aqueous solutions in man, *Toxicol. Lett. 10*:367–372 (1982)].
14. H.S. Brown, D.R. Bishop, and C.A. Rowan, The role of skin absorption as a route of exposure for volatile organic compounds (VOCs) in drinking water, *Am. J. Public Health 74*:479–484 (1984).
15. U.S. Environmental Protection Agency (EPA), *Health Assessment Document for Chloroform*, EPA/600/8-84/004F, U.S. EPA, Office of Health and Environmental Assessment, Environmental Criteria and Assessment Office, Washington, D.C., 1985.
16. U.S. Environmental Protection Agency (EPA), *Health Assessment Document for Trichloroethylene*, EPA/600/8-82-006F, U.S. EPA, Office of Health and Environmental Assessment, Environmental Criteria and Assessment Office, Washington, D.C., 1985.
17. U.S. Environmental Protection Agency (EPA), *Health Assessment Document for Tetrachloroethylene (Perchloroethylene)*, EPA/600/8-82-005F, U.S. EPA, Office of Health and Environmental Assessment, Environmental Criteria and Assessment Office, Washington, D.C., 1985.
18. U.S. Environmental Protection Agency (EPA), *Addendum to the Health Assessment Document for Tetrachloroethylene (Perchloroethylene), Updated Carcinogenicity Assessment for Tetrachloroethylene (Perchloroethylene, PERC, PCE) (Review Draft)*, EPA/600/8-82-005FA, U.S. EPA Office of Health and

Environmental Assessment, Environmental Criteria and Assessment Office, Washington, D.C., 1986.
19. U.S. Environmental Protection Agency (EPA), *Addendum to the Health Assessment Document for Trichloroethylene: Updated Carcinogenicity Assessment for Trichloroethylene (External Review Draft)*, EPA/600/8-82-006F, U.S. EPA Environmental Criteria and Assessment Office, Washington, D.C., 1987.
20. R.C. Cothern, W.A. Coniglio, and W.L. Marcus, Estimating risk to human health: Trichloroethylene in drinking water is used as the example, *Environ. Sci. Technol. 20*:111–116 (1986).
21. K.T. Bogen, L.C. Hall, L. Perry, R. Fish, T.E. McKone, P. Dowd, S.E. Patton, and B. Mallon, *Health Risk Assessment of Trichloroethylene in California Drinking Water*, Report prepared for the California Public Health Foundation (UCRL-21007). Lawrence Livermore National Laboratory, Livermore, CA, 1988.
22. R.J. Scheuplein and I.H. Blank, Permeability of the skin, *Physiol. Rev. 51*:702–747 (1971).
23. R.C. Wester and P.K. Noonan, Relevance of animal models for percutaneous absorption, *Int. J. Pharmaceut. 7*:99–110 (1980).
24. R.C. Wester and H.I. Maibach, Cutaneous pharmacokinetics: 10 steps to percutaneous absorption, *Drug Metab. Rev. 14*:169–205 (1983).
25. H.I. Maibach and R.C. Wester, Percutaneous absorption: In vivo methods in humans and animals, *J. Am. Coll. Toxicol. 8*:803–813 (1989).
26. J. Kao, J. Hall, and G. Helman, In vitro percutaneous absorption in mouse skin: Influence of skin appendages, *Toxicol. Appl. Pharmacol. 9*:93–103 (1988).
27. J.V. Wade, M.M. Mershon, L.W. Mitcheltree, and C.L. Woodard, The hairless guinea pig model and vesicant vapor exposures for bioassay purposes, *Proceedings of the 1989 Medical Defense Bioscience Review*, August 15–17, 1989 (U.S. Army Medical Research Institute of Chemical Defense, U.S. Army Medical Research and Development Command, eds.) Fort Detrick, Frederick, MD, 1989, pp. 569–575.
28. A.H. Conney, Pharmacological implications of microsomal enzyme induction, *Pharmacol. Rev. 19*:317–366 (1967).
29. R.J. Scheuplein and I.H. Blank, Mechanism of percutaneous absorption. IV. Penetration of nonelectrolytes (alcohols) from aqueous solutions and from pure liquids, *J. Invest. Dermatol. 60*:286–296 (1973).
30. K. Schmidt-Nielsen, *Scaling: Why Is Animal Size So Important?* Cambridge University Press, Cambridge, England, 1984, pp. 78–82.
31. K.T. Bogen, Pharmacokinetics for regulatory risk analysis: The case of trichloroethylene, *Regul. Toxicol. Pharmacol. 8*:447–466 (1988).
32. K.T. Bogen, Risk extrapolation for chlorinated methanes as promoters vs. initiators of multistage carcinogenesis, *Fund. Appl. Toxicol. 15*:536–557 (1990).
33. International Commission on Radiological Protection (ICRP), *Report of the Task Group on Reference Man*, ICRP No. 23. Pergamon Press, New York, 1975.

34. M.S. Roberts, R.A. Anderson, and J. Swarbrick, Permeability of human epidermis to phenolic compounds, *Pharm. Pharmacol. 29*:677–683 (1977).
35. T.A. Loomis, Skin as a portal of entry for systemic effects, *Current Concepts in Cutaneous Toxicity* (V.A. Drill and P. Lazar, eds.), Academic Press, New York, 1980, pp. 153–169.
36. E.A.C. Crouch, R. Wilson, and L. Zeise, The risks of drinking water, *Water Resources Res. 19*:1359–1375 (1983).
37. U.S. Environmental Protection Agency (EPA), *Techniques for the Assessment of the Carcinogenic Risk to the U.S. Population Due to Exposure from Selected Volatile Organic Compounds from Drinking Water Via the Ingestion, Inhalation and Dermal Routes*, NTIS Publ. No. PB84-213941, Office of Drinking Water, Washington, D.C., 1984.
38. J.B. Andelman, Human exposures to volatile halogenated organic chemicals in indoor and outdoor air, *Environ. Health Perspect. 62*:313–318 (1985).
39. U.S. Environmental Protection Agency (EPA), *Superfund Exposure Assessment Manual*, EPA/540/1-88/001, U.S. EPA Office of Remedial Response, Washington, D.C., 1988, pp. 121–133.
40. A.J. Beech, Estimated worst case trihalomethane body burden of a child using a swimming pool, *Med. Hypotheses 6*:303–307 (1980).
41. A.T. Shehata, A multi-route exposure assessment of chemically contaminated drinking water, *Toxicol. Ind. Health. 1*:277–298 (1985).
42. California Department of Health Services (CDHS), Notice of proposed rulemaking: Maximum contaminant level for tetrachloroethylene (PCE) in drinking water, R-75-87, Sacramento, CA, 1988.
43. T.E. McKone, Human exposure models, *Toxicol. Lett. 49*:321–339 (1989).
44. Brown and Hattis (1989).
45. R.J. Scheuplein, I.H. Blank, G.J. Brauner, and D.J. MacFarlane, Percutaneous absorption of steroids, *Invest. Dermatol. 54*:63–70 (1969).
46. C.A. Frankin, D.A. Somers, and I. Chu, Use of percutaneous absorption data in risk assessment, *J. Am. Coll. Toxicol. 8*:815–827 (1989).
47. I.H. Blank, J. Moloney, III, A.G. Emslie, I. Simon, and C. Apt, The diffusion of water across the stratum corneum as a function of its water content, *Invest. Dermatol. 82*:188–194 (1984).

15

Predictive Model for Dermal Exposure Assessment

Annette L. Bunge

Colorado School of Mines, Golden, Colorado

Gordon L. Flynn

The University of Michigan, Ann Arbor, Michigan

Richard H. Guy

University of California, San Francisco, California

I. INTRODUCTION

Dermatological disease and systemic illness resulting from exposure of the skin to toxic chemicals are currently major occupational and environmental health concerns in the United States. Prediction of detrimental effects upon exposure of the skin to hazardous chemicals is difficult, however, because of the complexity of the percutaneous absorption process and the lack of any established relationship(s) between transport rate and chemical properties. For cutaneous toxicity to be manifest, transdermal penetration must occur. The unique permeation barrier properties of skin ensure that the kinetics of percutaneous absorption to a very large extent determine the severity and time course of any potential hazard. It follows that one objective of risk estimation is to develop a predictive model for skin penetration and for dermal exposure assessment.

II. BACKGROUND

The rate at which a chemical enters the skin depends on the concentration difference ΔC across the skin according to

$$\frac{d}{dt}\frac{M_{\text{in}}}{A} = K_{p,s}\,\Delta C \tag{1}$$

where M_{in}/A is the cumulative mass entering into the stratum corneum per area exposed and the proportionality parameter $K_{p,s}$ is the skin permeability coefficient. Strictly speaking, $K_{p,s}$ unambiguously characterizes the skin barrier only when there is no chemical- or vehicle-related skin damage, and the rate of mass absorption has reached steady state such that the masses of material entering and leaving the skin are identical. For clarity, we refer to $K_{p,s}$ calculated from the steady-state mass flux as the steady-state permeability of skin. If the concentration of chemical in the vehicle remains constant at C_v° and the chemical concentration at the point of capillary uptake is zero, then ΔC is equal to C_v°.

The skin barrier consists of the stratum corneum and viable epidermis. Consequently, the steady-state permeability for the combined stratum corneum and viable epidermis is related to the steady-state permeabilities of each as follows:

$$\frac{1}{K_{p,s}} = \frac{1}{K_{p,sc}} + \frac{1}{K_{p,ve}} \tag{2}$$

Equation (2) therefore indicates that resistance in series are additive. The steady-state permeability across the stratum corneum from a given vehicle into an infinite sink depends on the diffusivity of the chemical in the stratum corneum D_{sc}, the diffusion path length through the stratum corneum L_{sc}, and the equilibrium partition coefficient of the chemical between the stratum corneum and the vehicle $K_{sc/v}$, as

$$K_{p,sc} = \frac{K_{sc/v}D_{sc}}{L_{sc}} \tag{3}$$

where $K_{sc/v}$ is defined as the concentration of chemical in the stratum corneum (mass/volume of stratum corneum at absorbing conditions) divided by the equilibrium chemical concentration in the vehicle (mass/volume) (1).

The steady-state permeability of the viable epidermis $K_{p,ve}$ is defined as

$$K_{p,ve} = \frac{K_{ve/v} D_{ve}}{L_{ve}} \tag{4}$$

Consequently, the steady-state permeability for the stratum corneum/viable epidermis skin composite can be written as

$$K_{p,s} = \frac{K_{p,sc}}{1 + B} \tag{5}$$

where the parameter B, defined as

$$B = \frac{K_{p,sc}}{K_{p,ve}} = \frac{D_{sc} L_{ve} K_{sc/v}}{D_{ve} L_{sc} K_{ve/v}} \tag{6}$$

measures the permeability of the chemical in the stratum corneum ($K_{p,sc}$) relative to that in the viable epidermis ($K_{p,ve}$). If the vehicle does not alter the physical properties of the stratum corneum, then thermodynamic equilibrium requires that the ratio of the stratum corneum and viable epidermis partition coefficients with the vehicle equal the equilibrium partition coefficient between the two layers; that is,

$$K_{sc/ve} = \frac{K_{sc/v}}{K_{ve/v}} \tag{7}$$

B can then be related to the physicochemical parameters of the stratum corneum and viable epidermis only:

$$B = \frac{D_{sc} L_{ve} K_{sc/ve}}{D_{ve} L_{sc}} \tag{8}$$

As such, B will be independent of the vehicle, provided that the vehicle has not altered the stratum corneum. The B parameter measures the ease with which a chemical absorbs into the (functionally lipoidal) stratum corneum compared to the viable epidermis. Consequently, B increases with increasing chemical lipophilicity. For many chemicals B is relatively small, and the permeability of skin is essentially equal to the permeability of the stratum corneum [as indicated in Equation (5)]. In this situation, the stratum corneum is said to be the controlling or rate-limiting resistance.

Several assumptions are usually made concerning rate control by the stratum corneum (2):

The full thickness of the stratum corneum contributes to the diffusion barrier.
No active transport occurs.

The stratum corneum is a pseudohomogeneous medium.
Penetrant and vehicle molecules diffuse across the stratum corneum as individual entities; that is, there is no carrier effect.
There are no size-limiting pores to affect absorption.
The stratum corneum is not changed progressively by the vehicle or penetrant.
Penetrant concentration changes do not alter stratum corneum or vehicle properties.

Many of these premises are sound; others, such as the last two, represent an ideality that is rarely encountered in reality.

III. DERMAL ABSORPTION RATE EQUATION PARAMETERS

A. Partition Coefficient

The partition coefficient $K_{sc/w}$ defines the equilibrium ratio of the concentration of the compound in the stratum corneum divided by that in the adjacent aqueous vehicle. The use of the stratum corneum/vehicle partition coefficient in Equation (3) acknowledges that attainment of equilibrium at the stratum corneum/water interface is rapid (implying that the thermodynamic activity of the permeant at the stratum corneum's surface is essentially identical to that in the adjacent bulk solution).

In principle, $K_{sc/w}$ values can be determined (3) by (a) allowing a known quantity of stratum corneum to equilibrate with a solution of known concentration of the compound under study; (b) determining the concentration of the compound in the solution after equilibrium is reached; and (c) using the difference between the initial and final solution concentrations to determine the amount of the compound partitioned into the stratum corneum. The ratio of the concentration in the stratum corneum (SC; in mass per unit volume of SC) to the concentration in the vehicle (in mass per unit volume of vehicle, which in this case is water) equals the experimental partition coefficient. This value is based upon chemical (free and unbound) present within all phases of the stratum corneum (lipid and protein). It follows that $K_{sc/w}$ may not faithfully reflect the chemical's concentration in that phase of the stratum corneum through which the diffusive process takes place.

B. Path Length of Chemical Diffusion

The diffusion path length L_{sc} has often been assumed to be equal to the average thickness of the stratum corneum, that is, 10–40 μm. However, there is increasing acceptance that permeants spanning a very diverse range of physicochemical properties transport through the stratum corneum via a tortuous pathway confined to the intercellular lipid channels. Such a pathway has been estimated to have a diffusion path length of several hundred μm (4–6).

C. Diffusion Coefficient

The other key determinant of the permeability coefficient and the flux is the diffusion coefficient. Diffusion coefficients of relatively small (<500 daltons) nonelectrolytes in simple liquids (e.g., water and light organic solvents) are of the order of 10^{-6}–10^{-5} cm^2/s at ambient temperatures. Solutes of much higher molecular weight (10^3–10^4 daltons) have considerably smaller diffusion coefficients (10^{-8}–10^{-7} cm^2/s) in nondense liquid media.

Experimentally derived, apparent diffusion coefficients across the stratum corneum have been reported to be as small as 10^{-13} cm^2/s (2,3,7,8), values that translate into lag times of several days. Such estimates assume an "as the crow flies" diffusion path length (typically, L_{sc} = 10 μm) and calculation of D_{sc} from the measured lag time (see later). Introducing tortuosity and recalculation of D_{sc} values (with $L_{sc} \sim 500$ μm) results in diffusion coefficients in the normal range for diffusion through lipid bilayer structures (i.e., 10^{-9}–10^{-7} cm^2/s). It must also be recognized that binding phenomena, concomitant with transport, delay the onset of steady states and lengthen the lag times.

Because solute diffusion through a solvent requires displacement of the solute through a continuum of solvent molecules, it follows that the ease of diffusion should be inversely related to solute size (i.e., the D of a large molecule is less than that of a small molecule). The dependence of D_{sc} on solute size [as measured by molecular weight (MW) or molecular volume (MV)] has been modeled in a number of ways. Kasting et al. (8), following Cohen and Turnbull (9), proposed an exponential dependence:

$$D_{sc} \propto e^{-\text{constant} \times \text{MV}} \tag{9}$$

Equation (9) can be used in the prediction of permeability coefficients (see later).

D. Concentration Gradient

There is good evidence that capillary clearance of most transcutaneously absorbed compounds in the upper dermis is highly efficient. Consequently, when there are no blood flow limitations to the removal of penetrating compounds, given the massive volume of distribution of the body, the concentration of the penetrant at the papillary vasculature is assumed to be zero. Therefore, the concentration differential of solute across the skin ΔC is fairly represented by the concentration of the solute in the contacting phase C_v°, which in the discussion here is assumed to be constant.

IV. ESTIMATION OF STEADY-STATE PERMEABILITIES FROM AQUEOUS VEHICLES

Two types of structure-activity models, empirical and theoretical, have been used to estimate the skin permeability coefficients of chemicals applied in aqueous solution. Theoretical skin permeability models take into account the contributions of the possible routes of penetration through the skin, and the interactions of the constituent materials of these routes with the permeating chemicals. Empirical models rely upon measured, experimental permeability coefficients of structurally related chemicals. Typically, the permeability coefficients of a series of compounds are measured and are then correlated with physicochemical properties, such as partition coefficient and molecular size. The derived correlations are then employed to predict the permeability coefficients of other structurally related compounds within the congeneric family.

Interestingly, the most successful predictive models of percutaneous permeability have blended the ingredients of these two strategies described. Scheuplein et al. (2,3,7,10) were the first to model the skin barrier. Permeability of the epidermis was viewed as a two-step process involving transport across (a) the stratum corneum and (b) the living epidermal tissue beneath. A parallel, independent follicular pathway, bypassing the stratum corneum, was included in the model to explain situations in which therapeutic responses occurred more quickly than expected. In this model, the stratum corneum (although described as a lipoprotein mosaic) was assumed to be a uniform diffusion medium. Using alkanols as test permeants, Scheuplein and coworkers clearly demonstrated the partitioning dependence of skin permeability and, for the first time, established the viable tissue between the stratum corneum and the upper dermal capillaries as a limiting barrier to chemicals of very high lipophilicity.

Subsequently, Michaels et al. (5) modeled the stratum corneum as a "brick-and-mortar" structure. It was assumed that molecules diffused passively through the stratum corneum via either its intracellular keratin or its interstitial lipid phase. The model depended upon the permeant's diffusivities in the lipoidal and (presumably polar) proteinaceous phases of the stratum corneum and its lipid-phase/protein-phase partition coefficient. To use the model prospectively, however, requires exacting information that is simply unavailable, rendering the model a conceptual tool rather than an algorithm for forecasting K_p values.

A more recent model proposed by Berner and Cooper (12) also assigns parallel lipoidal and polar pathways to the stratum corneum. Assuming that the fluxes through the polar and lipophilic pathways are independent and additive, $K_{p,sc}$ can be estimated from this model providing appropriate information is available for the area fractions of the two pathways (A_P and A_L), for the corresponding diffusion coefficients, and for the effective thickness of the stratum corneum. A third (heterogeneous oil-water multilaminate) pathway was subsequently added to the model, a refinement that only compounded the basic problem of the original concept: namely, that the necessary parameters lacked a physiological and/or experimental base.

The Berner and Cooper model (12) was closely related to an approach described earlier by Albery and Hadgraft (4), who accounted for the diffusive penetration of stratum corneum by transcellular and intercellular pathways. Area fractions for the specific routes are again required to use the model predictively; it is therefore subject to the same concerns. Perhaps the most important contribution of this model was its application to deduce the importance of the intercellular lipid pathway of permeation across the stratum corneum several years before general acceptance of this concept.

Kasting et al. (8) also considered skin permeation as a simple passive diffusion process and treated the stratum corneum as a homogeneous membrane responsible for the barrier function of the skin. New features incorporated into their model were a solubility limitation for chemical in the stratum corneum and a molecular size dependence for diffusivity. Their statement of maximum steady-state flux $J_{\max}$ is given by Equation (10):

$$J_{\max} = \frac{D_{sc}}{L_{sc}} C_{sc}^{\text{sat}} \tag{10}$$

where C_{sc}^{sat} is the saturation solubility of the chemical in the stratum corneum, D_{sc} is its diffusivity, and L_{sc} is the diffusion path length. Recognizing that D_{sc} is inversely dependent upon penetrant size (molecular volume) according to the relationship

$$D_{sc} = D^0 e^{-\beta \text{MV}} \tag{11}$$

where β is a constant and D^0 is the diffusion coefficient of a (hypothetical) molecule having MV = 0, Kasting et al. (8) arrived at

$$\log \frac{J_{\max}}{C_{sc}^{\text{sat}}} = \log \frac{D^0}{L_{sc}} - \frac{\beta \text{MV}}{2.303} \tag{12}$$

The model was used to analyze the skin penetration rates of 35 chemicals delivered across human skin from saturated propylene glycol solutions. The data conformed, with a high degree of statistical certainty, to the relationship

$$\log \frac{J_{\max}}{C_{sc}^{\text{sat}}} = 1.13 - 0.008\text{MV} \tag{13}$$

Replacement of MV with molecular weight did not significantly alter the statistical significance of the correlation.

Recently, Flynn (13) compiled a large data base (~90 values) of permeability coefficients through human skin following exposure to aqueous solutions. The results were analyzed empirically using the rationale articulated in the approach of Kasting et al. (8). Flynn also recognized the importance of molecular size in his analysis and divided the data set, somewhat arbitrarily, into "low-molecular-weight" and "high-molecular-weight" groupings. The permeability coefficients were also classified according to chemical polarity: (a) a large intermediate group, for which an essentially linear dependence of permeability upon octanol-water partition coefficient (K_{oct}) was apparent; (b) a subset of very polar chemicals, the $K_{p,sc}^{w}$ values of which were similar and were greater than those predicted by a linear extrapolation of the $K_{p,sc}^{w}$ versus K_{oct} dependence exhibited by the first group; and (c) a collection of very hydrophobic substances of high (but, again, essentially K_{oct} independent) $K_{p,sc}^{w}$. Empirical algorithms were then set out to predict permeability coefficients from partition coefficients and molecular weights.

Based on this work and the concepts first articulated by Scheuplein (3), Flynn (13) has argued that the perceived sigmoidal dependence of permeability upon partitioning is biophysically instructive and provides evidence that (a) small, highly water-soluble compounds have access to a limited "polar" pathway across the stratum corneum ($K_{p,sc}^{w}$ is therefore uninfluenced by the chemical's K_{oct} value), and (b) the rate-determining barrier for very hydrophobic substances is no longer the lipophilic stratum corneum but now resides in the underlying, and more aqueous in nature, viable tissue. The latter contention seems eminently reasonable given that it has been established experimentally (e.g., Reference 14) that skin stripped of its stratum corneum retains a residual, essentially fixed, resistance because of the diffusional barrier of the underlying strata. The residual permeability of human skin, stripped of its stratum corneum, is of the order of 0.1 cm/h.

The hypothesis of a polar, "pore" pathway is not as easily substantiated. Because the permeabilities of very polar compounds are extremely small, the establishment of steady-state transport can be difficult to measure because of analytical sensitivity limitations. Guy and Hadgraft (15) challenged the arbitrary separation of the $K_{p,sc}^{w}$ versus K_{oct} dependence on evidential and statistical grounds, pointing out that there was no direct experimental support for a polar pathway and that there was no statistical basis for the division of the data into three subsets. Most recently, Potts and Guy (16) have taken this argument one step further and have reanalyzed the Flynn data set (13) with a refinement of the Kasting et al. (8) approach. This interpretation led to the correlation

$$\log K_{p,sc}^{w}(\text{cm/s}) = -6.36(\pm 0.18) + 0.74(\pm 0.07) \times \log K_{oct} - 0.0060(\pm 0.0006) \times \text{MW} \quad (14)$$

with $r = 0.82$ and $s = 0.75$. Interestingly, other investigators (Bronaugh, and Amidon and Flynn, personal communications) arrived at functionally similar relationships at about the same time.

It follows that this simple analysis, based on a single lipid-based route of penetration, is sufficient to explain the percutaneous permeability of a very large and diverse group of penetrants ($-3 < \log K_{oct} < 6$ and $18 < \text{MW} < 750$). Furthermore, the regression coefficients derived have physicochemical significance and are consistent with the results of earlier investigations. Specifically,

1. The coefficient premultiplying $\log K_{oct}$ is in good agreement with that determined experimentally by Anderson et al. (17). The stratum corneum lipids are apparently more polar (and can act as more efficient hydrogen bond donors and acceptors) than bulk octanol.
2. The value of the coefficient multiplying MW is comparable to those that have been obtained in similar analyses of transport across phospholipid-based biomembranes. A common mechanism of diffusive mobility through the SC and these simpler structures is implied, therefore, and a model based upon lipid conformational fluctuations (so-called rotational isomerization) appears to be most plausible (9,16,18–20).
3. The constant term corresponds to $\log(D^0/L_{sc})$ as defined in Equation (12). The value derived from the regression analysis is several orders of magnitude smaller than that for simple biomembranes. Given the preceding discussion, however, large differences in D^0 seem unlikely, leading one to the conclusion that D^0/L_{sc} in the SC is very small because L_{sc} is so large. Once again, this conclusion is consistent with a tortuous diffusion path through the intercellular lipids of the SC that requires a molecule to travel several hundred micrometers to reach the dermal side of the barrier.

Thus, a single interpretation, which correlates $K_{p,sc}^{w}$ with K_{oct} and MW, is able to account satisfactorily, at a high level of statistical significance, for a large data set of human skin permeability coefficients. The model accounts for the permeabilities of diverse nonelectrolytes across the SC purely on the basis of the lipid properties of the membrane. Transport via the intercellular, lipid-filled domains of the SC is therefore strongly implicated. Molecular polarity and molecular size are the dominant factors that determine skin permeability.

V. ESTIMATION OF DOSE ABSORBED

For the purpose of risk assessment, the total amount of chemical that ultimately becomes systematically available following an exposure is the parameter of crucial importance. Figure 1 shows the cumulative mass of chemical entering the stratum corneum through the exposed (outside) surface as a function of time, when the stratum corneum is the controlling resistance and the chemical concentration on the exposed surface remains

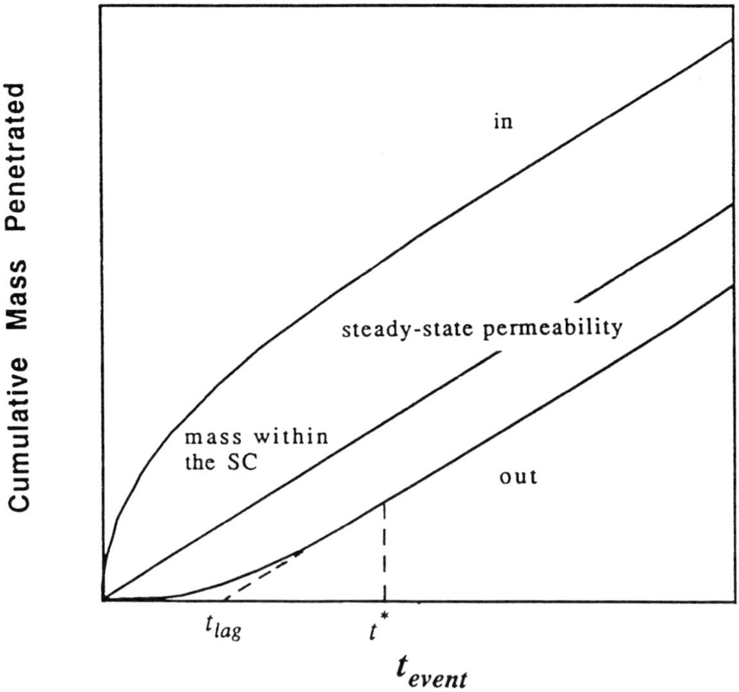

Figure 1 Cumulative mass of chemical penetrating into and out of the stratum corneum as a function of t_{event}.

unchanged. Figure 1 also shows the mass of chemical exiting from the stratum corneum's underside into the body.

The difference between the curves labeled *in* and *out* corresponds to the mass of chemical within the stratum corneum at any time. As illustrated in Figure 1, because some chemical resides in the stratum corneum itself, the cumulative mass entering the stratum corneum is always greater than that exiting from that skin. We argue that chemical absorbed into the stratum corneum invariably continues to transport into the viable tissue even after the chemical exposure ends. This is so because desquamation of the stratum corneum is too slow to compete with diffusion of chemical into the body.

The stratum corneum of normal skin desquamates at a rate of approximately 1 cell layer per day; there are typically 15–20 cell layers, so it follows that the stratum corneum is turned over once every 2–3 weeks. Thus, a chemical that has been taken up by the stratum corneum and is well mixed throughout the tissue is cleared extremely slowly by desquamation (the effective rate constant works out to be $\sim 10^{-3}$ h^{-1}). By contrast, the rate of clearance by simple diffusion is $\sim 10^2/K_{sc/w}$ h^{-1} (assuming rather conservative values for diffusivity and diffusion path length). Clearance by desquamation can only be equated with diffusional clearance, therefore, for exceedingly hydrophobic chemicals (those for which $K_{sc/w} \sim 10^5$) that are thermodynamically held back in the stratum corneum. Thus, if there is no loss of the chemical present in the skin by metabolism, irreversible binding, or evaporation, then all the chemical that entered the skin during the exposure period eventually becomes available to the body. Therefore, an appropriately conservative estimate of the total chemical exposure should be calculated based on the cumulative mass passing into the skin during the exposure event.

The rate at which mass enters the stratum corneum decreases with increasing exposure time. During short exposure times, the stratum corneum contains very little chemical; consequently, the chemical must diffuse only a short distance into the stratum corneum. Eventually the stratum corneum capacity is filled, and chemical entering the outer surface is offset by an equivalent amount of chemical exiting the tissue's underside. When this occurs, the concentration gradient across the entire stratum corneum becomes fixed. This situation is called steady state and corresponds to the linear regions of the curves labeled *in* and *out* in Figure 1.

The straight line through the origin of Figure 1 represents the steady-state permeability calculation from

$$DA_{event} = AC_v^o K_{p,sc} t_{event} \tag{15}$$

which assumes that the mass absorption rate is always at steady state. Thus, the cumulative masses into and out of the stratum corneum are equal. The slope of the steady-state permeability line is the same as the linear regions of the *in* and *out* curves. As seen in Figure 1, estimates of chemical absorption made using the steady-state permeability equation underpredict the actual amount absorbed into the stratum corneum, especially for short exposures when the rate of uptake is particularly rapid.

Extrapolation of the linear portion of the cumulative mass *out* curve to zero yields the lag time t_{lag}, which is related theoretically (for an isotropic membrane with no hydrodynamic boundary layers) to D_{sc} and L_{sc} by

$$t_{lag} = \frac{L_{sc}^2}{6D_{sc}} \tag{16}$$

In Equation (16), D_{sc} is the apparent diffusion coefficient of the penetrant consistent with the assumption that L_{sc} reflects the average thickness of the stratum corneum. Practically speaking, t_{lag} approximately measures how much time elapses before chemical begins to exit from the stratum corneum. Importantly, the lag time depends only on the diffusivity of the chemical in the stratum corneum and the effective stratum corneum thickness (which includes tortuosity). When the stratum corneum is the controlling resistance of the skin barrier, the time required to reach steady state t^* is about $2.4t_{lag}$ (21).

For highly lipophilic compounds, the viable epidermis contributes a significant resistance to chemical penetration of the skin barrier. The resulting chemical absorption becomes a function of the physicochemical properties of the chemical in both layers. The effect of the viable epidermis on the cumulative mass entering the stratum corneum is characterized by the B parameter defined in Equation (8). The epidermis is functionally aqueous in nature; hence, $K_{sc/ve}$ values may be expected to correlate with partition coefficients, such as the octanol-water partition coefficient K_{oct}. Direct experimental values of $K_{sc/ve}$ do not exist, although various dependencies upon K_{oct} have been proposed based on indirect measurements. Based on literature values of the magnitude of the variables in Equation (6) and reasonable expectations of the effect of MW on diffusivity in the viable epidermis, B can be approximated by (21,22)

$$B = \frac{K_{p,sc}^{w}(\mathrm{MW})^{1/2}}{0.72 \times 10^{-3}\ \mathrm{cm/s}} \tag{17}$$

where $K_{p,sc}^{w}$ is calculated from Equation (14). Figure 2 shows the cumulative mass absorbed (the dose absorbed, DA_{event}) into the stratum corneum per unit area, normalized by the quantity $K_{sc/v}C_v^\circ L_{sc}$, as a function of the dimensionless time ($t_{event}D_c/L_c^2 = t_{event}/6t_{lag}$). Plotting the cumulative mass

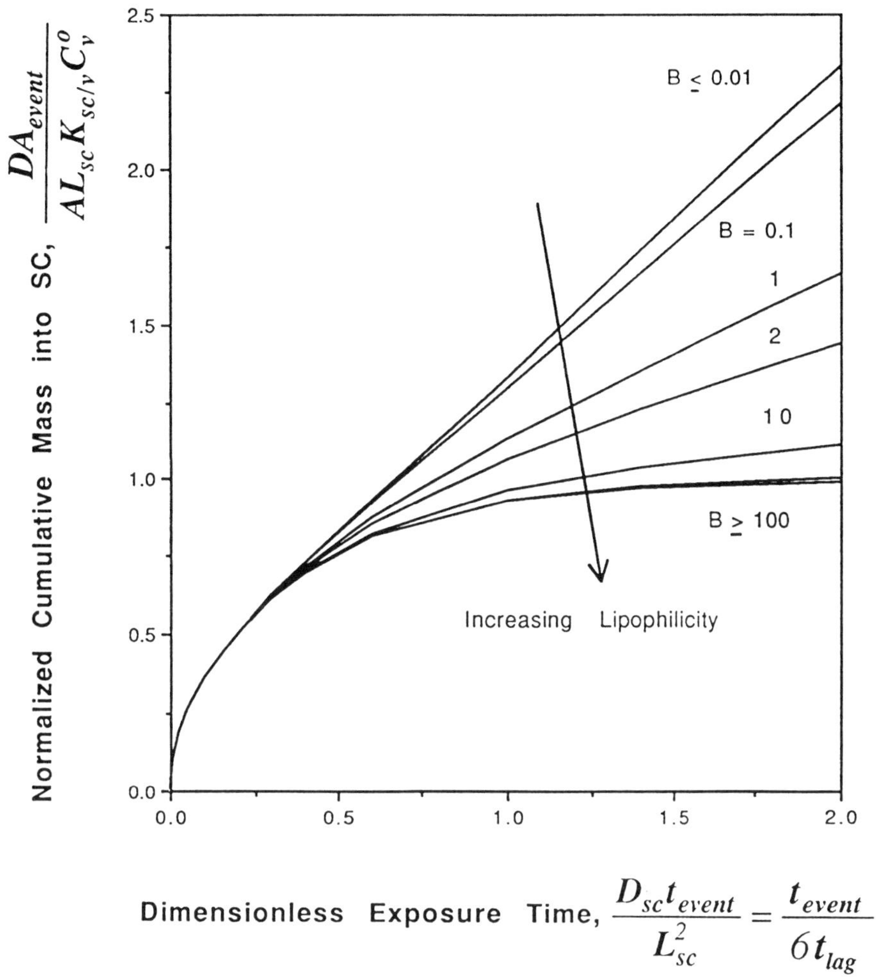

Figure 2 Cumulative mass of chemical entering the stratum corneum as a function of t_{event} when the skin is considered a two-layer composite comprising SC and viable epidermis (21).

absorbed in this way permits chemicals with different partition coefficients and effective diffusivities to be displayed on the same graph.

Figure 2 illustrates several important features that should be considered when estimating the amount of chemical absorbed during an exposure event. First, the normalized mass of chemical taken up by the stratum corneum per unit area ($DA_{\text{event}}/AK_{sc/v}C_v^{\circ}L_{sc}$) during an exposure shorter than the time required to reach steady state is independent of B. During these shorter exposure periods, the chemical has not reached the viable epidermis and the stratum corneum alone controls uptake. Following this initial non–steady-state period, the presence of the viable epidermis is felt and the normalized cumulative mass absorbed depends on B (see Figure 2). For moderately lipophilic or hydrophilic compounds (i.e., log K_{oct} less than about 3), B is less than 0.1 and the cumulative mass of chemical absorbed is controlled by the stratum corneum. As compounds increase in lipophilicity, the viable epidermis further restricts their flux out of the stratum corneum. If K_{oct} is very large ($B > 100$ or log $K_{oct} \simeq 7$), the viable epidermis essentially takes total control of the steady-state flux of the chemical.

Based on Figures 1 and 2, we conclude that DA_{event} estimates for risk assessment should be based on the total amount absorbed into the stratum corneum during the exposure event; furthermore, faster uptake rates during short exposure periods and the hydrophilic barrier presented by the viable epidermis to lipophilic chemicals must also be taken into account. The following two equations reasonably describe the family of curves in Figure 2 (21,22):

$$DA_{event} = 2AC_v^{\circ}K_{sc/v}\sqrt{\frac{D_{sc}t_{event}}{\pi}} \qquad \text{for } t_{event} \leq t^* \tag{18}$$

$$DA_{event} = \frac{AC_v^{\circ}K_{sc/v}L_{sc}}{1 + B}\left[\frac{D_{sc}t_{event}}{L_{sc}^2} + \frac{1 + 3B(1 + B)}{3(1 + B)}\right] \qquad \text{for } t_{event} > t^* \tag{19}$$

where t^*, which represents, essentially, the time required to reach steady state, depends on B as follows:

$$t^* = 0.4\,\frac{L_{sc}^2}{D_{sc}} \qquad \text{for } B \leq 0.60 \tag{20}$$

$$t^* = [b - \sqrt{(b^2 - c^2)}]\,\frac{L_{sc}^2}{D_{sc}} \qquad \text{for } B > 0.60 \tag{21}$$

where b and c are defined by Equations (23) and (24):

$$b = \frac{2(1 + B)^2}{\pi} - c \tag{22}$$

$$c = \frac{1 + 3B + 3B^2}{3(1 + B)} \tag{23}$$

For multiple, sequential exposures (e.g., showering daily), the maximum possible dose absorbed is obtained by assuming that the stratum corneum is completely cleared of all chemical before each new exposure.

To predict the mass absorbed using these equations requires estimates for D_{sc}, $K_{sc/ve}$, and L_{sc} rather than the steady-state skin permeability, which lumps all three parameters together. The stratum corneum thickness varies with age, gender, and location on the body. Nevertheless, 10–20 μm is a reasonable estimate for L_{sc} (2).

Most experiments on human skin have measured the stratum corneum or combined stratum corneum and viable epidermis (and sometimes the dermis) permeability, not the separate components of permeability, D_{sc} and $K_{sc/v}$. However, some experimental partition coefficient data are available. Based on these values, Cleek and Bunge (21) divided the correlation for steady-state stratum corneum permeability from water, Equation (14), into parts separately representing D_{sc}/L_{sc} and $K_{sc/w}$ as follows:

$$\log K_{sc/w} = 0.74 \log K_{\text{oct}} \tag{24}$$

$$\log \frac{D_{sc}}{L_{sc}} \text{ (cm/s)} = -6.36 - 0.006\text{MW} \tag{25}$$

Consistent with its derivation, the stratum corneum/vehicle partition coefficient obtained from Equation (14) is for a water vehicle and is indicated as such in Equation (24). Dermal absorption estimates using $K_{sc/w}$ apply for absorption from aqueous solutions. Approaches to adjust $K_{sc/w}$ for calculations of absorption from nonaqueous vehicles are discussed later. Based on data available at present, we recommend that Equation (24) not be used for highly hydrophilic chemicals with K_{oct} smaller than about 1.8 (i.e., $\log K_{\text{oct}} = -0.26$).

VI. ESTIMATION OF ABSORBED DOSE FROM NONAQUEOUS VEHICLES

Equations (18) and (19) assume that the chemical is being absorbed into the skin from an aqueous vehicle. In some environmental exposures, however, the absorbing chemical is present as a vapor, as a neat liquid, or in a nonaqueous vehicle. To estimate the dose absorbed in these situations, the appropriate values must be used in Equations (18) and (19) for C_v° and $K_{sc/v}$. If the vehicle does not significantly alter the physicochemical properties of skin, D_{sc} and B should not vary with the vehicle.

If the vehicle does not alter the skin, then the rate of uptake into the stratum corneum from saturated solutions should be independent of the nature of the vehicle. That is, the absorption rate from a saturated vapor, saturated water, or saturated nonaqueous phase should be the same as the absorption rate from the neat liquid chemical. However, there is no doubt that some vehicles interact with the skin. In such cases, it is difficult to predict the mass absorbed using absorption parameters from aqueous vehicles. The approach described here therefore represents a first approximation that may underestimate the actual absorption if vehicle-skin interactions are large.

For a nonaqueous liquid solution, C_v° is simply the concentration of the absorbing chemical in the vehicle (mass of chemical per volume of liquid). For exposure to a neat liquid chemical, C_v° is the density of the neat liquid. When the absorbing chemical is present as a vapor, the ideal gas law can be used to represent the gas-phase concentration in units of mass of chemical per gas volume:

$$C_v^\circ = \frac{p\mathrm{MW}}{RT} \tag{26}$$

where p is the partial pressure of the absorbing chemical in the vapor, T is absolute temperature, and R is the real gas constant. The partial pressure p should not exceed the saturation (i.e., vapor) pressure (p^{sat}) of the pure chemical at the temperature of the gas.

When partition coefficient data between the stratum corneum and the vehicle of interest are not available, approximate $K_{sc/v}$ values can be estimated by appropriately adjusting estimates of $K_{sc/w}$. These procedures assume that the vehicle of interest does not interact significantly with the stratum corneum, thereby altering its thermodynamic properties. If this is true, then partitioning between the stratum corneum and any vehicle must be related to the partition coefficient between the stratum corneum and water and the partition coefficient between the vehicle and water ($K_{v/w}$); that is,

$$K_{sc/v} = \frac{K_{sc/w}}{K_{v/w}} \tag{27}$$

If the vehicle and water do not interact significantly and $K_{v/w}$ is essentially independent of the chemical's concentration, then $K_{v/w}$ can be approximated from the saturation concentrations of the chemical in the water and the vehicle of interest. Then $K_{sc/v}$ can be estimated from

$$K_{sc/v} = K_{sc/w} \frac{C_w^{\mathrm{sat}}}{C_v^{\mathrm{sat}}} \tag{28}$$

where $K_{sc/w}$ is estimated using Equation (24). For neat liquids, the saturation concentration in the vehicle is represented by its liquid density.

Saturation concentrations are temperature dependent, and if possible, it is preferable to use saturation values at the skin surface temperature. If these values are not available, saturation concentrations at another temperature can be used provided the same temperature is used for both the vehicle and water. For a gaseous vehicle, C_v^{sat} is calculated from Equation (26) using the saturation (vapor) pressure (p^{sat}) for the absorbing chemical at the same temperature as the saturation concentration for water.

If desired, Equation (3) can be used to estimate a chemical's steady-state permeability through the stratum corneum following contact via a nonaqueous vehicle. Because the permeability coefficient includes the stratum corneum/vehicle partition coefficient, permeability values are different for different vehicles. Consequently, aqueous permeability coefficients should not be used to estimate absorption from nonaqueous vehicles.

VII. EXAMPLE CALCULATIONS

Dermal absorption of volatile organic compounds, specifically chloroform, during showering, bathing, and swimming in chlorinated water was discussed in several recent articles (23–25). Consequently, we have chosen to illustrate application of the predictive model described earlier using chloroform. Figure 3 shows the chloroform dose absorbed (normalized by the contacting concentration) per exposed area into the stratum corneum as a function of the exposure time. The dashed curve was calculated using Equations (18) and (19). These simplified equations closely represent the exact theoretical solution (21), which is shown as the solid curve. Table 1 lists the parameters [B, $K_{sc/w}$, and D_{sc} estimated by Equations (17), (24), and (25), respectively] used to calculate the curves plotted in Figure 3.

The right-hand axis of Figure 3 is the drinking water volume that must be orally ingested to result in an exposure equivalent to the predicted dermal absorption of chloroform. Assuming 100% absorption of all ingested chloroform, a material balance easily shows that the equivalent ingestion volume V_{ing} is given by

$$V_{ing} = \frac{DA_{event}}{A} C_w^{\circ} A \tag{29}$$

where the area of exposure A is assumed to be the entire skin surface area of 18,000 cm^2 (26). These equivalent ingestion volumes may be compared to public water supply regulations, which assume a daily ingestion of 2 L tap water for normal adults (23).

Figure 3 also shows the mass of chloroform leaving the stratum corneum to enter the body system during the same exposure time. As discussed earlier, much more chloroform will have entered the stratum corneum than

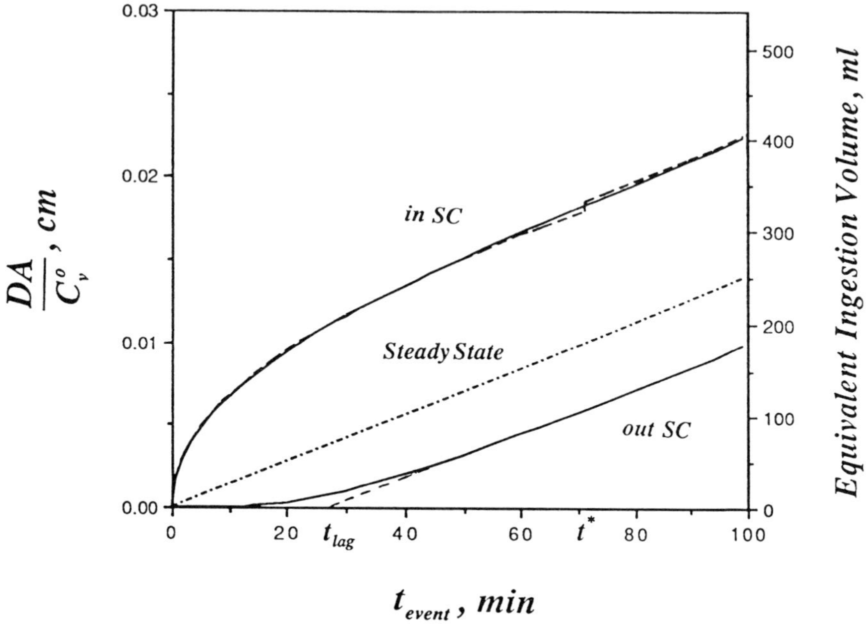

Figure 3 Chloroform dose absorbed into and out of the stratum corneum as a function of t_{event}.

Table 1 Parameters Used to Estimate Chloroform Dermal Absorption

Parameter		Equation
MW	119.4	—
log K_{oct}	1.97[a]	—
B	0.037	17
$K^{w}_{p,sc}$	0.0087 cm/h	14
$K_{sc/w}$	28.7	25
D_{sc}/L_{sc}	8.5×10^{-8} cm/s	26
L_{sc}	0.001 cm	—
t_{lag}	32.9 minutes	16
t^*	78.6 minutes	20

[a]Reference 27.

will have diffused across the stratum corneum to enter the systemic circulation during a brief exposure. For example, during a 10 minute shower, the predictive model estimates that the amount absorbed into the stratum corneum is equal to ingesting 131 ml of tap water. This is almost 900 times more than the mass of chloroform (ingestion volume equivalent of 0.15 ml) that would have diffused across the stratum corneum to enter the system during that same 10 minute shower. Ultimately, all the chloroform absorbed by the stratum corneum during the exposure may enter the body. We therefore reiterate that dermal exposure estimates should be based on the mass taken up into the stratum corneum.

Estimates of chloroform absorption from the steady-state permeability equation [Equation (15)], indicated in Figure 3 by the dotted line, underpredict the actual amount absorbed into the stratum corneum. For example, for a 10 minute shower, the predicted mass of chloroform absorbed is more than five times larger than the amount estimated by the steady-state permeability equation. As shown in Figure 3, however, even after a 79 minute exposure, when steady state has been reached, Equation (15) continues to underpredict the total amount absorbed. For a 100 minute exposure, the steady-state estimate of the mass absorbed is still only about 60% of the actual amount.

Calculations like those illustrated in Figures 1 and 3 raise concerns about the methods used to estimate permeability values from in vivo experiments. In vivo cumulative mass absorption data are sometimes analyzed by rearranging the steady-state permeability Equation (15) to calculate an apparent steady-state permeability,

$$K_{p,sc}^{\text{app}} = \frac{DA_{\text{event}}}{AC_v^{\circ}t_{\text{event}}} \tag{30}$$

Unless the in vivo experiment is conducted over a relatively long exposure time, a permeability estimate from mass absorption data will be larger than the true steady-state permeability. Again, this is because the steady-state treatment does not account for the additional mass initially absorbed to fill the stratum corneum reservoir. Figure 4 shows the scale of this overestimation for chloroform, in which the apparent permeability ($K_{p,sc}^{\text{app}}$) has been calculated by substituting the mass absorption from Equations (18) and (19) into Equation (30). The difference between the apparent and actual steady-state permeabilities decreases with increasing exposure period. As shown in Figure 4, however, the permeability coefficient calculated from the mass absorbed in 100 minutes still exceeds (by a factor of almost 1.7) the actual steady-state permeability value (0.0087 cm/h), even though steady-state was reached almost 20 minutes earlier.

In vivo determinations of $K_{p,sc}^{\text{app}}$ reflect $K_{p,s}$ only if the vehicle concentration remains unchanged over a sufficiently long exposure time for the

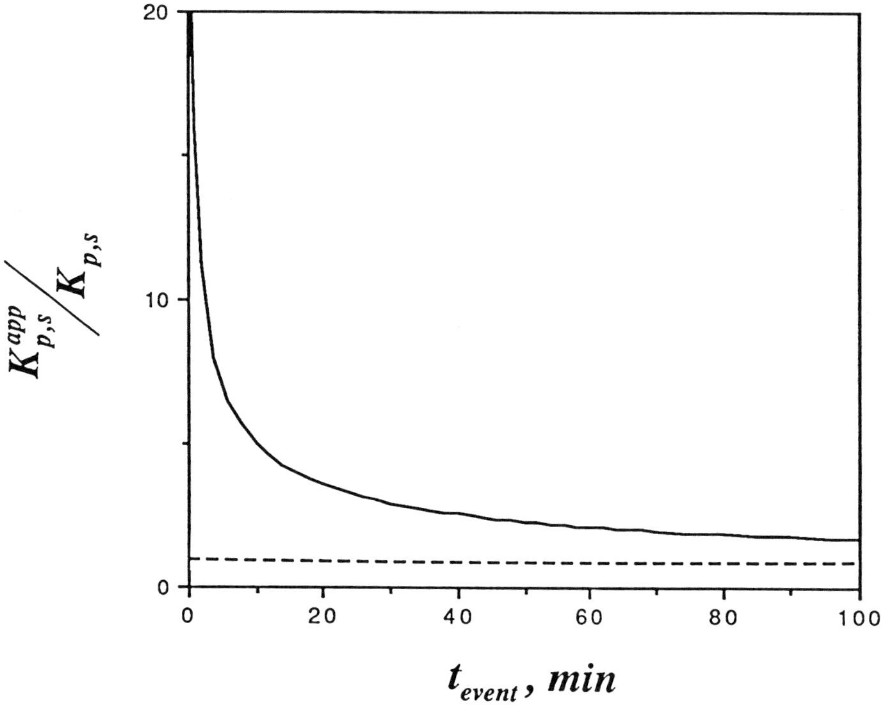

Figure 4 Apparent chloroform permeability calculated from cumulative mass absorption compared with the steady-state stratum corneum permeability as a function of time.

steady-state estimate of the mass absorbed [Equation (15)] approximately to equal the actual amount absorbed [including the unsteady-state period as described by Equation (19)]. The time required ($t^{\#}$) for the steady-state permeability Equation (15) to estimate 90% of the actual mass absorbed [Equation (19)] is given by

$$t^{\#} = 3\,\frac{D_{sc}}{L_{sc}^2}\,\frac{1 + 3B(1 + B)}{1 + B} \tag{31}$$

Using Equation (16), $t^{\#}$ can be related to t_{lag},

$$t^{\#} = 18t_{\text{lag}}\,\frac{1 + 3B(1 + B)}{1 + B} \tag{32}$$

For chloroform, the steady-state calculation predicts 90% or more of the actual mass absorbed only after an exposure time longer than $18t_{\text{lag}}$ (~ 10 h). That is, permeability values calculated from the measured cumulative mass absorbed for exposure times less than 10 h overpredict the actual steady-state permeability. However, meaningful permeability values can be obtained from shorter exposure times using the mass absorption rate. Once steady state has been established ($t^* \simeq 80$ minutes for chloroform), the steady-state permeability can be calculated from the rate of mass absorption as described by Equation (1).

We should also be cautious about in vivo permeability values estimated from data reflecting the mass that has entered the body during or just following the chemical exposure (e.g., blood levels or breath analysis). Since the mass taken up by the stratum corneum is larger than the mass leaving the stratum corneum to enter the body, these permeability values underestimate the actual permeability. Experiments that measure the mass absorbed into the stratum corneum are therefore preferred.

Jo et al. (23) estimated dermal exposure relative to inhalation exposure during showering using human subjects who were either unclothed or wearing a wet suit to prevent dermal exposure. It was shown that chloroform breath levels 5 minutes after a 10 minute shower correlated linearly with chloroform concentrations in the shower water. Chloroform concentrations in the breath of unclothed subjects were about 90% higher than those of the participants wearing wet suits. Based on these findings, Jo et al. (23) claimed that exposure via the dermal route was comparable to that by inhalation. Assuming that, in the unclothed subjects, 90% of the systemically available chloroform was dermally absorbed through 80% of the total skin area (18,000 cm^2), Bogen et al. (24) estimated an apparent permeability coefficient of 0.16 cm/h from the Jo et al. data, assuming that 100% of the mass inhaled (calculated from the vapor concentration and a respiration rate of 378 L/h) was absorbed. In a second paper, Jo et al. (25) estimated an average dermal absorption of 15.2 μg (0.22 μg/kg for a 70 kg individual) based on a similar calculation but assuming that the respiration rate was higher (840 L/h) and that only 77% of the inhaled dose was absorbed. This corresponds to 620 ml water ingested. Using the average concentration of chloroform in water (24.5 μg/L) (23) and exposure of the entire body for 10 minutes, we deduce an apparent permeability coefficient of 0.21 cm/h.

The Jo et al. data (23) raise some important concerns. Chloroform breath levels clearly correspond to systemic levels, specifically blood concentrations. If this is the case, their data indicate significant systemic concentrations within 5 minutes of a 10 minute exposure, specifically ingestion volume equivalents of 375 or 620 ml based on the Bogen et al. (24) or Jo et al. (23) assumptions, respectively. This is a much more rapid systemic response than expected from the predictive model estimate of ~30 minutes for the lag time. In fact, the predictive model indicates that an ingested volume equivalent of 131 ml will have absorbed into the stratum corneum during a 10 minute shower but that less than 1% of the amount would have diffused out in 15 minutes. If one assumes a very short lag time of only 5 minutes, then the amount of chloroform, that will have left the stratum corneum after a 10 minute shower, followed by a further period (without exposure) of 5 minutes (ingested volume equivalent of ~173 ml), is only about 50% of the chemical taken up during the shower (ingested volume equivalent, therefore, of ~337 ml). Yet, despite that the Bogen et al. (24) and Jo et al. (23) permeabilities were calculated from the amount of chemical that apparently *left* the SC, their estimates are 20–25 times *larger* than the value (0.0087 ± ~0.004 cm/h) calculated from Equation (14) (16).

The reasons for the discrepancy between the experimentally deduced value and that predicted by the model are not known at this time. One key issue, which must be examined, is the extent to which chloroform absorption can occur via routes other than transdermally or via inhalation when showering unclothed (e.g., oral, buccal, nasal, rectal, ocular, and vaginal). This control is lacking from the Jo et al. experiment (23) and could lead to a significant overestimate of exposure mediated via the skin.

More recently, Bogen et al. (24) attempted to determine chloroform permeability in vivo in hairless guinea pigs. The animals were immersed (up to the neck) in an aqueous chloroform solution for 70 minutes, and dermal absorption was determined (after correction for volatile losses) from the residual amount of chemical in the exposure solution. The permeability coefficient was calculated from the slope of cumulative mass absorbed versus time. The value obtained (0.13 ± 0.04 cm/h) in this way is 15 times larger than that predicted by Equation (16) for human skin (i.e., 0.0087 cm/h). As discussed, during the early period of dermal exposure absorption occurs more rapidly than the rate predicted using the true steady-state permeability coefficient. Consequently, the permeability coefficient reported by Bogen et al. (24) is larger than the actual steady-state value.

Figure 5 illustrates this point. The solid line is the cumulative absorption of chloroform predicted (in human) by Equations (18) and (19). The dashed line is a linear regression forced through the predicted cumulative absorption curve for the same time period (i.e., 70 minutes). The slope of the dashed line corresponds to an apparent permeability coefficient of 0.0208 cm/h, a value 2.4 times larger than the steady-state permeability coefficient predicted by Equation (16) of 0.0087 cm/h. If the data analysis of Bogen et al. (24) also overestimates the true permeability coefficient by 2.4-fold, then the actual steady-state permeability of chloroform through guinea pig skin should be about 0.054 cm/h. This value is approximately 6 times larger

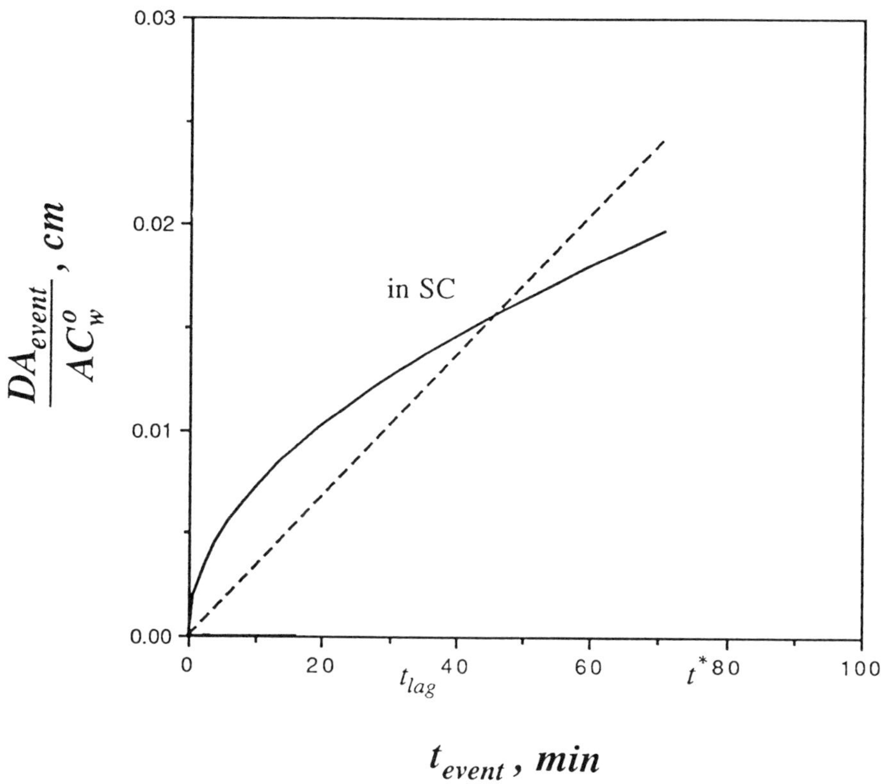

Figure 5 Linear analysis of the chloroform dose absorbed into the stratum corneum during a 70 minute exposure event.

than that predicted for human skin. Reasons for this difference may include interspecies variation in barrier function (e.g., stratum corneum thickness or lipid composition).

VIII. CONCLUSIONS

This chapter has outlined a simple description of skin permeability and has presented straightforward algorithms by which the amount of chemical absorbed into the body may be calculated following a variety of exposure conditions. A model with which to predict the steady-state permeability coefficient of a chemical across the skin (from an aqueous vehicle) has been described and physicochemically justified. Application of the model to an in vivo exposure scenario has been illustrated, and the utility of the approach with respect to improving future experimental designs has been demonstrated.

GLOSSARY

A	surface area of chemical exposure
b	parameter in t^* calculation, Equation (22)
B	parameter for the stratum corneum–viable epidermis skin composite measuring stratum corneum permeability relative to viable epidermis permeability
c	parameter in t^* calculation, Equation (22)
C_v°	concentration of the absorbing chemical in the vehicle; assumed to remain constant during the exposure period t_{event}
C_w°	concentration of the absorbing chemical in water; assumed to remain constant during the exposure period t_{event}
C_{sc}^{sat}	saturation concentration of the absorbing chemical in the SC
C_v^{sat}	saturation concentration of the absorbing chemical in the vehicle
C_w^{sat}	saturation concentration of the absorbing chemical in water
D^0	diffusivity of a hypothetical solute having MV = 0
DA_{event}	mass (dose) absorbed during an exposure event
D_{sc}	effective diffusivity of the absorbing chemical in the stratum corneum
D_{ve}	effective diffusivity of the absorbing chemical in the viable epidermis
J_{max}	maximum steady-state flux across the SC
$K_{sc/ve}$	equilibrium partition coefficient between the stratum corneum and the viable epidermis for the absorbing chemical
$K_{sc/v}$	equilibrium partition coefficient between the stratum corneum and vehicle for the absorbing chemical
$K_{sc/w}$	equilibrium partition coefficient between the stratum corneum and water for the absorbing chemical
$K_{ve/v}$	equilibrium partition coefficient between the viable epidermis and the vehicle for the absorbing chemical
K_{oct}	octanol-water partition coefficient for the absorbing chemical
$K_{p,sc}$	steady-state permeability coefficient of the absorbing chemical through the stratum corneum from a specified vehicle
$K_{p,sc}^w$	steady-state permeability coefficient of the absorbing chemical through the stratum corneum from water

$K_{p,ve}$	steady-state permeability coefficient of the absorbing chemical through the viable epidermis from a specified vehicle
$K_{p,s}$	steady-state permeability coefficient of the absorbing chemical through the skin (stratum corneum–viable epidermis composite) from a specified vehicle
L_{sc}	effective thickness of the stratum corneum
L_{ve}	effective thickness of the viable epidermis
M_{in}	cumulative mass absorbed into the stratum corneum during an exposure event t_{event}
MV	molecular volume of the absorbing chemical
MW	molecular weight of the absorbing chemical
p	partial pressure of a chemical vapor in a gaseous vehicle
p^{sat}	saturation (vapor) pressure of a chemical in a gaseous vehicle
r	correlation coefficient
R	real gas constant
s	standard error
sc	stratum corneum
t^*	time to reach approximately steady state
$t^\#$	time required for the steady-state estimate of mass absorption [Equation (15)] to equal 90% of the actual mass absorbed [Equation (32)]
t_{event}	time period of exposure event
t_{lag}	lag time
T	absolute temperature
ve	viable epidermis
β	parameter relating D_{sc} to the absorbing chemical's molecular volume [Equation (9)]
ΔC	concentration difference across the skin

ACKNOWLEDGMENTS

This work was supported in part by the U.S. Environmental Protection Agency under Assistance Agreement Nos. CR-817451 and CR-816785 and by Sandia National Laboratories (Contract No. 69-5518). We thank R.L. Cleek for his assistance, Kim Hoang, Professor Jonathan Hadgraft, and Dr. Russ Potts for their helpful comments, and Andrea Mazel for manuscript preparation.

REFERENCES

1. G.E. Parry, A.L. Bunge, G.D. Silcox, L.K. Pershing, and D.W. Pershing, Percutaneous absorption of benzoic acid across human skin. I. In vitro experiments and mathematical modeling, *Pharm. Res. 7*:230–236 (1990).
2. R.J. Scheuplein and I.H. Blank, Permeability of the skin, *Physiol. Rev. 51*:702–747 (1971).
3. R.J. Scheuplein, Mechanism of percutaneous adsorption. I. Routes of penetration and the influence of solubility, *J. Invest. Dermatol. 45*:334–346 (1965).
4. W.J. Albery and J. Hadgraft, Percutaneous absorption: Theoretical description, *J. Pharm. Pharmacol. 31*:140–147 (1979).
5. A.S. Michaels, S.K. Chandrasekaran, and J.E. Shaw, Drug permeation through human skin: Theory and in vitro experimental measurement, *Am. Inst. Chem. Eng. J. 21*:985–996 (1975).
6. R.O. Potts and M.L. Francoeur, The influence of stratum corneum morphology on water permeability, *J. Invest. Dermatol. 96*:495–499 (1991).
7. R.J. Scheuplein, Mechanism of percutaneous absorption. II. Transient diffusion and the relative importance of various routes of skin penetration, *J. Invest. Dermatol. 48*:79–88 (1967).
8. G.B. Kasting, R.L. Smith, and E.R. Cooper, Effect of lipid solubility and molecular size on percutaneous absorption, *Skin Pharmacokinetics* (B. Shroot and H. Schaefer, eds.), Karger, Basel, 1987, pp. 138–153.
9. M.H. Cohen and D. Turnbull, Molecular transport in liquids and gases, *J. Chem. Phys. 31*:1164–1169 (1959).
10. R.J. Scheuplein, On the application of rate theory to complex multibarrier flow coordinates: Membrane permeability, *J. Theor. Biol. 18*:72–89 (1968).
11. R.J. Scheuplein and I.H. Blank, Mechanisms of percutaneous absorption. IV. Penetration of nonelectrolytes (alcohols) from aqueous solutions and from pure liquids, *J. Invest. Dermatol. 60*:286–326 (1973).
12. B. Berner and E.R. Cooper, Models of skin permeability, *Transdermal Delivery of Drugs*, Vol. 3 (B. Berner and A.F. Kydonieus, eds.), CRC Press, Boca Raton, FL, 1987, pp. 41–56.
13. G.L. Flynn, Physicochemical determinants of skin absorption, *Principles of Route-to-Route Extrapolation for Risk Assessment* (T.R. Gerrity and C.J. Henry, eds.), Elsevier, New York, 1990, pp. 93–127.
14. H. Durrheim, G.L. Flynn, W.I. Higuchi, and C.R. Behl, Permeation of hairless mouse skin. I. Experimental methods and comparison with human epidermal permeation by alkanols, *J. Pharm. Sci. 69*:781–786 (1980).
15. R.H. Guy and J. Hadgraft, Physicochemical aspects of percutaneous penetration and its enhancement, *Pharm. Res. 5*:753–758 (1988).
16. R.O. Potts and R.H. Guy, Predicting skin permeability, *Pharm. Res. 9*:663–669 (1992).
17. B.D. Anderson, W.I. Higuchi, and P.V. Raykar, Heterogeneity effects on permeability-partition coefficient relationships in human stratum corneum, *Pharm. Res. 5*:566–573 (1988).

18. W.R. Lieb and W.D. Stein, Non-stokesian nature of transverse diffusion within human red cell membranes, *J. Membr. Biol. 92*:111–119 (1986).
19. R.O. Potts and M.L. Francoeur, Lipid biophysics of water loss through the skin, *Proc. Natl. Acad. Sci. USA 87*:3871–3873 (1990).
20. H. Trauble and D.H. Haynes, The volume change in lipid bilayer lamellae at the crystalline-liquid crystalline phase transition, *Chem. Phys. Lipids 7*:324–335 (1971).
21. R.L. Cleek and A.L. Bunge, A new method for estimating dermal absorption from chemical exposure. I. General approach, *Pharm. Res. 10*:497–506 (1993).
22. R.L. Cleek and A.L. Bunge, A new method for estimating dermal absorption from chemical exposure. II. Effect of molecular weight and octanol-water partitioning, *Pharm. Res.* (in press, 1994).
23. W.K. Jo, C.P. Weisel, and P.J. Lioy, Routes of chloroform exposure and body burden from showering with chlorinated tap water, *Risk Anal. 10*:575–580 (1990).
24. K.T. Bogen, B.W. Colston, and L.K. Machiaca, Dermal absorption of dilute aqueous chloroform, trichloroethylene, and tetrachloroethylene in hairless guinea pigs, *Fund. Appl. Toxicol. 18*:30–39 (1992).
25. W.K. Jo, C.P. Weisel, and P.J. Lioy, Chloroform exposure and the health risk associated with multiple uses of chlorinated tap water, *Risk Anal. 10*:581–585 (1990).
26. International Commission on Radiological Protection (ICRP), *Report of the Task Group on Reference Man*, IRCP No. 23, Pergamon Press, New York, 1975.
27. United States Environmental Protection Agency, *Dermal Exposure Assessment: Principles and Applications*, EPA/600/8-91/011B, 1992.

16

Establishing Concepts and Developing Methodologies in Dealing with Various Routes of Exposure to Water Contaminants: Dose Versus Health Risk

Rhoda G. M. Wang*

U.S. Environmental Protection Agency, Washington, D.C.

Ronald C. Wester

University of California School of Medicine, San Francisco, California

I. INTRODUCTION

With increasing awareness of the potential inhalation exposure hazard to water contaminants, notably the contribution of radon and other volatile organic chemicals (VOC) to the overall body burden, identifying appropriate methods for dose estimate has become an important task. However, not all water contaminants that may contribute significantly to the inhaled dose from daily household water uses behave like VOC. Furthermore, the volatilization potential differs among VOC, making the task of differentiating various water contaminants that may or may not be potential inhalants more difficult.

Data provided by McKone in the assessment of inhalation exposure on several VOC by way of a three-compartment model support the conclusion

**Present affiliation*: California Environmental Protection Agency, Sacramento, California

that for those chemicals with high volatility the dosages contributed via inhalation may exceed those via ingestion (1). Individual patterns of exposure may vary significantly from household to household depending on room size and ventilation conditions, as well as individual water use patterns. Thus, in addition to developing methodologies to compute the volatilized fraction of individual water contaminants, microenvironmental monitoring of human activities would certainly improve dose estimate. Sound regulatory decision making requires the availability of solid data that encompass a wide range of exposure scenarios. Human exposure studies are the most relevant for dose estimates and risk assessment. However, existing data are few or incomplete for water contaminants; those reported are preliminary but valuable. Many reports claiming human studies at best may involve the use of human skin but rarely involve in vivo human volunteer studies. The committees for human research established in university-related medical centers provide an appropriate means to evaluate the virtues of each study and serve as governing bodies for study approval, as well as ensure the safety of the volunteers entering these studies. Valuable information may be generated from these studies that can be used effectively for dose estimates and health risk assessment.

Federal and state regulatory agencies are involved in regulating drinking water contaminants. In dealing with numerous potential water contaminants that may adversely affect human health, regulatory scientists need to develop methods to examine the potential dose resulting from all routes of exposure. Concern over inhalation exposure to water contaminants requires prioritizing the contaminants according to their volatilization potential and their intrinsic toxicity (the former measures dose and the latter measures effect).

Dermal exposure to water contaminants during bathing and showering may also contribute to the total exposure dose. This route of exposure has not been fully explored. It is generally recognized that with the exception of occupational workers who may have excessive dermal contact with industrial and agricultural chemicals, such as during handling and accidental spills, the dermal dose derived from skin contact with household water contaminants may be insignificant in general. However, preliminary results from a recently conducted human study in which chloroform was monitored during showering demonstrate that the combined dose from inhalation and dermal exposure to chloroform during showering was comparable to the computed daily oral dose derived from water ingestion (2,3). We do not think this is a shocking discovery. Each chemical must be carefully evaluated for its potential dermal, inhalation, and oral dose under a given exposure scenario.

II. BRIEF OVERVIEW ON THE TRANSPORT AND UPTAKE OF TOXICANTS THROUGH INHALATION AND PERCUTANEOUS ROUTES

A. Inhalation

Inhalation exposure to volatilized water contaminants shares common ground with inhaling air contaminants from other sources. First, the lower respiratory tract is physiologically most vital in the transport of inhaled toxicants to the bloodstream. A network of branching occurs distally, and with the decrease of airway size the number of airways increases. At the site of alveoli, molecular diffusion becomes the dominant airspace transport. Simply stated, molecular diffusion results in a net flow from regions of high concentration to regions of low concentration. Absorption by airspace walls to the surrounding vascular tissues and entering into the systemic bloodstream tends to lower the concentration gradient in the airspaces and creates a new concentration gradient. The absorbed dose (internal dose) is proportional to the inhaled dose, which in turn relates to the air concentration of the contaminant. Volatilized water contaminants reaching the air sporadically depend on individual household water use patterns and other house ventilation parameters.

Factors influencing the transport and delivery of an inhaled compound to the bloodstream may include the blood/air partition coefficient, which reflects the solubility of the compound in blood, the rate of blood flow through the lungs, and the partial pressures of the contaminants in arterial and mixed venous blood. The following pharmacokinetic parameters are essential for estimating the uptake of an inhaled compound: the time after initial exposure, the first-order chemical rate constant for the air-blood compartment, the tissue/blood partition coefficient, and the rate constant for the elimination of the inhaled compound from the system.

In essence, the transport and uptake of an inhaled toxicant involves complicated physiological processes. Inhalation exposure to water contaminants has the unique feature that such contaminants must first be volatilized into the air and then mixed with the air. The volatilization of water contaminants is temperature dependent. Showering and cooking may result in higher concentrations of volatilized contaminants than household water usage at room temperature. Moreover, when water contaminants evaporate and mix with the air they do not possess the physical barrier, that is, particle size variations, present predominantly in synthetic toxic chemicals when dispersed into the air for intended—air application or spray of a pesticide—or nonintended situations—accident. The exposure scenario of water contaminants in one's own dwelling classifies it as an indoor exposure condition. Unlike toxic gases, which may react directly with lung tissues to produce major effects, water contaminants that volatilize into the air normally do not exist in the concentration range that produces direct pulmonary effects. However, their potential chronic effects, such as those produced via indoor exposure to radon, should not be overlooked. Thus, the health concern of exposure to water contaminants is likely to be focused on potential chronic effects as a result of cumulative, long-term, low-dose exposure conditions. While emphasizing on the physiological and anatomical aspects of the respiratory tract and the pathway of inhalation exposure to water contaminant, one should not lose sight of the toxicological aspects of an inhaled compound. After all, it is the pharmacological and toxicological nature of the chemical that evokes an effect and causes health concerns that may range from nonspecific to specific tissue or target organ effects.

B. Percutaneous Route

The uniqueness of percutaneous penetration of water contaminants due to recreational events, such as swimming, and daily water uses, such as showering and bathing, places these activities in a category that differs from dermal contact with chemicals under nonaqueous conditions (i.e., dermal delivery system of drugs or dermal contact with industrial chemicals in work-related settings). Traditionally, regulatory agencies are concerned with dermal exposure and absorption of pesticides in workers and in residents who live or work close by pesticide application sites. Over the years, the techniques for evaluating dermal exposure and absorption under various use conditions have been improved and refined. However, most of the data points are derived from animal studies that repeatedly demonstrate that percutaneous penetration rates differ among species (4). Newer in vitro techniques appear to present a parallelism between in vitro and in vivo percutaneous absorption studies for certain chemicals (5), but it is clearly indicated that there may be a distinct difference in human dermal absorption in vivo and in vitro (6). This is likely a result primarily of the lack of intact subcutaneous tissues and vascular apparatus of surgically removed or cadaver skin, and without these vital structures the physiological functions of normal skin cannot take place. Thus, in vivo human percutaneous penetration data, whenever feasible, is still the best source of information that can be used for realistic dermal dose estimate. In a recent paper, Wester and Maibach recommended criteria for choosing animal models from which data may be extrapolated to humans (7). The contribution to the total absorbed dose through dermal contact with contaminated soil has been considered a potentially significant route of exposure,

Table 1 In Vivo Percutaneous Absorption of *p*-Nitroaniline in the Rhesus Monkey Following 30 Minutes Exposure to Surface Water: Comparisons with In Vitro Binding and Absorption

Phenomenon	% Dose absorbed/bound
In vivo percutaneous absorption, rhesus monkey	4.1 + 2.3
In vitro percutaneous absorption, human skin	5.2 + 1.6
In vitro binding, powdered human stratum corneum	2.5 + 1.1

Source: From Reference 5.

especially in young children. An extensive review of dermal uptake of contaminants from water and soil contact can be found in the dermal exposure assessment document issued by the U.S. Environmental Protection Agency (EPA) (8).

The main skin barrier that prevents water loss, as well as limiting xenobiotics from entering the skin, is the stratum corneum. Extensive descriptions of the anatomy and physiology of the skin can be found in the literature (9–11). Since the stratum corneum is the essential skin barrier, attempts have been made to study powdered stratum corneum and relate its binding to various chemicals. Using similar amounts of powdered stratum corneum generated from in vitro excised skin and in vivo rhesus monkey's skin, Wester and colleagues compared the dermal absorption of several chemicals (5). For nitroaniline, a good correlation exists between stratum corneum binding, in vitro percutaneous absorption of human skin, and in vivo rhesus monkey skin for 30 minutes of exposure to surface water (Table 1). For other chemicals, this correlation may not be well defined. More intensive studies are needed further to define the correlation and predictability of each type of study on a chemical-specific basis.

In general, the dose contributed from dermal exposure to household water may not be sufficient to cause an acute toxicological concern. On a chronic water use basis, however, a toxicant that penetrates the skin sufficiently and repeatedly may contribute to the overall cumulative dose that one should not overlook, especially if known chronic adverse effects exist.

III. METHODOLOGY DEVELOPMENT IN DOSE ESTIMATES

A. Complexity of Human Activities and Household Factors

Exposure to water contaminants is related to individual water use pattern as well as the size and construction of the house. Human activities predominantly dictate the quantity of contaminants released into the air and the extent of dermal contact. House factors may include the size and construction of the house and air ventilation. For instance, a smaller house with more residents invariably involves more water usage and higher levels of air contaminants, as opposed to a larger house with fewer residents. By the same token, choices of shower or bathing and the time spent in each activity are strictly personal. In both cases deriving a combined dose from inhalation and dermal routes of exposure is possible.

B. Volatilization Models

1. Equilibrium and Semiequilibrium Models

The extent to which a chemical volatilizes from drinking water into the air is ultimately determined by the net rate of chemical release from water into the air. Assuming that drinking water contaminants are always in equilibrium between the water in which they are dissolved and the boundary layer of air immediately above the water, an equilibrium equation may be expressed as

$$(\mathrm{O}_{wa})^i = \frac{RT}{H^i} = \frac{C^i_{ew}}{C^i_{ea}} \tag{1}$$

where $(\mathrm{O}_{wa})^i$ is the chemical-specific Ostwald water/air partition coefficient, R is the universal gas constant, 8.2×10^{-5} atm-m^3/mol°K, T is the absolute temperature, 293°K at 20°C, H^i is the chemical specific Henry's law constant, atm-m^3/mol, C^i_{ew} is the equilibrium concentration of i in water (mg/L), and C^i_{ea} is the equilibrium concentration of i in air (mg/L). Since the total amount of contaminants initially found in water is partially volatilized in air,

$$C^i_w V_w = C^i_{ew} V_w + C^i_{ea} V_a \tag{2}$$

where C^i_w is the initial concentration of i in water (mg/L), V_a is the daily effective volume of air in the house (L), and V_w is the daily volume of water usage in the house (L/day). After introducing the fraction of chemical volatilized into the air, F^i, into the equation, $F^i = (C^i_w - C^i_{ew})/C^i_w$, substituting, and rearranging these equations, the following equations can be derived:

$$F^i = \frac{1}{1 + C^i_{ew} V_w / C^i_{ea} V_a} \tag{3}$$

$$F^i = \frac{1}{1 + \mathrm{O}^i_{wa} R_w / V_h R_v} \tag{4}$$

Equation (4) represents the equilibrium method by which O_{wa} can be estimated from RT/H, and house-specific parameters, such as water usage R_w, house volume V_h, and air exchange rate R_v, can be assumed.

Because complete equilibrium is never attained, and using volatilization of radon as an example, the volume-weighted average volatilization fraction for radon has been observed to be 0.55 (U.S. EPA) (12); thus, Equation (4) overestimates the volatilization by an approximate factor of 2. Hence, we propose a semiequilibrium equation that may be obtained by applying 0.5. (In the latter part of this chapter we demonstrate the validity of this proposal.) The semiequilibrium equation may be expressed as

$$F'^i = \frac{0.5}{1 + O^i_{wa}R_w/V_hR_v} \tag{5}$$

Equation (5) is the semiequilibrium equation used to compute the realistic volatilized fraction of water contaminants, and F'^i represents the volatilized fraction of chemical i of the semiequilibrium equation.

Although the factor applied to account for nonattainment of equilibrium is not expected to be a constant across the board for all chemicals, taking into account of all possible variables and uncertainties, it is believed that the semiequilibrium method provides useful information in estimating the volatilization potential of a water contaminant and can be applied effectively in routine screening. Table 2 applies the equilibrium methods to estimate nine chemicals ranging from low to high volatilization (13).

Table 2 Calculation of F^i by the Equilibrium and Semiequilibrium Methods for Nine Chemicals

Chemical	H (atm)	O_{wa}	F^i	F'^i
Benzene	5.60E-03	4.29E+00	0.9993	0.4996
Chloroform	3.16E-03	7.6OE+00	0.9987	0.4994
Ethylene dibromide	6.70E-04	3.58E+01	0.9941	0.4970
Naphthalene	4.60E-04	5.22E+01	0.9914	0.4957
Dibromochloropropane	2.37E-04	1.01E+02	0.9834	0.4917
DDT	2.80E-05	8.57E+02	0.8750	0.4375
Phenol	4.50E-07	5.33E+04	0.1011	0.0506
Formaldehyde	3.27E-07	7.34E+04	0.0756	0.0378
DEHP	4.10E-08	5.85E+05	0.0101	0.0051

F'^i is a correction factor of 0.5 that was applied to convert F^i to F'^i for nonattainment of equilibrium.

Source: From Reference 13.

2. Two-Resistance Model

This method was developed by Mackay (14) for calculating relative air-water exchange coefficients. Taking into account the extreme complexity of the volatilization processes at the air/water interface, only approximate empirical correlation can be obtained. Using radon as an example, the overall mass transfer can be calculated from liquid-phase resistance K_l^i and the air-phase resistance K_a^i by the equation

$$\frac{1}{K^i} = \frac{1}{K_l^i} + \frac{RT}{K_a^i H^i} \tag{6}$$

Relating the ratio K_l^i/K_a^i to known parameters, such as the viscosity in air and water, the following equation can be derived:

$$\frac{K_l^i}{K_a^i} = \left(\frac{\mu_i}{\mu_a}\right)^{1/3} \left(\frac{D_l^i}{D_a^i}\right)^{2/3} \tag{7}$$

where μ_i is the viscosity in water (9.82×10^{-7} m^2/s), μ_a is the viscosity in air (1.56×10^{-5} m^2/s, D_l^i is the diffusion coefficient of i in water, and D_a^i is the diffusion coefficient of i in air.

Substituting the values of μ_l and μ_a and combining Equations (6) and (7) yields

$$K^i = \frac{b}{2.5/D_l^i + RT/D_a^i H^i} \tag{8}$$

where b is a function of the chemical-independent factors that influences the ratio K_l/K_a (15).

To compare the volatilized fraction of chemical i with that of a known chemical, radon (Rn), the following equation can be derived:

$$\frac{F^i}{F^{\mathrm{Rn}}} = \frac{K^i}{K^{\mathrm{Rn}}} = \frac{(2.5/D_l^{\mathrm{Rn}} + RT/D_a^{\mathrm{Rn}} H^{\mathrm{Rn}})}{(2.5/D_l^i + RT/D_a^i H^i)} \tag{9}$$

This equation contains parameters, such as diffusivity of the contaminant i in water and air phases (D_l^i and D_a^i), that are not readily available in the literature. These computations are somewhat complicated and require knowledge of the molecular structure. A thorough discussion of the computation of diffusion coefficients may be found in the literature (13,16). To facilitate comparison with the equilibrium method, Table 3 displays the computation of volatilized fractions of a similar series of chemicals using the two-resistance McKone method.

Table 3 Calculation of F^i by the Two-Resistance McKone Method for Nine Chemicals[a]

Chemical	BP	MW	MV	Diffusion in H_2O	Diffusion in air	Henry's constant	F^i/F^{Rn}	F'^i
Benzene	353.25	78.11	96	8.99E-10	9.36E-06	5.60E-03	0.7420	0.4081
Chloroform			NC[b]	9.20E-10	8.70E-06	3.16E-03	0.7509	0.4130
Ethylene dibromide	404.75	187.88	98.4	8.87E-10	8.31E-06	6.70E-04	0.7145	0.3930
Naphthalene	491.05	128	147.6	6.98E-10	6.96E-06	4.60E-04	0.6019	0.3310
Dibromochloropropane			NC	7.20E-10	6.90E-06	2.37E-04	0.5891	0.3240
DDT	642	354.5	333.5	4.32E-10	4.33E-06	2.80E-05	0.2628	0.1445
Phenol	455	94.11	103.4	8.61E-10	8.44E-06	4.50E-07	0.0152	0.0084
Formaldehyde	253.5	30.03	29.6	1.80E-09	1.18E-05	3.27E-07	0.0185	0.0102
DEHP	696	390.6	524.8	3.31E-10	3.42E-06	4.10E-08	0.0008	0.0004

[a]Diffusivity values were obtained from Reference 1.
[b]Not calculated.
Source: From Reference 13.

C. Percutaneous Penetration: Theoretical and Physical-Chemical Versus Experimental Measurements

Traditionally, the K_p value, a chemical-specific physical constant, may be derived theoretically or experimentally through in vitro skin penetration studies. Other physical constants, such as the octanol/water partition coefficient ($K_{o/w}$), are also important parameters. The octanol/water partition coefficient ($K_{o/w}$) has been adopted as a measure of the skin permeability coefficient K_p. This is based on the assumption that, as the lipophilicity of the compound increases, the skin permeability also increases. For certain classes of chemicals, such as phenols, $K_{o/w}$ may not represent lipophilicity, and this type of correlation does not exist. More recently, Flynn observed that the best correlation of log K_p and log $K_{o/w}$ occurs in molecules that are of intermediate polarity. This may include middle chain-length alkanols and more polar phenols. Partitioning into the lipid component of the stratum corneum by these molecules appears to be an essential phenomenon for percutaneous absorption. Roughly, the log $K_{o/w}$ values of this group of chemicals range from -0.5 to 3.0 (17). The large number of chemicals with which we deal in our daily lives—drugs or environmental toxicants—differ in molecular weight. Attempts have been made to provide a simple means for estimating the permeability coefficient by lipophilicity and molecular weight. The latter is roughly divided into low and high molecular weight (18). Thus, it appears a first-hand estimate of K_p can be derived as as function of $K_{o/w}$ and molecular weight (MW). This would be a simplistic mathematical approach to screening chemicals for their respective K_p (Table 4). A thorough review of hundreds of compounds selected with respect to log $K_{o/w}$ (-1.0 to 4.0), K_p (-5.0 to -0.08), and molecular weight (18–800) indicates that this approach may provide an important method for

Table 4 Correlation Between Skin Permeability (K_p), Octanol-Water Partition Coefficient ($K_{o/w}$), and MW in Aqueous Solution

Octanol-Water Partition Coefficient	Low-MW compounds (<150)	High-MW compounds (>159)
Log $K_{o/w} < 0.5$	Log $K_p = -3$	Log $K_p = -5$
$0.5 \leq \log K_{o/w} \leq 3.0$	Log $K_p = -3.5 + \log K_{o/w}$	
$0.5 \leq \log K_{o/w} \leq 3.5$		Log $K_p = -5.5 + \log K_{o/w}$
log $K_{o/w} > 3.0$	Log $K_p = -0.05$	
log $K_{o/w} > 3.5$		Log $K_p = -1.5$

Source: From Reference 17.

initial screening of chemicals that may be of dermal penetration concern. It is the authors' opinion that for preliminary screening purposes, a first-hand rough estimate is that when log K_p is greater than -2.5 (log $K_p > -2.5$), irrespective of molecular weight, there may be a need for a closer look for potential dermal absorption during household water use. For illustration, the following few chemicals are among those having log $K_p > -2.5$: chloroform, benzene, tetrachloroethylene, and toluene.

The mathematical approach to screening chemicals using generated K_p values has two current restrictions. The first is that these K_p-generated values are restricted to water vehicles and are not applicable to other vehicles, such as soil. More importantly, the K_p values are usually derived from a large surface water volume (1 ml/cm^2 skin surface area). This makes it difficult to compare with in vivo results, for which the surface vehicle is literally a few micrometers of water per cm^2 skin surface area.

The second restriction to the mathematical approach to generating K_p values is that the K_p values have not been validated for humans (or relevant animal models) in situations in vivo. This is of concern because K_p values were derived from in vitro results, and in vitro results with highly lipophilic compounds do not correlate with in vivo results. The in vitro system cannot handle lipophilic compounds because of nonsolubility in the receptor fluid under the skin. This includes many compounds of environmental concern (polychlorobiphenyls, PCB, DDT, benzo[*a*]pyrene, and chlorodane) for which the in vitro results show minimal percutaneous absorption whereas the in vivo results demonstrate that these compounds are readily absorbed through skin. To demonstrate this point, Wester et al. carried out in vivo and in vitro studies on PCB and found that in vitro absorption of the PCB by the cadaver skin did not correlate with in vivo findings in the rhesus monkey. This was because of the low water solubility of this highly lipophilic chemical and the lack of a PCB partition from skin into the water receptor fluid (18). Therefore, when mathematically derived K_p values or K_p values derived from in vitro studies are used to determine percutaneous absorption for human health hazard, a cautionary flag must be raised. The K_p must be kept within its derived boundaries (such as water vehicles), and it must be ascertained whether the K_p value has been validated for human in vivo.

Many of the K_p values were derived from in vitro studies. It should be emphasized that such methodologies are inadequate to generate reliable K_p values because (1) the species from which the skin was derived from for in vitro studies could affect the results, and (2) the site from which the skin was removed also affects the outcome, even when human skin was used for deriving the skin permeability values. This is because of the lack of in vivo structure and functions. The results are at best a "hopeful close estimate." On the other hand, dermal absorption of certain water contaminants can be monitored in humans through noninvasive collection of biological samples, notably urine and air samples. This provides insight into the dermal penetration potential of the subject chemical, but not every chemical is a candidate. Generally speaking, chemicals should be tested in animals for their pharmacokinetic behavior first, and the test compounds or their metabolites should be fairly polar and readily excrete in the urine. With respect to the in vivo percutaneous absorption on various chemicals between humans and their closely related kin, rhesus monkeys, Table 5 clearly demonstrate the similarity of the dermal absorption rate in rhesus monkeys for a variety of chemicals. Other than humans, the rhesus monkey may be the most appropriate animal model for dermal absorption studies (7).

Numerous mathematical models that have been developed over the past years attempted to compute K_p from various physical chemical constants (19–22). These models were developed by either including a new parameter or modifying an existing theory. A good review of this subject may be found in the dermal exposure assessment document recently compiled by the U.S. EPA (8). It is the authors' opinion that this is an area in which research is much needed and that a study purposely designed for in vivo, in vitro, and mathematical modeling with chosen parameters should be conducted simultaneously for water or soil contaminants of concern to reveal the true nature of the dermal dose computed from all three approaches. When compiled with adequate numbers of environmental contaminants, a valuable correlation could be generated. The experimental design is very crucial in these types of studies. Instead of picking up bits and pieces of information derived from different investigators with very different study designs and approaches, which does not allow fair comparison, a few well-designed and controlled studies are warranted to build a solid data base.

D. Update on the Application of PBPK Methods to Estimate Exposure of Water Contaminants via Various Routes

The physiologically based pharmacokinetic model (PBPK) is designed to simulate the disposition of a toxicant within the body. Physiological functions, such as blood flow, tissue distribution, and metabolism, are grouped into various physiological compartments that conform with the general concept of the kinetic behavior of a chemical under study. Because of the availability of these physiological and chemical parameters in various species, it is fair to say that interspecies extrapolation is made easy. In essence, the most striking difference between PBPK modeling and other means of estimating dose is that the former offers estimation of the internal dose, the dose of tissues and organs. For instance, Travis et al. reported a five-compartment model to study the disposition of benzene (23). The model accurately predicted the experimental results, such as blood and expired air concentrations, as well as the excretion of phenol, a major metabolite in urine, for several animal species, including human.

This may be translated to the successful development of a method that can utilize animal data to predict human dose under various exposure conditions. This technique, however, although exciting, needs further refinement and more vigorous testing to ensure its usefulness. For example, other authors working on the same compound expressed their concern over the disparity of goodness of fit for benzene over three physiologically based pharmacokinetic models (24).

A simple gas uptake inhalation technique has been proposed by Gargas and Clewell (25). This is a very useful technique to study the inhalation uptake of VOC. By exposing animals to various VOC at targeted chamber concentrations, metabolic and kinetic constants can be determined. Taking into consideration any enzymatic reactions that may have occurred to describe adequately the observed VOC uptake behavior, one may construct an appropriate PBPK model. This technique can accommodate a large number of VOC and allow direct human dose extrapolation. Thus, it has the potential as a most powerful tool for predicting human dose. It would be an ideal situation to apply this method in conjunction with human showering exposure studies and construct appropriate dose prediction models. Obviously, the prerequisite is to have a thorough understanding of the metabolic pathway of a chemical under study, and when needed, animal studies can be easily incorporated.

The PBPK approach is applicable to modeling all routes of exposure. A skin compartment is usually created. With the input of percutaneous absorption data observed in an appropriate species, prediction of dermal dose is attainable for humans. However, choosing the animal species is of utmost importance. An inappropriate species not only does not provide comparable human dose prediction, but such extrapolation may prove to be erroneous and misleading. Recent studies on isofenphos, an insecticide, conducted in rats and humans separately is a good example. The results indicated that rat percutaneous absorption (PBPK model) is approximately 20-fold greater than humans (26,27).

IV. HEALTH CONCERNS FOR INHALED OR DERMALLY EXPOSED CONTAMINANTS THROUGH DOMESTIC WATER USES

A. Historical Background

It has been known for over a century that chemicals can volatilize from water to air, but only more recently have investigators begun studying the volatilization of drinking water contaminants. Over the last decade, the concern of cancer risk from indoor exposure to radon has stimulated a great deal of studies over radon and other volatile organic chemicals (28–30). That volatilization of drinking water contaminants is of interest because of the large amount of water usage in a household (approximately 200 L/person/day), as opposed to water ingestion, which is typically 2 L/person/day, and because drinking water standards are established without considering the inhalation and dermal routes of exposure to water contaminants.

B. Dose, Toxicology, and Health Risks

Dose is a function of the concentration of water contaminants, the duration of exposure (inhalation and dermal), and the actual amount absorbed through various routes of exposure.

When a contaminant volatilizes from water, the concentration of the particular contaminant in water is reduced accordingly; this in turn reduces the ingested dose. Theoretically, it also reduces the dermal dose, but in reality, since dermal contact with household water does not contribute substantially to the overall dose in general, this effect is considered minor with respect to the total dose.

The intrinsic toxicity of a water contaminant plays a major role in health risk assessment. That the health risk advisories differ over a wide range is a good example. With the development of various methodologies to assess the overall dose through all routes of exposure of water contaminants, it is conceivable that the setting standards of drinking water would be impacted. Although the overall dose of water contaminants, when all potential routes of exposure are taken into account, may or may not differ substantially from those previously estimated based on ingested dose alone, it nevertheless brings into focus the concept and practice of applying all routes of exposure for dose estimates. Thus, the total dose contributed from various routes of exposure may remain relatively unchanged in some instances, even when individual contributory sources vary, that is, increasing inhalation dose of VOC from volatilization with corresponding decrease in ingested dose. In other cases, incorporating the noningestion dose may alter the total dose significantly, which warrants reconsideration of existing drinking water standards.

V. DISCUSSION

The intent of this paper is to stimulate discussion on the pros and cons of including all routes of exposure to water contaminants for dose estimates. This approach, when used properly, would increase the awareness of the regulators and other interested parties of those chemicals that may contribute significantly to total dose through noningestion water uses and thus require regulators to reset the drinking water standards to reduce the maximum contamination levels of contaminants.

The benefit would be reducing the health hazard of water-using consumers and protecting the public health. In carrying out this mandate, however, a significant increase in work load may be required to recompute the hundreds of chemicals and their respective dosages. In so doing, the federal agency charged with computing these exposure estimates would undoubtedly be overwhelmed. Thus, *work smarter not harder* appears to be absolutely necessary. Consequently, the exploration of appropriate methodologies to prioritize the various water contaminants is essential. Based on these concerns, we are proposing a simple strategy to screen chemicals with inhalation and dermal penetration potential. At the time of writing, there is no known method that could carry out this project and fulfill all expectations. The approach we have adopted here is at the very preliminary stage and subject to vigorous testing. However, based on the nine chemicals that were chosen, which represent a range of volatilization potentials, irrespective of which of the two methods applied to compute the volatilized fraction of each chemical, the outcome showed very little difference. Thus, the simplicity of the semiequilibrium method appears to offer some promise to the risk assessor.

The contribution to total dose from dermal contact of household water uses is not known at this time. However, a methodical way would allow each chemical to be evaluated on a case by case basis. Perhaps for initial screening, one needs to develop some guidance to separate those that are highly permeable to the skin from those that are not. Since household water use is encountered by everybody on a daily basis, it would be prudent to suggest more human monitoring studies be conducted during all natural use situations. Since data derived from human studies are very scanty, these studies would definitely enrich our understanding of the dose contributed from various routes of exposure in the most natural state of human exposure. An exercise of this kind, using experimentally derived numbers, would undoubtedly improve the end product of health risk assessment to water contaminants.

ACKNOWLEDGMENTS

The authors thank Drs. Nancy Chiu and Ken Bailey of the U.S. EPA and Dr. Fanny Ennever of the Bowman Gray School of Medicine for their inspirational comments. This work is in part supported by the Intergovernmental Personnel Act (IPA) of the U.S. EPA.

REFERENCES

1. T.E. McKone, House exposure models, *Toxicol. Lett. 49*:321–339 (1989).
2. W.K. Jo, C.P. Weisel, and P.J. Lioy, Routes of chloroform exposure and body burden from showering with chlorinated tap water, *Risk Anal.* 10:575–580 (1980).
3. W.K. Jo, C.P. Weisel, and P.J. Lioy, Chloroform exposure and the health risk associated with multiple uses of chlorinated tap water, *Risk Anal. 10*:581–585 (1990).
4. M.J. Bartek and J.A. LaBudde, Percutaneous absorption in vitro, *Animals Models in Dermatology* (M. Maibach, ed.), Churchill, Livingston, New York, 1975, pp. 103–120.
5. R.C. Wester, M. Mobayen, and H.I. Maibach, In vivo and in vitro absorption and binding to powered stratum corneum as methods to evaluate skin absorption of environmental chemical contaminants from ground and surface water. *J. Toxicol. Environ. Health 21*:367–374 (1987).
6. H.I. Maibach and R.C. Wester, Percutaneous absorption: In vivo methods in humans and animals, *J. Am. Coll. Toxicol.* 8:803–813 (1989).
7. R.C. Wester and H.I. Maibach, Animal models for percutaneous absorption, *Health Risk Assessment: Dermal and Inhalation Exposure and Absorption of Toxicants* (R. Wang, J. Knaak, and H. Maibach, eds.), CRC Press, Boca Raton, FL, 1992.
8. U.S. EPA, *Dermal Exposure Assessment: Principle and Application*, Office of Health and Environmental Assessment, Washington, D.C. Interim Report, EPA/600/8-91/011B, 1992.
9. A.S. Michaels, S.K. Chandrasekaran, and J.E. Shaw, Drug permeation through human skin: Theory and in vitro experimental measurement, *AICE J. 21*(5): 985–996 (1975).
10. P.M. Elias, K.R. Feingold, G.K. Menon, S. Grayson, M.L. Williams, and G. Grubauer, The stratum corneum two-compartment model and its functional implications. *Pharmacology and the Skin*, Vol. 1, *Skin Pharmacokinetics* (B. Shroot and H. Schaefer, eds.), Karger, Basel, 1987, pp. 1–9.
11. R.M. Marks, S.P. Barton, and C. Edwards (eds.), *The Physical Nature of the Skin*, MTP Press, Ltd., Lancaster, 1988.
12. U.S. EPA, *Criteria Document for Radon*, Office of Water, Washington, D.C., 1991.
13. U.S. EPA, *Guidelines for Incorporation of Inhalation and Dermal Exposures from Drinking Water in the Calculation of Health Advisories and DWEL Values*, Subcontract LSI-8700, 1991.
14. D. Mackay, Air/water exchange coefficients, *Environmental Exposure from Chemicals*, Vol. 1 (W.B. Neeley and G.E. Blan, eds.), CRC Press, Boca Raton, FL, 1989, pp. 96–108.
15. T.E. McKone, Human exposure to volatile organic compounds in household tap water: The indoor inhalation pathway, *Environ. Sci. Technol. 21*:1194–1201 (1987).

16. W.J. Lyman, W.F. Reehl, and D.H. Rosenblatt, *Handbook of Chemical Property Determination Methods. Environmental Behavior of Organic Compounds*, McGraw-Hill, New York, 1982.
17. G.L. Flynn, Physiochemical determinants of skin absorption, *Principles of Route-to-Route Extrapolation for Risk Assessment* (T.R. Gerity and C.J. Henry, eds.), Elsevier, Amsterdam, 1990, pp. 93–127.
18. R.C. Wester, H.I. Maibach, D.A.W. Bucks, J. McMaster, and M. Mobayan, Percutaneous absorption and skin decontamination of PCBs: In vitro studies with human skin and in vivo studies in the Rhesus monkey, *J. Toxicol. Environ. Health 31*:235–246 (1990).
19. A.S. Michaels, S.K. Chandrasekaren, and J.E. Shaw, Drug permeation through human skin: Theory and in vitro experimental measurement, *AIChE J. 21*(5):985–996 (1975).
20. W.J. Albery and J. Hadgraft, Percutaneous absorption: Theoretical description, *J. Pharm. Pharmacol. 31*:140–147 (1979).
21. B.B. Berner and E.R. Cooper, Models of skin permeability, *Transdermal Delivery of Drugs*, Vol. II (A.F. Kydonieus and B.B. Berner, eds.), CRC Press, Boca Raton, FL, 1987, pp. 41–56.
22. G.B. Kasting, R.L. Smith, and E.R. Cooper, Effects of lipid solubility and molecular size on percutaneous absorption, *Pharmacol. Skin 1*:138–153 (1987).
23. C.C. Travis, J.I. Quillen, and A.D. Arms, Pharmacokinetics of benzene, *Toxicol. Appl. Pharmacol. 102*:400–420 (1990).
24. F.Y. Bois, T.J. Woodraff, and R.C. Spear, Comparison of three physiologically based pharmacokinetic models of benzene disposition, *Toxicol. Appl. Pharmacol. 110*:79–88 (1991).
25. M.L. Gargas and H.J. Clewell, III, Gas uptake inhalation techniques and the rates of metabolism of chloromethanes, chloroethanes, and chloroethylenes in the rat, *Inhal. Toxicol. 2*:295–319 (1990).
26. J.B. Knaak, M. Al-Bayato, O.G. Raabe, and J.N. Blancato, In vivo percutaneous absorption studies in the rat: Pharmacokinetics and modelling of isofenphos absorption, *Prediction of Percutaneous Penetration: Methods, Measurements and Modelling* (R.C. Scott, R.H. Guy, and J. Hadgraft, eds.), IBC Technical Service, London, 1991.
27. R.C. Wester, H.I. Maibach, J. Melendres, L. Sedik, J.B. Knaak, and R. Wang, In vivo and in vitro percutaneous absorption and skin evaporation of isofenphos in man, *Toxicol. Appl. Pharmacol. 19*:521–526 (1992).
28. D. Crawford-Brown and C.A. Cothern, A Bayesian Analysis of scientific judgment of uncertainties in estimating risks due to Rn-222 in U.S. public drinking water supplies, *Health Phys. 53*:11–21 (1987).
29. National Academy of Science, *Health Risk of Radon and Other Internally Deposited Alpha Emitters*, BEIR IV, National Academy Press, Washington, D.C., 1988.
30. J.B. Andelman, Human exposure to volatile halogenated organic chemicals in indoor and outdoor air. *Environ. Health Perspect. 62*:443–460 (1985).

17

Estimating Human Health Risks for Trichloroethylene, a Water Contaminant, Using Physiological Models

Jeffrey W. Fisher

Armstrong Laboratory, Wright-Patterson Air Force Base, Ohio

I. INTRODUCTION

Trichloroethylene (TCE) is a common and persistent environmental contaminant found in groundwater near most large cities and at Superfund landfill sites. In many cases, the groundwater is the primary source for water consumption. Because of the widespread distribution of TCE in the environment, a significant fraction of the population may ingest or inhale TCE over an extended period of time. Typically, water concentrations of TCE are in the ppb range (1). Health concerns for environmental exposure to TCE stem largely from positive outcomes in laboratory cancer bioassay studies with rodents at relatively high exposure concentrations (2). Epidemiological evidence that TCE is a human carcinogen is judged to be equivocal by the U.S. Environmental Protection Agency (EPA) (2). Recent epidemiological evidence for TCE (3,4) was not considered by the U.S. EPA.

Metabolic activation is apparently required for TCE to exert its carcinogenic effect in rodents. Recent cancer studies with rodents have linked two metabolites of TCE (dichloroacetic acid, DCA, and trichloroacetic acid, TCA) with hepatocellular carcinoma formation (2,5,6). TCE is metabolized by the cytochrome P_{450} system. Trichloroacetaldehyde is formed, possibly via a transient epoxide that rapidly undergoes an intramolecular rearrangement (7). Trichloroacetaldehyde is either oxidized to TCA or reduced to trichloroethanol. Other minor metabolites are formed via dechlorination reactions (oxalic acid, carbon dioxide, and DCA) (8).

The U.S. EPA (2) has estimated the health risks of cancer in humans exposed to low concentrations of TCE by using data from rodent cancer bioassay studies. In their risk analysis approach, a nonthreshold extrapolation model (linearized multistage model) was used in the estimation of an excess cancer risk of 1 in 1 million based on the amount of TCE metabolized. Various target organs and types of cancer were included in their calculations. More recently, to estimate the health risks from inhalation exposure to TCE in humans, a classic compartmental model was developed by the U.S. EPA and used to estimate the amount of TCE metabolized for chronic TCE vapor exposures in rodents (9). Both these risk analyses for TCE were conducted without the advantage of proper laboratory studies with rodents to characterize quantitatively the metabolism of TCE. That is, metabolic capacity (e.g., V_{max}, the maximum metabolic rate) was not determined for each strain and species of bioassay rodent. In addition, no attempts were made by the U.S. EPA to develop a human kinetic model for TCE and estimate the metabolic capacity for TCE in humans.

In this chapter, human liver and lung cancer risks are presented for lifetime ingestion of drinking water containing TCE or inhalation of TCE vapors. A generic, multiroute physiologically based pharmacokinetic (PBPK) exposure model with TCE was constructed for mice and humans and used in conjunction with a linearized multistage model to calculate lung and liver cancer risks. The linearized multistage model was used for low-dose extrapolation to maintain consistency with U.S. EPA methodology. These cancer risk calculations were based on quantitative metabolic studies carried out in the laboratory using TCE-exposed B6C3F1 mice (10) and published cancer bioassay findings in which B6C3F1 mice developed liver and lung cancer (2,9).

In reviewing the cancer bioassay studies reported for TCE (2,9), two cancer studies, a gavage dosing study (11) and an inhalation exposure study (12), both with B6C3F1 mice, were selected for risk assessment purposes. Both routes of TCE exposure produced liver and lung cancer, which provided a "pharmacokinetic" route-to-route comparison of the predicted amount of TCE metabolized with incidences of liver and lung cancer.

A. PBPK Model Development in Mice

PBPK models for volatile organics, such as TCE, require chemical-specific information on tissue solubility and metabolic capacity (V_{max}) and species-specific physiological information on respiratory rates, blood flows, and organ volumes. In this chapter, a brief overview of the laboratory experiments and model simulation results are given for TCE in mice and humans. These research findings are described elsewhere (10,14–16). The description of the PBPK model used to simulate the uptake, distribution, metabolism, and elimination of TCE in mice is reported in Fisher et al. (10). This PBPK model is a generic four-compartment model (Figure 1) as described by Ramsey and Andersen (13). Physiological parameters for the TCE model were taken from the literature and tissue solubilities and metabolic constants for mice determined by experimentation.

Tissue solubilities (partition coefficients) for TCE were determined using vial equilibration. This in vitro gas-phase method has been used by Gargas et al. (17) to estimate partition coefficients for a large number of volatile organic chemicals. For TCE, tissue/air partition coefficients were determined for fat, muscle, liver, and blood (10).

Metabolic rate constants were determined by gas uptake techniques (10). In brief, to conduct gas uptake studies, either male or female mice were placed in a glass jar with a recirculating closed atmosphere. A measured amount of TCE was introduced into the glass chamber, and loss of TCE from the bell jar was monitored for 1–6 h. Loss of TCE from the closed glass jar atmosphere represents systemic uptake and metabolism of TCE by the mice. A series of chamber loss curves for TCE were generated, and a V_{max} value was estimated for male and female mice by optimization (Simusolv, Dow Chemical Company) using a five-compartment PBPK model in which the gas uptake chamber represented one compartment (10). A detailed description of the gas uptake system is reported in Gargas et al. (18).

At this point, a PBPK model for TCE was developed, but not validated, for male and female B6C3F1 exposed to TCE vapors. To validate the model for TCE, 4 h constant concentration TCE inhalation exposures were conducted over a wide range of vapor concentrations with male and female B6C3F1 mice (10). These TCE inhalation concentrations spanned the TCE exposure concentrations used in the Maltoni inhalation cancer bioassay (12). TCE blood concentrations were measured during and after exposure and compared with model-predicted blood concentrations. Measured blood concentrations of TCE did not closely agree with prediction. To fine-tune the model's ability to characterize metabolism of TCE, measured blood concentrations of TCE (not predicted blood concentrations) were used to estimate the amount of metabolized TCE.

With a completed vapor exposure model for TCE, the PBPK model was expanded by adding a first-order rate constant to describe the gastrointestinal uptake of TCE into systemic circulation via the liver (Figure 1). This approach is a nonphysiological method for describing oral uptake of TCE using classic compartmental modeling techniques. Experimentally, to determine the first-order uptake rate constant for TCE, male and female

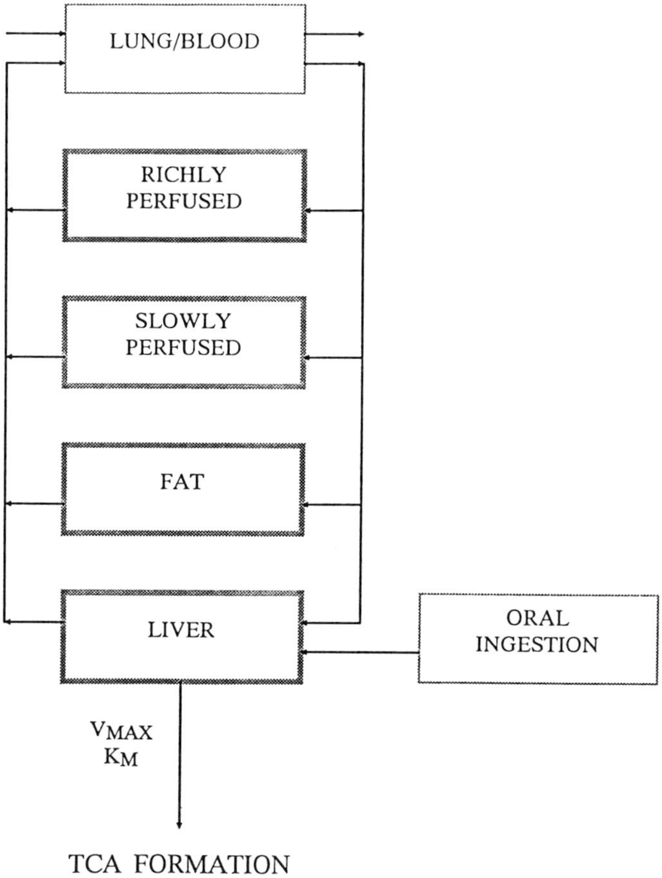

Figure 1 Physiologically based pharmacokinetic four-compartment model for trichloroethylene (TCE) in mice and humans. The lung and gastrointestinal tract (for oral ingestion of TCE) are not included in the set of differential equations that constitutes the PBPK model but represent routes of TCE exposure. Systemic clearance of TCE in mice and humans is characterized by lung exhalation and metabolism of TCE.

mice were gavaged with large doses of TCE in corn oil, and TCE-blood and TCA-plasma time course data were collected. These doses were similar to the cancer bioassay dose concentrations (11). Systemic uptake of TCE in corn oil was multiphasic and prolonged and could not be described by simple first-order kinetics (16,19). Again, to increase the model's predictive ability to characterize the metabolism of TCE, an "effective" first-order uptake rate constant (for TCE) was determined by optimization of the TCA plasma concentration profiles.

B. PBPK Model Development in Humans

The generic PBPK model structure for TCE in humans was identical to the PBPK model structure used for mice (Figure 1). The development of the human PBPK model for TCE is described elsewhere in detail (14,15). In summary, physiological parameters, such as blood flows, and volume of organs and chemical parameters, such as partition coefficients (tissue/air) for TCE, were taken from the literature. Human partition coefficient values for TCE were found in the literature for blood and fat; those for other tissue groups were estimated from published rodent partition coefficient values. Metabolic capacity for TCE in humans was estimated by calibrating the PBPK model using blood and urine time course data from controlled experiments in which human volunteers were exposed to either TCE or TCA. To describe the kinetic behavior of the stable metabolite, TCA, kinetic constant values were refined by computer optimization with a data-driven, one-compartment pharmacokinetic model (14,15). The production, systemic clearance, and urinary excretion data for TCA were then used in conjunction with limited blood time course data for TCE to estimate the metabolic capacity for oxidation of TCE in humans by computer optimization. In general, for repeated TCE vapor exposures, there was good agreement between computer-predicted and observed TCE-blood and TCA-plasma concentrations, but for single TCE vapor exposures, TCA plasma concentrations were slightly underestimated.

II. METHODS

The guiding philosophy for this TCE risk analysis was that with the development of a generic model for TCE in mice and humans, metabolism of TCE could be quantitatively characterized in both species and used in the assessment of liver and lung cancer risks. To estimate the liver and lung cancer risk for humans exposed to TCE, a PBPK model for TCE in B6C3F1 mice was used to simulate the cancer bioassay mice exposures for oral ingestion and inhalation of TCE. The lifetime average amount of TCE metabolized each day (AMET, mg/kg/day) was determined for each B6C3F1 mouse exposure dose or concentration per sex and used with the liver and lung cancer incidence data in a linearized multistage model (20) to predict AMET values that correspond to excess cancer risks of 1 in 1 million for lung and liver cancer in mice. These AMET values then became the "internal target doses" that, assuming equal sensitivity between mice and humans, would produce excess cancer risks of 1 in 1 million in humans. The human model was exercised by varying the drinking water or vapor concentrations of TCE until the human AMET value was equivalent to the linearized multistage model-derived mouse AMET value.

A. Bioassay Simulations

PBPK model parameters for the mouse and human are presented in Tables 1 and 2. The National Cancer Institute (NCI) TCE gavage bioassay (11) dosing schedule was simulated for male and female B6C3F1 mice over a 7 day period. Animals in this bioassay were dosed 5 days/week with weekends off. TCE was predicted to clear systemic circulation between each daily dosing. However, TCA was predicted to accumulate during the week from daily TCE dosing and then to clear the systemic circulation over the weekend. Exposure duration and concentrations from the Maltoni group's (12) TCE inhalation bioassay exposure schedule were simulated for male and female B6C3F1 mice over a 7 day period. Animals in this study were exposed for 7 h/day, 5 days/week, with weekends off. Similar to the NCI gavage bioassay, TCE was predicted to clear the systemic circulation between vapor exposures and TCA was predicted to accumulate during the week and to clear the systemic circulation over the weekend. The average daily amount of TCE metab-

Table 1 Physiological Model Parameters for Humans and Mice

	Female mice	Male mice	Humans
Tissue group (fraction of body weight, BW)			
Liver	0.04	0.04	0.026
Richly perfused	0.05	0.05	0.050
Slowly perfused	0.72	0.78	0.620
Fat	0.10	0.04	0.190
Flows (L/h)			
Alveolar ventilation	$30BW^{0.74}$	$30BW^{0.74}$	$12.6BW^{0.74}$
Cardiac output (CO)	$30BW^{0.74}$	$30BW^{0.74}$	$14.9BW^{0.74}$
Tissue group (fraction of CO)			
Liver	0.24	0.24	0.26
Richly perfused	0.52	0.52	0.44
Slowly perfused	0.19	0.19	0.25
Fat	0.05	0.05	0.05

Table 2 Kinetic Constants for Modeling TCE and TCA in Humans and Mice

	Female mice	Male mice	Humans
Partition coefficients per tissue group			
Liver/blood	1.62	2.03	6.82
Richly perfused/blood	1.62	2.03	6.82
Slowly perfused/blood	0.48	1.00	2.35
Fat/blood	31.4	41.3	73.3
Blood/air	14.3	13.2	9.20
TCE metabolic rate constants			
V_{max}, mg/kg/h	23.2	32.7	14.9
K_m, mg/L	0.25	0.25	1.5

olized for both routes of TCE exposure was calculated by multiplying the computer-predicted cumulative 7 day amount of TCE metabolized by $1/7$. The average daily amount of TCE metabolized was then multiplied by the fraction of lifetime exposure to TCE (78 weeks/104 weeks) to obtain the lifetime average AMET value on a per day basis (2).

B. Linearized Multistage Model and Human PBPK Model

A linearized multistage model (20) was used to estimate the lower bound of the 95% confidence interval for the lifetime average amount of TCE metabolized per day that corresponds to an excess risk of 1 in 1 million risk for liver and lung cancer in the bioassay mice. The multistage model calculations were determined for extra risk using the Monte Carlo method. These calculations were performed in male and female mice for two routes of TCE exposure, inhalation and oral ingestion. These AMET values for oral and inhalation exposures to TCE in male and female mice then became "target" internal dose measures for assessing the risk of liver and lung cancer in humans exposed to TCE in drinking water and air.

C. Human Equivalent Exposure

To obtain the target internal AMET values for humans, a human PBPK model for TCE and TCA was exercised by varying the exposure concentration for inhalation of TCE vapors and for oral ingestion of TCE in water. Because the dose rate is not well defined for oral ingestion of TCE in drinking water, the drinking water exposure model was configured to simulate a single, 2 L ingestion of water per day or four separate ingestions of water, totaling 2 L, over a 12 h period per day. For modeling purposes, the rate of absorption of TCE into the liver was described as first order with a rate constant value of 5.5/h (19,21). AMET values for human exposure to TCE in drinking water were calculated on a per day basis at steady state by conducting a PBPK model simulation for 44 days and another simulation for 45 days and determining the difference between the cumulative time-dependent AMET values. Oral ingestion of TCE was assumed to occur 7 days per week. For vapor exposure to TCE, two exposure conditions were considered: continuous exposure, 24 h/day, 7 days/week, and an intermittent exposure, 7 h/day, 5 days/week. Again, PBPK model simulations were conducted for 44 and 45 days to determine an average daily AMET value at steady state for the continuous exposure. PBPK model simulations were conducted for 42 and 49 days to determine an average weekly AMET value at steady state for the intermittent exposure. The average weekly AMET value for intermittent exposure was adjusted (multiplied by 1/7) to obtain an average daily AMET value.

III. RESULTS

Tables 3 and 4 list the simulated AMET values and corresponding liver and lung incidences for the inhalation and gavage bioassays, respectively. The lifetime average amount of TCE metabolized in mice (mg/kg/day)

Table 3 Liver and Lung Tumor Incidences in B6C3F1 Mice Exposed to TCE by Inhalation (12) and Corresponding AMET Values

TCE exposure (ppm)	Hepatoma incidence[a]	Lung tumor incidence[b]	AMET (mg/kg/day)
Female mice			
600.0	9/87	14/87	285.7
300.0	4/89	7/89	249.7
100.0	4/90	6/90	111.5
0.0	3/90	2/90	—
Male mice			
600.0	6/88	1/88	355.9
300.0	3/88	2/88	301.3
100.0	1/86	2/86	108.5
0.0	1/85	2/90	—

[a]Hepatomas include all malignant tumors of hepatic cells. Hepatoma incidence in male mice was statistically insignificant (9), and a large fraction of these mice died during the bioassay (12).

[b]Lung tumors include early adenomas (adenomatous hyperplasia), adenomas, and adenocarcinomas. Lung tumors in male mice were not used in this risk analysis.

Table 4 Liver and Lung Tumor Incidences in B6C3F1 Mice Gavaged with TCE (9) and Corresponding AMET Values

TCE dose (mg/kg)	HC incidence[a]	Lung tumor incidence[b]	AMET (mg/kg/day)
Male mice			
2339.0	31/48	2/48	211.4
1169.0	26/50	5/50	176.5
0.0	1/20	0/20	—
Female mice			
1739.0	11/47	7/47	196.2
869.0	4/50	4/50	158.7
0.0	0/20	1/20	—

[a]Hepatocellular carcinoma (HC) formation.
[b]Lung tumors include adenomas and carcinomas. Lung tumors in male mice were not used in this risk analysis.

appear to correlate with liver and lung tumor incidences. That is, as liver and lung tumor incidences increased, corresponding AMET values increased, although the correspondence was not 1:1. Male mice dosed orally with TCE appeared to be very sensitive to liver tumor formation (Table 5). Both routes of TCE exposure produced similar risks for lung tumors in female mice (Table 5).

The inhaled concentrations of TCE required to produce an excess risk for liver and lung tumors of 1 in 1 million in humans are given in Table 6 for the dose metric AMET, under continuous and intermittent exposure conditions. For ingestion of drinking water, the 1 sip per day and 4 sips per day model conditions did not result in a significant difference in liver cancer risks. The drinking water concentrations of TCE required to pro-

Table 5 Linearized Multistage Model Calculations Using AMET

Route of exposure	Sex of mice	95% Lower bound on AMET at 10^{-6} risk (mg/kg/day)	
		Liver	Lung
Inhalation	Male	5.200E-3	—
	Female	3.360E-3	1.754E-3
Gavage	Male	1.875E-4	—
	Female	1.024E-3	1.179E-3

Table 6 TCE Concentrations in Drinking Water and Air That Correspond to a Lifetime Risk Of 10^{-6} for Lung and Liver Cancers in Humans

	Lifetime exposure concentration for 10^{-6} risk		
		Liver tumors	
Route of exposure	Lung tumors, female mice	Male mice	Female mice
Inhalation, ppb			
Continuous	10.0	15.0	10.0
Intermittent	23.0	69.0	47.0
Drinking water, μg/L	46.0	7.0	39.0

duce an excess risk of liver and lung tumor formation equal to 1 in 1 million in humans for the dose metric AMET are given in Table 6.

IV. HEALTH RISKS

The U.S. EPA health assessment document on TCE (2) reported a calculated 1 in 1 million excess risk of liver cancer in humans for a lifetime ingestion of 2 L water/day containing 3.1 μg TCE/L water and for a lifetime continuous inhalation of TCE vapors at 0.14 ppb. These risk calculations were based on two gavage bioassays (11,22). In this study, the calculated drinking water and atmospheric concentrations of TCE associated with a 1 in 1 million excess risk liver cancer are larger than the U.S. EPA calculated values by a factor of about 2 for ingestion of TCE in drinking water (using the gavage-dosed male mice from the NCI 1976 bioassay) and a factor of 71 for continuous inhalation of TCE (using the vapor-exposed female mice from the Maltoni bioassay). In a more recent draft addendum to the health assessment document on TCE (9), the U.S. EPA, using a compartmental model and inhalation vapor exposure bioassay data (12), estimated that a continuous lifetime vapor exposure of 0.11 ppb TCE would correspond to an excess lung cancer risk of 1 in 1 million, which is 91 times lower than the vapor exposure concentration calculated in this study using female mice from Maltoni's bioassay (Table 6).

Oral ingestion and inhalation are both important routes of TCE exposure for the assessment of cancer risks. Based on this analysis, the risk of developing lung cancer is slightly greater than that of developing liver cancer for inhalation of TCE, and for oral ingestion of TCE in drinking water, the risk of developing liver cancer is greater than that of developing lung cancer.

The gavage bioassay with TCE was used to assess cancer risks for ingestion of TCE in drinking water, and the vapor exposure bioassay with TCE was used to assess inhalation cancer risks of TCE in humans (16). From a pharmacokinetic perspective, the gavage bioassay kinetic profile for TCE in mice is very different from the kinetic profile for mice or human oral ingestion of low concentrations of TCE dissolved in water. The dose rate of delivery of large doses of TCE to the liver in the gavaged bioassay animals is much greater than ad libitum ingestion of TCE in drinking water. For oral ingestion of TCE in drinking water, the dose rate for delivery of TCE to the liver is more similar to an inhalation exposure than a gavage-dosing exposure (19), which suggests that for metabolized systemic toxins, dose rate of toxicant delivery may be more important than route of exposure in assessing health risks. Also, the use of vehicles (e.g., corn oil) in performing bolus intubations has a significant effect on the uptake kinetics of solvents, such as TCE (19), which adds to the difficulty in interpreting gavage bioassay studies.

Exposing mice to TCE vapors for assessing human cancer risks from exposure to TCE vapors is less problematical from a pharmacokinetic perspective. For example, in the lowest bioassay concentration (12), mice were exposed to 100 ppm TCE, which is near the actual exposure concentrations used for pharmacokinetic studies with human volunteers (50–70 ppm TCE). This does not ensure that similar internal exposures occur between mice and humans, but major factors that control the dose rate of delivery of TCE to the liver, such as blood/air partition coefficient values and breathing and perfusion rates, can be readily accounted for in a PBPK model.

Lung tumor data for female mice from the NCI 1976 bioassay and liver tumor data for male mice from the Maltoni bioassay were used in this risk assessment. The U.S. EPA elected not to use either of these data sets in their risk analysis of TCE. The U.S. EPA concluded that in the Maltoni bioassay a large portion of the male mice died and the hepatoma incidences were not statistically significant (12). Although this data set was compromised, male mice were included in this risk analysis to allow route-to-route comparison of cancer risks. It was not clear why the U.S. EPA did not use the female mice from the NCI bioassay study (11); the female mice from the NCI bioassay were included in this risk analysis because survival rates were high for these mice and a dose-related response was observed for lung tumors. Neither of these bioassay data sets, however, provided the most conservative estimates of cancer risks in humans (Table 6).

V. CONCLUSIONS

PBPK modeling provides a framework for attempting to "make sense" of multiple bioassay data sets for different routes of exposure. It also allows the use of an internal measure of dose (e.g., amount of TCE metabolized) rather than an external measure of dose (e.g., atmospheric concentration of TCE) in the calculation of cancer risks. In this study, the cancer risks calculated by the U.S. EPA for human exposure to TCE were refined by using PBPK models to estimate the metabolized dose. PBPK models offer promise as a useful tool in the chemical risk assessment process and may provide insight into mechanisms of toxicological action for chemicals.

The mechanism of action by which TCE exerts its carcinogenic effect is not well understood, which limits our ability to calculate cancer risks that are free from troublesome uncertainties, such as, is there a dose-response threshold for TCE carcinogenicity? Are B6C3F1 mice reasonable surrogates for humans? Recent studies by Bull and colleagues (5) suggest that the acid metabolites of TCE (TCA and DCA) play an important role in carcinogenesis. Laboratory studies, designed to elucidate key events at a cellular and molecular level, are needed to understand better the nature of TCE exposure and tumor formation.

REFERENCES

1. W.E. Coleman, R.D. Lingg, R.G. Melton, and F.C. Kopfler, The occurrence of volatile organics in five drinking water supplies using gas chromatography/mass spectrometry, *Identification and Analysis of Organic Pollutants of Water* (L.H. Keith, ed.), Ann Arbor Science Pub., Ann Arbor, MI, 1976, pp. 305–327.
2. U.S. EPA, *Health Assessment Document for Trichloroethylene*, Office of Health and Environmental Assessment, Washington, D.C., EPA/600/8-82/006F, PB-249696, 1985.
3. R. Spirtas, P. Stewart, and S. Lee, Retrospective cohort mortality study of workers at an aircraft maintenance facility. I. Epidemiological results, *Br. J. Ind. Med. 48*:515–530 (1991).
4. R. Spirtas, S. Lee, and D. Marano, Retrospective cohort mortality study of workers at an aircraft maintenance facility. II. Exposures and their assessment, *Br. J. Ind. Med. 48*:531–537 (1991).
5. R.J. Bull, I.M. Sanchez, M.A. Nelson, J.L. Larson, and A.J. Lansing, Liver tumor induction in B6C3F1 mice by dichloroacetate and tricholoroacetate, *Toxicology 63*:341–359 (1990).
6. S.L. Herren-Freund, M.A. Pereira, M.D. Khoury, and G. Olson, The carcinogenicity of trichloroethylene and its metabolites, trichloroacetic acid and dichloroacetic acid, in mouse liver, *Toxicol. Appl. Pharmacol. 90*:183–189 (1987).

7. J.W. Daniel, The metabolism of ^{36}Cl-labelled trichloroethylene and tetracholothylene in the rat, *Biochem. Pharmacol. 12*:795–802 (1963).
8. W. DeKant, A. Schultz, M. Metzler, and D. Henschler, Absorption, elimination, and metabolism of trichloroethylene: A quantitative comparison between rats and mice, *Xenobiotica 16*:143–152 (1986).
9. U.S. EPA, *Addendum to the Health Assessment Document for Trichloroethylene: Updated Carcinogenicity Assessment for Trichloroethylene, Draft*, Office of Health and Environmental Assessment, Washington, D.C., EPA/600/8-82/006FA, 1987.
10. J.W. Fisher, M.L. Gargas, B.C. Allen, and M.E. Andersen, Physiologically based pharmacokinetic modeling with trichloroethylene and its metabolite, trichloroacetic acid, in the rat and mouse, *Toxicol. Appl. Pharmacol. 109*:183–195 (1991).
11. National Cancer Institute, *Carcinogenesis Bioassay of Trichloroethylene*, CAS No. 79-01-6, DHEW Publ. No. (NIH) 76-802, 1976.
12. C. Maltoni, G. Lefemine, and G. Cotti, Experimental research on trichloroethylene carcinogenesis, *Archives of Research on Industrial Carcinogenesis* (C. Maltoni and M.A. Mehlman, eds.), Princeton Scientific, Princeton, NJ, 1986.
13. J.C. Ramsey and M.E. Andersen, A physiologically based description of the inhaled pharmacokinetics of styrene monomer in rats and humans, *Toxicol. Appl. Pharmacol. 73*:159–175 (1984).
14. Clement International Corporation, *Development and Validation of Methods for Applying Pharmacokinetic Data in Risk Assessment*, Vol. II, *Trichloroethylene*, Harry G. Armstrong Aerospace Medical Research Center, Human Systems Division, Air Force Systems Command, Wright-Patterson AFB, OH, AAMRL-TR-90-072, AD A237366, 1990.
15. B.C. Allen and J.W. Fisher, Pharmacokinetic modeling of trichloroethylene and trichloroacetic acid in humans, *Risk Anal. 13*(1):71–86 (1993).
16. J.W. Fisher and B.C. Allen, Estimating liver cancer risks in humans exposed to trichloroethylene using physiological models, *Risk Anal 13*(1):87–95 (1993).
17. M.L. Gargas, R.J. Burgess, D.E. Voisard, G.H. Cason, and M.E. Andersen, Partition coefficients of low-molecular-weight volatile chemicals in various liquids and tissues, *Toxicol. Appl. Pharmacol. 98*:87–99 (1989).
18. M.L. Gargas, M.E. Andersen, and H.J. Clewell, III, A physiologically based simulation approach for determining metabolic constants from gas uptake data, *Toxicol. Appl. Pharmacol. 86*:341–352.
19. J.W. Fisher and M.E. Andersen, Using inhalation kinetic data in the development of physiological models for oral absorption: A case study with trichloroethylene, *Principles of Route-to-route Extrapolation for Risk Assessment* (T.R. Gerrity and C.J. Henry, eds.), Elsevier, New York, 1990, pp. 297–311.
20. R.B. Howe and C. Van Landingham, *Global 86*, Clement Associates, Ruston, LA, 1986.

21. J.W. Fisher, T.A. Whittaker, D.H. Taylor, H.J. Clewell, III, and M.E. Andersen, Physiologically based pharmacokinetic modelling of the pregnant rat: A multiroute exposure model for trichloroethylene and its metabolite trichloroacetic acid, *Toxicol. Appl. Pharmacol.* *99*:395–414 (1989).
22. National Toxicology Program, *Carcinogenesis Bioassay of Trichloroethylene*, CAS No. 79-01-6, NTP 81-84, NIH Publication No. 82-176, 1982.

18

Risk Assessment Based on Data from the Toxic Chemical Release Inventory

Andrew Battin, Nancy Chiu, and James Darr

U.S. Environmental Protection Agency, Washington, D.C.

I. INTRODUCTION

A key provision of the Emergency Planning and Community Right-to-Know Act is Section 313, the Toxic Chemical Release Inventory (TRI), which requires companies who manufacture, process, or use certain chemicals to report annually the quantities of those chemicals that are released to the environment. In 1987, the first year covered by TRI reporting requirements, the U.S. Environmental Protection Agency (EPA) received over 70,000 reports on the 309 chemicals and 20 chemical categories subject to the requirements (chemicals can be added or removed by EPA, so the list usually changes slightly from year to year).

The TRI provided EPA with quantitative emissions data that were not generally available in the past. Furthermore, EPA has compiled the data in a computerized form amenable to analysis. The Office of Pollution Prevention and Toxics (OPPT; formerly the Office of Toxic Substances) in EPA sought to make use of this unique information source in its chemical screening program.

The OPPT screening program has two major objectives:

1. Identify those chemicals that present the greatest health or environmental risks and propose risk reduction and pollution prevention options.
2. Identify testing needs for commercial chemicals that lack an adequate base of information for risk assessment.

The opinions expressed in this chapter are those of the authors. These opinions do not necessarily reflect the position of the U.S. Environmental Protection Agency.

Utilization of the TRI data in the OPPT screening program presented both an opportunity and a challenge. The TRI could enable OPPT to perform more sophisticated analysis and make better decisions, but the assessment methods employed would have to satisfy the fundamental need of the screening program to review a fairly large number of chemicals quickly and at a relatively low unit cost. The approach employs quantitative methods within an overall screening framework that is more qualitative in nature. This is a significant departure from traditional approaches, in which quantitative methods are typically reserved for more rigorous, detailed risk assessment. Although the screening results include numerical risk estimates, the results are properly interpreted as qualitative indicators of the nature and level of risk.

TRI data on chemical releases to water, when used in conjunction with relevant data on receiving streams and downstream drinking water utilities, provide a valuable tool for identifying potential drinking water exposures. Because the TRI covers releases to all environmental media, it allows the user to place exposures and risks from one medium, such as drinking water, into perspective with risks from other environmental exposures.

A total of 30 TRI chemicals were evaluated in the screening program. Of those, 4 chemicals, representing a range of hazard and exposure concerns, are discussed in this paper.

II. CHEMICAL CHARACTERIZATION

The four chemicals reported here include acrylonitrile (AN; CAS RN 107-13-1), 1,2-dichloroethane (DCE; CAS RN 107-06-2), hydrazine, and *N,N*-dimethylaniline (NNDMA; CAS RN 302-01-2). All four chemicals were identified during TRI screening as toxicants and/or carcinogens with high production and environmental releases. The first three chemicals are highly toxic to humans and animals if inhaled, ingested, or dermally absorbed. They are classified by EPA as B1 or B2 "probable human carcinogens" based on epidemiological studies of workers and/or sufficient evidence from animal studies. In addition, hydrazine is extremely toxic to aquatic organisms, such as algae. NNDMA is a less well characterized toxicant.

The environmental fate properties relevant to the exposure assessment are listed in Table 1.

III. TRI ENVIRONMENTAL RELEASES

All release figures are based on the reported 1987 TRI data recorded on the Toxic Release Inventory System (TRIS) developed by OPPT contractor

Table 1 Key Environmental Fate Characteristics

	Air		Water		
Chemical	Volatility	Persist	Biodeg	Bioconc	Soil/Sediment
AN	High	$t_{1/2}$ = 10–50 h	$t_{1/2}$ = 2–3 weeks	Low	Leaches readily
DCE	High	$t_{1/2}$ = 12–120 days	$t_{1/2}$ = 3–6 months	Low: log BCF = 0.85	Leaches readily
Hydrazine	High	$t_{1/2}$ = 6 h	$t_{1/2}$ = 7–28 days	Low	Mobility varies with CEC
NNDMA	Moderate	$t_{1/2}$ = 3–21 h	$t_{1/2}$ = 1–6 months	Low: BCF = 33.5	Moderately leachable

Source: From Reference 1.

Table 2 1987 TRI Releases[a] (Pounds)

Chemical	Air (stack + fugitive)	Water		Land	
		Direct	Indirect	Underground	Surface
AN	6,140,000	21,500	173,000	4,400,000	38,500
DCE	6,140,000	70,700	139,000	1,600,000	445,000
Hydrazine	41,500	32,600	1,600		161,000
NNDMA	130,000	17,600	179,000		250

[a]Includes both direct releases and estimated indirect releases following treatment. Figures are from 1987 TRI data available at the time the modeling was done. Slightly different results may be obtained from the final version of the 1987 data base because of later additions or corrections.
Source: From Reference 2.

Sycom, Inc. The environmental releases to specific media of individual chemicals are tabulated in Table 2.

IV. ENVIRONMENTAL EXPOSURE MODELING

A. Water Modeling

The routing graphical display system (RGDS) model was used in estimating dissolved chemical concentrations in stream reaches nationwide because of the point releases from facilities and public-owned treatment works (POTW) obtained in TRIS (3,4). The model runs under the STORET environment on the EPA IBM mainframe computer system and uses data from the EPA REACH file (stream segment identification file), GAGE file (stream flows), and IFD file (locations of discharging facilities and POTW on reaches). The RGDS model produces an output file for each chemical, which consists of chemical concentrations in each reach for approximately 59,000 stream reaches in the continental United States. The model uses the average daily TRIS release from each facility or POTW as identified by the NPDES (National Pollutant Discharge Elimination System) number and estimates the dissolved concentrations of a chemical of each reach segment by accounting for dilution and first-order decay in surface water. In this study the mean annual flow was used for drinking water analysis, and low flow was used to assess aquatic ecotoxicity.

In calculating the effluent concentrations from POTW and other non-POTW plants, removal percentages via biodegradation, adsorption, and volatilization (Toxics Data Base, Office of Water, U.S. EPA) were applied. The estimated removal percentages at treatment plants were 75, 90, 90, and 10% for AN, DCE, hydrazine, and NNDMA, respectively. Removal of chemicals via biodegradation (5) based on half-lives in surface water were 23, 180, 7, and 80 days for AN, DCE, hydrazine, and NNDMA, respectively.

A postprocessing program was developed for this study, which creates two output tables for each chemical. The first table consists of stream reaches ranked by descending chemical concentration at either mean or low annual stream flow and the number of facilities discharging the specified chemical that are located on the reach. The low limit of precision of model output is 0.001 μg/L (1 ppt). Computed reach concentrations below this limit were output as zero (6).

The second type of output table summarizes the calculated concentration and either dose or estimated excess cancer cases (over a 70 year period) in all reaches having drinking water utility intakes. The locations of the utility intakes and populations served were obtained from the EPA Water Supply Data Base file. The lifetime average daily dose (LADD) was calculated as

$$\text{LADD} = \frac{C \times \text{DR}}{W1000}$$

where
LADD = lifetime average daily dose, mg/kg/day
C = concentration in reach, μg/L
DR = average daily drinking water consumption for an adult assumed, 2 L/day
W = average adult body weight, assumed to be 70 kg

The number of excess cancer cases was calculated as

$$\text{Can} = \text{LADD} \times \text{Pop} \times Q_1^*$$

where:
Can = number of excess cancer cases
Pop = population served by the utility
Q_1^* = oral cancer potency-slope factor for the chemical, $(\text{mg/kg/day})^{-1}$

B. Air Modeling

The industrial source complex long-term (ISCLT) model implemented in the atmospheric modeling subsystem (GAMS) of graphical exposure modeling system was used to estimate ambient atmospheric concentrations of the released chemicals (5,7,8).

The results should be considered screening level because of the lack of site-specific release characteristics. The following assumptions were made in building the required ISCLT input files:

The 1980 population density in people per square kilometer within 3 km of the facility determined urban or rural modeling for releases at each facility. If the density was 750 or greater, urban dispersion modeling was used; otherwise rural dispersion modeling was used.

Ring distances in the range of 50–50,000 m were used with three concentration estimates per ring in 16 directions.

A 99.9% destruction efficiency was applied to off-site transfer quantities reported to TRIS to obtain release amounts.

No decomposition or transformation of chemical in the ambient air after releases since the travel time (~4 h) to reach 50 km is smaller than the half-life of the chemicals modeled in this exercise.

A release height of 10 m was used for point and off-site transfer releases, and 3 m height was used for area source releases.

The area sources size was set at 10 by 10 m.

Effects of buoyancy and momentum were essentially eliminated by setting the exit gas temperature to 293 K and the exit gas velocity to 0.01 m/s.

The off-site releases were modeled at the off-site zip code location.

The 1980 population data base was created from the 1980 Master Area Reference File (MARF) (9) released by the Bureau of Census. Donnelly Marketing Information Services has added the latitude and longitude coordinates to all BG/ED (block group and enumeration districts) in MARF within the United States. The dose and risk estimates were obtained from ISCLT based on the population distribution data.

The LADD for a BG/ED is calculated as

$$\text{LADD}(i) = \text{Conc}(i) \times 0.001 \frac{\text{IR}}{\text{BW}}$$

where:

(i) = index for a given BG/ED
LADD(i) = lifetime average daily dose, mg/kg/day
Conc(i) = annual average concentration, $\mu g/m^3$
IR = daily inhalation rate, assumed to be 20, m^3/day
BW = human body weight, assumed to be 70 kg

Lifetime excess risk over background for a BG/ED using a cancer potency slope factor is calculated using the equation

$$\text{Risk}(i) = \text{LADD}(i) \times Q_1^*$$

where:
Risk(i) = lifetime excess cancer risk over background
Q_1^* = inhalation cancer potency slope factor, $(\text{mg/kg/day})^{-1}$

The expected number of excess cancer cases for a BG/ED is calculated using the expression

$$\text{Cases}(i) = \text{Risk}(i) \times \text{Pop}(i)$$

where:
Cases(i) = excess lifetime cancer cases
Pop(i) = population at the BG/ED (persons)

The concentration, dose, and risk values are based on sector segment averages assigned to BG/ED. This occurs only if the centroid of the BG/ED is located within the sector segment (a combination of distance and direction from the source) and is based on average concentrations.

V. HAZARD ASSESSMENT

OPPT sought information that would allow a rapid evaluation of the hazards a chemical may present to human health or the environment. The best information sources for this type of assessment are those that

1. Are readily available
2. Reach "bottom-line" conclusions through peer review or a comparable formal evaluation
3. Include quantitative estimates of toxicological potency

The information source that best met the needs of the screening effort is the EPA Integrated Risk Information System (IRIS) (10). IRIS contains quantitative estimates of potency for both carcinogenic effects and other toxicological end points. For chemicals not in IRIS, we used sources that included the National Library of Medicine Hazardous Substances Data Bank and review documents prepared by EPA and other institutions.

Because thc TRI data allow quantitative, albeit crude estimates of exposure, OPPT sought to incorporate quantitative indicators of hazard potential into the screening approach. Carcinogenic risk was evaluated using EPA "slope factors" that express carcinogenic potency in terms of lifetime risk per mg/kg/day average dose. Most potency values are determined from animal studies that provide sufficient dose-response data for quantitative analysis.

The screening exercise used the most relevant potency factor available for the exposure under consideration. If the only potency factor available was derived from an animal study employing a different route of exposure than the primary route of exposure for humans, we performed a simple route-to-route conversion of the potency factor unless there was a strong reason to believe that the carcinogenic response would be so much different between routes as to render such a conversion completely invalid. In fact, there were no cases for which we rejected a potency factor because of concerns over route-to-route extrapolation. For some chemicals, we used potency factors that have not been verified by the agency. Again, it is important to remember that this exercise was intended as an initial screen, not as a definitive determination of risk. For the three carcinogens discussed in this chapter (acrylonitrile, 1,2-dichloroethane, and hydrazine), the agency has derived and formally approved both oral and inhalation slope factors. We used these slope factors along with exposure estimates to determine both individual risk to highly exposed populations and overall population risk. Chemicals and facilities associated with either high individual risks or significant population risks were flagged for further assessment.

EPA expresses potency for toxicological effects other than cancer as a reference dose (RfD). The reference dose is defined as "an estimate (with uncertainty spanning perhaps an order of magnitude) of a daily exposure to the human population (including sensitive subgroups) that is likely to be without appreciable risk of deleterious effects during a lifetime" and is expressed as a mg/kg/day dose. The RfD is usually based on the most sensitive known effect. Although some RfD are based on actual human data, RfD are most often calculated from results obtained in oral chronic or subchronic animal studies. EPA calculates a comparable measure of potency for inhalation exposures, called a reference concentration (RfC). At the time of the screening exercise, very few RfC values were available so we typically used the corresponding RfD value as a rough indicator of inhalation toxicity. In making these judgments, we took into account considerations of metabolism and pharmacokinetics in judging whether the oral RfD was a valid indicator for the inhalation toxicity of the subject chemical. Of the four chemicals discussed in this chapter, the only RfD/RfC value available at the time of the screening was an RfD for *N,N*-dimethylaniline.

The RfD was used in the screening exercise as a benchmark for comparison to the estimated exposure levels. Chemicals whose projected exposures exceeded the RfD were flagged for further assessment. For chemicals lacking RfD, we attempted to identify a no observed adverse effect level (NOAEL) or lowest observed effect level (LOAEL) from a well-conducted, peer-reviewed study. NOAEL and LOAEL were used in margin of exposure (MOE) calculations. The MOE is defined as the NOAEL or LOAEL divided by the exposure level. Thus, the higher the MOE, the lower the risk. The general guideline employed in interpreting the MOE was to flag a chemical for further assessment if an MOE based on a LOAEL was <1000 or if an MOE based on a NOAEL was <100.

Ecological risk in aquatic environments was evaluated by comparing projected exposure levels to ecotoxicity "concern levels" estimated by EPA scientists from data in the literature or from structure-activity predictions.

A. Chemical-Specific Health Effects

1. Acrylonitrile

Acrylonitrile is readily absorbed via inhalation, oral, and dermal exposure. AN is classified by EPA as a B1 probable human carcinogen based on the observation of a statistically significant increase in incidence of lung cancer in exposed workers 1980) and on the sufficient evidence of carcinogenicity from animal studies (6,10,11). EPA has developed an inhalation cancer potency factor of 0.24 $(mg/kg/day)^{-1}$ for AN and an oral potency factor of 0.54 $(mg/kg/day)^{-1}$. AN has also demonstrated developmental toxicity and reproductive toxicity in animal studies.

2. 1,2-Dichloroethane

DCE is readily absorbed by humans and animals via all exposure routes and is rapidly distributed to body tissues (10,12). EPA has classified DCE as a B2 probable human carcinogen based on the observation of sufficient evidence of carcinogenicity from animal studies. EPA has developed an oral and inhalation cancer potency factor of 9.1×10^{-2} $(mg/kg/day)^{-1}$.

3. Hydrazine

A dermal application study suggests that hydrazine is readily absorbed by animals (10,13). Hydrazine is classified by EPA as a B2 probable human carcinogen based on the sufficient animal carcinogenicity data. Based on the hematoma incidence data in a mouse gavage study, EPA developed an oral slope factor of 3.0 $(mg/kg/day)^{-1}$. Based on nasal cavity adenoma or adenocarcinoma in a rat inhalation study, EPA developed an inhalation slope factor of 17.1 $(mg/kg/day)^{-1}$.

4. *N*,*N*,-Dimethylaniline

Effects on the spleen were the most sensitive toxicological end points observed in oral subchronic studies in rats and mice (10). Based on dose-related increases in the incidence and severity of splenomegaly in mice, EPA established an oral reference dose of 2×10^{-3} mg/kg/day.

B. Chemical-Specific Environmental Effects

Lowest observed effect levels or concern levels established by EPA for the four chemicals studied are given in Table 3.

Table 3 LOAEL or Concern Levels for Aquatic Organisms

Chemical	Acute freshwater	Chronic freshwater	Acute marine	Chronic marine
AN (14)	2.6 ppm (fish)	7.5 ppm (*D. magna*)		
DCE (15)	118 ppm (fish)	4 ppm (fish)	113 ppm (fish)	
		2 ppm (invertebrate)		
Hydrazine (16)	30 ppm (fish)	3.4 ppm (fish)		
	6 ppt (algae)	1 ppt (algae)		
NNDMA (17)	5 ppm (fish, *D. magna*)	0.3 ppm (*D. magna*)		

VI. ENVIRONMENTAL EXPOSURE AND RISK ESTIMATES

Using the ISCLT model in GAMS as described in Section IV.B, the air concentrations of each chemical within a 50 km diameter of all air release sources were estimated. Surface water and drinking water concentrations downstream from the water release sites were also estimated using the RGDS as described in Section IV.A.

A. Acrylonitrile

1. Ambient Air

Risk levels and lifetime excess cancer cases due to AN air emission was estimated using the EPA-established cancer risk slope factor of 0.24 (mg/kg/day)$^{-1}$, model-estimated ambient air concentrations, and population data within GAMS. The estimates can be summarized as follows:

Concentration (μg/m^3)	Risk level	Population at risk	Lifetime cases
3.46E1	2.37E-3	207	4.91E-1
3.46E1–1.46E1	2.37E-3–1E-3	1,574	3.17E0
1.46E1–1.46E0	1E-3–1E-4	41,457	9.16E0
1.46E0–1.46E-1	1E-4–1E-5	871,895	2.22E1
1.46E-1–1.46E-2	1E-5–1E-6	6,612,942	1.76E1
1.46E-2–1.46E-3	1E-6–1E-7	11,266,397	4.18E0
1.46E-3–1.46E-4	1E-7–1E-8	33,013,050	1.17E0
1.46E-4–1.46E-5	1E-8–1E-9	20,161,329	9.48E-2
1.46E-5–1.46E-6	1E-9–1E-10	9,214,621	4.06E-3

The model estimates that there are an 58.1 excess lifetime cancer cases among a total 81,183,472 people exposed. The annual incidence is 0.829 cases per year. The geographical distribution of the excessive cancer cases is presented in Figure 1.

2. Drinking Water

Lifetime excess cancer risk estimates of populations served by water supplies containing the highest AN concentrations using the model estimates, the EPA-established slope factor (oral) of 0.54 $(mg/kg/day)^{-1}$, and census data within REACHSCAN are summarized as

Concentration (μg/L)	Risk level	Population at risk	Lifetime cases
0.673	1.12E-5	200	0.002
0.651	1.09E-5	105,000	1.029
0.564	0.94E-5	2,200	0.019
0.563	0.94E-5	47,091	0.381
0.551	0.92E-5	15,000	0.124
0.492	0.82E-5	50,000	0.370
0.454	0.76E-5	9,900	0.067
0.451	0.75E-5	37,103	0.252
0.438	0.73E-5	760,000	5.016
0.428	0.71E-5	52,000	0.333

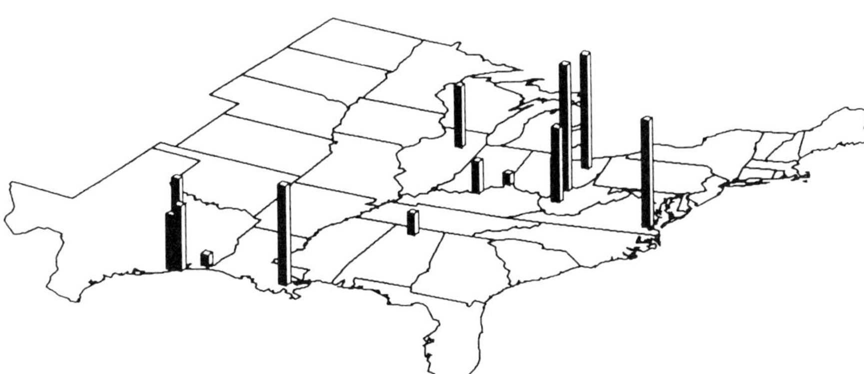

Figure 1 Excess cancer cases from acrylonitrile. GAMS estimates using ISCLT and TRI release amounts of air. Facilities with over 1.0 modeled lifetime cases presented. Range of lifetime cases presented is 1.3–10.1. Total estimated cases from 94 facilities releasing to air, 58.1.

From a total of 1,078,494 population consuming drinking water containing the 10 highest estimated AN concentrations, the model estimates that an excess of 7.6 cancer cases may result. Moreover, it is possible that more drinking water utilities have lower AN concentrations but serve much larger populations. This could result in additional excess cancer cases.

Potential of AN leaching to the groundwater from the landfill sites is high because of the (1) high water solubility and low absorbability of AN, and (2) large releases to land. In addition, the high vapor pressure of AN indicates that evaporation from soil surfaces may also occur.

B. 1,2-Dichloroethane

1. Ambient Air

For risk levels and lifetime excess cancer cases due to DCE air emissions, data used in the estimations included the EPA-established cancer risk slope factor of 9.1×10^{-2} $(mg/kg/day)^{-1}$, model-estimated ambient air concentrations, and population data within GAMS. The model estimates that there are 65.7 excess lifetime cancer cases among a total 72 million persons exposed. The annual incidence is 0.939 cases per year.

2. Drinking Water

The lifetime excess cancer risk of populations served by water supplies containing DCE was estimated using the model estimates, the EPA-established slope factor (oral) of 9.1×10^{-2} $(mg/kg/day)^{-1}$, and census data within RGDS. One drinking water utility that serves approximately 40,000 persons was originally identified as a possible source of DCE exposure. However, follow-up assessment that more accurately placed the location of the drinking water intake eliminated this concern.

3. Surface Water

The 10 highest mean flow DCE estimates range from 3.26 to 798.05 ppb. The estimates are lower than the freshwater chronic concern level of 2 ppm recommended by the EPA OPPT.

The potential of DCE leaching to the groundwater from the landfill sites can be high because of the high water solubility and low absorbability of DCE.

C. Hydrazine

1. Ambient Air

Based on the modeled air concentration estimates resulting from the TRI air releases and the inhalation slope factor (17.1/mg/kg/day), an estimated

79.4 lifetime cancer cases will result from about 39 million persons exposed to various concentrations of contaminated air.

2. Drinking Water

Estimated hydrazine concentrations at various drinking water intakes range from 0.001 to 0.024 μg/L. This translates into daily doses of 6.8^{-4} to 2.9×10^{-6} mg/kg/day to exposed populations. Applying the oral slope factor of 3/mg/kg/day established by EPA, there are 0.36 estimated lifetime cancer cases among the 1.56 million exposed persons.

3. Surface Water

The 10 highest model mean flow stream hydrazine concentration estimates range from 0.443 to 3.636 μg/L. The surface water estimates are lower than the EPA OPPT-recommended fish concern levels (acute 3.4 ppm; chronic 30 ppb). However, these estimates are three orders of magnitude higher than the algae aquatic concern level (acute 6 ppt; chronic 1 ppt). Moreover, it is estimated that there are 380 reach segments in eight states, totaling 7000 miles in river length, with mean stream concentrations equal to or over 1 ppt (Figure 2).

D. *N,N*-Dimethylaniline

1. Ambient Air

The modeled exposure concentrations for the highest exposed population of 2368 persons corresponds to a lifetime average daily dose of 2.9×10^{-3} mg/kg/day. This dose exceeds the EPA-recommended RfD value of 2×10^{-3} mg/kg/day.

2. Drinking Water

Estimates for the 10 highest NNDMA concentrations at various utility sites range from 0.161 to 2.358 μg/L. This translates into a daily intake dose range of 4.6×10^{-6} to 6.7×10^{-5} mg/kg/day. These daily intake estimates are lower than the EPA-established oral RfD value by two orders of magnitude.

3. Surface Water

The 10 highest model mean flow stream concentration estimates range from 1.4 to 136.8 ppb. The surface water estimates are lower than the fish concern levels (acute 5 ppm; chronic 0.3 ppm) recommended by EPA OPPT.

Figure 2 Hydrazine concentrations > 1 ppt.

VII. ASSESSMENT SUMMARY AND DISCUSSION

Acrylonitrile, 1,2-dichloroethane, and hydrazine all present significant levels of cancer risk from ambient air exposures. All three chemicals are undergoing further evaluation in the Office of Pollution Prevention and Toxics to better characterize these risks and to identify options for reducing exposures. OPPT has also sent letters to companies who release large quantities of the chemicals, encouraging them to reduce releases.

Water-modeling results showed possibly high levels of acrylonitrile at the intake of one large drinking water utility. Follow-up discussion with this utility, however, indicated that treatment effectively removes acrylonitrile from finished drinking water. Water releases of hydrazine may present a significant risk to aquatic species, especially algae. Modeled concentrations of hydrazine exceeded the aquatic toxicity "concern level" in a large number of stream segments. This risk to aquatic species will be included in further OPPT assessment of hydrazine.

Modeled air exposures of *N*,*N*-dimethylaniline for the highest exposed population correspond to an average daily intake slightly higher than the current oral RfD. Toxicological data on NNDMA are limited, and EPA confidence in the current RfD is low. NNDMA will be considered for additional toxicological testing in the OPPT chemical testing program. Final results of a cancer bioassay on NNDMA done by the National Toxicology Program (NTP) were not available at the time of the TRI screening exercise. Future assessment in OPPT, however, will take into account the results of the NTP study.

VIII. CONCLUSIONS

We believe that the screening exercise was very useful in discriminating between chemicals that present risks warranting further assessment and chemicals that present minimal risk. The approach was particularly well suited to evaluating carcinogenic risks from air exposures. The TRI data provided quite adequate input to the standard models for estimating lifetime exposure and cancer risk. Risk estimates for drinking water exposures, however, must be regarded as much cruder indicators because they are based on concentrations estimated at the utility intake. Estimating concentrations in drinking water after treatment was beyond the scope of the screening exercise. Because all aspects of the screening exercise employed conservative assumptions, we are comfortable with the conclusion of "low risk" reached for a number of the chemicals.

The TRI data allowed us to reach judgments about the relative contribution to risk from specific facilities and the relative risks in different geographical areas. We believe this to be an important enhancement over previous screening approaches for which site-specific environmental release data were not available. OPPT has taken advantage of this more sophisticated analysis in targeting specific facilities for further assessment and voluntary risk reduction activities.

The major limitations encountered in the exercise stem from limitations in the TRI data itself. Chief among these is that the current TRI contains only annual release estimates. It provides no data whatsoever on peak levels or the time course of releases. Consequently, one must be very cautious in using TRI data to assess toxicological risks, such as developmental toxicity or acute ecological effects, when short-term exposures must be considered. EPA is taking steps to address this shortcoming by considering a "peak release" amendment that will require facilities to report peak releases and other parameters that will allow better exposure estimates.

Another significant limitation is in the range of facilities subject to TRI reporting. At present only manufacturing sector facilities (Standard Industrial Classification Codes 20–39) are required to report on the TRI. This poses problems in assessing chemicals with releases from facilities that fall outside the specified classifications. For example, TRI data alone may be inadequate for assessing exposures to widely used solvents, such as trichloroethylene. EPA is considering an expansion of the range of standard industrial classification codes subject to TRI reporting.

ACKNOWLEDGMENTS

The authors gratefully acknowledge the contributions of the following individuals: Dr. J. Vincent Nabholz of the Office of Pollution Prevention and Toxics, EPA, for his assistance in evaluating the ecotoxicity of the compounds studied and Kevin Terpie and Scott Rheingrover of General Sciences Corporation and Stuart Wollman of Israel Hydrological Services for their assistance in the exposure modeling phase of the study.

REFERENCES

1. Syracuse Research Corp., *Chemical Fate Rate Constants of SARA § 313 Chemical and Superfund Health Evaluation Manual Chemicals*, EPA Contract 68-02-4254 and EPA 68-08-004, 1989.
2. Sycom, Inc., *Toxic Release Inventory System*, Office of Toxic Substances, Environmental Protection Agency, Washington, D.C., 1988.
3. W.M. Grayman, *Routing and Graphical Display System (RGDS) User's Manual—Draft*, Research Triangle Institute, Monitoring and Data Support Division, Office of Water Regulations and Standards, U.S. EPA, Work Assignment #32, Contract No. 68-01-6904, 1985.
4. W.M. Grayman, *Modifications of the Routing Graphical Display System (RGDS) to Support Hot Spot Analysis of Chemical Releases to Surface Waters*, General Sciences Corporation, Work Assignment #1-53, Contract No. 68-02-4281, 1988.
5. General Science Corp., *Graphical Exposure Modeling System (GEMS) User's Guide*, U.S. EPA Office of Toxic Substances, Contract No. 68-02-4281, 1989.
6. U.S. EPA, *Chemical Hazard Information Profile on Acrylonitrile*, Office of Toxic Substances, Washington, D.C., 1989.
7. General Science Corp., *Graphical Exposure Modeling System*, EPA Contracts 68-02-42-81 and 68D00080, 1990.
8. H.E. Cramer Corp., *Industrial Source Complex (ISC) Dispersion Model User's Guide*, U.S. EPA Office of Air Quality Planning and Standards, Contract No. 68-02-3886, 1986.
9. Donnelly Marketing Information Services, *Master Area Reference File (MARF)*.
10. U.S. EPA, *Integrated Risk Information System*, Office of Health and Environmental Assessment, Environmental Criteria and Assessment Office, Cincinnati, OH, 1989.

11. Agency for Toxic Substances and Disease Registry and U.S. EPA, Atlanta, GA, *Toxicological Profile for Acrylonitrile*, 1989.
12. Agency for Toxic Substances and Disease Registry and U.S. EPA, Atlanta, GA, *Toxicological Profile for 1,2-Dichloroethane*, 1989.
13. U.S. EPA, *Health and Environmental Effects Profile for Hydrazine and Hydrazine Sulfate*, Office of Research and Development, Cincinnati, OH, 1984.
14. U.S. EPA, *Ambient Water Quality Criteria for Acrylonitrile*, Office of Water Regulations and Standards, Washington, D.C., NTIS Report No. PB-81-117285, 1981.
15. U.S. EPA, *Aquatic Toxicity Evaluation for 1,2-Dichloroethane*, Office of Toxic Substances, Washington, D.C., 1989.
16. E.W. Odenkirchen, J.V. Nabholz, *Generic Environmental Hazard Assessment of Hydrazines and Related Compounds*, Dynamic Corp., Office of Toxic Substances, Washington, D.C., 1989.
17. U.S. EPA, *Aquatic Toxicity Evaluation for* N,N-*Dimethylaniline*, Office of Toxic Substances, Washington, D.C., 1990.

19

Empirical Comparisons of Carcinogen Potencies Across Species and Their Bearing on Appropriate Scaling of Doses for Risk Assessment

Lorenz Rhomberg

U.S. Environmental Protection Agency, Washington, D.C.

I. INTRODUCTION

The assessment of potential health risks to people drinking water contaminated with traces of chemicals must include (in addition to characterization of the levels of intake of toxins) an investigation of how health effects are expected to vary among levels of exposure. For chemicals regulated as potential carcinogens, the presumption is usually made for regulatory purposes that there is no level in the drinking water that is absolutely safe but that the probability of the induction of cancer decreases with decreasing levels of contamination. The increment in presumed risk per unit of increased concentration in the drinking water—the so-called unit risk—is most often determined indirectly from the results of lifetime cancer bioassays in laboratory rodents. Applying such laboratory results to the characterization of potential human health effects involves several extrapolations: from the high exposure levels tested in rodents to the low levels that humans would consume in the drinking water, from the controlled, regular, lifelong pattern of exposure in the bioassay (often by routes of uptake other than in drinking water) to the more variable patterns characteristic of humans in the real world as they are exposed to the agents in the drinking water, and finally, from the toxicological reactions of laboratory rodents to those expected in humans. The present chapter addresses the last of these, in particular the question of how to define dose levels presumed of equivalent carcinogenicity in rodents and humans, so that the observations on rodents can be used to characterize the potential magnitude of human risk at the levels of water contamination in question.

As a matter of necessity, the potential for a chemical agent to cause toxic reactions in humans is often investigated by observing the reactions of exposed experimental animals, usually rats and mice. This practice rests on the high degree of physiological, biochemical, and anatomical similarity among mammalian species; the biological reactions in the experimental animals may be taken as evidence that humans might show similar responses to the same chemical exposures. When the objective is to use the animal data to predict the degree or probability of response in humans—that is, when the aim is quantitative extrapolation—one must define the dose levels for humans and animals expected to produce the same degree of effect. For this, it is necessary to take into account the pronounced difference in *scale* between the tested model organisms and humans. That is, even if fundamental similarity is presumed, one must allow for the fact that humans are much larger than experimental rodents and experience chronic exposure to a toxicant for longer lifetimes.

Defining such "toxicologically equivalent" doses has been problematical. Alternatives that have found use include scaling daily administered amounts by body weight or by body surface area, scaling cumulative lifetime intake by body weight, and equating exposures to contaminated air, food, or water according to the concentration of toxic agent. Despite considerable study and debate (1–13), no alternative has emerged as clearly preferable on either empirical or theoretical grounds. The various federal agencies conducting chemical risk assessments have developed their own preferences and precedents for cross-species scaling methodology. This variation stands among the chief causes of variation among estimates of a chemical's potential human risk, even when assessments are based on the same data.

In some sense, the diversity of approaches to dose scaling is appropriate since there is considerable uncertainty about the best approach. The resulting disagreement in risk estimates, however, results in some awkwardness in the regulatory arena. Increasingly, regulatory procedures are being mandated that establish decision points contingent on whether a certain human risk level is expected according to "generally accepted" risk assessment procedures. Variation in methodology frequently leads to ambiguity about whether regulatory action should take place. It has therefore become important to resolve differences in cross-species scaling assumptions.

A second impetus for reexamining the scaling question comes from the increasing availability of comparative pharmacokinetic information on toxic agents. Pharmacokinetic analysis uses data on absorption of agents into the body, distribution among the tissues, metabolic activation or detoxification, and elimination to develop a picture of the disposition of a dose by the body and consequent exposure of the actual target tissues of toxic action. Pharmacokinetic differences among species clearly contribute to the magnitude of equipotent doses. However, the appropriate use of such information for the dose equivalency question hinges on resolving the role of pharmacokinetics compared to that of species differences in the magnitude of toxic reaction to a given degree of target-tissue exposure (i.e., "pharmacodynamics"). Distinguishing the roles of these two aspects of potency scaling has been hampered by imprecisely articulated rationales for the various methods.

Discussions and debate on the issues have been held under the auspices of the Interagency Pharmacokinetics Group (IPG), an ongoing work group of federal scientists from the U.S. Environmental Protection Agency, the Consumer Product Safety Commission, the Food and Drug Administration, and the Occupational Safety and Health Administration that deals with issues of common interests arising in the application of pharmacokinetics to risk assessment.

Recently, the IPG advanced a proposal (14), based on these discussions, advocating the adoption by the participating federal regulatory agencies of a unified method: scaling daily administered amount of a carcinogenic agent in proportion to the ¾ power of body weight. This dose-scaling approach was proposed as a default assumption to be used when case-specific data are inadequate, to achieve presumed equal lifetime cancer risks in different mammalian species following lifetime exposure. The IPG proposal applies only to the scaling of carcinogen doses; the further issue of scaling doses for noncarcinogenic end points was not addressed.

The present chapter is adapted from the report containing the IPG dose-scaling proposal. In this chapter, empirical investigations of how carcinogen potencies vary among species are reviewed, and the implications for choice of a general risk assessment methodology for carcinogen dose scaling are discussed.

II. APPROACHES TO CHOOSING A CROSS-SPECIES SCALING FACTOR

There are two broad and complementary approaches to choosing a cross-species scaling factor. The first is empirical: one may seek cases in which human epidemiological data allow a direct estimate of an agent's potency and then investigate the success of various scaling methods in predicting this potency from animal data. The second approach is theoretical and is grounded in the principles of *allometry*, which is the study of the regular variation in features of anatomy and physiology as a function of overall body size. The strategy for this second approach is to develop a scientific rationale for a particular scaling factor by investigating the allometric variation in the biological features and processes that influence and underlie carcinogenic potency.

Clearly, in many cases there are agent-specific ways in which humans and experimental animals differ in a nonsystematic fashion. These may include metabolic activation or detoxification and interaction with key receptors or target molecules. Such factors create unpredictable deviation from the general pattern of scaling and must be discovered and accounted for on a case-by-case basis. The discussion in this chapter is about a *default scaling factor*, one to be applied in the absence of adequate case-specific information. Lacking such information, one provisionally assumes that the agent in question is an example of a "typical" or "average" chemical that follows a general pattern of cross-species potency differences. This presumption may be modified as information becomes available, but the default assumptions still serve as the benchmark against which the new information is evaluated.

The search for better definition and support of such a default scaling factor is aided by an understanding of the data and principles that may be applied to the question. This chapter attempts to lay some appropriate groundwork. The primary focus is on the empirical data on carcinogen potencies in different species, including a review of studies that have attempted to glean insight into the dose-scaling question through analysis of such information. A brief discussion of the theoretical, allometric approach is also offered.

III. EMPIRICAL INVESTIGATIONS OF CARCINOGEN POTENCIES ACROSS SPECIES

This approach attempts to find a factor value that is empirically successful in producing good estimates of potency in humans from data on potencies in other species. The underlying reason such a factor works is a secondary consideration. The advantage of an empirical approach is that by directly examining carcinogenic potencies (rather than influences on potency, such as pharmacokinetics), all relevant factors are included. The disadvantage is that the data are few and of low resolution. One must hope that the agent-specific factors just mentioned average to yield a good estimate of the general relationship.

A. Studies of Acute Toxicity

A number of studies have sought general scaling factors empirically. Freireich et al. (2), testing and extending the suggestion of Pinkel (1), examined maximum tolerated doses (MTD) of 18 antineoplastic drugs in mice, rats, hamsters, dogs, monkeys, and humans. LD_{10} (doses causing 10% lethality) were used for rodents and were presumed to be a level of toxicity equivalent to an MTD. Doses from experiments of different lengths were reexpressed in terms of an exposure regimen of 5 consecutive days on the assumption that cumulative dose is proportional to effect. The authors concluded that, when doses were expressed as mg/m^2 body surface area per day, good predictions of human MTD were obtained from all animal species, but that body weight scaling of doses overpredicted human MTD (i.e., underpredicted potency in humans) by a margin that increased as one extrapolates from smaller and smaller species. Since an MTD is intended to be a dose causing no lethality and an LD_{10} causes 10% lethality, the equivalence of these two end points can be questioned. Antineoplastic drugs typically have very steep dose-response curves, however, and survival near the MTD is maintained by close monitoring and intervention, which the rodent LD_{10} determinations lack.

Collins et al. (15,16) found that the human MTD for 16 antineoplastic drugs is well predicted on average by the mouse LD_{10} when doses are expressed as mg/m^2 of body surface area. That is, if these end points of acute toxicity are taken as equivalent, scaling doses in proportion to surface area tends to equalize toxicity across species. Moreover, Collins et al. (16) compared the blood levels (in terms of the areas under the curve of concentration in plasma as it declines over time, or $C \times T$) that correspond to equally toxic administered doses and found that these were an even better predictor in that they displayed less case-by-case variation. These results illustrate three important points: (1) scaling administered doses in this way tends to equalize blood levels across species; (2) areas under the curve of blood concentration can serve as a predictive measure of the toxic response to a dose, even across species; and (3) obtaining pharmacokinetic data on internal dose measures can increase the precision of the cross-species prediction of equivalently toxic doses by accounting for case-by-case variation.

Travis and White (17) reanalyzed the Freireich et al. (2) data set and nearly doubled the number of drugs by adding a similar data set of Schein et al. (18). Instead of simply examining the success of previously proposed scaling methods, they used regression techniques empirically to determine the optimal power of body weight to achieve the best-fitting allometric relationship of MTD across species. For both data sets individually and for the combined data set, a power of 0.72–0.74 led to the best cross-species predictions. In the analysis of the combined data, a power of unity (body weight scaling) was clearly rejected at the 95% level of significance, and a power of ⅔ (surface area scaling) was barely rejected. The authors discuss the history of empirical studies of allometric variation in a number of physiological features, primarily basal metabolism, and argue that their result is part of a general empirical support for scaling by the ¾ power of body weight. Recently, Watanabe and coworkers (19) argued that Travis and White underestimate the variance around the estimate of optimal power of body weight. By explicitly accounting for measurement error in the original values, they estimate a wider confidence interval that, although excluding a power of unity, does not exclude the ⅔ power. Travis and Morris (20), in reply, point out that the ¾ power is found to be optimal in both analyses and that their recommendation for this value rests not only on this test but on the rationale provided by the allometric scaling of physiological features underlying pharmacokinetics (21). Goddard and Krewski (22) argue that, owing to the complexity of the proper characterization of variation in scaling behavior among chemicals, conclusions about the bearing of summary statistics on individual instances of cross-species extrapolations should be tempered.

The difficulty with applying such studies to the present question is that they address acute systemic toxicity of a rather narrowly defined type rather than carcinogenesis. Although dose scaling for different toxic end points should have some features in common (notably pharmacokinetics), it is not altogether clear how lifelong risks that accumulate over time (such as cancer risk) should relate to short-term toxicity dependent only on immediate insults to target tissues.

Some empirical studies of comparative potencies of carcinogens in different species have been done. Such studies face the difficulty of precisely determining potencies in humans based on epidemiological data. There is also some ambiguity in defining potencies in animals, owing to the variations in route of exposure, sex and strain differences, varying experimental designs, and so on. Nonetheless, such studies represent the direct investigation of the question at hand.

B. National Academy of Sciences Study

The National Academy of Sciences (NAS) (23) examined the potencies of six carcinogenic agents in bioassays using mice and rats and from human epidemiological studies. They recommended as a dose measure cumulative lifetime amount of agent administered (mg) per kg body weight. Such scaling is more "conservative" (i.e., predictive of higher human risk from animal results) than either surface area scaling or body weight scaling (from which it differs by a factor of 35, owing to the lack of adjustment for differences in length of lifetime). The NAS conclusion was not based on formal quantitative comparison with surface area scaling ($mg/kg^{2/3}/day$) or body weight scaling.

C. Rat and Mouse Studies

The paucity of carcinogen potencies in humans known directly from epidemiological data limits the precision of such comparisons. Crouch and Wilson (24) instead investigated dose scaling between rats and mice in about 70 ingestion cancer bioassays from the National Cancer Institute (NCI) testing program. They measured potency by the parameter of a fitted one-hit dose-response model (in units of risk per mg/kg/day), focusing on the tumor site and type producing the greatest potency (excluding testicular tumors in Fischer 344 rats and skipping cases in which potency was less than twice sensitivity in either species). A geometric mean of potencies in each sex (which were highly correlated) was used. Interspecies comparisons were based on the best-fitting line of unit slope on a plot of the logarithm of potency in rats against the logarithm of potency in mice. The intercept of such a line gives the geometric mean of the factor by which the rat potency must be divided to give the mouse potency. Body weight scaling predicts a factor of 1 (i.e., equal risk per mg/kg/day in both species), but surface area scaling predicts a factor of about 2.1–2.3, depending on the exact body weights. (For comparison, the scaling by $mg/kg^{3/4}/day$, as advocated in the Interagency Pharmacokinetics Group proposal (14), predicts a ratio of about 1.8 or 1.9.) The results depend on the strain of rat used. In the 17 cases of comparison between Osborne-Mendel rats and B6C3F1 mice, the mean ratio of potencies was 0.40; these rats were somewhat less sensitive than mice, contrary to the expectations of both scaling methodologies. When Fischer 344 rats were compared to the same mouse strain (18 cases), a mean ratio of 4.5 was obtained, indicating that rats were even more sensitive than surface area scaling would expect. A geometric mean of these two ratios is 1.3.

Crouch and Wilson (24) also examined ratios of rodent potency to epidemiologically derived human potency, comparing "insofar as possible" studies with the same route of exposure and duration in fraction of a lifetime. Owing to imprecision in the epidemiologically based human estimates, no precise curve fitting was attempted, but the authors state that humans appear to be more sensitive to a mg/kg/day dose by about a factor of 5 compared to either rats or mice. (Surface area scaling predicts a ratio of about 6 for rat-human comparisons and 12 for mouse-human comparisons.)

A similar comparison of rats and mice, based on an expanded base of 187 NCI bioassays, was conducted by Crouch (25). (Despite the larger original data base, there were only a few more chemicals in the final analysis, apparently owing to more stringent requirements for significance of potency estimates.) Again, the rat strain influenced the results: for Osborne-Mendel rats the mean ratio was 0.63; for Fischer 344 rats it was 2.29. (A geometric mean of these two ratios is 1.20.) Separate analysis of males and females changed these ratios only slightly. An analysis irrespective of rat strain yielded a ratio of 1.62.

Gaylor and Chen (26) examined data on rats, mice, and hamsters in the extensive data base of Gold et al. (27) on TD_{50}, the dose (mg/kg/day) leading to a halving of the actuarially adjusted percentage of tumor-free animals at the end of a standard life span. The tumor site and type showing the highest potency (i.e., lowest TD_{50}) was chosen to represent the species, and only agents with responses in both species were included. For 190 compounds administered in the diet, the geometric mean ratio of TD_{50} in rats and mice was 0.455 = 1/2.20. That is, rats were on average about 2.2-fold more sensitive. Ratios for other routes of exposure varied somewhat, although based on much lower sample sizes than the ingestion results just cited. By gavage, 32 compounds had a mean ratio of 1:1.32, in drinking water 10 compounds had a mean ratio of 1.45 (i.e., rats were less sensitive), and by inhalation 7 compounds had a mean ratio of 1:11.2 (i.e., rats were much more sensitive).

Chen and Gaylor (28) investigated cancer bioassays conducted by the NCI and the National Toxicology Program (NTP) on compounds administered orally to rats and mice. They compared "virtually safe doses" (VSD), defined as doses associated with a lifetime cancer risk of 1 in 1 million. These were determined by the method of Gaylor and Kodell (29): that is, a linear extrapolation was conducted from an upper bound on a fitted multistage model dose-response curve. Thus, both the rat and mouse VSD are in some sense "upper bounds." Chemicals were included if judged by the NTP to be positive in at least one species and, when in only one, if there was at least a positive trend in the other species for the same tumor site or type. Unlike the studies mentioned earlier, Chen and Gaylor (28) focused on the correspondence of VSD at the same site and sex across species. VSD were expressed in terms of concentration (parts per million, ppm). Since daily intakes of contaminated media (air, food, and water) tend to be proportional to body surface area, the expectation from surface area scaling is that VSD expressed in ppm would be about equal across species; body weight scaling would expect a ratio of rat to mouse VSD to be slightly greater than 2. Again, the results depend on the strain of rat used: for Fischer 344 rats the mean ratio is 1.15, for Osborne-Mendel rats it is 1.68, and for Sprague-Dawley rats it is 1.78. Ignoring rat strain gives a mean ratio of 1.27. These results are intermediate between the expectations of surface area and body weight scaling. For ease of comparison with other studies, one may convert these ratios from a ppm basis to a mg/kg/day basis using empirically based daily food and water consumption patterns in rats and mice [for food, 5 and 13% of body weight for rats and mice, respectively, and, for water, 7.8 and 17% (30)]. On a mg/kg/day basis, the rat-mouse VSD ratios are 0.44–0.53 for Fischer rats, 0.647–0.771 for Osborne-Mendel rats, and 0.69–0.82 for Sprague-Dawley rats. (The range reflects using rat-mouse ratios of water and food consumption, respectively, which differ slightly.)

Metzger et al. (31) expanded Crouch's (25) earlier data set by including all 264 cases from the Gold et al. (27) data base in which a significant TD_{50} was obtained in an oral study of rats and mice (of any strain), that is, including studies that were not in the NCI/NTP data base. A best-fitting line of unit slope showed a TD_{50} ratio of 1.46 between mice and rats. This is intermediate between the ratio of 1.0 expected from body weight scaling and that of 2.5 from surface area scaling (using the authors' assumptions about body weights).

Recently, a lively discussion has developed around the observed correlation over chemicals of carcinogenic potency and maximum tolerated dose (32–36). Some cite the relationship as evidence of underlying commonality between carcinogenesis and other chronic toxic effects (37). Others believe the correlation is an artifact of the limited range of tumor responses possible in the standard cancer bioassay; such responses are artificially tied to noncancer effects because testing is at or near the maximum dose that can be tolerated chronically without overt toxicity (38–40). The outcome of this debate could affect the interpretation of the mouse-to-rat comparisons of carcinogenic potencies. If these potencies are artifactually related to maximum tolerated doses (which are known to be correlated between rodent species), then the performance of dose-scaling methods may be saying more about chronic noncancer toxicity than about carcinogenicity. This issue is much less likely to affect comparisons of animal and human cancer potencies, however, which are discussed later.

D. Animal-to-Human Studies

A major study of animal-to-human extrapolation of cancer potencies was carried out by Allen et al. (41) and reported in several publications (42–

44). A total of 23 chemicals were identified that permitted quantitative evaluation of potency in humans and in animals. Risk-related doses (RRD) were calculated, defined as the average daily dose per kg body weight expected to result in an extra cancer risk of 25% over a lifetime. Chemicals were included even if RRD estimates were "infinite" for one species, as happens when no carcinogenic effect is observed. Unlike the studies reviewed earlier, the Allen et al. (41) study considered a large number of alternative ways of representing the potency in animals, as well as various methods for extrapolating the resulting RRD to humans. Alternative sets of "risk assessment assumptions" restricted the animal data base according to various criteria of experimental design, route of exposure, and tumor type. Different levels of averaging results over experiments, sex, and species were tried. Finally, different methods for combining the multiple animal results on a given chemical into a single measure of its "potency in animals" were examined. This complexity allows an admirably comprehensive look at animal-to-human extrapolation, but it also makes manifest a problem latent in the other extrapolation studies: the performance of a scaling factor depends on how the animal potency is characterized. A factor that tends to overpredict human risk can be "rescued" by a method for characterizing animal potency that tends to produce a low estimate, and vice versa.

When the objective is to examine alternative dose-scaling factors, it seems that the best approach is to examine analyses that aim at broadly based and unbiased estimates of the potency in animals. Certain risk assessment practices, such as using upper bounds on dose-response curves and extrapolating from the most sensitive sex and species of animal, are explicitly conservative; they may be appropriate science policies for regulatory purposes, but when the issue is empirically to choose a best-performing scaling factor, they introduce a bias, favoring a less conservative factor to compensate for their conservatism and restore a good prediction of the known human potency.

To compare potencies, Allen et al. (41) fit a line of unit slope to the data of epidemiologically observed log RRD in humans plotted against the predicted human log RRD based on the animal data and the chosen scaling methodology. The intercept of this line gives an average ratio of the observed to the predicted potency, a ratio of unity indicating unbiased prediction. The analyses discussed prominently in the publications and reports of this study (41–44) show that body weight scaling leads to a ratio of approximately 1 to somewhat less than 1 depending on the particular suite of risk assessment assumptions chosen (i.e., slightly underpredicting human risk); surface area scaling overpredicts human risk severalfold.

These results are sometimes cited as tending to support mg/kg/day scaling, but such a conclusion should be tempered. The particular choice of risk assessment assumptions (among many examined) in the widely cited analysis is that with results least favorable to surface area scaling; most alternatives discussed in the full report (41) show that body weight scaling underestimates human risks by about the degree to which surface area overestimates it. Moreover, these analyses contain a bias of the sort just outlined above: the animal potency for a chemical is characterized by the median of the *lower bounds* on the RRD for the various animal data sets rather than on best estimates. At present it is unresolved how much the use of central estimates of animal risk to predict central estimates of human risk—a more appropriate analysis for resolving the scaling factor—would shift the results toward favoring surface area scaling.

Two additional studies of comparative cancer potencies should briefly be mentioned, both favoring a somewhat more conservative scaling factor. Raabe et al. (45) compared bone cancer risks from radium in watch dial painters (who ingested radium by tipping brushes on their tongues) and in beagle dogs exposed to radium by injection. Doses were measured as dose to bone of deposited radium, so this comparison can be seen as lacking the pharmacokinetic component of cross-species differences. Potency was measured by the relative mean degree of life shortening as a function of dose. The authors argued that a cumulative lifetime radiation dose per unit of bone seemed to give good correspondence between human and dog. This result could be related to mg/kg per lifetime scaling for chemical agents.

Kaldor et al. (46) examined the carcinogenic potency of five antineoplastic drugs using potencies derived from bioassays in rodents and from the secondary tumors the drugs caused in human cancer patients. They argued that potency seemed to be related to total cumulative lifetime exposure per kg body weight. Recently, Dedrick and Morrison (47) investigated three of these agents more deeply by adding a consideration of their pharmacokinetics to the calculations. They corroborated the apparent success of mg/kg per lifetime scaling for these agents. The authors point out that, although they examined few agents, these are agents for which rigorous human dose estimates are available and for which human and animal dosing are at similarly high levels, just below those causing unacceptable toxicity.

E. Summary of Empirical Evidence

The empirical evidence on cross-species scaling of carcinogen potencies can be summarized as follows. The correlation of agent potencies across species is clearly and strongly demonstrated. This correlation extends to humans, so far as is ascertainable from the limited number of agents for which potencies can be estimated epidemiologically. There is a remarkable agreement among studies that the dose-scaling methods in current use span a range that appears approximately correct. The resolution of the data available at present, however, does not permit a clear choice between surface area and body weight scaling. Empirically chosen scaling factors tend to fall between these two choices in most cases, but the specific results depend on the laboratory strains used, route of administration, details of the methods for characterizing the carcinogenic potency in animals, and the statistical methods used in curve fitting. The data seem consistent in indicating that body weight scaling somewhat underestimates risks in larger species. The exception is when Osborne-Mendel or Sprague-Dawley rats are compared to B6C3F1 mice, in which comparison the rats are seen to be less affected even by doses scaled to body weight. The preponderance of data are from Fischer 344 rats, however, and this is the strain used in most modern bioassays.

Several points should be borne in mind while interpreting the empirical scaling data. First, although several studies are reviewed, they overlap considerably in their data bases: the individual studies are not independent tests. Second, the specific results of a study depend on details of the methodology. The Allen et al. (41) study showed that whether potencies were averaged over sexes, whether both benign and malignant tumors were counted, whether projections were made for specific tumor sites or for the most potent site, and other such factors could swing the analysis toward favoring one scaling method or another. It is hard confidently to identify and isolate the specific contribution of dose scaling among the many factors that contribute to the final predictions of human risk. Third, the epidemiologically based human potencies that serve as "targets" for the animal-based extrapolations are themselves very uncertain and, as in the animal data, dependent on the specifics of the methodology used in their estimation. As a result of this and of the previous point, the comparability of animal- and human-based potencies may be problematical. (For example, potencies calculated from human data are usually based on cancers that were the cause of death following partial lifetime exposure, but animal-based estimates usually reflect incidental as well as fatal tumors arising after full lifetime exposure.) A final point to be borne in mind is that the reported empirically derived factors represent averages over large numbers of cases. Although the means vary over a narrow range, the individual chemicals show ratios of potencies in different species that span orders of magnitude. Most of the rat-to-mouse comparisons were within an order of magnitude of the average scaling relationship, but several agents showed a 100-fold difference. Variances of rodent-to-human potency ratios were higher, reflecting the uncertain determination in humans and the lack of standardized experimental design. The existence of this scatter of cases around the mean helps to define the limits to the resolution of any scaling method and emphasizes the importance of case-to-case variation. Moreover, it provides some insight into the distribution of uncertainty in the cross-species dose extrapolation step of risk assessment.

Despite these shortcomings, the empirical data support the general practice of scaling rodent potencies to humans and show that, on average, current methods perform satisfactorily. Certainly, any method that produces average results an order of magnitude higher or lower than the range represented by body weight and surface area scaling contradicts the empirical data. The data suggest that a scaling factor between the surface area and body weight scaling can be considered to have empirical support.

IV. AN ALLOMETRIC APPROACH TO DOSE SCALING

The complement to the empirical investigation of potency scaling is a more theoretical approach that seeks to identify the biological factors whose variation underlies the variation in a carcinogen's potency across species and then attempts to adjust for their effects. Clearly, these factors are numerous and, for the most part, poorly understood. Fortunately, there are some rather simple and general quantitative patterns in the variation of many features of anatomy and physiology across differently sized mammalian species, representing broad trends in the way the essentially similar mammalian system operates in large and small editions. Although specific processes acting on specific chemicals can (and do) deviate from these broad trends, it is argued here that the general patterns can provide a benchmark that expresses the expectation about a chemical's carcinogenic potency in small mammals, such as experimental rodents, and larger animals such as humans. This expectation can be refined (or refuted) by case-specific biological and mechanistic data, when available, showing how the actual processes of metabolism and carcinogenesis differ from the presumptions of the broad trend analysis that serves as the default.

Clearly, pharmacokinetic differences across species constitute a major influence on the relative potency of toxic agents. One of the major aims of pharmacokinetic analysis is to aid in cross-species extrapolation of effects by being able to identify and take into account differences between rodents and humans in their uptake and metabolism of a toxic agent. The basic physiological factors that mediate the uptake, distribution, metabolism, and excretion of chemicals by the body are relatively well understood. Moreover, many of these factors are qualitatively similar among mammalian species, and they vary quantitatively in a regular and well-studied way, primarily as a function of body size. (The study of such variation patterns is know as allometry.) It is therefore illuminating to approach the cross-species dosimetry problem by examining how the sizes of body components and the rates at which they function vary among species and how such variation in scale influences the "internal" exposure of the actual target organs of toxicity to the active agent.

The question is made difficult, however, by a lack of a clear expectation about how doses, even internal doses, should be expressed to yield equivalent risks across species. Species are different in more than pharmacokinetics: among other things, they have organs of different sizes (and hence number of cells to act as targets for transformation), they may have different inherent rates of repair and turnover, and they experience daily dosing for different numbers of days in a full lifetime. The question of how properly to define tissue doses of equivalent toxic effect remains problematical (14,48,49).

A full treatment of this approach is beyond the scope of the present chapter. The arguments have been well discussed by a number of authors (50–63) and are elaborated in the Interagency Pharmacokinetics Group proposal (14). A brief sketch, however, should serve to convey the flavor of the analysis.

Basically, one can note that the physiological processes that underlie pharmacokinetic phenomena display, for the most part, regular and well-known patterns of variation across mammalian species. Organ volumes, capacities, and similar measures tend to maintain proportionality to body weight, and the rates of many physiological processes (including intakes of air, food, and water, cardiac output, glomerular filtration, basal metabolism, and clearance of many compounds) tend to maintain proportionality to roughly the ¾ power of body weight. (On a per kg basis, the rates vary with the $-\frac{1}{4}$ power of body weight—that is, they are some sevenfold slower in humans than in mice.)

The regularity of this variation suggests that, to a first approximation, mammals represent differently sized versions of the same basic anatomical and physiological "machinery," with larger editions operating at a relatively slower rate. This leads to longer persistence of compounds in the tissues of larger organisms. Compared with a mouse, a human given the same mg/kg dose should take about seven times longer to clear it from the body, leading to a correspondingly larger area under the curve of concentration versus time. Such $C \times T$ measures are often used as indices of expected toxic effect. If instead the human is given a smaller initial dose, one scaled not to body weight but to the ¾ power of body weight (which would be some seven times smaller), the area under the curve will be similar to that of the mouse. In other words, the approach is to scale the dose not to body size but to the capacity of the body to process the incoming material.

V. DISCUSSION

The approaches outlined here aim to arrive at a very broad generalization about carcinogen exposures that can be considered of equal risk in experimental animals and humans, one that can be applied to potentially carcinogenic chemicals lacking adequate information on pharmacokinetics and mechanisms of action. They attempt to provide a rational basis for a prima facie characterization of potential risks in humans, consistent with our empirical knowledge of carcinogen potencies in animals and humans and with the known general consequences of species variation in body size and the rates of physiological processes.

To achieve this wide applicability and generality, it is necessary to rely on simplified, broad patterns and trends of biological variation while bypassing many details and causes of case-by-case variation. This is not to deny the importance of these details nor to denigrate the value of case-specific data that show species- or dose-related differences in uptake, metabolism, or physiological actions of putative carcinogenic agents. To the contrary, the intention is to provide a framework for the use of such data, allowing (and, indeed, encouraging) one to go beyond the prima facie case based on overall trends to address the impact of specific knowledge about the chemical and its actions.

The empirical data on carcinogen potencies estimated in various animal species and in humans demonstrate the large variability involved. Although scaling doses by $W^{3/4}$, as proposed by the Interagency Pharmacokinetics Group (14), characterizes the trend fairly well, individual chemicals may deviate from this overall pattern by two orders of magnitude or more in either direction. The existence of such underlying variation means that the extrapolation of chemically induced risks observed in one circumstance (say, in a mouse lifetime cancer bioassay) to another (say, to people exposed to environmental pollutants) must be carefully and properly interpreted. Clearly, the projection of an equivalent dose is not merely a conversion of units, with the resulting human dose achieving an equal factual standing to the original animal observation. The projection is an hypothesis formulated in the face of uncertainty. In the most basic case, when little additional information may be brought to bear, this hypothesis represents a best guess based on general principles and the recognition of overall trends. This best guess is surrounded by an envelope of considerable uncertainty, owing to the dozens of particulars that make each chemical's disposition and toxic effects in various species unique, despite the overall trends. When applicable pharmacokinetic and mechanistic insights into the particular chemical and its actions are available, they can (and should) be used to refine the projections by identifying and accounting for these chemical-specific factors.

Every projection of human equivalent dose, no matter how sophisticated, is associated with both uncertainty and variability. The uncertainty concerns whether the scaling method employed has correctly embodied and utilized the information at hand (be it general cross-species trends over all chemicals or case-specific insights from pharmacokinetics and mechanistic studies). The variability arises because even a sophisticated projection, when applied to a population of cases, at best predicts the mean of an array of actual values that reflect the myriad individual factors that no analysis can completely take into account. The "true" dose of equivalent risk varies among exposed humans according to how each individual deviates form the overall human norm, owing to genetic factors, environmental influences, age, sex, life-style, and countless details of personal history.

The goal of a cross-species scaling methodology, then, is not to arrive at true values of equivalent doses under all circumstances (for this is impossible, even in principle). Rather, it is to embody correctly and without bias the impact of the information at hand, providing rational estimates that take into account what is known, recognizing that true values vary around this estimate as a result of case-by-case particulars, many of which are either unknown or are known to vary among the individuals for whom the projections are being made.

It must be acknowledged that there is considerable uncertainty about the best scaling method to achieve this unbiased projection. In particular, the empirical data on comparative carcinogen potencies are also compatible with both body weight and surface area scaling. It is sometimes suggested that there should be more than one "default" scaling methodology, with different generalized procedures to be applied to different classes of chemical carcinogens. At present, it is not clear how such division of cases would be made, however, nor what the consequences on a generalized method should be.

VI. CONCLUSIONS

The empirical data on comparative carcinogenic potencies in different species support the general practice of scaling rodent potencies to humans and show that, on average, current methods perform rather well. The data are not of sufficient resolution, however, to distinguish between surface area and body weight dose scaling. The empirical investigations of variation in carcinogen potency across species, considered as a whole, are fully consistent with the proposal of the Interagency Pharmacokinetics Group (14): to adopt as a default methodology (to be used when case-specific pharmacokinetic and other data are inadequate) the scaling of daily lifetime carcinogen doses by body weight to the ¾ power to be presumed equal in carcinogenic effect.

Theoretical support for scaling carcinogen doses by the ¾ power of body weight is available from analysis of the allometric variation in key physiological parameters across mammalian species. Such an analysis has the benefit of providing an articulated rationale for the scaling methodology and of setting out the underlying assumptions explicitly.

ACKNOWLEDGMENTS

This chapter is adapted from a report of the Interagency Pharmacokinetics Group (IPG), which I wrote on the group's behalf. That report benefited considerably from discussion and comment by IPG members. In particular, I acknowledge the contributions of Robert Brown, Murray Cohn, Robert Scheuplein, Ronald Lorentzen, and Christine Whittaker. I also thank the numerous people who commented on an earlier, widely distributed draft of the IPG report.

REFERENCES

1. D. Pinkel, The use of body surface area as a criterion of drug dosage in cancer chemotherapy, *Cancer Res. 18*:853–856 (1958).
2. E.J. Freireich, E.A. Gehan, D.P. Rall, L.H. Schmidt, and H.E. Skipper, Quantitative comparison of toxicity of anticancer agents in mouse, rat, hamster, dog, monkey and man, *Cancer Chemother. Rep. 50*:219–244 (1966).
3. N. Mantel and M.A. Schneiderman, Estimating safe levels, a hazardous undertaking, *Cancer Res. 35*:1379–1386 (1975).
4. D.P. Rall, Species differences in carcinogenesis testing, *Origins of Human Cancer, Book C, Human Risk Assessment*, Cold Spring Harbor Conferences on Cell Proliferation, Vol. 4, (H.H. Hiatt, J.D. Watson, and J.A. Winsten, eds.), Cold Spring Harbor Laboratory, Cold Spring Harbor, NY, 1977.
5. D.G. Hoel, Some problems in low-dose extrapolation, *Origins of Human Cancer, Book C, Human Risk Assessment*, Cold Spring Harbor Conferences on Cell Proliferation, Vol. 4, (H.H. Hiatt, J.D. Watson, and J.A. Winsten, eds.), Cold Spring Harbor Laboratory, Cold Spring Harbor, NY, 1977.
6. M. Hogan and D.G. Hoel, Extrapolation to man, *Principles of Toxicology* (A.W. Hayes, ed.), Raven Press, New York, 1982.
7. E.J. Calabrese, *Principles of Animal Extrapolation*. John Wiley & Sons, New York, 1983.
8. E.J. Calabrese, Extrapolation from animal data, *Toxic Substances and Human Risk: Principles of Data Interpretation* (R.G. Tardiff and J.V. Rodricks, eds.), Plenum Press, New York, 1987.

9. K.S. Crump, A. Silvers, P.F. Ricci, and R. Wyzga, Interspecies comparison for carcinogenic potency to humans, *Principles of Health Risk Assessment* (P.F. Ricci, ed.), Prentice-Hall, Englewood Cliffs, NJ, 1985.
10. I.W.F. Davidson, J.C. Parker, and R.P. Beliles, Biological basis for extrapolation across mammalian species, *Regul. Toxicol. Pharmacol. 6*:211–237 (1986).
11. J.R. Gillette, Dose, species, and route extrapolation: general aspects, *Pharmacokinetics in Risk Assessment*: *Drinking Water and Health*, Vol. 8 (National Research Council), National Academy Press, Washington, D.C. (1987).
12. F. Vocci and T. Farber, Extrapolation of animal toxicity data to man, *Regul. Toxicol. Pharmacol. 8*:389–398 (1988).
13. T.A. Hill, R.C. Wands, and R.W. Leukroth, Jr. (eds.), *Biological Bases for Interspecies Extrapolation of Carcinogenicity Data*, Federation of American Societies for Experimental Biology, FDA Contract No. 223-83-2020, Food Safety and Applied Nutrition, Food and Drug Administration, Department of Health and Human Services, Washington, D.C., 1986.
14. U.S. EPA, A cross-species scaling factor for carcinogen risk assessment based on equivalence of mg/kg$^{3/4}$/day, *Federal Register 57*:42152–42173 (June 5, 1992).
15. J.M. Collins, D.S. Zaharko, R.L. Dedrick, and B.A. Chabner, Potential role for preclinical pharmacology in phase I clinical trials, *Cancer Treat. Rep. 70*:73–80 (1986).
16. J.M. Collins, C.K. Grieshaber, and B.A. Chabner, Pharmacologically guided phase I clinical trials based upon preclinical drug development, *J. Natl. Cancer Inst. 82*:1321–1326 (1990).
17. C.C. Travis and R.K. White, Interspecific scaling of toxicity data, *Risk Anal. 8*:119–125 (1988).
18. P.S. Schein, R.D. Davis, S. Carter, J. Newman, D.R. Schein, and D.P. Rall, The evaluation of anticancer drugs in dogs, and monkeys for the prediction of quantitative toxicities in man, *Clin. Pharmacol. Ther. 11*:3–40 (1979).
19. K. Watanabe, F.Y. Bois, and L. Zeise, Interspecies extrapolation: A reexamination of acute toxicity data, *Risk Anal 12*:301–310 (1992).
20. C.C. Travis and J.M. Morris, On the use of 0.75 as an interspecies scaling factor, *Risk Anal. 12*:311–313 (1992).
21. C.C. Travis, R.K. White, and R.C. Ward, Interspecies extrapolation of pharmacokinetics, *J. Theor. Biol. 142*:285–304 (1990).
22. M.J. Goddard and D. Krewski, Interspecies extrapolation of toxicity data, *Risk Anal. 12*:315–317 (1992).
23. National Academy of Sciences (NAS), *Pest Control*, Vol. 1, *An Assessment of Present and Alternative Technologies*, National Academy Press, Washington, D.C. (1975).
24. E. Crouch and R. Wilson, Interspecies comparison of carcinogenic potency, *J. Toxicol. Environ. Health 5*:1095–1118 (1979).
25. E.A.C. Crouch, Uncertainties in interspecies extrapolations of carcinogenicity, *Environ. Health Perspect. 50*:321–327 (1983).

26. D.W. Gaylor and J.J. Chen, Relative potency of chemical carcinogens in rodents, *Risk Anal.* *6*:283–290 (1986).
27. L.S. Gold, C.B. Sawyer, R. McGaw, G.M. Buckman, M. DeVecidna, R. Levinson, N.K. Hooper, W.R. Hevendor, L. Bernstein, R. Peto, M.C. Pike, and B.N. Ames, A carcinogenic potency database of the standardized results of animal bioassays, *Environ. Health Perspect.* *58*:9–319 (1984).
28. J.J. Chen and D.W. Gaylor, Carcinogenic risk assessment: Comparison of estimated safe doses for rats and mice, *Environ. Health Perspect.* *72*:305–309 (1987).
29. D.W. Gaylor and R.L. Kodell, Linear extrapolation Algorithm for low dose risk assessment of toxic substances, *J. Environ. Pathol. Toxicol.* *4*:305–312 (1980).
30. U.S. EPA, Health assessment document for epichlorohydrin, EPA-600/8-83-03F, National Technical Information Service, Springfield, VA, PB85-132363/AS (1984).
31. B. Metzger, E. Crouch, and R. Wilson, On the relationship between carcinogenicity and acute toxicity, *Risk Anal.* *9*:169–177 (1989).
32. L. Zeise, R. Wilson, and E. Crouch, Use of acute toxicity to estimate carcinogenic risk, *Risk Anal.* *4*:187–198 (1984).
33. C. Whipple, Toxicity and carcinogenicity, *Risk Anal.* *5*:261 (1985).
34. L. Bernstein, L.S. Gold, B.N. Ames, M.C. Pike, and D.G. Hoel, Toxicity and carcinogenic potency, *Risk Anal.* *5*:263–264 (1985).
35. L. Zeise, E.A.C. Crouch, and R. Wilson, Reply to comments: On the relationship of toxicity and carcinogenicity, *Risk Anal.* *5*:265–270 (1985).
36. L. Zeise, E.A.C. Crouch, and R. Wilson, A possible relationship between toxicity and carcinogenicity, *J. Am. Coll. Toxicol.* *5*:137–151 (1986).
37. E. Crouch, R. Wilson, and L. Zeise, Tautology or not tautology? *J. Toxicol. Environ. Health* *20*:1–10 (1987).
38. L. Bernstein, L.S. Gold, B.N. Ames, M.C. Pike, and D.G. Hoel, Some tautologous aspects of the comparison of carcinogenic potency in rats and mice, *Fund. Appl. Toxicol.* *5*:79–86 (1985).
39. J.P. Rieth and T.B. Starr, Experimental design constraints on carcinogenic potency estimates, *J. Toxicol. Environ. Health* *27*:287–296 (1989).
40. J.P. Rieth and T.B. Starr, Chronic bioassays: Relevance to quantitative risk assessment of carcinogens, *Regul. Toxicol. Pharmacol.* *10*:160–173 (1989).
41. B.C. Allen, A.M. Shipp, K.S. Crump, B. Kilian, M.L. Hogg, J. Tudor, and B. Keller, *Investigation of Cancer Risk Assessment Methods*, Research Triangle Institute, U.S. EPA Contract No. 68-01-6807, National Technical Information Service, No. PB88-127113, 1987.
42. K. Crump, B. Allen, and A. Shipp, An investigation of how well human carcinogenic risk from chemical exposures can be predicted by animal data, with emphasis upon selection of dose measure for extrapolation from animals to humans, Twenty-Sixth Hanford Life Sciences Symposium, *Modeling for Scaling to Man: Biology, Dosimetry, and Response*, Richland, WA, October 20–23, 1987.

43. K. Crump, B. Allen, and A. Shipp, Choice of dose measure for extrapolating carcinogenic risk from animals to humans: An empirical investigation of 23 chemicals, *Health Phys. 57*(Suppl. 1):387–393 (1989).
44. B.C. Allen, K.S. Crump, and A.M. Shipp, Correlation between carcinogenic potency of chemicals in animals and humans, *Risk Anal. 8*:531–561 (1988).
45. O.G. Raabe, S.A. Book, and N.J. Parks, Lifetime bone cancer dose-response relationships in beagles and people from skeletal burdens of ^{226}Ra and ^{90}Sr, *Health Phys. 44*(Suppl. 1):33–48 (1983).
46. J.M. Kaldor, N.E. Day, and K. Hemminki, Quantifying the carcinogenicity of antineoplastic drugs, *Eur. J. Can. C., 24*:703–711 (1988).
47. R.L. Dedrick and P.F. Morrison, Carcinogenic potency of alkylating agents in rodents and humans, *Cancer Res. 52*:2464–2467 (1992).
48. U.S. EPA, Technical analysis of new methods and data regarding dichloromethane hazard assessment, EPA/600/8-87/029A (Review Draft, June 1987), National Technical Information Service, Springfield, VA, PB87-228557/AS, 1987.
49. U.S. EPA, Update to the health assessment document and addendum for dichloromethane (methylene chloride): Pharmacokinetics, mechanism of action, and epidemiology, EPA/600/8-87/030A (Review Draft, July 1987), National Technical Information Service, Springfield, VA, PB87-228565/AS, 1987.
50. R.L. Dedrick, K.B. Bischoff, and D.S. Zaharko, Interspecies correlation of plasma concentration history of methotrexate (NSC-740), *Cancer Chemother. Rep. 54*(1):95–101 (1970).
51. K. Schmidt-Nielsen, Energy metabolism, body size, and problems of scaling, *Fed. Proc. 29*:1524–1532 (1970).
52. K. Schmidt-Nielsen, Scaling in biology: The consequences of size, *J. Exp. Zool. 194*:287–308 (1975).
53. K. Schmidt-Nielsen, *Scaling: Why Is Animal Size So Important?* Cambridge University Press, Cambridge, 1984.
54. R.L. Dedrick, Animal scale-up, *J. Pharmacol. Biopharmacol. 1*:435–461 (1973).
55. S.L. Lindstedt and W.A. Calder, Body size, physiological time, and longevity of homeothermic animals, *Q. Rev. Biol. 56*:1–16 (1981).
56. H. Boxenbaum, Interspecies scaling, allometry, physiological time, and the ground plan of pharmacokinetics, *J. Pharmacol. Biopharmacol. 10*:201–227 (1982).
57. H. Boxenbaum, Evolutionary biology, animal behavior, fourth-dimensional space, and the raison d'être of drug metabolism and pharmacokinetics, *Drug Metab. Rev. 14*:1057–1097 (1983).
58. H. Boxenbaum, Interspecies pharmacokinetic scaling and the evolutionary-comparative paradigm, *Drug Metab. Rev. 15*:1071–1121 (1984).
59. H. Boxenbaum, Time concepts in physics, biology, and pharmacokinetics, *J. Pharmacol. Sci. 75*:1053–1062 (1986).
60. J. Mordenti, Man versus beast: Pharmacokinetic scaling in mammals, *J. Pharmacol. Sci. 75*:1028–1040 (1986).

61. S.L. Lindstedt, Allometry: Body size constraints in animal design, *Pharmacokinetics in Risk Assessment: Drinking Water and Health*, Vol. 8 (National Research Council), National Academy Press, Washington, D.C., 1987.
62. E.L. O'Flaherty, Interspecies conversion of kinetically equivalent doses, *Risk Anal. 9*:587–598 (1989).
63. R.J.M. Ings, Interspecies scaling and comparisons in drug development and toxicokinetics, *Xenobiotica 20*:1201–1231 (1990).

20

Modeling Risks from Water Contaminants: The Application of Concentration-Response Models

A. John Bailer

Miami University, Oxford, Ohio, and
National Institute for Occupational Safety and Health, Cincinnati, Ohio

Christopher J. Portier

National Institute of Environmental Health Sciences,
Research Triangle Park, North Carolina

I. INTRODUCTION

Developing exposure standards in the United States has traditionally been divided into two major components: risk assessment and risk management. Risk management covers a broad range of topics, including developing regulatory policies and the overall evaluation of public health. Included in this process is the evaluation of the economic, social, and political consequences of certain regulatory options. Risk management has been discussed in other reports (e.g., Reference 1) and is not discussed further in this chapter.

Risk assessment can be divided into three components: the identification of agents that lead to adverse human and ecosystem responses, the assessment of the level of exposure in a population to these agents, and the evaluation of the relationship between exposure and response. The "responses" range from cancer onset or increases in genetic damage in humans occupationally exposed to a chemical to the loss of diversity in ecosystems following the dumping of a pollutant. The "agent" or hazard ranges from a chemical or mixture of chemicals to excessive noise levels in the workplace or temperature increases in a stream as a result of some industrial process.

The first component of a risk assessment, "hazard identification," is a qualitative determination of possible adverse effect associated with a hazard. This can be based on the chemical properties of the hazard: for example, structure-activity relationships may suggest that an agent could be hazardous because other structurally related agents are known to be hazardous (2). Alternatively, a hazard can be identified by observing adverse health outcomes in a toxicological or epidemiological study. A second component of any risk assessment is to identify the extent of exposure in a population to an identified hazard. The identification of which populations are exposed to the hazard and to what levels or doses of exposure must also be considered (3). The relationship between hazard exposure and adverse health outcome is then addressed in the concentration-response phase of a risk assessment. Given the assessment of exposure and a concentration-response model for an identified hazard, an overall assessment of the level of concern for the hazard can be derived in a "risk characterization," the final phase of a risk assessment.

Many chapters in this book have addressed issues in the identification of a hazard and the assessment of exposure. Discussion in previous chapters ranged from evaluating pharmacokinetic data for exposure assessment (4) to determining total dose (concentration) over various routes of exposure (5,6). This chapter focuses on an overview of the dose-response assessment, highlighting statistical and mathematical details in the context of risk estimation. Dose-response assessment at its most basic level involves the application of mathematical models to experimental or observational results relevant to the end point under study. Many of the techniques used to develop risk estimates are seldom recognized as mathematical models. Even the simplest techniques, such as applying safety factors to no observed adverse effect levels (NOAEL), are in fact a form of applying a mathematical model to the data. Following this overview, a detailed discussion of concentration (dose)-response models is presented in which various responses, dose metrics, and models are considered.

We assume in this discussion that a hazard has been identified and its exposure levels have been assessed. At this stage, a critical assessment of the relationship between exposure and adverse outcome begins.

II. MODELING CONCENTRATION-RESPONSE RELATIONSHIPS

Before discussing concentration-response models, we first consider possible responses and relevant dose scales or metrics for use in risk assessment. Note that *concentration* and *dose* are used interchangeably throughout the text. *Response* and *end point* are also used interchangeably.

A. Thresholds

There are wide variety of end points or responses of interest in risk assessment. At the most general level, a distinction might be made between end points that exhibit threshold responses versus those end points that exhibit nonthreshold responses. A *threshold response* is a response that does not occur unless the concentration (dose) of a hazard exceeds a particular level, the so-called threshold for the hazard. In this context, any dose below the threshold is considered "safe." A *nonthreshold response* is associated with a hazard for which no level of the hazard is considered absolutely safe. In risk assessment, carcinogenesis or mutagenesis are often considered nonthreshold end points. Teratogenesis, systemic toxicity, and acute toxicity are often considered threshold end points.

The concept of dose-response modeling using thresholds receives less and less support as we begin to understand the mechanisms by which agents produce toxic effects. In pharmacology, biochemistry, molecular biology, and other sciences, mathematical modeling focuses on the use of complicated mathematical relationships that allow for nonlinear relationships between dose and response, but not for thresholds. Thus, in discussing the types of models that may be appropriate for a certain toxic response, we focus on both linear and nonlinear models. The nonlinear models are divided into two broad groups: models exhibiting convex nonlinearity (this is similar to a threshold model in which the slope of the response in the low-dose region approaches zero) and models exhibiting concave nonlinearity (this is the opposite model, in which the slope of the dose-response curve increases as a dose of zero is approached). Convex (concave) nonlinear models are sometimes called sublinear (supralinear) models. Examples of concave, convex, and linear curves are presented in Figure 1.

B. Issues in Extrapolation

As noted in other chapters, risk assessment frequently must address extrapolating a dose-response function across species, from high experimental doses to low dose levels, and across different routes of exposure. For the extrapolation across species, a dose metric is chosen that has some realistic biological basis as a measure of the potential for toxic response (in equivalent units of the dose metric) in the two species. This includes such simple measures as daily dose expressed in units of mg/kg/day to a measure as complicated as the steady-state concentration of some protein induced by this compound in both species. The issue of extrapolation to exposure doses outside the range of the data is accomplished by the use of mathematical models like those discussed in detail later in this chapter. Finally, the issue of different routes of exposure can be handled via simple phys-

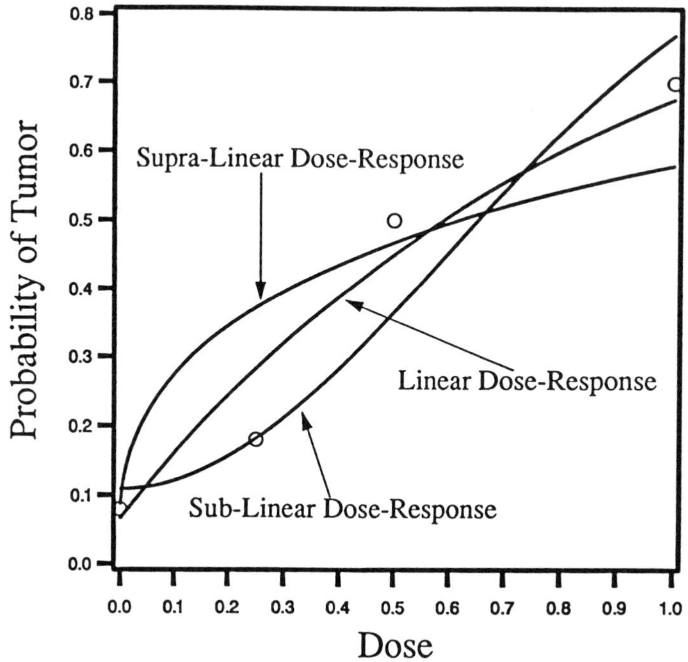

Figure 1 Sublinear, linear, and supralinear dose-response curves.

iological models concerning uptake via the route of exposure. Before we consider these approaches in detail, let us consider the data available to serve as the basis for these extrapolations. In this context, we discuss the gaps or deficiencies commonly encountered when using such data.

1. Available Data

Human epidemiological studies and animal experimental studies are the primary source of data for conducting dose-response studies in the course of a risk assessment. Human studies are typically observational in nature, with no explicit experimental intervention. A common source of such data is an occupational cohort that has been exposed to some hazard in the workplace. These data are especially relevant in that human health can be directly assessed; however, they also experience many of the shortcomings of observational studies, as described in the following section. Animal data are generated most often in controlled experimental settings. Long-term lifetime exposure rodent studies are commonly used to study cancer incidence. For other end points, such as teratogenicity or other systemic effects, shorter studies are conducted.

2. Data Gaps

When the protection of human health is of concern, data on the responses of humans exposed to some hazard is clearly of great interest. This type of data is often unavailable. If it is available, the route of exposure to the hazard may be different from the route of exposure being studied for possible regulation of the hazard. Another problem is that humans are often exposed simultaneously to many potential hazards. Additional problems arise when evaluating exposure to mixtures of hazards (7). Occupational studies must also address the problems related to the worker population not being the most sensitive of all human populations. In addition to all of these difficulties, one of the major drawbacks of the use of epidemiological data in the dose-response assessment of environmental hazards is in the description of the pattern or level of exposure for each human or group of humans. Without accurate exposure information, the development of dose-response models to predict "safe" exposure levels or to predict levels of risk at particular exposure levels may not be justified.

A considerable effort is underway to improve the utility of epidemiological data for risk estimation. Recent work in epidemiology has focused on the measurement of biomarkers in human populations as surrogates of activity in the target cell or the target organ of exposed humans. Commonly used markers have been those that measure the concentrations of a chemical or its metabolites in hair, milk, placenta, amniotic fluid, semen, and fat. Examples of the types of biomarkers being quantified are given in References 8–10.

The use of these biomarkers in risk assessment requires validation of the relationship between the marker and the effect being studied in humans. When extrapolation from an animal species to humans is necessary, validation that the animal marker and the human marker are similarly related to the exposure under study is needed. Other major problems include the existence of susceptible subpopulations of humans and the overall problem of interindividual variability in humans. Current research efforts will help quantify this variability and clarify the implications of interindividual variability on risk estimates (11,12).

In the absence of human data of sufficient quality to develop and fit dose-response models, data from animal experiments are used in the risk assessment modeling exercise. Accurate exposure assessment is seldom an issue because of the laboratory nature of these studies. The uncertainties associated with exposure route (e.g., rodent gavage study versus human inhalation exposure), exposure duration (e.g., lifetime rodent exposure versus intermittent occupational exposure), and site relevance (e.g., dose-related increase in tumors in mouse Zymbal glands versus human lymphoma) are commonly encountered. The relevance of an animal model to human outcome is another common area of concern. Differences in metabolic activity between species, as well as different target sites for toxic effects, must be considered. The relevance of high-dose studies in animal species for low-dose human exposure is also a concern. This is a problem due to possible nonlinearities in the relationship between metabolism at high doses and that at low doses.

As noted earlier, a general strategy that has emerged for addressing many of these concerns is to determine some dose metric that is equivalent between species. In other words, what dose of the hazard in an animal species is equivalent to the dose of the hazard in humans? *Equivalence* means that these doses lead to the same level of risk (probability of adverse response). One strategy for determining equivalent doses is to use physiologically based pharmacokinetic models that explicitly take into account the physiological differences in these species. These models, although better representations of the disposition of toxins, are still subject to uncertainty since many of the model components are unknown and are estimated using allometric formulas estimated from in vitro studies for use in an in vivo model or reflect calibration of the model to fit a particular data set. Additionally, use of these models may introduce even greater variability in the risk assessment process (13). Another strategy is to use information on the relationship between the administered dose and certain metabolites to estimate an internal dose. This is relevant if the toxicity of a hazard arises from its metabolites. Another dose metric may be found when the toxic effect is the result of receptor binding to nuclear or microsomal proteins to alter cell homeostasis (14,15). Finally, dose metrics can be constructed for other biomarkers, including DNA adducts (16).

The implications of the choice of dose metric for extrapolating to low-dose risks are illustrated in Figure 2. In Figure 2, three different dose metrics have been applied to data on the carcinogenicity of 2,3,7,8-tetrachlorodibenzo-*p*-dioxin (TCDD) (17,18). The dose curve refers to the application of a one-stage model of carcinogenesis (see later) to these data using the administered dose metric. The CYP1A1 curve is the application of a dose metric (the steady-state concentration of the microsomal protein cytochrome P450 1A1) other than the administered dose (15) and a one-stage model of carcinogenesis to these tumor incidence data. The third is the use of another dose metric, CYP1A2 (the steady-state concentration of the microsomal protein cytochrome P450 1A2), and the one-stage model. It can be seen that in the observable response range (Figure 2, left) the

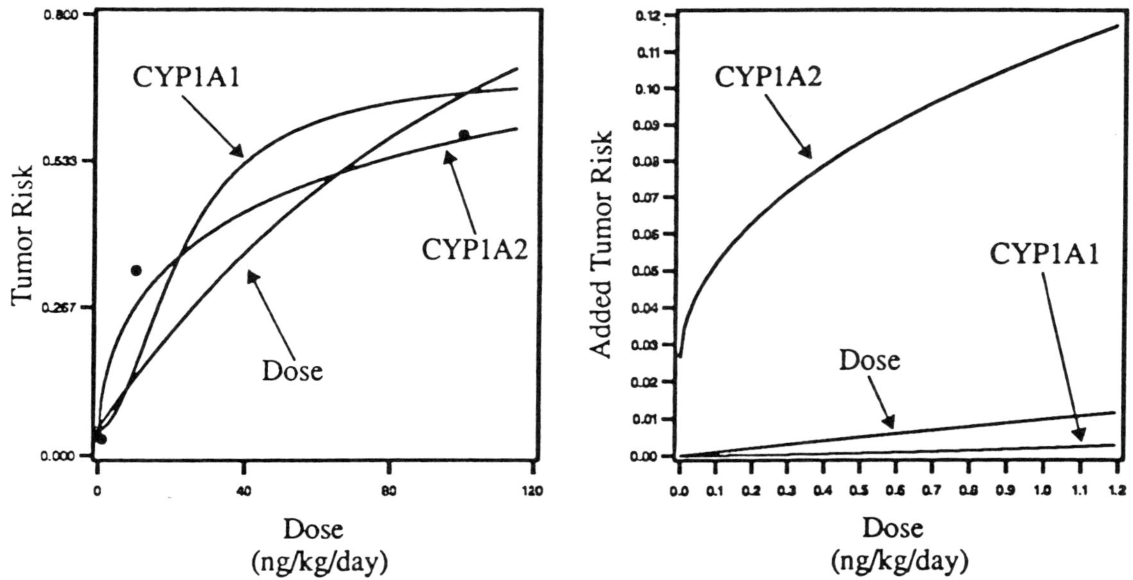

Figure 2 Impact of the selection of a dose metric on extrapolated low-dose risk estimates. Three model fits to data from an experiment examining TCDD-induced tumorigenicity are displayed in the two plots. Observed tumor incidence is plotted as points on a plot on the left. The three curves represent one-stage model fits using the administered dose scale (dose) and two internal dose metrics (CYP1A1 and CYP1A2). The plot on the right represents model fits in the low-dose region.

three models all adequately describe these data, the two dose metrics doing slightly better than the administered dose metric. However, in the low-dose region (Figure 2, right) the behavior of the three models is considerably different, the models incorporating CYP1A1 and CYP1A2 as dose surrogates giving orders of magnitude differences in risk for specified small doses. The reason for these large differences is that the model using CYP1A1 is convex (threshold looking) and that using CYP1A2 is concave in the low-dose region. The model utilizing the administered dose metric is linear.

All these uncertainties can be reduced with greater understanding of the mechanism by which hazards induce their toxic effects. This understanding will lead to dose-response models that reflect the mechanistic process by which a hazard acts.

C. Common Models

Models should reflect the various responses to be evaluated and, preferably, be based upon some mechanistic understanding of the process being modeled. In this section, we first introduce notation that will be used when formally presenting dose-response models. For risk assessment modeling, there has been very little use of mechanistic modeling, especially for noncancer end points. Thus, our discussion of mechanism is limited. We divide our discussion of risk assessment dose-response models into subsections describing models that are commonly applied for threshold (or noncancer) responses and nonthreshold (or cancer) responses.

1. Notation

We first introduce some basic notation. Let d_i correspond to the dose or concentration administered (or the dose in terms of some calculated or constructed metric) in the ith treatment group ($i = 1, \ldots, k$). Typically, d_1 is a zero-dose control, and d_k corresponds either to a maximally tolerated dose (MTD) or to the highest concentration in a serial dilution experiment. For a given subject or animal in each treatment group, a variety of responses may be measured (as discussed previously).

In carcinogenicity studies, the presence or absence of tumors of a particular type, along with a measure of survival time, is usually recorded for each experimental unit. In cancer response models, we describe models relating the dose d_i to the probability of tumor onset by study termination, denoted $P(d_i)$, or by some time t, denoted $P(d_i, t)$. For noncancer studies, we analyze the relationship between some response Y_i and the dose d_i.

2. Noncancer Response Models

Noncancer or threshold responses generally have been modeled by two methods. The most common and historically most frequently used method has been an uncertainty or safety factor approach in which an estimate of the threshold has been modified to reflect degrees of uncertainty using so-called safety or uncertainty factors. The second approach, the so-called benchmark dose approach, derives dose estimates from a dose-response model that then may be modified by safety factors. These two approaches are now discussed in turn.

a. Safety Factor

The safety factor approach is based on a fairly simple experimental idea. First, determine the largest experimental dose that is not significantly different from the zero-dose control group. This experimental dose is often called the no observed adverse effect level. The NOAEL represents an estimate of the threshold for a particular hazard. The next step is to modify the NOAEL to reflect various levels of uncertainty. If the NOAEL is based on experimental animal studies, one uncertainty factor (UF) might reflect that responses in the human population may be more variable than responses in inbred animal populations. Another UF might reflect the uncertainty in extrapolating from animals to humans. Typically, a factor of 10 is associated with each UF. Experimental justification for the choice of these values is discussed by Dourson and Stara (19). The NOAEL is then divided by the product of the UF to construct a reference dose (RfD) or concentration (RfC).

The construction of the RfD or the RfC is commonly encountered in noncancer quantitative risk assessment (20,21). The noncancer end points analyzed include neurotoxic or teratogenic effects. An end point with a similar basis for construction is encountered in aquatic toxicology testing of effluents. The chronic value (ChV) is viewed as a point estimate of the presumably safe (no effect) concentration or threshold of an aquatic hazard (22). The ChV is constructed as the geometric mean between the no observed effect concentration (NOEC) and the lowest observed effect concentration (LOEC).

The RfD and the ChV are both sensitive to experimental design issues. The NOAEL, NOEC, and LOEC are all experimental levels. Thus, the RfD and ChV are constrained to simple functions of the experimental levels. Thus, the experimental design (number of doses or concentrations, spacing of doses, and replications at each dose group) is a critical determinant of these quantities (see, e.g., Reference 23). In addition, the RfD and ChV do not account for information on the shape or form of the concentration-response relationship. For both these measures, it is impossible to state the level of risk that might be associated with exposure to a hazard at concentration levels that exceed the RfD or ChV. It is also possible to manipulate the experimental design to achieve a particular RfD or ChV. For example, reducing the sample size at a particular concentration could change a response from statistically significant to nonsignificant. These difficulties have led to alternative methods for examining concentration-response relationships for noncancer risk assessment.

b. Benchmark Dose

The benchmark dose was proposed as an alternative to the NOAEL for use in the calculation of a reference dose (24). The benchmark dose (BD) is based on first fitting a dose-response model to experimental data in which the excess risk of an adverse outcome is modeled as a function of dose. An upper bound is derived for this mathematical relationship. Given a specified level of excess risk (e.g., 10% above background), the lower confidence limit on the dose associated with this excess risk level can be obtained from the upper bound estimate of the dose-response relationship. This derived quantity is the benchmark dose. Unlike the use of NOAEL in the calculation of a RfD, the BD uses the dose-response relationship (and hence uses more data). This approach has its limitations for risk assessment. Risk estimates are obtained by applying safety factors to some high level of risk estimated from the model. Also, the form of the model has an impact on the BD. Finally, the choice of the amount of excess risk used to derive the BD is arbitrary and can influence the exposure standard.

c. Aquatic Toxicology Risk Assessment Alternatives

Alternatives to safety factor approaches for assessing the toxicity of effluents have also been proposed. Norberg-King (25) suggested an alternative to the ChV for examining reproductive toxicity in a *Ceriodaphnia dubia*. This alternative is the 50% inhibition concentration (IC_{50}), which corresponds to the concentration of a toxin required to reduce the number of young produced in three broods to 50% of the control number. This method proceeds by identifying the concentration with average total young immediately above 50% of the control reproduction rate and the concentration immediately below 50% of the control rate. The IC_{50} is calculated using linear interpolation between these two concentrations. Inhibition concentration for other levels of inhibition are calculated similarly. A dose-response model-based analog of the IC_{50} has been proposed by Bailer and Oris (26). They suggest modeling the number of young produced as an exponential function of some polynomial of the concentrations, such as $Y_i = e^{b_0 + b_1 d_i + b_2 d_i^2} +$ error. As an initial suggestion, these authors propose using a Poisson model for the response distribution. An estimate of the concentration that leads to a specified level of productive inhibition can be obtained from the parameter estimates of b_1 and b_2. The extension of this work to low-dose risk estimation has not been addressed.

d. *Mechanistic Models*

To improve methods for risk estimation for noncancer end points, there must be a concerted effort to develop mechanistic models. This is not a task for the mathematician alone since it involves the development of novel bioassay designs and the collection of multiple types of data to derive a mechanism and then estimate parameters in any model that might result from the studies. Until such developments, we must rely upon the statistical models discussed earlier.

Several issues from the modeling discussed earlier can be studied in greater detail. The first is the statistical sensitivity of the approach to the design of the study. This has been studied for threshold approaches in teratology (27) but needs to be discussed for other end points. Also, the basic assumption that these end points generally follow a nonlinear response needs additional verification. For teratology, a linear concentration-response relationship may be appropriate for many compounds (28). Other end points could be studied in a similar fashion.

3. Cancer Response Models

Cancer response models are commonly applied in risk assessments, and many comprehensive surveys of these methods have been published (see, e.g., Reference 29). We review the common classes of dose-response models for carcinogenicity data (tolerance distribution models and multihit or multistage models) and conclude this section with a discussion of mechanistic models that are now being applied with greater frequency in risk assessment.

a. *Safety Factor: Threshold*

The safety factors approach and other methods for evaluating noncancer responses could be applied to carcinogenic end points. This is rarely, if ever, done in the United States since it is generally believed that any amount of a carcinogenic hazard has a nonzero probability of inducing cancer onset. However, other countries routinely apply safety factors in their cancer risk assessments, often ending with estimates of "safe" exposure levels that are orders of magnitude larger than estimates derived by U.S. agencies. There is a growing belief that carcinogenic compounds can be grouped into two broad classes, initiators and promoters, with safety factors applied to promoters (30). There is little evidence to support this approach (31,32), however, and several examples run counter to this observation (e.g., Reference 15). Thus, this issue needs additional research before it can be resolved.

b. *Tolerance Distribution Models*

It is assumed in the models that follow that there exists a minimum dose, say d_T, such that any exposure above this dose induces tumor formation

(d_T is the so-called threshold). In addition, this threshold varies over the population of individuals at risk (implying biological variability among individuals). The models in this class differ in terms of which probability distribution is used to describe the variability in thresholds in the population.

Probit or log-normal model. As the name of this model suggests, it is assumed that thresholds in the population follow a log-normal distribution. This implies that the probability of response (e.g., tumor onset) can be written as

$$P(d) = \Phi\frac{\log d - \mu}{\sigma}$$

where Φ is the cumulative normal probability. This relationship is often expressed in terms of a probit transformation:

$$Y(d) = \Phi^{-1}(P(d)) = \frac{\log d - \mu}{\sigma} = a + b \log d$$

Logistic model. A second commonly used threshold model assumes that the population threshold distribution is logistic. This written in our notation as

$$P(d) = \frac{1}{1 + e^{-\log\theta_0 - \theta_1 \log d}}$$

where θ_0, $\theta_1 > 0$. This logit transformation of the response probability is linearly related to the log of the dose:

$$Y(d) = \log\frac{P(d)}{1 - P(d)} = \log\theta_0 + \theta_1 \log d$$

Probit and logistic curves differ in that the probit curve approaches zero much faster than the logistic curve. Both curves are based on a simplistic, deterministic threshold assumption (deterministic for each person but varying among persons). These models can be easily implemented in most common statistical computer packages.

c. *Multistage and Multihit Models*

Hit theory models. The hit theory class of models is based on the assumption that cells experience transforming and irreversible events—*hits*. When sufficient hits are experienced, the carcinogenic process continues independently of dose. The motivation for these models is the interaction between particles and target sites necessary to induce a response.

The single-hit theory suggests that a carcinogenic response can be induced after a single susceptible target is hit by a single biologically effective unit of dose. If we assume constant exposure at dose rate d for duration

t, then the probability of carcinogenic response in the single-hit model can be written as

$$P(d) = 1 - e^{-\alpha t d}$$

where $\alpha > 0$. This expression is often reparameterized in terms of $\theta = \alpha t$ since θd can be interpreted as the expected number of hits by time t at dose d. The reparameterization leads to the formula

$$P(d) = 1 - e^{-\theta d}$$

Note that $P(d)$ is approximately equal to θd for small d (i.e., linear response in the low-dose region).

The gamma multihit model generalizes the single-hit model by requiring multiple hits on a single susceptible target to induce tumor formation. An additional parameter is needed to represent the minimum number of hits a target requires to obtain a carcinogenic response.

The multihit hit model is derived by assuming that the probability of observing x hits on the susceptible target follows a Poisson distribution with rate parameter equal to θd; that is, for $x = 0, 1, 2, \ldots,$

$$\Pr(x \text{ hits}) = \frac{e^{-\theta d}\,(\theta d)^x}{x!}$$

Suppose m hits are required to induce a carcinogenic response. Then the probability of a carcinogenic response is simply the probability of observing m or more hits on the target. This can be written as

$$\Pr(\text{response}) = \Pr(\text{at least } m \text{ hits}) = \sum_{x=m}^{\infty} \frac{e^{-\theta d}\,(\theta d)^x}{x!} = \int_0^{\theta d} \frac{y^{m-1}\,e^{-y}\,dy}{(m-1)!}$$

From this formula it can be seen that gamma tolerance distribution is implicitly associated with a multihit model. This interpretation does not require m to be an integer.

Multihit models are generally no longer used in risk assessment. The basic reason is that they are highly nonlinear in the low-dose range. In many cases, this results in these models being unable to fit the available tumor incidence data. However, recent research suggested that the multihit theory of carcinogenesis still has mechanistic support in basic biology (33) and should not be abandoned. Instead, this class of models should be incorporated into the most commonly applied class of models, the multistage models of carcinogenesis.

Multistage models. Multistage models are similar to multihit models in that probabilistic models are suggested to reflect various mathematical models of the carcinogenic process. These models start with the assumption that cancer is initiated after a certain number, say k, of heritable changes have been induced. Usually it is assumed that a system starts with all normal cells and that a specific order for progress of cellular change is required for the onset of tumor formation. Also, it is generally assumed that the time until the formation of a detectable tumor after a cell progresses through all k stages is "short" (relative to life span). If a chemical can have a linear effect on the transition rate between successive stages, then the probability of tumor onset by some time t can be written as

$$P(d) = 1 - e^{-\prod_{i=1}^{k}(\alpha_i + \beta_i d)t}$$

with α_i, $\beta_i \geq 0$, or as

$$P(d) = 1 - e^{-\sum_{i=0}^{k}(\theta_i d^i)t^k}$$

with $\theta_i \geq 0$. The exponent in the second $P(d)$ formula is often written as a polynomial of dose:

$$\sum_{i=0}^{k} q_i d^i \tag{34}$$

where $q_i = \theta_i t^k$

The multistage model is approximately linear in low doses, but the multihit model allows low-order polynomial terms in low doses. It is believed that the low-dose linearity of the multistage model provides a degree of "conservativeness" in the low-dose risk extrapolation process. This explains, in part, the popularity of the multistage model for risk assessment purposes.

d. Time to Response Models

The general idea in time to response models is that the time to the occurrence of the response of interest provides additional information that allows a "clearer" definition of the dose-response relationship. A benefit of such a model formulation is that the expected tumor prevalence at various points of time in some population at risk can be estimated. These models are comprised of two components, including some probability distribution of response times (e.g., written as a cumulative probability distribution function $F(t) = \Pr(T \leq t)$, where T is a random variable denoting the time until tumor onset), and a mathematical model relating dose level to response time. A common probability distribution for response times is the Weibull distribution. This probability distribution has been observed to provide a reasonable description for the distribution of tumor onset in a large historical control data base of animal carcinogenicity studies (35). In this model, the probability of tumor onset before time t is written as

$$P(t) = F(t) = 1 - e^{-\theta_1 t^{\theta_2}}$$

If θ_1 is related to dose raised to some power (i.e., $\theta_1 = \alpha d^{\beta}$) and θ_2 is independent of dose, then the probability of tumor onset before time t for experimental units receiving dose d is

$$P(d) = 1 - e^{-\alpha d^{\beta} t^{\theta_2}}$$

Another time to response model that has been frequently used in risk assessment is the Hartley-Sielkin model (36). The original formulation of the multistage model of carcinogenesis is similar to the Hartley-Sielken model.

e. Models Including Cellular Proliferation

With few exceptions, the mathematical modeling of carcinogenesis has concentrated on the use of multistage models. The first practical application of this class of models was done by Armitage and Doll (37). The Armitage-Doll (AD) model of carcinogenesis assumes that cells progress from a state in which they display normal cellular function into a malignant state via a fixed number of stages that must occur in a particular sequence. For this model adequately to describe cancer data, this progression generally required between four and seven stages. Each stage represents an additional critical mutation in the genome, the final mutation leading to the formation of a single malignant cell that progresses rapidly to an observable neoplasm. The model was applied in a variety of contexts, including the description of early- and late-stage carcinogens (38), the design and analysis of carcinogenicity experiments (39,40), and the estimation of low-dose cancer risks (41).

Several researchers (42–44) noted a major inadequacy in the theoretical basis of the AD model. The AD model failed to take into account the growth kinetics of cell populations. A new model, the two-stage (TS) model of carcinogenesis, was developed that allows for the birth and death or differentiation of cells. Specifically, cells progress through a first stage in which normal cells experience one critical mutation before a second stage in which a second mutation occurs, leading to the formation of malignancies. Both types of cells, normal cells and initiated cells (the cells that have experienced the first mutational event), undergo a birth-death process that allows the population of cells in a particular stage either to expand in number through mitosis or to reduce in number through cell death or terminal differentiation. The TS model has been applied to tumor incidence data, capturing features of the data the AD model failed to describe (44,45). Newer techniques in toxicology, cell biology, biochemistry, and other fields allow the application of the TS model to a much broader range of experimental data (46,47). When properly applied, this model has done an adequate job of describing the experimental results, with few exceptions (48).

The advent of newer information on carcinogenic mechanisms and the availability of faster computers has led to the creation of more complicated mathematical models of carcinogenesis (33,49). This will continue to be the case in the future.

To date, few risk assessments have applied these multistage models of carcinogenesis to include clonal expansion. There is an ongoing effort to support their application in this area. Like all other models used in risk assessment, these models also have their limitations (50). However, with the ability to incorporate mechanistic information and multiple forms of data directly, it is expected that these models will be more appropriate than nonmechanistic models for risk assessment.

f. Low-Dose Extrapolation and Model Selection

Low-dose extrapolation refers to a common problem in risk assessment. Often the dose-response modeling uses experimental studies in which the experimental doses far exceed doses experienced in the environment or in the workplace. High experimental doses are selected to increase the sensitivity of the hypothesis tests to detect a significant dose-response relationship if it is present. The problem encountered in the low-dose extrapolation exercise is that many models fit data equally well but lead to low-dose estimates that may differ by several orders of magnitude. In fact, the further the extrapolation, the more divergent the various models become. In a modeling study of 19 different data sets, Krewski and Van Ryzin (51) observed the following low-dose risk ordering:

Log-normal $<$ logistic and multihit $<$ multistage $<$ one hit

We see that the linear model in low doses (one-hit or multistage) provides an upper bound on the true dose-response (conservative estimate). This occurs because the dose-cumulative incidence curve is S shaped, so the response curve is bounded above by a line in the low-dose region. This is not necessarily true in general; this is discussed and illustrated by Bailar et al. (52). Other factors can severely impact the dose-response shape in the low-dose region, such as pharmacokinetics (53) and receptor binding (15). A carefully constructed extrapolation to low-dose risks must consider all sources of nonlinearity.

III. DISCUSSION AND FUTURE DIRECTIONS

We have reviewed some of the quantitative issues arising in risk assessments, focusing on the many dose-response models that are commonly used. Most of these models are at best gross depictions of the underlying process by which the toxic response is generated. Basic biological research clarifying the biological mechanisms associated with the toxic response will be of great importance in determining the direction of the field of risk assessment. The greater use of pharmacokinetic models for dose calculations and the multistage models of carcinogenesis for dose-response modeling are testaments to this trend.

ACKNOWLEDGMENTS

We thank Drs. Beth Gladen, Kristen Meier, James Oris, and Steve Simon for their careful reading of early versions of this manuscript.

REFERENCES

1. National Academy of Sciences, *Risk Assessment in the Federal Government: Managing the Process*, National Academy Press, Washington, D.C., 1983.
2. J. Ashby and R.W. Tennant, Chemical structure, *Salmonella* mutagenicity and extent of carcinogenicity as indicators of genotoxic carcinogenesis among 222 chemicals tested in rodents by the U.S. NCI/NTP, *Mutat. Res. 204*:17–115 (1988).
3. U.S. Environmental Protection Agency, Guidelines for exposure assessment, *Federal Register 57*:22888–22938 (1992).
4. J. Blancato and N. Chiu, Use of pharmacokinetic models to estimate internal dose from exposure, *Water Contamination and Health: Integration of Exposure Assessment, Toxicology, and Risk Assessment* (R. Wang, ed.), Marcel Dekker, New York, 1994, pp. 217–239.
5. R. Wester and H. Maibach, Percutaneous absorption of chemicals from water simulating swimming and bathing and from vapor exposure, *Water Contamination and Health: Integration of Exposure Assessment, Toxicology, and Risk Assessment* (R. Wang, ed.), Marcel Dekker, New York, 1994, pp. 149–165.
6. R. Wang and R. Wester, Establishing concepts and developing methodologies in dealing with various routes of exposure to water contaminants: dose versus health risk, *Water Contamination and Health: Integration of Exposure Assessment, Toxicology, and Risk Assessment* (R. Wang, ed.), Marcel Dekker, New York, 1994, pp. 375–391.
7. U.S. Environmental Protection Agency, Guidelines for health risk assessment of chemical mixtures, *Federal Register 51*:34014–34025 (1986).
8. M.A. Fingerhut, W.E. Halperin, D.A. Marlow, L.A. Piacitelli, P.A. Honchar, M.H. Sweeney, A.L. Greife, P.A. Dill, K. Steenland, and A.J. Suruda, Cancer mortality in workers exposed to 2,3,7,8-tetrachlorodibenzo-p-dioxin, *N. Engl. J. Med. 324*:212–218 (1991).
9. F. Perera, R. Santella, H. Fischman, A. Munshi, M. Poirier, D. Brenner, H. Mehta, and J. Van Ryzin, DNA adducts, protein adducts ans sister chro-

matid exchange in cigarette smokers and non-smokers, *J. Natl. Cancer Inst.* *79*:449–456 (1987).
10. T. Wong, T. Sloop, and G. Lucier, Nondectable concentrations of human placental Ah receptor are associated with potent induction of microsomal BaP hydroxylase in individuals exposed to polychlorinated biphenyls, quarterphenyls and dibenzofurans, *Toxicol. Appl. Pharmacol.* *85*:60–68 (1986).
11. G. Lucier and C. Thompson, Issues in biochemical applications to risk assessment: When can lymphocytes be used as surrogate markers? *Environ. Health Perspect.* *76*:187–191 (1987).
12. F. Perera, The potential usefulness of biological markers in risk assessment, *Environ. Health Perspect.* *76*:141–145 (1987).
13. C. Portier and N. Kaplan, The variability of safe dose estimates when using complicated models of the carcinogenic process. A case study: Methylene chloride, *Fund. Appl. Toxicol.* *13*:533–544 (1989).
14. M. Kohn, G. Lucier, G. Clark, C. Sewall, A. Tritscher, and C. Portier, A mechanistic model of the effects of dioxin on gene expression in the rat liver, *Toxicol. Appl. Pharmacol.* *120*:138–154 (1993).
15. C. Portier, A. Tritscher, M. Kohn, C. Sewall, G. Clark, L. Edler, D. Hoel, and G. Lucier, Ligand/receptor binding for 2,3,7,8-TCDD: Implications for risk assessment, *Fund. Appl. Toxicol.* *20*:48–56 (1993).
16. S.A. Belinsky, C.M. White, T.R. Devereux, and M.W. Anderson, DNA adducts as a dosimeter for risk estimation, *Environ. Health Perspect.* *76*:3–8 (1987).
17. R.J. Kociba, D.G. Keyes, J.E. Beyer, R.M. Carreon, C.E. Wade, D.A. Dittenber, R.P. Kalnins, L.E. Frauson, C.N. Park, S.D. Barnard, R.A. Hummel, and C.G. Humiston, Results of a two-year chronic toxicity and oncogenicity study of 2,3,7,8-tetrachlorodibenzo-p-dioxin in rats, *Toxicol. Appl. Pharmacol.* *46*:279–303 (1979).
18. R.M. Sauer, Pathology working group 2,3,7,8-tetrachlorodibenzo-*p*-dioxin in Sprague-Dawley rats, Pathco Incorporated Technical Report, Ijamsville, MD, 1990.
19. M.L. Dourson and J.F. Stara, Regulatory history and experimental support of uncertainty (safety factors), *Regul. Toxicol. Pharmacol.* *3*:224–238 (1983).
20. D.G. Barnes and M. Dourson, Reference dose (RfD): Description and use in health risk assessments, *Regul. Toxicol. Pharmacol.* *8*:471–486 (1988).
21. C.A. Kimmel, Quantitative approaches to human risk assessment for noncancer health effects, *Neurotoxicology* *11*:189–198 (1990).
22. C.I. Weber, W.H. Peltier, T.J. Norberg-King, W.B. Horning, F.A. Kessler, J.R. Menkedick, T.W. Neiheisel, P.A. Lewis, D. Klemm, Q.H. Pickering, E.L. Robinson, J.M. Lazorchak, L.J. Wymer, and R.W. Freyberg, Short-term methods for estimating the chronic toxicity of effluents and receiving waters to freshwater organisms, 2nd ed., EPA/600/4-89/001A, U.S. Environmental Protection Agency, Cincinnati, OH, 1989.

23. J.T. Oris and A.J. Bailer, Statistical analysis of the *Ceriodaphnia* toxicity test: Sample size determination for reproductive effects, *Environ. Toxicol. Chem. 12*:85–90 (1993).
24. K. Crump, A new method for determining allowable daily intakes, *Fund. Appl. Toxicol. 4*:854–871 (1984).
25. T.J. Norberg-King, An interpolation estimate for chronic toxicity: The ICp approach, National Effluent Toxicity Assessment Center Technical Report 05-88, U.S. Environmental Protection Agency, Environmental Research Laboratory-Duluth, Duluth, MN, 1988.
26. A.J. Bailer and J.T. Oris, Modeling reproductive toxicity in *Ceriodaphnia* tests, *Environ. Toxicol. Chem. 12*:787–791 (1993).
27. N. Kaplan, M. Hogan, C. Portier, and D. Hoel, An evaluation of the safety factor approach in risk assessment, *Developmental Toxicology: Mechanisms and Risk* (J. McLachlan, ed.), Cold Spring Harbor Laboratory, Banbury Report 26, Cold Spring Harbor, NY, 1988, pp. 335–346.
28. G. Carr and C. Portier, An evaluation of the Rai and Van Ryzin dose-response model in teratology, *Risk Anal. 11*:111–120 (1991).
29. C. Brown and J. Koziol, Statistical aspects of the estimation of human risk from suspected environmental carcinogens, *SIAM Rev. 25*:151–181 (1983).
30. R. Kroes, Contributions of toxicology towards risk assessment of carcinogens, *Arch. Toxicol. 60*:224–228 (1987).
31. R.P. Perera, The genotoxic/epigenetic distinction: Relevance to cancer policy, *Environ. Res. 34*:175–191 (1984).
32. D. Hoel and C. Portier, Nonlinearity of carcinogenesis dose-response functions, *Environ. Health Perspect.* (in press).
33. C. Portier, A Kopp-Schneider, and C. Sherman, Using cell replication data in mathematical modelling in carcinogenesis, *Environ. Health Perspect.* (in press).
34. K.S. Crump, H.A. Guess, and K.L. Deal, Confidence intervals and tests of hypotheses concerning dose response relations inferred from animal carcinogenicity studies, *Biometrics 33*:342–451 (1977).
35. C. Portier, J. Hedges, and D. Hoel, Age-specific models of mortality and tumor onset for historical control animals in the National Toxicology Program's carcinogenicity experiments, *Cancer Res. 46*:4372–4378 (1986).
36. R.L. Sielken, Use of the Hartley-Sielken model in low-dose extrapolation, *Toxicological Risk Assessment*, Vol. I (D.B. Clayson, D. Krewski, and I. Munro, eds.), CRC Press, Boca Raton, FL, 1985, pp. 165–186.
37. P. Armitage and R. Doll, The age distribution of cancer and a multistage theory of cancer, *Br. J. Cancer 8*:1–12 (1954).
38. N. Day and C. Brown, Multistage models and primary prevention of cancer, *J. Natl. Cancer Inst. 64*:977–989 (1980).
39. C. Portier and D. Hoel, Optimal design of the chronic animal bioassay, *J. Toxicol. Environ. Health 12*:1–19 (1983).
40. C. Portier and D. Hoel, Design of animal carcinogenicity studies for goodness-of-fit of multistage models, *Fund. Appl. Toxicol. 4*:949–959 (1984).

41. H. Guess, K. Crump, and R. Peto, Uncertainty estimates for low-dose-rate extrapolations of animal carcinogenicity data, *Cancer Res. 37*:3475–3483 (1977).
42. P. Armitage and R. Doll, A two-stage theory of carcinogenesis in relation to the age distribution of human cancer, *Br. J. Cancer 11*:161–169 (1957).
43. J. Neyman and E. Scott, Statistical aspects of the problem of carcinogenesis, Fifth Berkeley Symposium on Mathematical Statistics and Probability, University of California Press, Berkeley, CA, 1967, pp. 745–776.
44. S. Moolgavkar and D. Venzon, Two-event models for carcinogenesis: Incidence curves for childhood and adult tumors, *Math. Biosci. 47*:55–77 (1979).
45. S. Moolgavkar, N. Day, and R. Stevens, Two-stage model for carcinogenesis: Epidemiology of breast cancer in females, *J. Natl. Cancer Inst. 65*:559–569 (1980).
46. S.H. Moolgavkar, E.G. Luebeck, M. de Gunst, R.E. Port, and M. Schwarz, Quantitative analysis of enzyme-altered foci in rat hepatocarcinogenesis experiments. I. Single agent regimen, *Carcinogenesis 11*:1271–1278 (1990).
47. A. Kopp-Schneider and C.J. Portier, Birth and death/differentiation rates of papillomas in mouse skin, *Carcinogenesis 13*:973–978 (1992).
48. C. Portier and A.J. Bailer, Two-stage models of carcinogenesis for historical control animals from the National Toxicology Program, *J. Toxicol. Environ. Health 27*:21–45 (1989).
49. C.J. Portier and A. Kopp-Schneider, A multistage model of carcinogenesis incorporating DNA damage and repair, *Risk Anal. 11*:535–543 (1991).
50. C.J. Portier, Statistical properties of a two-stage model of carcinogenesis, *Environ. Health Perspect. 76*:125–131 (1987).
51. D. Krewski and J. Van Ryzin, Dose response models for quantal response toxicity data, *Current Topics in Probability and Statistics* (M. Csörgö, D. Dawson, J.N.K. Rao, and E. Saleh, eds.), North-Holland, New York, 201–231 (1981).
52. J. Bailar, E. Crouch, D. Spiegelman, and R. Shaikh, One-hit models of carcinogenesis: Conservative or not? *Risk Anal. 8*:485–497 (1988).
53. D. Hoel, N. Kaplan, and M. Anderson, Implication of nonlinear kinetics on risk estimation in carcinogenesis, *Science 219*:1032–1037 (1983).

21

Problematic Risk Assessments for Drinking Water Contaminants: Selenium, Aldicarb, and Nickel

Susan F. Velazquez and Kenneth A. Poirier

U.S. Environmental Protection Agency, Cincinnati, Ohio

I. INTRODUCTION

As mandated by the Safe Drinking Water Act, the U.S. Environmental Protection Agency (EPA) is charged with setting national drinking water standards for chemicals with known or anticipated occurrence that may pose a threat to human health. This includes many different types of contaminants, including inorganics (both those that are purely toxic, such as lead, and those that are essential at low doses yet toxic at high doses, such as selenium), radioisotopes, volatile organics, aromatic hydrocarbons, and various classes of pesticides. The process of performing risk assessments of chemicals of such different natures and with such varying data bases is inherently fraught with uncertainty. To provide consistent and scientifically defensible risk assessments, the U.S. EPA has developed strategies for both cancer and noncancer risk assessments. This chapter provides a brief overview of the noncancer risk assessment process (described in more detail elsewhere in this book), followed by three examples of chemicals that identify particular aspects of this process that must be addressed on an individual chemical basis: selenium, aldicarb, and nickel.

The risk assessment process consists of four components: hazard identification, dose-response assessment, exposure assessment, and risk characterization (1). Although elements from each of these four components may be identified, the discussion on problematical risk assessments is focused primarily on the step of dose-response assessment. For such an assessment, data from humans (epidemiological, case studies and others) or laboratory animals may be used. Data from human studies are indisputably preferred because an interspecies extrapolation need not be made. However, adequate human data are often not available. For these chemicals, the U.S. EPA relies on studies in laboratory animals for performing a dose-response assessment. For the three risk assessments described in this chapter, data from both humans and laboratory animals were integral components of the final quantitative risk estimate.

In the quantification of noncarcinogenic effects, a reference dose (RfD) is calculated. The RfD is an estimate (with uncertainty spanning perhaps an order of magnitude) of a daily exposure to the human population (including sensitive subgroups) likely to be without an appreciable risk of deleterious health effects during a lifetime. The RfD is generally derived from a no observed adverse effect level (NOAEL) or lowest observed adverse effect level (LOAEL), identified from a subchronic or chronic study and divided by an uncertainty factor(s) (UF) and a modifying factor (MF). The RfD is expressed in mg/kg body weight (BW) per day, and the calculation is as follows:

$$\text{RfD} = \frac{\text{NOAEL or LOAEL}}{\text{UF} \times \text{MF}}$$

Selection of the UF to be employed in the calculation of the RfD is based upon professional judgment considering the entire data base of toxicologic effects for the chemical. To ensure that UF are selected and applied in a consistent manner, the U.S. EPA employs a modification to the guidelines proposed by the National Academy of Sciences (2,3). In short, UF are applied to account for extrapolation of data from animals to humans, to account for the variation in sensitivities of humans, to extrapolate from a subchronic study to a lifetime exposure, to extrapolate from a LOAEL to a NOAEL, and to account for an incomplete data base. These UF are generally 10-fold each, but when the total UF exceeds 1000, more than one area of uncertainty may be expressed by a single factor of 10. In addition to the total uncertainty factor, a modifying factor (no greater than 10; default = 1) may be used in the risk assessment to account for uncertainties that are not otherwise explicitly covered.

As illustrated in the examples to follow, the RfD for a chemical is often calculated using data from a single study that identifies an appropriate toxicological end point and demonstrates a dose-response relationship, preferably including a NOAEL and LOAEL. However, it is important to understand that many other studies usually support the RfD, both qualitatively and quantitatively. The RfD is generally conservative, based upon the study that identifies the *critical effect*, which is the effect that occurs at the lowest dose of administered compound. However, as illustrated later for nickel, there are cases in which humans with a documented hypersensitivity to a chemical may not be adequately protected. This is particularly true of sensitizing metals, such as nickel, chromium, and cobalt. Furthermore, although exposures at or below the RfD should not categorically be considered safe, exposures above the RfD should likewise not be considered unsafe. Other factors must be taken into consideration in making such a judgement; these include the degree of imprecision of the RfD (those with a larger UF inherently having less precision) and possible differences in bioavailability of a chemical in different media.

Risk assessments that have been agreed upon by a consensus of the U.S. EPA RfD/RfC Work Group may be accessed through an on-line data base termed the Integrated Risk Information System (IRIS) (4). The risk assessments described later for selenium, aldicarb, and nickel are all verified and available through IRIS.*

II. SELENIUM

Selenium is found in varying regional concentrations with an average abundance of 0.09 ppm in the earth's crust (5). It is found in greatest abundance in igneous rock with sulfide ores. Elemental selenium can be reduced or oxidized to other more commonly found oxidation states of −2, 4, or 6 (6).

A. Essentiality and Physiological Roles

The essentiality for selenium has been well documented in livestock based upon the alleviation of specific deficiency conditions by selenium supplementation of the diet (6,7). Selenium has been clearly demonstrated to be a cofactor of glutathione peroxidase, a hydrogen and lipid peroxide reducing enzyme (8). Human requirements for selenium were not conclusively established until 1979, when an association was made between low-selenium status and cardiomyopathy (Keshan disease) in China for young children and women of child-bearing age (9,10). More recently, iatrogenic episodes of selenium deficiency were reported in patients receiving intravenous total parenteral administration of feeding solutions devoid of selenium. Findings included low glutathione peroxidase activity and low se-

*These assessments are subject to change if new information becomes available or if risk assessment methodologies change. The reader is referred to the IRIS database for U.S. EPA's most recent assessment.

lenium levels in erythrocytes (11), muscular weakness and discomfort (12), and cardiomyopathy (13). It is important to note that glutathione peroxidase activity is a valid indicator of human selenium status only in populations with relatively low selenium intakes, since the enzyme activity plateaus at adequate selenium intake levels (14), thereby precluding the use of this biochemical indicator under excessive selenium intake situations.

B. Nutritional Requirements and Dietary Levels

The National Academy of Sciences has determined the recommended dietary allowance for selenium to be 0.87 μg/kg, or approximately 70 and 55 μg/day for the reference adult North American male and female, respectively (15). Requirements for selenium increase during pregnancy to 65 μg/day and, for lactation, to 75 μg/day. Selenium requirements for infants and children vary according to age. However, based on the reference weights of the National Health and Nutrition Examination Survey II (NHANES II), these populations demonstrate an increased requirement per unit weight relative to adults. For infants, the selenium requirement is 1.67 μg/kg, and for children the requirement ranges from 1.07 to 1.53 μg/kg. The establishment of the recommended dietary allowance for selenium was based primarily on reported clinical signs of selenosis (i.e., loss of hair and nails) in approximately 50% of a population living in Enshi County, Hubei Province, People's Republic of China (16). These estimates, however, were based upon estimates of intake from only six and three inhabitants in the high- and low-contamination areas, respectively.

In addition to the recognized nutritional requirements for selenium, there is evidence that selenium plays a role in inhibiting carcinogenesis. A discussion of this literature is outside the scope of this chapter, and the reader is referred to a comprehensive review by Milner and Fico (17).

C. Selenium Toxicity

Selenium toxicity has been clinically described according to three types: acute selenosis, subacute selenosis, and chronic selenosis. The acute condition is caused by consuming relatively large amounts of selenium over a short period of time. This condition is characterized by the development of an unsteady gait, cyanosis of the mucous membranes, and labored breathing, sometimes resulting in death. Pathological findings include congestion of the liver, endocarditis and myocarditis, degeneration of the smooth musculature of the gastrointestinal tract, gallbladder, and bladder, and erosion of the long bones (18).

Subacute selenosis occurs from exposure to large doses of selenium over a longer period of time, resulting in neurological dysfunction (impaired vision, ataxia, and disorientation) and respiratory distress. It is typically seen in grazing livestock feeding upon selenium-accumulating plants and had been referred to as "blind staggers" (19).

Prolonged exposure of animals to more moderate levels of selenium results in skin lesions involving alopecia, hoof necrosis and loss, emaciation, and increased serum transaminase and alkaline phosphatase. In humans, the condition is characterized by chronic dermatitis, fatigue, anorexia, gastroenteritis, hepatic degeneration, enlarged spleen, and increased concentrations of selenium in the hair and nails (20).

D. Effect of Speciation of Selenium on Toxicity

The essentiality and toxicity of selenium varies according to the valence state of selenium when incorporated into biomolecules and the form in which selenium is fed or administered. Although it is difficult to make as assessment for several selenium compounds by a similar mode of administration in a common animal species, there is general agreement that sodium selenite, sodium selenate, selenomethionine, and selenodiglutathione are among the more toxic species (6). The relative potency of systemic toxicity for selenium compounds is also similar in both in vivo and in vitro experiments examining potency of antitumorigenic activity (21).

E. Toxicity of Selenium in Humans

Yang and coworkers (22,23), in a follow-up to their earlier study (16), studied a population of approximately 400 individuals living in an area of China with unusually high environmental concentrations of selenium. The subjects were evaluated for clinical and biochemical signs of selenium intoxication. Three geographical areas with low, medium, and high selenium levels in the soil and food supply were chosen for comparison in the studies. The earlier study was conducted in response to endemic selenium intoxication in two separate areas with sample sizes of only six and three. Comparisons were then made to a selenium-adequate area ($n = 8$) and low-selenium area ($n = 13$). The later studies by Yang et al. (22,23) provide a much larger sample size and include additional analysis of tissue selenium levels. This allows a more accurate estimation of the dose-response relationship observed for selenium toxicity. Selenium levels in soil and approximately 30 typical food types commonly eaten by the exposed population showed a positive correlation with blood and tissue selenium levels. The daily average selenium intakes based on lifetime exposure were 70, 195, and 1438 μg for adult males and 62, 198, and 1238 μg for adult females in the low-, medium- and high-selenium areas, respectively. Significant correlations between selenium concentrations of serum and various tissues were used to estimate daily selenium intake. In this manner, selenium intake was estimated from serum concentrations of selenium at which various alterations in biochemical parameters indicative of possible selenium-induced liver dysfunction (e.g., prolongation of clotting time and serum glutathione titer) and clinical signs of selenosis (e.g., hair or nail loss and morphological changes in the nails) were noted. Likewise, a marginal safe level of daily selenium intake was estimated from serum selenium levels in subjects without biochemical or morphological signs of toxicity.

Analysis of the results indicated that persistent clinical signs of selenosis were observed in only 5 in 349 adults, a potentially sensitive subpopulation. The blood selenium concentration in this group ranged from 1.054 to 1.854 mg/L, with a mean of 1.346 mg/L (Yang, personal communication). Clinical signs observed included the characteristic "garlic odor" of excess selenium excretion in the breath and urine, thickened and brittle nails, hair and nail loss, lowered hemoglobin levels, mottled teeth, skin lesions, and central nervous system abnormalities (peripheral anesthesia, acroparesthesia, and pain in the extremities). Alterations in the measured biochemical parameters occurred at dietary intake levels of 750–850 μg/day. These alterations were described as a delay in prothrombin time (i.e., increase in blood coagulation time and reduction in blood glutathione concentration). These indicators were poorly characterized, however, and are not typically used as an index for clinical selenosis resulting from chronic exposure to selenium (15).

Glutathione is an important conjugate-forming compound for the detoxification and excretion of electrophilic metabolites. If glutathione is depleted or markedly reduced in the liver, the hepatotoxicity of these compounds is likewise expected to be enhanced (24). The significance of decreased serum glutathione is not well characterized, however, and should not be used in this context as a biochemical marker of selenium toxicity. Likewise, there is no indication that prothrombin activity is affected by excess selenium administration (25). The description of this effect by Yang et al. (23) was based on the general Chinese population, for which there is insufficient documentation of normal clotting times.

Based upon the blood selenium levels shown to reflect clinical signs of selenium intoxication, a while-blood selenium concentration of 1.35 mg/L corresponding to 1.261 mg daily selenium intake is indicative of the lowest correlative selenium intake causing overt signs of selenosis. The next lowest whole-blood selenium concentration of 1.0 mg/L, corresponding to 0.853 mg selenium per day, produces no clinical signs of selenosis. The NOAEL for this study is 0.85 mg Se/day, and the LOAEL is 1.26 mg Se/day. Using the average adult body weight of 55 kg in this population (23), the NOAEL and LOAEL can be expressed as 0.015 and 0.023 mg Se/kg BW/day, respectively.

An epidemiological study by Longnecker et al. (25) evaluated 142 volunteers from South Dakota and Wyoming chosen at random from households listed in a telephone directory or from ranches with suspected high selenium intake based on previous cases of livestock selenosis. The geographical areas were chosen because of known seleniferous topsoil and high concentrations of selenium in plants and food. The subjects were followed for 1 year and completed health questionnaires, underwent physical examinations, provided blood samples for clinical assessment, and provided blood, urine, toenails, and duplicate-plate food collections for selenium analysis. The average selenium intake was 239 μg/day, approximately two to three times higher than the national average. The concentration of selenium in whole blood, serum, urine, and toenails and the amount in diet were highly correlated. Blood selenium concentration was highly correlated with selenium intake. The correlation was very similar to that reported by Yang et al. (22). Liver function, hematological function, and clinical chemistry were not found to be altered as a result of selenium intake. High regression coefficient predictor variables for selenium toxicity (muscle twitching, paresthesia, nail loss, nail lines, hair loss, and garlic breath) were not found in increased frequency for this population. No signs of selenium toxicity were found in this population, including individuals whose selenium intake was as high as 724 μg/day. This report corroborates that of Yang et al. (23), which showed that a selenium intake of up to 853 μg/day is not associated with the nail or hair loss typical of selenium intoxication.

F. Calculation of the RfD

The study by Yang et al. (23) was chosen as the critical study from which the RfD is derived as follows:

$$\text{RfD} = \frac{0.015 \text{ mg Se/kg/day}}{3} = 0.005 \text{ mg Se/kg/day}$$

where 0.015 mg Se/kg/day = NOAEL identified for chronic human consumption (23) and 3 = uncertainty factor to account for sensitive individuals.

In the derivation of this RfD, an uncertainty factor of 3 was applied to the NOAEL to account for sensitive individuals. A full factor of 10 was not deemed necessary since similar NOAEL were identified in two moderately sized human populations exposed to selenium levels in excess of the RDA throughout a lifetime without apparent clinical signs of selenosis.

Because the specific forms of selenium ingested by the Chinese cohort could not be identified, it is assumed that many of the naturally occurring and nutritionally important forms of selenium were present in the food and water supply. Hence, the RfD was derived for compounds of selenium found in the food chain.

G. Reproductive and Developmental Toxicity

Selenium has been reported to cause growth retardation, decreased fertility, embryotoxicity, fetotoxicity, and teratogenic effects in animals. Yang et al. (23) observed malformations in chickens hatched from eggs produced in areas of China with high selenium levels; however, teratogenic effects in human infants were not reported.

The developmental toxicity of selenomethionine was investigated by Tarantal et al. in nonhuman primates (26). A group of 40 pregnant long-tailed macaques were dosed daily by nasogastric intubation with 0, 0.025, 0.150, or 0.3 mg selenium/kg as selenomethionine on gestational days 20–50. Dams were examined clinically and followed to term (gestational day 165). No significant maternal or fetal developmental effects or teratogenesis were found up to 0.3 mg/kg selenium, the highest dose tested.

Schroeder and Mitchener administered 3 ppm selenium as selenate (390 μg/kg/day) to CD mice through four generations (27). Maternal effects were not observed. There was a significant increase in young deaths in the F_1 generation and an increase in numbers of runts in generations F_1–F_3. By the F_3 generation there was also a decrease in breeding events.

Rosenfeld and Beath (28) administered selenium as potassium selenate to sires and pregnant rats through five breeding cycles at 1.5, 2.5, or 7.5 ppm selenium (75, 125, or 375 μg/kg/day). No effect was observed on reproduction, the number of young reared, or the reproduction of two successive generations of dams and sires in groups receiving 1.5 ppm selenium. In the group receiving 2.5 ppm selenium, there was a 50% reduction in the number of young reared. At 7.5 ppm there was a decrease in the fertility of the females but not the males, a decrease in the number of survivors, and a reduction in the rate of growth in the young.

Nobunaga et al. (29) administered 3 or 6 ppm selenium (390 or 780 μg/kg/day, respectively) as selenite to IVCS mice for 30 days before mating and throughout gestation. On day 18 of gestation, maternal mice were sacrificed and the embryos removed. Number of litters, total implants, total implants per dam, dead fetuses, dead embryos, resorptions, surviving fetuses (% of total implants), litter size, gross malformations, and skeletal anomalies were not significantly different between selenium-treated and control mice. The only significant effect noted was a decrease in the body weight of surviving fetuses in mice given 6 ppm selenium.

Based on the strength of these animal studies, it was determined that reproductive and developmental toxicity would not be of concern at or below the identified NOAEL and derived RfD for selenium. Therefore, no additional uncertainty factor application was deemed necessary. Any concern for safety is adequately addressed by the uncertainty factor applied for sensitive human subpopulations.

H. Confidence in the RfD

Confidence in the critical study from which the RfD for selenium is derived is medium. Although this is a human epidemiological study in which a sizable population with sensitive subpopulations was studied, several possible interactions were still not fully accounted for (e.g., fluoride intake and protein status). Also, except for clinical signs of selenosis, there are no other reliable indicators, biochemical or clinical, of selenium toxicity. Confidence in the data base is high because many animal studies and epidemiological studies (reviewed in Reference 6) support the principal study. The human study by Longnecker et al. (25) with a freestanding NOAEL strongly corroborates the NOAEL identified in the principal study. Therefore, high confidence is placed on the RfD based on support of the critical study and the high level of confidence in the data base.

IV. ALDICARB

Aldicarb (also known by the trade name Temik) is a synthetic carbamate ester pesticide for which there are no natural sources. Approved for use on a variety of crops, aldicarb is applied below the surface of soil and is taken into the plant systemically to help control insects, mites, and nematodes.

Exposure to the general population may occur either through residues in food or through contaminated water. In addition, workers may be exposed to aldicarb through dermal and inhalation routes. Several cases of human poisonings have been documented, primarily by way of contaminated watermelons and cucumbers, crops not approved for the use of aldicarb. The information gleaned from these accidental poisonings is used in conjunction with a controlled study in humans and a laboratory study in dogs to support a quantitative risk assessment for aldicarb.

The risk assessment problems presented by aldicarb include the question of how to deal with a compound known to be metabolized to two compounds (aldicarb sulfoxide and aldicarb sulfone), which are also quite toxic, and second, how to do a risk assessment for a chemical for which the toxic effects are acute and reversible and have a very steep dose-response relationship. Perhaps more problematical is how to assess the effect of acetylcholinesterase inhibition in the absence of actual clinical manifestation of toxicity. These issues are highlighted here.

A. Toxic Metabolites of Aldicarb

Two metabolites of aldicarb, the sulfoxide and the sulfone, elicit the same toxicological effects as the parent compound. The sulfoxide is about equally toxic, but the sulfone is considerably less toxic than aldicarb. The oral LD_{50}* in rats for aldicarb and its sulfoxide are both roughly 1 mg/kg BW (30,31); the LD_{50} for the sulfone is about 25 mg/kg BW (31). However, although recognizing this difference in potency demonstrated in acute exposures, the U.S. EPA determined that data for clinical effects and long-term exposures are insufficient to support a separate quantitative risk assessment for aldicarb sulfone. To be protective of human health, the data the U.S. EPA used in promulgating the drinking water maximum contaminant level goal for aldicarb and aldicarb sulfoxide are also used to support the regulations for aldicarb sulfone.

B. Mechanism of Toxicity

The toxicity of aldicarb results from its inhibition of the enzyme acetylcholinesterase (AChE). This is due to the close structural similarity between aldicarb (a methyl carbamate) and acetylcholine, the physiological substrate for AChE. Acetylcholinesterase is responsible for the hydrolysis (and hence inactivation) of the neurotransmitter acetylcholine at synaptic and myoneural junctions. Although the toxic effect of aldicarb are due to its effect on AChE in nerve synapses, AChE activity in erythrocytes may be used as a measurement of a biochemical effect resulting from aldicarb exposure (32).

The signs and symptoms of aldicarb poisoning are cholinergic in nature. Typically, the acute toxicity of aldicarb is manifested by excessive sweating, rapid pulse, lacrimation, salivation, nausea, vomiting, nonreactive contracted pupils, and respiratory distress. If the dose is not high enough to result in death, these symptoms generally last only a few hours at most, and complete recovery is achieved.

The inhibition of AChE by aldicarb results from carbamylation of the enzyme, rendering it inactive. There is generally a sufficient level of AChE

---

*LD_{50} is the dosage (mg/kg BW) causing death in 50% of the exposed animals.

activity that no clinical effects are manifested until a significant degree of inhibition is achieved. However, the relationship between AChE inhibition and clinical effects has not been clearly established, in part because there is some controversy about whether erythrocyte AChE activity is a good predictor of brain AChE activity.

Risk assessment scientists have difficulty in determining whether a statistically significant inhibition of AChE in plasma, red blood cells, and/or the brain should be considered an adverse effect in the absence of clear signs of clinical toxicity. This problem has been discussed for a number of years by both U.S. EPA and others. Some scientists consider the inhibition of AChE in plasma and red blood cells a biomarker of exposure but clinical signs are required to define an adverse effect. Other scientists consider AChE inhibition to be a precursor to adverse (or potentially adverse) effects, thereby fitting the U.S. EPA definition of a critical effect in and of itself.

Vandekar et al. (33) performed studies in both rats and humans with a carbamate pesticide (propoxur) with a mechanism of toxicity similar to that of aldicarb. In rats, no noticeable symptoms resulted from an intramuscular dose of propoxur that caused a reduction in both brain and plasma cholinesterase by as much as 40%. When enzyme activity was reduced to 50% of normal, a slight tremor was observed, and at higher doses, the degree of AChE inhibition was closely correlated with the severity of the resulting symptoms. Also reported in this study were effects on human volunteers; only one dose was tested that resulted in a reduction in erythrocyte AChE inhibition to 50% of normal. Symptoms were first noticed 15 minutes after ingestion of the carbamate pesticide. The symptoms, including nausea, vomiting, and profuse sweating, grew progressively worse for 45 minutes following exposure, after which time recovery began. The subject felt completely recovered within 2 h from ingestion of the chemical. Such rapid recovery following exposure to carbamate pesticides is typical.

For aldicarb, red blood cell and brain cholinesterase inhibition and clinical signs of toxicity have been found in numerous studies. The following highlights the U.S. EPA evaluation for aldicarb in the face of this uncertain science.

C. Human Studies

There have been several documented cases of aldicarb poisonings in humans who have ingested contaminated produce. Goldman et al. reported three separate incidents involving poisoning of 28 humans exposed to aldicarb and aldicarb sulfoxide in contaminated watermelons and cucumbers (34). Although it is difficult to determine the actual level of exposure, the dose range associated with ensuing symptoms (nausea, vomiting, and diarrhea) was 0.002–0.0086 mg/kg BW based on self-reports of food consumption and estimates of body weight. The authors noted that the dosage calculations were uncertain and a wide range of sensitivity was observed in the individuals exposed.

In a controlled human study by Union Carbide, adult male volunteers (four per group) were administered a single dose of aldicarb at 0.025, 0.05, or 0.1 mg/kg BW (35). The subjects were given the aldicarb dissolved in 200 ml water to drink in the morning, after which normal activities resumed. Blood and urine sampling and clinical observations were reported at 1, 2, 3, 4, and 6 h following the administration of aldicarb. Observations included the measurement of pulse, blood pressure, and pupil size. Subjects' complains were also recorded.

Significant dose-related cholinesterase inhibition in whole blood was reported for all three groups. At 1 h after dosing, the cholinesterase activity was reduced to about 53, 39, and 35% of normal levels for the low dose through the high dose. Peak inhibition occurred between 1 and 2 h, with almost complete recovery by 6 h. All four subjects receiving the highest dose manifested clinical signs, including nausea and vomiting, pinpoint nonreactive pupils, malaise, weakness, epigastric pain, air hunger and yawning, excessive sweating, salivation, and slurred speech. None of these symptoms were severe enough to require treatment. Most of the volunteers in the lower two dose groups had no signs or symptoms of aldicarb toxicity. Based on significant inhibition of cholinesterase in whole blood, the lowest dose of 0.025 mg/kg is considered the LOAEL for this study.

D. Animal Studies

Rhône-Poulenc administered technical-grade aldicarb in the diet to groups of beagle doses (five per sex per dose) for 1 year at levels of 0, 1, 2, 5, and 10 ppm (36). Animals were evaluated for mortality, body weight, food and water consumption, hematology, histopathology, and clinical chemistry. The only compound-related effect was inhibition of plasma cholinesterase, with a significant dose-related decrease reported in male dogs given 2, 5, or 10 ppm aldicarb at all testing periods (weeks 5, 13, 26, and 52). There was also a significant reduction in plasma cholinesterase at week 52 in three of the male dogs receiving 1 ppm aldicarb (mean inhibition of 30.9%; range 1.4–34.4%). Plasma sampling at weeks 5, 13, and 26 weeks did not reveal any significant effects in the 1 ppm dose group. Brain cholinesterase was inhibited in male dogs at the 10 ppm dose level. There was an increase in the combined incidences of soft and mucoid stools and diarrhea in male and female dogs given diets containing 2 ppm aldicarb or higher.

The Office of Pesticide Programs of the U.S. EPA has accepted a level of 25% cholinesterase inhibition as indicative of a level that may be associated with health effects as opposed to merely an indicator of exposure (37). Therefore, based on the inhibition seen in the male dogs, the dietary level of 1 ppm (about 0.02 mg/kg/day) is the LOAEL for this study.

In chronic laboratory studies performed in rats, NOAEL were identified that were higher than the levels at which effects were seen in the dogs (38,39). Therefore, the data from dogs is used in the quantitative risk assessment for aldicarb because it appears that this species may be more sensitive than rodents. There are no data to suggest that aldicarb poses any health threat of carcinogenicity or reproductive effects.

E. Calculation of the RfD

After extensive review of the data, it was determined by the RfD/RfC Work Group that the RfD for aldicarb is best derived using a weight of evidence approach incorporating data from more than one study.* The principal studies used in the developing the RfD are Rhône-Poulenc (36), Goldman et al. (34), and Union Carbide (35). In the Rhône-Poulenc study, beagle dogs exposed to 1 ppm aldicarb (0.02 mg/kg/day) experienced a significant reduction in plasma cholinesterase activity. This dose level is identified as a LOAEL. Union Carbide identified a LOAEL of 0.025 mg/kg in humans based on a significant reduction in erythrocyte cholinesterase activity. The reports from accidental human poisonings by Goldman et al. suggested limited effects in more sensitive individuals resulting from an intake as low as 0.002 mg/kg/day. The RfD for aldicarb that would be derived using any one of these three studies are all supportive of each other; therefore, the final RfD is considered the result of an assessment utilizing all three studies (subscripts RP, UC, and G following RfD refer to the three supporting studies):

$$\text{RfD}_{(\text{RP})} = \frac{0.02 \text{ mg/kg}}{100} = 0.0002 \text{ mg/kg/day}$$

where 0.02 mg/kg = LOAEL based upon significant reduction in cholinesterase in plasma of dogs (36) and 100 = one 10-fold uncertainty factor to account for both interspecies differences and a minimal LOAEL. The

*Subsequent to the preparation of this chapter, the RfD/RfC Work Group verified a new assessment for aldicarb (available on IRIS as of November 1, 1993).

other factor of 10 is used to account for sensitive subpopulations of humans.

$$RfD_{(UC)} = \frac{0.025 \text{ mg/kg}}{100} = 0.00025 \text{ mg/kg/day}$$

where 0.025 mg/kg = LOAEL based on significant AChE inhibition in whole blood in a controlled human study (35) and 100 = uncertainty factor to account for use of a minimal LOAEL in humans and variability within humans. Although the acute toxicity of aldicarb is the end point of concern, this UF also incorporates a factor to account for the fact that this study was based on a single exposure, after which time the subjects were allowed to recover fully.

$$RfD_{(G)} = \frac{0.002 \text{ mg/kg}}{10} = 0.0002 \text{ mg/kg/day}$$

where 0.002 mg/kg = minimal LOAEL (estimated) in humans consuming contaminated cucumbers and watermelons (34) and 10 = uncertainty factor to account for minimal effects in humans who may have comprised a more sensitive subgroup (based on the fact that not all exposed individuals responded to the same degree).

Based on the clinical observations and acetylcholinesterase inhibition reported in the three studies just described, an RfD of 0.0002 mg/kg/day for aldicarb was verified by the RfD/RfC Work Group in July of 1990 (37).* Confidence in the critical studies from which the RfD is derived is considered medium to low, primarily because a NOAEL could not be established for plasma or whole-blood cholinesterase inhibition and the NOAEL for clinical signs in humans is equivocal. The supporting data base is given a high level of confidence because of the numerous studies demonstrating the same effects. Confidence in the RfD overall is considered medium to high.

V. NICKEL

Nickel is a ubiquitous component of the environment, predominantly found in valence state 2. The primary health concern posed by nickel to humans has been in an occupational setting by the route of inhalation. However, the risk assessment described in this chapter pertains to nickel as a drinking water contaminant; therefore, little mention is made of the more extensive inhalation data base.

*This assessment has been superceded by the verification of a new RfD of 0.001 mg/kg/day, based on a weight-of-evidence approach using three human studies.

In the process of performing a risk assessment for nickel in drinking water, the U.S. EPA considered many factors, among them the following:

Identification of the most relevant toxicological end point(s)
Differing potencies of various compounds of nickel
Bioavailability of ingested nickel from different sources
Role of ingested nickel in aggravating dermal eczematous lesions in hypersensitive individuals
Levels of nickel found in an average diet
Possible essentiality of nickel for humans
Relationship of the carcinogenicity of inhaled nickel to an oral exposure

Each of these issues, addressed in the U.S. EPA risk assessment for nickel in drinking water, is discussed here.

A. General Toxicity

Long-term ingestion of low levels of nickel does not appear to pose a health threat to humans; indeed, trace amounts of nickel may be required for health. Ingestion of larger amounts of nickel have been found to cause fetotoxic effects in laboratory animals and a generalized toxicity characterized by changes in organ weights. Other than acute poisonings, the only data showing nickel to be of concern for humans by the route of ingestion are from patients with a demonstrated hypersensitivity to nickel.

The data from chronic feeding studies in laboratory animals do not indicate that ingested nickel poses any carcinogenic risk. Therefore, the following discussion is for noncancer end points.

Ambrose et al. performed a 2 year feeding study in which rats were provided diets containing 0, 100, 1000, or 2500 ppm nickel as nickel sulfate hexahydrate (40). Assuming that rats consume feed equivalent to 5% of their body weight per day, these dietary levels correspond to nickel intakes of 0, 5, 50, or 125 mg/kg/day. Decreased body weight was reported for the 2500 ppm group and sporadically for the 1000 ppm group. No hematological or urinary effects were observed. Female rats exhibited a statistically significant increase in relative heart weight and decrease in relative liver weight at the middle and high doses. Upon this basis, the middle dose of 50 mg/kg/day was determined to be the LOAEL for this study and the lowest dose of 5 mg/kg/day was determined to be the NOAEL.

In addition to the study in rats, Ambrose et al. also exposed beagle dogs for 2 years to diets containing 0, 100, 1000, or 2500 ppm nickel (corresponding to about 0, 2.5, 25, or 62.5 mg Ni/kg/day) (40). The highest dose resulted in gastrointestinal irritation, decreased body weight gain, increased relative kidney and liver weights, and pathological changes in the lung and bone marrow. The middle dose of 25 mg/kg/day was the NOAEL for this study, suggesting that dogs are less sensitive than rats to the toxicological effects of nickel.

The results of the Ambrose et al. rodent study (40) are also supported by several subchronic studies. Whanger identified a NOAEL of 10 mg/kg/day and a LOAEL of 50 mg/kg/day (based on hematological effects) in a 6 week dietary study in weanling rats (41). Nation et al. (42) identified a NOAEL of 10 mg/kg/day and a LOAEL of 20 mg/kg/day (based on behavioral end points) following a 2 week dietary exposure to adult male rats. In a 91 day study by American Biogenics Corporation, CD rats were administered 0, 5, 35, or 100 mg/kg nickel (as nickel chloride hexahydrate) by gavage (43). The lowest dose of 5 mg/kg was determined to be the NOAEL and 35 mg/kg the LOAEL based primarily on hematological effects. These studies taken collectively support the dose-response relationship demonstrated by the Ambrose et al. study.

B. Reproductive Toxicity

Although exposure to nickel has not been shown to cause adverse reproductive effects in humans, studies have demonstrated reproductive and developmental toxicity in laboratory animals.

In a three-generation study using albino, Wistar-derived rats, Ambrose et al. administered nickel sulfate hexahydrate in the diet at levels of 0, 250, 500, or 1000 ppm nickel (0, 12.5, 25, or 50 mg/kg/day) (40). Decreased body weight for all three generations was observed at the highest dose. The incidence for stillborn pups was increased for both F_{1a} (4, 19, 12, and 15% for controls through high dose) and F_{1b} (2, 4, 20, and 25%). Statistical analysis for these data was not reported. Increased fetal mortality was not observed for F_2 or F_3 generations. This study is of limited use in the risk assessment process because of the small numbers of animals used (17–20 females mated per generation) and the use of pups rather than litters as the unit of comparison. The results are also equivocal and do not clearly identify a NOAEL or LOAEL.

A small-scale three-generation rat study by Schroeder and Mitchener indicated an increase in neonatal mortality and the number of runts born to dams exposed to 0.72 mg Ni/kg/day in drinking water (44). This was the only dose tested; therefore no dose-response relationship can be established.

In a two-generation study by the Research Triangle Institute (RTI), nickel chloride was administered in drinking water to male and female CD rats (30 per sex per dose) at dose levels of 0, 50, 250, and 500 ppm (0, 7.3, 30.8, and 51.6 mg/kg/day) for 90 days before breeding (45). At the 500 ppm dose level there was a significant decrease in the P_0 maternal body weight, along with absolute and relative liver weights. The dose of 250 ppm (30.8 mg/kg/day) was a NOAEL for P_0 breeders.

For two sequential P_0 matings there was a dose-related decrease in live litter size, an increase in neonatal mortality, and a decrease in pup weight, but statistically significant effects were seen at the high dose only. There was also dose-related mortality between 3 and 7 weeks of age for F_1 rats exposed to 250 or 500 ppm nickel. Several organ weight changes were reported in the F_1 generation, but these were not associated with any pathological lesions.

For the F_1 matings there were also dose-related decreases in live litter size and increased mortality per litter. These were significant only in the high-dose group.

Although the RTI study provides evidence for the fetotoxicity of nickel, establishing a dose-response relationship is difficult. In addition to statistical concerns, other factors may have influenced the outcome of the study. Most significantly, the room temperature was up to 10°F higher than normal at certain times during gestation and the early postnatal days. Much lower levels of humidity were also recorded. As evidenced in the literature, temperatures that are 10°F above normal during fetal development can cause adverse effects (46). Because of the problems with temperature and humidity control, it is difficult to make a direct association between the effects reported in the RTI study and nickel exposure.

Smith et al. exposed groups of 34 female Long-Evans rats to drinking water containing 0, 10, 50, or 250 ppm nickel as nickel chloride (47). The authors calculated the average ingested doses of nickel to be 0, 1.3, 6.8, or 31.6 mg/kg/day. Exposure commenced 11 weeks before mating and continued throughout two successive gestation (G_1 and G_2) and lactation (L_1 and L_2) periods. Pups were observed until weaning; breeder males were unexposed.

Dams in the 250 ppm group consumed significantly less water in the prebreeding and gestational periods and significantly less food in the prebreeding and G_2 and L_2 periods. Weight gain was significantly reduced during G_1 in the high- and middle-dose groups. Throughout the study there were no overt clinical signs of toxicity in any of the groups.

Indices of reproductive performance were comparable across all groups. These include mating success, rate of impregnation, number of fetuses per litter, and gestation length. Pup birth weight was not affected by treatment. The frequency of perinatal mortality reflects the most significant finding of the study. The proportion of dead pups per litter was significantly elevated only at the high dose in L_1 and at 10 and 250 ppm in L_2 (not statistically significant at middle dose of 50 ppm; $p = 0.076$). The number of dead pups per litter was significantly increased at each dose in L_2. The lowest dose in this study (10 ppm, about 1.3 mg/kg/day) is generally considered to represent a LOAEL based on an increase in perinatal mortality.

C. Dermatotoxicity Related to Hypersensitivity

Chronic dermatitis, especially of the hands, is the most prevalent effect of nickel in the general population. The North American Contact Dermatitis Group found that 11% of 1200 patients were sensitive to nickel sulfate applied as a 2.5% solution in a dermal patch test (48). Peltonen (49) and Prystowsky et al. (50) surveyed subjects more representative of the general population and concluded that 1.5–5% of the population is sensitive to nickel, the prevalence being much higher in females. Menne et al. determined the frequency of nickel allergy in a sample of Danish women from the general population as 14.5% (51). Of the allergic women, 57% experienced hand eczema.

Chronic hand dermatitis is a frequent and often incapacitating manifestation of nickel hypersensitivity. Initial sensitization to nickel occurs from dermal contact, but it has been demonstrated that recurring flares may be triggered by the ingestion of nickel. Although recurrent vesicular eruption on the hands is most common, eczematous flares are also reported on the feet, trunk, eyelids, and other regions of the body that may have been sensitized through contact with nickel. Unfortunately, it is very difficult to establish a good dose-response relationship, primarily because of the large variation in the level of sensitivity of individuals to nickel. The human studies described here are difficult to interpret for several reasons: very small numbers of subjects (mostly women already determined to be sensitive to nickel by a patch test) were used in the studies; many investigators reported a placebo effect; many studies were not conducted in a double-blind manner, thereby introducing investigator bias; and it was often not specified whether subjects had fasted overnight, whether there were other dietary restrictions, or the typical intake of nickel for the individuals under study. As described in more detail later, the source of nickel (e.g., drinking water versus food) may have a profound effect upon the degree to which it is absorbed by the gastrointestinal tract. It is also important to note that most of these studies involved the ingestion of capsules containing nickel, resulting in a bolus dose. The absorption and pharmacokinetics following such an exposure are likely to be quite different from an exposure given incrementally throughout the day. These points should be borne in mind when interpreting the findings of the following studies.

Following an overnight fast, groups of five nickel-sensitive women were given 100 ml water along with one oral dose of nickel sulfate containing 0.6, 1.25, and 2.5 mg nickel (52). Worsening of hand eczema was reported in two in five, three in five, and five in five women (receiving 0.6, 1.25, and 2.5 mg nickel, respectively). Erythema was observed in one in five, four in five, and four in five women. There appears to be a good dose-response relationship, but this study did not report controls. The authors concluded that the amount of nickel in a typical diet (comparable to their dose of 0.6 mg) may be of significance in the hand eczema of some patients. However, this conclusion does not take into account that these patients received a bolus dose of nickel (in the fasted state) in addition to what the normal diet contributes.

Kaaber et al. reported that 17 of 28 patients experienced aggravation of dermatitis following ingestion of 2.5 mg nickel (as nickel sulfate), but not a placebo (53). Of these 17, the dermatitis of 9 improved during a period of 6 weeks on a low-nickel diet. When a normal diet was resumed, dermatitis flared in 7 of those 9. The only dose used in this study, 2.5 mg nickel, clearly posed a problem to patients with a hypersensitivity to nickel. In a subsequent study, 14 patients were challenged with single oral doses of 0.6, 1.2, or 2.5 mg nickel (54). Eczema was aggravated in 11 of 14 patients receiving nickel, but not a placebo. Only 1 patient responded to 0.6 mg, 1 patient to 1.2 mg, and 9 patients to 2.5 mg nickel.

The problem of false positives in studies with hypersensitive humans was demonstrated in a study by Jordan and King (55). A group of 10 nickel-sensitive women were given 0.5 mg nickel supplements or placebos on 2 consecutive days of a week. Each woman served as her own control, receiving nickel 1 week and a placebo a different week. Of the 10 patients, 6 experienced flares of hand eczema following both placebo and nickel administration. Only 1 patient demonstrated a reproducible flaring of vesicular dermatitis associated with nickel, not with the placebo. This study was conducted in a double-blind manner, with neither the patient nor the doctor knowing whether the capsules consumed during a particular week contained nickel or placebo. The high level of placebo response illustrates the importance of conducting studies in this manner.

Gawkrodger et al. reported similar results (56). They reported that 5 of 10 women responded to doses of both 0.4 and 2.5 mg nickel, but 10 of 26 also reacted to a placebo. They determined the LOAEL for their study to be 5.6 mg nickel, a dose at which 100% of the women responded. A similarly high placebo response was reported by Burrows et al. (57).

Nielsen described a study in which 12 nickel-sensitive women were challenged for a 4 day period with a diet providing 490 μg Ni/day (58). This level of nickel may be higher than the amount normally consumed by these individuals, but it is also within the range of what may be considered a "normal" diet. Unfortunately, this was the only dietary level administered. No changes were observed before the start of the nickel challenge to day 0 (start of challenge). On day 4, the eczema of 6 patients was considered worse according to both the patients' impressions and a dermatologist's evaluation. The author suggested that the delayed reaction in this study may be attributed to the fact that the dose of nickel was ingested in the diet throughout the day, as opposed to studies that employed a bolus dose. Such a difference in the manner by which the nickel is administered may greatly affect the pharmacokinetics of ingested nickel.

D. Calculation of the RfD

As with any quantitative risk assessment, the first step is to choose a toxicological end point of concern for humans for which there exists a sufficient data base. In performing a risk assessment for nickel, the potential for dermal toxicity in hypersensitive individuals must be addressed first because it represents the most substantial data base available for toxicological effects in humans resulting from the ingestion of nickel. In some regards, it seems to be the perfect end point upon which to base a risk assessment: the data were obtained from humans who comprise a sensitive subgroup. However, there are a sufficient number of problems with the studies that make up this particular data base to prevent their use. Most of these problems have already been described: small sample sizes, lack of controls, few studies available performed in a double-blind manner, and a data base that does not lend itself to identification of an NOAEL and/or LOAEL, for example. Despite these problems, it is important that this end point be acknowledged in any risk assessment for nickel and that any quantitative estimate based on another end point be considered protective for hypersensitive humans. Having stated this, it is also recognized that it may not be possible to protect the most sensitive individuals because these patients have been shown to react to levels of nickel that are present in a standard diet (58). Therefore, these individuals must take care to reduce their personal intake, this being most easily achieved by dietary modification. One level of nickel intake that was shown by several investigators to elicit a response in the majority of hypersensitive patients was 2.5 mg nickel (52–54,56). This dose is in addition to the normal dietary intake, which is probably of the order of a couple of hundred micrograms/day. Using an estimate of 300 μg/day, the total intake is about 2.8 mg/day. It is recognized that the dose of 2.5 mg was taken in bolus form, perhaps having a significant effect on the pharmacokinetic parameters following ingestion. However, no data appear to be appropriate for modifying such an intake to correspond to an intermittent intake as would occur from a drinking water source. Therefore, the total dose of 2.8 mg is considered a total daily intake. Assuming a 70 kg body weight, this is equivalent to 0.04 mg/kg/day. Although the clinical studies in humans were of short exposure duration, often a single dose, it appears that an uncertainty factor is necessary to account for extrapolation to a lifetime number. However, it is not an accumulation of nickel over time that is responsible for the elicitation of a response in these patients; rather, it appears to be a sudden spiking of nickel intake above the normal low level of intake that results in an eczematous response. Therefore, no modification for exposure duration is deemed appropriate. Because the dose of 0.04 mg/kg/day, calculated earlier, is associated with adverse dermatotoxicity in a majority of hypersensitive individuals, a quantitative risk assessment for nickel should be less than this amount.

The second toxicological end point considered for derivation of an RfD was that of fetotoxicity. Nickel has not been shown to cause any adverse reproductive or developmental effects in humans, but neither are there data to suggest that at a high enough dose this would not be a relevant toxicological end point for humans. Indeed, that several studies in rodents have reported fetotoxicity to result from nickel exposure suggest that this may be a critical target system for nickel in other species (44,45,47). Unfortunately, these studies also have deficiencies that appear to preclude their use for serving as the basis for a quantitative risk assessment. As with the studies in hypersensitive humans, the critical end point is clearly identified, but it is very difficult to establish a good dose-response relationship or to identify a NOAEL and/or LOAEL.

The study by Schroeder and Mitchener (44) is not acceptable because it is based on a total of only five matings and only one dose level, and also because it was determined that the level of dietary chromium in this study was inadequate and this may have contributed to the toxicity of nickel. The three-generation study by Ambrose et al. (40) is problematical because there was a higher incidence of stillborn pups in the first generation at all levels of nickel, but this was not observed in either of the two subsequent generations. There were also statistical limitations: pups were used as the statistical unit of comparison, as opposed to litters. The study by RTI (45) was considered for the basis for development of an RfD, but the possibly confounding effects from the increased temperature during gestation have raised serious doubts about the validity of the test results. Identifying a tentative NOAEL of 50 ppm (7.3 mg/kg/day) of nickel in drinking water from this study and applying an UF of 100 to account for inter- and intraspecies variability would result in an RfD of 0.07 mg/kg/day. The more recent study by Smith et al. (47) was also considered for use as the basis for a quantitative risk assessment. As described previously, it is difficult to establish clearly a dose-response relationship for this study because effects that were statistically significant at the lowest dose (1.3 mg/kg/day) were not at the middle dose (6.8 mg/kg/day).

The study the U.S. EPA chose to serve as the basis for the RfD for nickel is that by Ambrose et al. (40), which demonstrated a dose-response relationship and also identified a NOAEL (100 ppm diet; equivalent to 5 mg/kg BW/day) and a LOAEL (1000 ppm), as described previously. The RfD is calculated as

$$\text{RfD} = \frac{5 \text{ mg/kg/day}}{300} = 0.017 \text{ mg/kg/day}$$

(rounded to 0.02 mg/kg/day), where 5 mg/kg/day = NOAEL identified by chronic rat dietary study (40) and 300 = uncertainty factor to account for interspecies (× 10) and intraspecies (× 10) variability and to account for inadequacies in the reproductive data (× 3).

The RfD of 0.02 mg/kg/day, although based on one study, is considered protective of all end points discussed earlier. It is lower than the intake of 0.04 mg/kg/day, which is associated with effects in hypersensitive individuals, and it is lower than the RfD of 0.07 mg/kg/day, which may be estimated from the RTI study, yet it is not lower than the level to which humans are generally exposed in the diet (0.02 mg/kg/day corresponds to 1.4 mg/day for a 70 kg adult).

The RfD of 0.02 mg/kg/day was originally verified by the U.S. EPA RfD/RfC Work Group on July 16, 1987 (59). In light of new information regarding the issue of hypersensitivity and the reproductive study performed by Smith et al. (47), the risk assessment for nickel was presented to the Science Advisory Board in August 1990. The consensus from this meeting was that the RfD of 0.02 mg/kg/day based on the study by Ambrose et al. (40) is appropriate; however, it was also thought that the discussion leading to such a quantification must address the issues of reproductive toxicity and dermal effects in hypersensitive humans. This information is reflected in the risk assessment for the soluble salts of nickel, which is presented in the EPA on-line data base (IRIS).

E. Nickel Speciation

The issue of speciation is one of critical importance in determining the toxicity of virtually all inorganics, nickel being no exception. This is primarily because the degree to which nickel is absorbed in the gastrointestinal tract is directly related to the solubility of the particular compound. A dramatic illustration of the effect of speciation can be seen by observing the level of the LD_{50} for various nickel salts administered by gavage to rats (taken from Reference 60):

Nickel salt	LD_{50} (mg/kg BW)
Nickel sulfate hexahydrate	67
Nickel chloride	117
Nickel hydroxide	1013

Most of the toxicity testing for nickel by the oral route involves soluble species of nickel (generally the sulfate or chloride). Because the concern for exposure through drinking water is pertinent to the soluble salts only, the toxicity studies performed with these compounds are suitable for use.

Where this issue is perhaps more important is in evaluating the carcinogenicity of nickel. Some scientists contend that a carcinogen by one route may be considered to have carcinogenic potential by any route, but nickel challenges this notion. Certain forms of nickel are suspected or confirmed carcinogens in humans by the route of inhalation. These are nickel subsulfide, nickel refinery dust, and nickel carbonyl. It is further believed that it is the nickel ion that is eliciting the carcinogenic response; however, the soluble salts of nickel do not appear to elicit nearly the same level for concern for carcinogenicity (see Reference 61 for review). This is true for inhalation, but even more so for the route of ingestion. Indeed, in the multigenerational reproductive studies, as well as the lifetime study by Ambrose et al. (40) chosen for the derivation of the RfD, nickel was not shown to elicit a carcinogenic response. It is also known that humans consume low levels of nickel ($\sim$100–300 μg/day, depending on the diet) for a lifetime with no associated tumorigenic response. Therefore, although certain species of nickel have been identified as carcinogens, this does not appear to be relevant to a risk assessment for the ingestion of soluble nickel salts.

F. Essentiality of Nickel

To carry the previous discussion a step further, it is important to bear in mind that nickel has been shown to be an essential element for several species, including rats, sheep, cows, goats, and minipigs (62). No definitive physiological requirement for nickel has been demonstrated for humans, but that it has been shown to be essential at low levels for several other species suggests that the same is true for humans. Some studies have suggested that nickel is involved in the function of certain metalloenzymes and also may serve as a bioligand cofactor that facilitates the absorption of iron from the gastrointestinal tract (see Reference 63 for review). Most likely a human deficiency state has not been documented because of the more than adequate levels found in the diet. Based on data from chicks, Nielsen estimated that the daily nickel requirement of humans is about 35 μg/day, an amount that is easily attained (64).

G. Bioavailability

Although one may be comfortable with the RfD calculated for nickel, there is one more important factor to consider in using such a risk assessment for the basis of a drinking water standard: the relative bioavailability of nickel, particularly from drinking water compared with a dietary source. Sunderman et al. demonstrated a dramatic effect of the source of nickel upon its bioavailability (65). Two groups of healthy volunteers ingested 12, 18, or 50 μg nickel (as nickel sulfate) following a 12 h fast. One group ingested the nickel with water only; the other group ingested the dose with food. The absorption of nickel averaged 27 ± 17% when taken with water but only 0.7 ± 0.4% when taken with food, indicating that dietary constituents significantly reduced the uptake of nickel from the gastrointestinal tract. Other investigators have also demonstrated an effect of dietary constituents on the bioavailability of nickel (66,67).

VI. Summary

The prevailing message from this exercise is that, although guidelines are essential for ensuring scientific consistency in the risk assessment process, it is critically important that each chemical be viewed in light of its particular nature. The information contained in the guidelines is meant to be used on a chemical-specific basis rather than as a general formula.

For selenium, primary consideration must be given to the fact that although selenium is toxic at high doses, a deficiency of selenium can also result in adverse health effects. The risk assessment process must be modified to take both the toxicity and the essentiality of selenium into account. Because of the strength of the two epidemiological studies, a total uncertainty factor of 3 is applied in the derivation of the RfD for selenium. The calculated RfD is believed to be protective against toxicity while allowing a sufficient intake of selenium to guard against deficiency.

For aldicarb, identifying a NOAEL is difficult because an exposure to low levels of aldicarb may result in inhibition of acetylcholinesterase that is not associated with any actual health effects and is entirely reversible. However, there is a threshold for inhibition of this enzyme above which adverse health effects, ultimately death, occur. The quantitative risk assessment for aldicarb is based on both clinical effects and enzyme inhibition reported in both humans and laboratory animals. By using a more extensive data base as opposed to a single study to develop an RfD for aldicarb, there is greater confidence in the resulting number.

For nickel, a similar weight of evidence approach is used. For this chemical, however, the different studies incorporated into the RfD derivation represent different target organs for nickel: general toxicity manifested as organ weight changes, fetotoxicity in rodents, and dermal effects in hypersensitive humans. The final quantitative estimate for nickel is considered protective for all end points discussed.

ACKNOWLEDGMENTS

The authors gratefully acknowledge Drs. Michael Dourson and Moiz Mumtaz from the Office of Research and Development (U.S. EPA) and Dr. George Ghali from the Office of Pesticide Programs (U.S. EPA) for their scientific review and contributions and Mrs. Judi Olsen for her editorial review of this manuscript. The views in this paper are those of the authors and do not necessarily reflect the views of policies of the U.S. Environmental Protection Agency. The U.S. Government has the right to retain a nonexclusive royalty-free license in and to any copyright covering this article.

REFERENCES

1. National Research Council, *Risk Assessment in the Federal Government: Managing the Process*, National Academy Press, Washington, D.C., 1983.
2. National Academy of Sciences (NAS), *Drinking Water and Health*, Vol. 1, NAS, Washington, D.C., 1977, pp. 19–63.
3. National Academy of Sciences, *Drinking Water and Health*, Vol. 3, NAS, Washington, D.C., 1980, pp. 25–67.
4. U.S. EPA, *Integrated Risk Information System* (*IRIS*), Online, Office of Health and Environmental Assessment, Environmental Criteria and Assessment Office, Cincinnati, OH, 1992.
5. H.W. Lakin, Selenium accumulation in soils and its absorption by plants and animals, *Geol. Soc. Am. Bull. 83*:181 (1972).
6. G.F. Combs and S.B. Combs, Effects of selenium excesses, *The Role of Selenium in Nutrition* (G.F. Combs and S.B. Combs, eds.), Academic Press, New York, 1986, pp. 463–525.
7. O.A. Levander, Selenium, *Trace Elements in Human and Animal Nutrition*, 5th ed., Vol. 2 (W. Mertz, ed.), Academic Press, New York, 1986, pp. 209–279.

8. J.T. Rotruck, A.L. Pope, H.E. Ganther, A.B. Swanson, D.G. Hafeman, and W.G. Hoekstra, Selenium: Biochemical role as a component of glutathione peroxidase, *Science 179*:588–590 (1973).
9. Keshan Disease Research Group of the Chinese Academy of Medical Sciences, Observations on effect of sodium selenite in prevention of Keshan disease, *Chin. Med. J. 92*(7):471–476 (1979).
10. Keshan Disease Research Group of the Chinese Academy of Medical Sciences, Epidemiologic studies on the etiologic relationship of selenium and Keshan disease, *Chin. Med. J. 92*(7):477–482 (1979).
11. O.A. Levander and R.F. Burk, Report on the 1986 A.S.P.E.N. Research Workshop on Selenium in Clinical Nutrition, *J. Parenter. Enteral Nutr. 10*(6):545–549 (1986).
12. A.M. van Rij, C.D. Thomson, J.M. McKenzie, and M.F. Robinson, Selenium deficiency in total parenteral nutrition, *Am. J. Clin. Nutr. 32*:2076–2085 (1979).
13. R.A. Johnson, S.S. Baker, J.T. Fallon, E.P. Maynard III, J.N. Ruskin, Z. Wen, K. Ge, and H.J. Cohen, An occidental case of cardiomyopathy and selenium deficiency, *N. Engl. J. Med. 304*(20):1210–1212 (1981).
14. P.D. Whanger, M.A. Beilstein, C.D. Thomson, M.F. Robinson, and M. Howe, Blood selenium and glutathione peroxidase activity of populations in New Zealand, Oregon and South Dakota, *FASEB J. 2*(14):2996–3002 (1988).
15. NAS, *Recommended Dietary Allowances*, 10th ed., National Academy Press, Washington, D.C., 1989, pp. 217–224.
16. G. Yang, S. Wang, R. Zhou, and S. Sun, Endemic selenium intoxication of humans in China, *Am. J. Clin. Nutr. 37*:872–881 (1983).
17. J.A. Milner and M.E. Fico, Selenium and tumorigenesis, *Selenium in Biology and Medicine*, Proceedings of the Third International Symposium on Selenium in Biology and Medicine (G.F. Combs, Jr., J.E. Spallholz, O.A. Levander, and J.E. Oldfield, eds.), Van Nostrand-Reinhold, New York, 1987, pp. 1034–1043.
18. K.W. Francke and A.L. Moxon, A comparison of the minimum fatal doses of selenium, tellurium, arsenic and vanadium, *J. Pharmacol. Exp. Ther. 58*:454–459 (1936).
19. I. Rosenfeld and O.A. Beath, *Selenium: Geobotany, Biochemistry, Toxicity and Nutrition*, Academic Press, New York, 1964, pp. 198–208.
20. J.R. Harr and O.H. Muth, Selenium poisoning in domestic animals and its relationship to man, *Clin. Toxicol. 5*(2):175–186 (1972).
21. K.A. Poirier and J.A. Milner, The effect of various selenocompounds on Ehrlich ascites tumor cells, *Biol. Trace Elem. Res. 1*:25–34 (1979).
22. G. Yang, R. Zhou, S. Yin, L. Gu, B. Yan, Y. Liu, Y. Liu, and X. Li, Studies of safe maximal daily dietary selenium intake in a seleniferous area in China. I. Selenium intake and tissue levels of the inhabitants, *J. Trace Elem. Electrolytes Health Dis. 3*(2):77–87 (1989).
23. G. Yang, S. Yin, R. Zhou, L. Gu, B. Yan, Y. Liu, and Y. Liu, Studies of safe maximal daily dietary Se-intake in a seleniferous area in China. II. Re-

lation between Se-intake and the manifestation of clinical signs and certain biochemical alterations in blood and urine, *J. Trace Elem. Electrolytes Health Dis.* *3*(2):123–130 (1989).
24. B. Ketterer, B. Coles, and D.J. Meyer, The role of glutathione in detoxication, *Environ. Health Perspect.* *49*:59–69 (1983).
25. M.P. Longnecker, P.R. Taylor, O.A. Levander, M. Howe, C. Veillon, P.A. McAdam, K.Y. Patterson, J.M. Holden, M.J. Stampfer, J.S. Morris, and W.C. Willett, Selenium in diet, blood, and toenails in relation to human health in a seleniferous area, *Am. J. Clin. Nutr.* *53*:1288–1294 (1991).
26. A.F. Tarantal, C.C. Willhite, B.L. Lasley, C.J. Murphy, C.J. Miller, M.J. Cukierski, S.A. Book, and A.G. Hendrickx, Developmental toxicity of L-selenomethionine in *Macaca fascicularis*, *Fund. Appl. Toxicol.* *16*:147–160 (1991).
27. H.A. Schroeder and M. Mitchener, Toxic effects of trace elements on the reproduction of mice and rats, *Arch. Environ. Health* *23*(2):102–107 (1971).
28. I. Rosenfeld and O.A. Beath, Effect of selenium on reproduction in rats, *Proc. Soc. Exp. Biol. Med.* *87*:295–297 (1954).
29. T. Nobunaga, H. Satoh, and T. Suzuki, Effects of sodium selenite on methylmercury embryotoxicity and teratogenicity in mice, *Toxicol. Appl. Pharmacol.* *47*(1):79–88 (1979).
30. H. Martin and C.R. Worthing (eds.), *Pesticide Manual*, British Crop Protection Council, Croydon, 1977, p. 6.
31. C.S. Weil, EPA Pesticide Petition No. 9F0798, Mellon Institute Report No. 31-48, 1968.
32. World Health Organization (WHO), *Environmental Health Criteria 121: Aldicarb*, Geneva, 1991.
33. M. Vandekar, R. Plestina, and K. Wilhelm, Toxicity of carbamates for mammals, *Bull. WHO 44*:241–249 (1971).
34. L.R. Goldman, N. Beller, and R.J. Jackson, Aldicarb food poisonings in California, 1985–1988: Toxicity estimates for humans, *Arch. Environ. Health* *45*(3):141–147 (1990).
35. Union Carbide Corporation, MRID No. 00101911, FOI, EPA, Washington, D.C. 20460, 1971.
36. Rhone-Poulenc AG Company, MRID No. 40695901, FOI, EPA, Washington, D.C. 20460, 1988.
37. U.S. EPA, *Integrated Risk Information System* (*IRIS*), Online, Office of Health and Environmental Assessment, Environmental Criteria and Assessment Office, Cincinnati, OH, 1992.
38. C.S. Weil and C.P. Carpenter, Two-year feeding of Compound 21149 in the diet of rats, Mellon Institute Report Nl. 28-123, EPA Pesticide Petition No. 9F0798, 1965.
39. C.S. Weil and C.P. Carpenter, Aldicarb (A), aldicarb sulfoxide (ASO), aldicarb sulfone (ASO_2) and a 1:1 mixture of ASO:ASO_2. Two-year feeding study in the diet of rats, Mellon Institute Report No. 35-82, 1972.

40. A.M. Ambrose, P.S. Larson, J.R. Borzelleca, and G.R. Hennigar, Jr., Long-term toxicologic assessment of nickel in rats and dogs, *J. Food Sci. Technol. 13*:181–187 (1976).
41. P.D. Whanger, Effects of dietary nickel on enzyme activities and mineral content in rats, *Toxicol. Appl. Pharmacol. 25*:323–331 (1973).
42. J.R. Nation, M.F. Hare, D.M. Baker, D.E. Clark, and A.E. Bourgeois, Dietary administration of nickel: Effects on behavior and metallothionein levels, *Physiol. Behav. 34*:349–353 (1985).
43. American Biogenics Corporation, *Ninety-day gavage study in albino rats using nickel*, Draft Final Report, Research Triangle Institute, P.O. Box 12194, Research Triangle Park, NC 27709, 1986.
44. H.A. Schroeder and M. Mitchener, Toxic effects of trace elements on the reproduction of mice and rats, *Arch. Environ. Health 23*:102–106 (1971).
45. Research Triangle Institute (RTI), Two generation reproduction and fertility study of nickel chloride administered to CD rats in drinking water: Fertility and reproductive performance of the P_0 generation (Part II of III) and F_1 generation (Part III of III), Final Study Report, Office of Solid Waste Management, U.S. EPA, Washington, D.C., 1987.
46. M.J. Edwards, Hyperthermia as a teratogen: A review of experimental studies and their clinical significance, *Teratogenesis Carcinog. Mutagen. 6*:563–582 (1986).
47. M.K. Smith, E.L. George, J.A. Stober, H.A. Feng and G.L. Kimmel, Perinatal toxicity associated with nickel chloride exposure, *Environ. Res. 61*(2):200–211 (1993).
48. North American Contact Dermatitis Group, Epidemiology of contact dermatitis in North America: 1972, *Arch. Dermatol. 108*:537–540 (1973).
49. L. Peltonen, Nickel sensitivity in the general population, *Contact Dermatitis 5*(1):27–32 (1979).
50. S.D. Prystowsky, A.M. Allen, R.W. Smith, J.H. Nonomura, R.B. Odom, and W.A. Akers, Allergic contact hypersensitivity to nickel, neomycin, ethylenediamine, and benzocaine, *Arch. Dermatol. 115*:959–962 (1979).
51. T. Menne, O. Borgan, and A. Green, Nickel allergy and hand dermatitis in a stratified sample of the Danish female population: An epidemiological study including a statistic appendix, *Acta Derm. Venereol.* (*Stockh.*) *62*:35–41 (1982).
52. E. Cronin, A. DiMichiel, and S.S. Brown, Oral nickel challenge in nickel-sensitive women with hand eczema, *Nickel Toxicology* (S.S. Brown and F.W. Sunderman, Jr., eds.), Academic Press, New York, 1980, pp. 149–152.
53. K. Kaaber, N.K. Veien, and J.C. Tjell, Low nickel diet in the treatment of patients with chronic nickel dermatitis, *Br. J. Dermatol. 98*:197–201 (1978).
54. K. Kaaber, T. Menne, J.C. Tjell, and N. Veien, Antabuse treatment of nickel dermatitis. Chelation—a new principle in the treatment of nickel dermatitis, *Contact Dermatitis 5*:221–228 (1979).
55. W.P. Jordan and S.E. King, Nickel feeding in nickel-sensitive patients with hand eczema, *J. Am. Acad. Dermatol. 1*:506–508 (1979).

56. D.J. Gawkrodger, S.W. Cook, G.S. Fella, and J.A.A. Hunter, Nickel dermatitis: The reaction to oral nickel challenge, *Br. J. Dermatol. 115*:33–38 (1986).
57. D. Burrows, S. Creswell, and J.D. Merrett, Nickel, hands, and hip prostheses, *Br. J. Dermatol. 105*:437–444 (1981).
58. G.D. Nielsen, Oral challenge of nickel-allergic patients with hand eczema, *Nickel and Human Health: Current Perspectives*, Advances in Environmental Science and Technology (E. Nieboer and A. Aitio, eds.), John Wiley and Sons, New York, 1989, p. 1.
59. U.S. EPA, *Integrated Risk Information System (IRIS)*, Online, Office of Health and Environmental Assessment, Environmental Criteria and Assessment Office, Cincinnati, OH, 1992.
60. U.S. EPA, *Drinking Water Quantification of Toxicologic Effects (QTE) for Nickel*, Environmental Criteria and Assessment Office, Office of Health and Environmental Assessment, Cincinnati, OH, Office of Water, Washington, D.C., 1991.
61. P. Grandjean, O. Andersen, and G.D. Nielsen, Carcinogenicity of occupational nickel exposures: An evaluation of the epidemiological evidence, *Am. J. Ind. Med. 13*:193–209 (1988).
62. National Research Council, *Recommended Dietary Allowances*, 10th ed., National Academy Press, Washington, D.C., 1989, p. 267.
63. F.H. Nielsen, Nickel, *Trace Elements in Human Nutrition*, Vol. 1 (W. Mertz, ed.), Academic Press, San Diego, CA, 1987, pp. 245–273.
64. F.H. Nielsen, Ultratrace elements in nutrition, *Annu. Rev. Nutr. 4*:21–41 (1984).
65. F.W. Sunderman, Jr., S.M. Hopfer, K.R. Sweeney, A.H. Marcus, B.M. Most, and J. Creason, Nickel absorption and kinetics in human volunteers, *Proc. Soc. Exp. Biol. Med. 191*:5–11 (1989).
66. N.W. Solomons, F. Viteri, T.R. Shuler, and F.H. Nielson, Bioavailability of nickel in man: Effects of foods and chemically defined dietary constituents on the absorption of inorganic nickel, *J. Nutr. 112*:39–50 (1982).
67. E. Foulkes, Nickel absorption studies in rats, Report for Work Assignment No. 011302, Outreach Program Support (ODW-40), U.S. EPA, Office of Drinking Water, Washington, D.C. (unpublished), 1984.

22

Are There Carcinogens in Water for Human Consumption? An Epidemiologist's Viewpoint

Arthur O. S. Chiu and David L. Bayliss

U.S. Environmental Protection Agency,
Washington, D.C.

I. INTRODUCTION

Are there carcinogens in water for human consumption? How can we find out? What do we know about carcinogenesis and chemical carcinogens? How do we estimate human cancer risks if we have evidence of chemical carcinogenicity? Can we prevent cancer if we know the risks? What about other health risks? These are questions obviously of everyone's concern and for which we attempt to provide some answers in this discussion.

In a recent review, we stated that carcinogenesis is a complex and dynamic interaction of host and environment (1). Important host factors include genetic constitution and health status. Among environmental factors, diet, pollutants in air and water, occupation, and life-style characteristics, such as smoking, are known to affect cancer development in the human population. This perspective represents a synthesis of our accumulated knowledge from epidemiology and animal experiments (2–4).

Cancer epidemiology, in contrast to animal toxicological studies, requires no transference of the conclusions from the analysis of the animal data to indicate human carcinogenic potential. The long latency between the inception of cancer and its manifestation, and comparatively low incidence in human beings, as well as the difficulty of ascertaining exposure, have made an epidemiological approach demonstrating a cause-and-effect relationship between the suspected carcinogen and a specific cancer a very difficult if not impossible task. Long-term animal studies can provide tentative answers to questions posed by nonpositive but suggestive epidemiological studies. In addition, animal experiments have provided new perspectives on the interaction between the genome and the environment.

Ramazzini in the eighteenth century ascribed the prevalence of breast cancer among nuns to life-style factors. Percival Pott in 1775 published the now famous monograph from his clinical experience in St. Bartholomew Hospital and demonstrated implicitly the association of scrotal cancer with occupational exposure to soot in chimney sweeps (5). All these were looked at with rekindled interest in the twentieth century, and additional observations regarding the interaction of the genome and the environment were made. Doll and Peto, for example, commented in a recent National Cancer Institute monograph that "30% of current U.S. cancer deaths [were] due to tobacco in 1988" (4).

The recent flurry of activity in basic research in oncogenes also coincides with and reinforces clinical observations regarding genetic influences on cancer development (6). From his observation on the genetic and sporadic forms of retinoblastoma multiforme, Knudson pointed out not only a similar genetic basis between these two forms, but also highlighted the importance of the mutation in suppression of the oncogenes in the development of cancer in these cases (7–9). The emphasis and awareness of the dual importance of genetics and environment in carcinogenesis are further documented by other clinical and epidemiological studies mentioned later (10–13).

II. NATURE VERSUS NURTURE

Sellers et al. demonstrated that there is an increase in the risk of cancer among family members of a proband in 1988. Such a finding indicates the importance not only of genetics but also those effects from habits or life-style similarities in the same family. Meadows, in a study of the late effects of cancer treatment, observed in treated cancer patients that in addition to a genetic predisposition toward cancer development, alkylating agents used in chemical therapies for the "primary" increase the risk of a "second cancer" among the survivors years after termination of the treatment. The obvious conclusion is that chemical carcinogens increase cancer risks among those genetically predisposed individuals who have already developed a primary cancer years earlier (14–16).

On a first approximation, it appears that genetics and environment can be equally important in affecting cancer development. For public health

Table 1 Measurable Carcinogens in Drinking Water

Chemical	MW (D)	U.S. EPA Group	Concentration: (μg/L)
Chloroform	119.4	B2	366 (3×10^{-6} M)
Vinyl chloride	62.5	A	10 (1.6×10^{-7} M)
Benzene	78	A	10 (1.3×10^{-7} M)
Carbon tetrachloride	154	B2	5 (3.2×10^{-8} M)
Dieldrin	380.9	B2	8 (2×10^{-8} M)
PCB (Aroclor 1260)	324	B2	3 (9×10^{-9} M)
Diphenylhydazine	184	B2	1 (5×10^{-9} M)
Bis-(2-chloroethyl)ether	143	B1	0.42 (1×10^{-9} M)
Trichloroethylene	131.4	B2	0.05 (3.8×10^{-10} M)
Endrin	380.9	B2	0.08 (2×10^{-10} M)
Lindane (gBHC)	290.8	C	0.01 (3.4×10^{-11} M)
Chlordane	409.8	B2	0.01 (2.4×10^{-11} M)

Source: Modified from Wilkins et al. (37).

measures, prevention through avoidance of environmental exposures seems to be a sensible avenue to reduce human cancer incidence.

Preussman pointed out succinctly the following points about cancer incidence in different geographical areas in the ambient environment. "Site-specific cancer incidence differs in different geographic regions. There is a 50-fold higher stomach cancer incidence in Japan in contrast to USA. While USA has a 50-fold higher incidence in recto-colonic cancer in comparison to Japan" (17).

The genetic differences between Japanese and Americans may play a role in the carcinogenesis of some sites of the gastrointestinal tract in addition to the environment, such as described, it is incorrect to assign the cause of esophageal cancer to genetic influences, for example, in Iran: "A 100-fold difference in esophageal cancer rates are observed in areas only 100 km distant apart in Northern Iran. In Gonbad, eg the incidences of esophageal cancer per 100,000 are 165(m) and 195(f). In contrast, in Rudbar, the incidences are 12(m) and 3(f)." Presumably the two populations are similar in genetic makeup. Differences in cancer incidence thus reflect an environmental influence. Indeed, the drying of the littoral region around the Caspian Sea is suspected to have contributed to the building up of chemical "carcinogens" in that area (18).

There is reason to believe that chemical carcinogens may play a significant role in the development of esophageal carcinoma. Indeed, nitrosomethylacetonyl-3-methylbutylamine and nitrosomethylacetonyl-2-methylpropylamine were implicated as the carcinogenic metabolite produced in pickles by the fungus *Fusarium moniliforme* that may, if consumed, lead to esophageal cancer in China (19–21).

Emigrants, on the other hand, shed the prevalent cancer patterns of their native countries, and they and their offspring take on those of their new homes. For example, first-generation Japanese emigrating to the United States lose the characteristic high stomach cancer risk of Japan but develop a high risk of recto-colon cancer identical to that of their host country by the third generation (22). The change in life-style or diet responsible for this change is probably not through any conscious effort, even though the final result is beneficial for the reduction in the incidence of stomach cancer. Yet, the appreciation and understanding from epidemiology and animal experiments have helped us better to define the interaction of environmental influences (diet and life-style) with that of genetics in the process of carcinogenesis in these organs.

III. EXAMPLE OF SUCCESS IN CANCER PREVENTIVE MEASURES

Indeed, Swedish chimney sweeps who are educated and instructed to take showers after work suffered no scrotal cancer deaths in a recent epidemiological study (23). This serves in stark contrast to the classic example studied by Pott in London in the eighteenth century. It can be concluded as a corollary from Bayes' theorem that a cause-and-effect relationship is strongly implicated between epidermoid cancer and the exposure to soot in the rugae of the scrotum. What is important here, however, is that the

Table 2 Detectable but Not Measured Chemical Carcinogens in Water

	MW	U.S. EPA Group
Benzo[*a*]pyrene	252.3	A
Heptachlor	373	B2
DDT	354.5	B2
bBHC	290.8	—
aBHC	290.8	—
Aldrin	369.4	—
Heptachlor epoxide	389.3	—

Source: Modified after Wilkins et al. (37).

prevention of cancer can be accomplished through simple public health measures without the need to know a mechanism of causation.

On the other hand, these Swedish authors pointed out there was a significant excess risk of cancer of the lung and pleura (SMR 207, 95% confidence interval, CI, 144–290), esophagus (SMR 365, CI 146–751), and bladder cancer (SMR 236, CI 149–354) in these same educated chimney sweeps. They have perhaps, in contrast to Pott's English chimney sweeps, lived long enough to succumb to those diseases induced by the soot exposure. The total incidence of cancer was also significantly increased (SMR 132, CI 115–151). The excess lung cancers were predominantly squamous cell carcinoma and undifferentiated or small cell carcinoma (23). They concluded: "The excess risk of cancer of the esophagus, stomach, lung and bladder depended mainly on an excess during followup for more than 30 years from start of exposure." The median latency time (time from start of exposure to diagnosis) was 43 years for esophageal cancer, 47.5 years for lung cancer, 43.5 years for bladder cancer, and 42 years for all cancer in all sites.

In the examples just given, we intended to show that the route of exposure and the carcinogenicity of a chemical are equally important for fashioning various preventive public health measures. Chemical carcinogens via the oral route can induce cancer in the gastrointestinal tract, but the net caloric intake and other concomitant dietary supplement may affect the ultimate manifestation of cancer in the stomach or the colon. Prevention of exposure, with special emphasis on specific routes, such as the skin (scrotal folds) or dietary intake, can reduce cancer in those specific areas of most vulnerability. A concerted effort with an appreciation of all possible routes of exposure is then, we suggest, necessary for any successful approach to the prevention of cancer at any particular site.

The importance of a global approach in cancer prevention belies our approach in public health measures for cancer prevention. To illustrate this point, it is necessary to discuss cancer patterns in Swedish chimney sweeps *after* they have implemented the bathing procedure. If the removal by shower of a contact carcinogen, such as benzo[*a*]pyrene in the soot, has effectively eliminated scrotal cancer, it is likely that urinary bladder and other centrally located organ systems of these chimney sweeps are still under the influence of these carcinogen(s) through either ingestion or inhalation, or both.

The underlying assumption is that carcinogens in the soot are taken up into the body. The procarcinogen is further metabolized, and the ultimate carcinogen (23,24) finally exerts its influence on the distal urinary tract. This is our current understanding of carcinogenesis from animal experiments.

To recapitulate, the determination of the carcinogenicity of soot (such as benzo[*a*]pyrene) helps to explain the increase in cancer incidences of the scrotum and the bladder. The immediate tissue of concern was the skin of the scrotal folds. Prevention of direct contact by showers has probably eliminated the risk of scrotal cancer. Simultaneously, the urinary bladder in this same population becomes the next target at risk of cancer development and, therefore, our immediate concern for cancer preventive measures in public health.

IV. TARGET SITES VERSUS ROUTES OF EXPOSURE

Questions have been raised about whether there is a unique mechanism of carcinogenesis or whether a chemical can cause cancer in only one organ. If a particle can only enter the body via the respiratory tract through inhalation, is the lung the only organ to be affected? It might be either "absorption" or "activation" by the different organs that ultimately determines the carcinogenicity of that compound.

For the epidemiologist, asbestos is a classic example. This known carcinogen can cause bronchogenic carcinoma, pleural and peritoneal mesothelioma, and gastric carcinoma. The incidences of these cancers differ, reflecting the predilection of the carcinogen for one organ over the others. Nonetheless, even in this specific case in which the systemic involvement seems unlikely because of the particulate nature of the carcinogen, more than one organ is usually affected.

Thorotrast is another example in human beings. The carcinogen is taken up in the liver predominantly after intravenous administration but nevertheless causes cancer in the lung in addition to the liver after a prolonged latency (26). Ionizing radiation is another known human carcinogen. Survivors from the atomic bomb drops in Hiroshima and Nagasaki showed an initial rise in acute lymphocytic and acute myelocytic leukemia in the first 5 years, followed by a rise in carcinomas in other sites 25 years later (27).

Arsenic in drinking water is another example of a chemical carcinogen for which the target organs include not only the skin but visceral organs as well (28–30). Neumann et al. further concluded that the "promoting, presumably non-genotoxic effects" of the aromatic amines "determine tumor development in target tissues and the rate of tumor growth" (31). The promoting effect of arsenic is likely responsible for the various tumor incidences of the skin, bladder, lung, and liver.

It is not possible to predict accurately the target organs based upon the physical or chemical nature of a carcinogen. It is important to ascertain all routes of exposure to any suspected carcinogen to develop and implement a comprehensive plan for cancer prevention of all target organs. It is also important to note the rather long latency (20–40 years) derived from the observations on these environmentally induced cancers. That many of these cancers have such long latent periods should cause those persons with special interests other than public health to pause and reflect before reaching negative conclusions regarding a lack of carcinogenicity of a product produced by these same interest groups following an inadequate follow-up of only 5–10 years.

V. CANCERS ARISE ANTENATALLY OR FROM THE GERM PLASM

Yoshimoto et al. demonstrated that among the in utero A-bomb survivors (32):

1. The observed cancers occurred earlier in the 0.30^{+} Gy dose than in the 0 Gy dose group.
2. The incidence continues to increase over time.
3. The crude cumulative incidence rate 40 years after the A-bomb is 3.9-fold greater in the 0.30^{+} Gy group.
4. The susceptibility to radiation-induced cancer is higher in prenatally than postnatally exposed survivors.

Recall the observations by Gardner et al. (33) that an increased risk of leukemia and non-Hodgkin's lymphoma among children near Sellafield was associated with paternal employment and recorded external dose of whole-body penetrating radiation during work at the plant before conception. This result suggests an effect of ionizing radiation on the father's genes that may be leukemogenic in the offspring. Carcinogens like ionizing radiation may not only affect the adult but also impose a cancer risk on the offspring. The risk of cancer from exposure may not then be limited to the somatoplasm alone, even though the carcinogenicity on the germ plasm is obviously of a smaller magnitude according to the incidence in later life.

Water for consumption includes not only drinking water but the water used for washing and bathing as well. A chemical carcinogen with high vapor pressure can exert its deleterious effect not only in drinking water but through contact with the skin and inhalation in the lung also. Once the chemical has gained access to the body, cancer development in various target organs is a finite possibility. Quantitative estimates of cancer risks from these "routes" for a chemical must then be summarized to estimate the overall impact, even if there is no other multiplicative interaction between these areas during cancer development.

We have summarized our knowledge of human carcinogenesis from epidemiological studies with emphasis on how a chemical can affect multiple target organs, such as the metabolic (liver), transport (leukemia and lymphoma), depo (squamous cell carcinoma and sarcoma), and excretory organs (kidney and bladder), regardless of route of exposure. In the next section, we review briefly our risk assessment of two representative chemicals after outlining our approach. We also briefly discuss the noncarcinogenic risk assessment of toxic effects of the same chemicals. Because noncarcinogenic risk assessment is still in its infancy, we do not think it appropriate to dwell on these topics at this point, other than to point out that the basic rationale for prevention is the same.

VI. SHORT NOTE ON THE U.S. EPA APPROACH IN CARCINOGEN CLASSIFICATION: q_1^* AND THE REFERENCE DOSE

The U.S. Environmental Protection Agency (EPA) has developed cancer risk assessment guidelines to assist in the classification of chemicals according mainly to the available epidemiology or long-term animal experiments. All chemicals can be placed into one of five categories according to the level of evidence of carcinogenicity. If it is a human carcinogen based upon sufficient epidemiological evidence, it is placed in group A. When the epidemiology evidence is nonexistent or inadequate but there is sufficient evidence from long-term animal experiments, the chemical is placed into group B, probable human carcinogen. If long-term animal experiments provided limited or suggestive evidence, in the absence of epidemiology data the chemical is placed into group C, possible human carcinogen category. A classification as a group D chemical simply reflects the total lack of data in epidemiology and long-term animal experiments. In contrast, a group E classification indicates that there is sufficient evidence from long-term animal experiments and, of course, no human evidence, that the chemical is definitely not a carcinogen. From this qualitative risk assessment, we can then choose the appropriate data base to carry out a quantitative estimate of the possible human cancer risk using the computer modeling approach.

VII. CANCERS OR NONCANCERS: ARE THEY ALL PROBLEMS OF PUBLIC HEALTH CONCERN?

Are other toxic chemicals besides carcinogens in the water against which regulatory measures should be taken to protect public health?

Noncarcinogenic toxic effects produce, of course, more immediate impact on the populace exposed to the pollutants in the ambient environment. Quantitative risk assessment of human cancer of any chemical pollutant without considering its other noncarcinogenic effect is then incomplete until we have taken both areas into account. For the consideration of economics and human health, we discussed in the beginning the importance of prevention via public health measures in reducing cancer risks via environmental exposures. It is appropriate to point out that a similar consideration in quantitation of the magnitude of the noncancer problem has been undertaken for reducing these other risks.

The reference dose (RfD) used in a noncarcinogenic setting is a term based on a similar approach as the q_1^* we generated in the quantitative cancer risk assessment for the same chemical. An analysis of animal and human data is carried out to provide an estimate of human health risk based on these available data. The reference dose is based on the additional assumption that thresholds exist for certain noncancer toxic effects, such as tissue necrosis. In general, the RfD is an estimate, with uncertainty spanning perhaps an order of magnitude, of a daily exposure to the human population, including sensitive subgroups, that is likely to be without an appreciable risk of deleterious effect during a lifetime. "RfDs can also be derived for the non-carcinogenic health effects of compounds, which are also carcinogens."

Our interest in the prevention of exposure from cancer-causing agents in the environment stems from the fact that cancer is by far the disease that causes the greatest mortality and morbidity in the world. It is also given knowledge that public health preventive measures, through regulation or education, are far more effective in reducing illness than elaborate and painful treatment protocols. Nonetheless, it is from the experience of these measures, we believe, that comparative steps should now be taken in public health for the prevention of similar exposures to other chemical noncarcinogenic toxicants.

VIII. EXAMPLES OF TOXIC CHEMICALS IN WATER OF PUBLIC HEALTH CONCERN

A. Chloroform

Of all the suggested constituents found in drinking water that conceivably could carry a carcinogenic risk, chlorination products lead to most concern. However, we reflect first upon the importance of the treatment of drinking water by this simple "purification" procedure, which has a tremendous impact in bettering our public health. Then we discuss the possible carcinogenic effect of chloroform after years of continuous exposure.

Waterborne diseases brought about by the contamination of drinking water have been, through the centuries, one of the major scourges of humanity. The two most common infectious diseases in medical records are, first, gastrointestinal and, second, respiratory. In the past, whole populations in Europe, Asia, and now even today in South America have been wiped out by diarrhea, dysentery, amebiasis, and cholera brought about by the presence of microorganisms in water used for human consumption. The simple addition of chlorine to contaminated water brought an end to diseases transmitted in this manner.

In more recent years, however, it has been discovered that chlorine added to water that is contaminated with organics reacts to form a class of chemicals known as the trihalomethanes, of which chloroform is the main constituent. Over 300 volatile organic contaminants have been found in drinking water. Many have been identified as carcinogens (34) even before chlorine has reacted with them. However, chloroform has been shown to exceed peak concentrations of detected carcinogens by as much as 37 times that of the highest carcinogen found in drinking water, which is vinyl chloride. Peak concentrations of 266 μg/L have been recorded (34). The absolute need for chlorination around the world with the formation of heavy concentrations of chloroform have made this particular constituent found in drinking water one of utmost concern.

It has also been the subject of intense scrutiny because of the appearance in the literature of a whole host of epidemiological studies (34–37). The net import is that the chloroform formed in this way by the interaction of chlorine with organics in drinking water may be carcinogenic to humans (*Health Assessment Document for Chloroform*, EPA/600/8-84/004F, September 1985). However, one must hasten to add that the evidence is not conclusive, in that chloroform is not the sole or exclusive constituent of the drinking water that may be responsible for the seemingly moderate but consistent excess risks to increase certain site-specific cancers.

These cancers continue to "surface" in many studies done under different circumstances. These include cancer of the large intestine, rectum, and bladder. The moderately elevated risks observed in these studies could be explained by confounding effects of other uncontrolled influences, such as smoking, diet, air pollution, occupation, and life-style. Most of these studies cannot, indeed, control for all the possible confounding effects that have been identified for drinking water because of the purpose and/or design.

It is perhaps important to determine the magnitude of such risks, and an arbitrary decision must be made to decide how much chloroform in water for human consumption is acceptable for public health purposes.

B. Trichloroethylene

Another chemical that is beginning to appear more and more often in drinking water in recent years is trichloroethylene (TCI). This chemical is a degreaser of metals in industry. Although it has no natural sources, it has been found in ambient air and water all across the country. As much as 403 ppb have been measured in surface and subsurface waters. In finished drinking water, concentrations have ranged from 1 ppb to as much as 32 ppb. TCI when consumed via the oral route is absorbed completely. The principal urinary metabolites are trichloroethanol, its glucuronide, and trichloroacetic acid.

The animal evidence for carcinogenicity consists of elevated incidences of hepatocellular carcinomas in male and female B6C3F1 mice, lymphomas in female mice, renal adenocarcinomas in male Fischer 344 rats, and malignant liver tumors in male and female B6C3F1 mice. On the basis of these studies, the U.S. EPA decided that these studies constituted sufficient evidence of carcinogenicity in animals, according to the U.S. EPA guidelines, and it was classified as a probable human carcinogen.

With respect to the human studies the evidence is not as clear. No epidemiological studies implicate trichloroethylene directly as a significant risk factor in cancer. In the earlier studies, emphasis was placed on determining whether TCI is a hepatocellular carcinogen, as suggested by the animal data, or possibly a gallbladder carcinogen. However, confounding factors and insufficient follow-up have rendered these studies unhelpful in determining if such a risk is present in humans.

In later studies, the suggestion arose quite by accident that a leukemia risk might be associated with human consumption of TCI-contaminated drinking water. In areas around the country where significantly increased risks of leukemia were found and investigated, small amounts of TCI were found in the drinking water of the surrounding communities. This ecological connection has never been adequately investigated. Hence, because of confounding factors it was impossible to conclude based, upon the human studies alone, that TCI was causally related to any type of cancer. However, this is an avenue that needs further investigation. This should not prevent us from maintaining an awareness about possible deleterious effects brought on by the presence of this and other chemicals in drinking water.

ACKNOWLEDGMENTS

This article was written in the authors' private capacities and does not reflect the views of the Human Health Assessment Group, the Office of Health Environment Assessment, the Office of Research and Development, or the U.S. Environmental Protection Agency.

REFERENCES

1. M.M.L. Chu and A. Chiu, Environmental carcinogenesis and biotechnology, *J. Biotechnol. 16*:17–36 (1990).
2. J.W. Cullen, Design of cancer preventive studies, *Cancer Detect. Prevent. 9*:125–138 (1986).
3. C. Lerman, P. Rimer, and P.F. Engstro, Reducing avoidable cancer mortality through prevention and early detection regimens, *Cancer Res. 49*:4955–4962 (1989).
4. R. Doll and R. Peto, *The Causes of Cancer: Quantitative Estimates of Avoidable Risks of Cancer in the U.S. Today*, Oxford University Press, New York, 1981.
5. P. Pott, Chirurgical observations relative to the cataract, the polypus of the nose, the cancers of the scrotum, the different kinds of ruptures, and the mortification of the toes and feet, Hawes, Clark and Collins, London, 1775, pp. 7–13, reprinted in *Natl. Inst. Monogr. 10*:7–13 (1963).
6. National Academy of Sciences, *Risk Assessment in the Federal Government: Managing the Process*, National Academy Press, Washington, D.C., 1983.
7. R. Preussman, Carcinogenesis by environmental chemical carcinogens, *Cancer Campaign*, Vol. 9, *The Cancer Patient: Illness and Recovery* (E. Grundmann, ed.), Gustav Fischer Verlag, Stuttgart, 1985, pp. 23–28.
8. A.G. Knudson, Mutation and cancer: Statistical study of retinoblastoma, *Proc. Natl. Acad. Sci. USA 68*:820–823 (1987).
9. A.G. Knudson, A two-mutation model for human cancer, *Adv. Viral Oncol. 7*:1–17 (1987).
10. L. Tomatis, The value of long term testing for the implementation of primary prevention, *Origins of Human Cancer* (H.H. Hiatt, J.D. Watson, and J.A. Winsten, eds.), Book C, Cold Spring Harbor Laboratory, Cold Spring Harbor, NY, 1977, pp. 1339–1357.
11. J.M. Wainwright, A comparison of conditions associated with breast cancer in Great Britain and America, *Am. J. Cancer 15*:2610–2545 (1931).
12. V.E. Anderson, H.O. Goodman, and S.C. Reed, *Variables Related to Human Breast Cancer*, University of Minnesota, Minneapolis, 1958.
13. M.T. Macklin, Inheritance of cancer of the stomach and large intestine of man, *J. Natl. Cancer Inst. 24*:551–571 (1960).
14. E. Lovett, Family studies in cancer of the colon and rectum, *Br. J. Surg. 63*:13–18 (1976).
15. G.K. Tokuhata and A.M. Lillienfeld, Familial aggregation of lung cancers in humans, *J. Natl. Cancer Inst. 30*:289–312 (1963).
16. T.A. Sellers, R.C. Elston, C. Stewart, and H. Rothschild, Familial risks of cancer among randomly selected probands, *Gen. Epidemiol. 5*:381–391 (1988).
17. A.T. Meadows, Risk factors for second malignant neoplasms: Report from the late effects study group, *Bull. Cancer 75*:125–130 (1988).
18. H. Hormozdiari, N.G. Dag, B. Aramesh, and G. Mahboubi, Dietary factors and esophageal cancer in the Caspian littoral of Iran, *Cancer Res. 35*:3493–3498 (1975).

19. T.D. Hong, H.J.S. Huang, L.J.S. Young, A. Oro, R. Bookstein, E.Y.H.P. Lee, and W.H. Lee, Structure of the human retinoblastoma gene, *Proc. Natl. Acad. Sci. USA 86*:5502–5506 (1989).
20. M.H. Li, P. Li, and B. Li, Recent progress in research on esophageal cancer in China, *Adv. Cancer Res. 33*:173–249 (1980).
21. S. Lu, C. Ji, and M. Li, Studies on the synthesis of 2 new nitrosamines from isoamylamine and sodium nitrite by fungi, *Acta Acad. Ned. Sinica 8*(4):255–260 (1986).
22. W. Haenszel, W. Kuribara, W. Segi, and R.K.C. Lee, Stomach cancer among Japanese in Hawaii, *J. Natl. Cancer Inst. 49*:969–988 (1972).
23. P. Gustavsson, A. Gustavsson, and C. Hogstedt, Excess of cancer in Swedish chimney sweeps, *Br. J. Ind. Med. 45*:777–781 (1988).
24. J.A. Miller and C. Miller, The metabolic activation of chemical carcinogen to reactive electrophils (J.M. Yuhas, R.W. Tennant, and J.D. Regan, eds.), *Biology of Radiation Carcinogenesis*, Raven Press, New York, 1979.
25. E.C. Miller and J.A. Miller, Milestones in chemical carcinogenesis, *Semin. Oncol. 6*:445–460 (1979).
26. R. Morant and J.R. Ruettner, Thorotrast Spaetschaeden: Zuercher Erfahrungen, *Schweiz. Med. Wochenschr. 117*:952–957 (1987).
27. D.L. Preston, H. Kato, K.J. Kopecky, and S. Fujita, Study of the mortality of A-bomb survivors. VIII. Cancer mortality 1950–82, *Radiat. Res. 111*:151–178 (1987).
28. S. Yeh, S.W. How, and C.S. Lin, Arsenic cancer of skin, *Cancer 21*(2):312–339 (1968).
29. C.J. Chen, Y.C. Chuang, S.L. You, T.M. Lin, and H.Y. Wu, A retrospective study on malignant neoplasms of bladder, lung and liver in blackfoot disease endemic area in Taiwan, *Br. J. Cancer 53*:399–405 (1986).
30. C.J. Chen, Y.C. Chuang, T.M. Lin, and H.Y. Wu, Malignant neoplasms among residents of a black-foot disease endemic area in Taiwan, *Cancer Res. 45*:5895–5899 (1985).
31. H.G. Neumann, R. Hammerl, W. Hillesheim, and M. Wildschutte, Role of genotoxic and nongenotoxic effects in multistage carcinogenesis of aromatic amines, *Environ. Health Perspect. 83*: (1989).
32. Y. Yoshimoto, H. Kato, and W.J. Schull, Risk of cancer among children exposed in utero to A-bomb radiations 1950–84, *Lancet 2*:665–669 (1988).
33. M.J. Gardner, M.P. Snee, A.J. Hall, C.A. Powell, S. Downes, and J.D. Terrell, Results of case-control studies of leukemia and non-Hodgkin's lymphoma among young people near Sellafield nuclear plant in W. Cumbria, *Br. Med. J. 300*:423–434 (1990).
34. C.R. Bucher, R.J. Kuzma, and C.M. Forcade, Drinking water as an epidemiologic risk factor for cancer, *Origins of Human Cancer*, Book A (H.H. Hiatt, J.D. Watson, and J.A. Winsten, eds.), Cold Spring Harbor Laboratory, Cold Spring Harbor, NY, 1977, pp. 347–356.
35. M. Alavanja, I. Goldstein, and M. Susser, A case-control study of gastrointestinal & urinary tract cancer mortality and drinking water chlorination,

Water Chlorination: Environmental Impact and Health Effects, Vol. 2 (R.L. Jolley, H. Gorchev, and D.H. Hamilton, Jr., eds.), Ann Arbor Science Publishers, Ann Arbor, MI, 1978, pp. 395–409.

36. K.P. Cantor, R. Hoover, T.J. Mason, et al. Associations of cancer mortality with halomethanes in drinking water, *J. Natl. Cancer Inst. 61*:979–985 (1978).
37. J.R. Wilkins, III, N.A. Reiches, and C.W. Kruse, Organic chemical contaminants in drinking water and cancer, *Am. J. Epidemiol. 110*(4):420–448 (1979).

Index